D1693717

Edited by
Yoshinobu Kawai, Hideo Ikegami,
Noriyoshi Sato, Akihisa Matsuda,
Kiichiro Uchino, Masayuki Kuzuya,
and Akira Mizuno

Industrial Plasma Technology

Related Titles

Rauscher, H., Perucca, M.,
Buyle, G. (eds.)

Plasma Technology for Hyperfunctional Surfaces
Food, Biomedical and Textile Applications

2010
Hardcover
ISBN: 978-3-527-32654-9

d'Agostino, R., Favia, P., Kawai, Y.,
Ikegami, H., Sato, N., Arefi-Khonsari, F.
(eds.)

Advanced Plasma Technology

2008
Hardcover
ISBN: 978-3-527-40591-6

Heimann, R. B.

Plasma Spray Coating
Principles and Applications

2008
Hardcover
ISBN: 978-3-527-32050-9

Hippler, R., Kersten, H., Schmidt, M.,
Schoenbach, K. H. (eds.)

Low Temperature Plasmas
Fundamentals, Technologies and Techniques

2008
Hardcover
ISBN: 978-3-527-40673-9

Smirnov, B. M.

Plasma Processes and Plasma Kinetics
580 Worked-Out Problems for Science and Technology

580 Worked-out Problems for Science and
Technology
2007
Softcover
ISBN: 978-3-527-40681-4

Aliofkhazraei, M., Sabour Rouhaghdam, A.

Plasma Electrolysis
Fabrication of Nanocrystalline Coatings

2011
Hardcover
ISBN: 978-3-527-32675-4

Edited by Yoshinobu Kawai, Hideo Ikegami, Noriyoshi Sato, Akihisa Matsuda, Kiichiro Uchino, Masayuki Kuzuya, and Akira Mizuno

Industrial Plasma Technology

Applications from Environmental to Energy Technologies

WILEY-VCH Verlag GmbH & Co. KGaA

The Editors

Prof. Yoshinobu Kawai
Kyushu University, Grad.
School of Eng. Sciences
Kasugakoen 6-1, Kasuga
Fukuoka 816-8580
Japan

Prof. Hideo Ikegami
Technowave Inc.
13-18
Aoi 1-chome, Higashi-ku
Nagoya-shi Aichi 461-0004
Japan

Prof. Noriyoshi Sato
Graduate School of Engineering
Tohoku University
Kadan 4-17-113
Sendai 980-0815
Japan

Prof. Akihisa Matsuda
Osaka University, Grad. School
of Engineering Science
Machikaneyama-cho 1-3,Toyonaka
Osaka 565-0871
Japan

Prof. Kiichiro Uchino
Kyushu University, Grad.
School of Eng. Science
Kasugakoen 6-1, Kasuga
Fukuoka 816-8580
Japan

Prof. Masayuki Kuzuya
Matsuyama University
Fac. of Pharmaceutical Science
Bunkyo-cho 4-2, Matsuyama
Ehime 790-8578
Japan

Prof. Akira Mizuno
Toyohashi University of Techn.
Dept. of Ecologic. Engineering
Tempaku-cho
Toyohashi 441-8580
Japan

Library of Congress Card No.: applied for

British Library Cataloguing-in-Publication Data
A catalogue record for this book is available from the British Library.

Bibliographic information published by the Deutsche Nationalbibliothek
The Deutsche Nationalbibliothek lists this publication in the Deutsche Nationalbibliografie; detailed bibliographic data are available on the Internet at <http://dnb.d-nb.de>.

Cover Design Adam Design, Weinheim
Typesetting Laserwords Private Limited, Chennai, India
Printing and Binding Strauss GmbH, Mörlenbach

Printed in the Federal Republic of Germany
Printed on acid-free paper

ISBN: 978-3-527-32544-3

Contents

Industrial Plasma Technology. Edited by Yoshinobu Kawai, Hideo Ikegami, Noriyoshi Sato, Akihisa Matsuda, Kiichiro Uchino, Masayuki Kuzuya, and Akira Mizuno

ISBN: 978-3-527-32544-3

Preface

The first International School of Advanced Plasma Technology was held in 1992 at Villa Monastero, Varenna, Italy, with approximately 100 participants from all over the world. Most of the participants were enthused by the rapidly increasing needs of plasma applications in the area of nanoscale etching and patterning, amorphous solar-cell fabrication, surface modification and coating, as well as hydrophile/hydrophilic surface modifications.

The School was so well organized that tutorial speakers were encouraged to be pedagogical and to focus on the fundamentals of their work areas so that the nonspecialists could readily follow their presentation and take up synergetic collaboration with the workers in other specialties during the School activities.

Making a long-awaited debut, Japanese Association for Plasma Technology organized the Second International School in 2004. The special areas covered during these proceedings involved methods of controlling plasma parameters and their diagnoses (Advanced Plasma Technology (Wiley-VCH Verlag, 2008)). Some presentations in the proceedings dealt with biomedical applications and others with handling of hazardous environmental wastes.

The Third International School of Advanced Plasma Technology was also organized by Japanese Association for Plasma Technology again at the same venue, Villa Monastero, Varenna, from July 28 to July 31, 2008. The crucial purpose of the School was to enhance cross-disciplinary communication among the participants during the schooling period and after. Eventually, exchange of information between plasma researchers and industrial specialists was successfully achieved. With the rearranged proceedings, the present edition will provide graduate students, university researchers, and industrial specialists with useful, valuable, and indispensable information together with many hints and ready-to-use know hows.

Japanese Association for Plasma Technology
April 2009

Yoshinobu Kawai (editor-in- chief)
Hideo Ikegami
Masayuki Kuzuya
Akihisa Matsuda
Akira Mizuno
Noriyoshi Sato
Kiichiro Uchino

Industrial Plasma Technology. Edited by Yoshinobu Kawai, Hideo Ikegami, Noriyoshi Sato, Akihisa Matsuda, Kiichiro Uchino, Masayuki Kuzuya, and Akira Mizuno

ISBN: 978-3-527-32544-3

Yoshinobu Kawai works as Research Professor at the Interdisciplinary Graduate School of Engineering Sciences at Kyushu University, Japan, since 1984. He is an expert in the production of large diameter plasma in the frequency range from VHF to microwave for amorphous and microcrystalline silicon solar cell and etching. He is one of the editors of Advanced Plasma Technology (Wiley-VCH, 2007).

Hideo Ikegami is Professor Emeritus of both Nagoya University and National Institute for Fusion Science in Nagoya, Japan. His primary interest is basic plasma physics and various plasma applications. In 1996, he founded a consulting company of plasma application technology, Technowave Inc., and served as its president. He is one of the editors of Advanced Plasma Technology (Wiley-VCH, 2007).

Noriyoshi Sato has worked in research and education on plasma physics and is now Professor Emeritus at Tohoku University, Japan. His primary interest is basic plasma behavior related to various plasma applications. He is elevating his works in collaboration with companies. He is one of the editors of Advanced Plasma Technology (Wiley-VCH, 2007).

Akihisa Matsuda is Special Guest Professor at the Graduate School of Engineering Science of Osaka Univeristy, Japan, since 2007, after being Director of Research Initiative for Thin Film Silicon Solar Cells at the National Institute of Advanced Industrial Science and Technology (AIST), Japan. His main field of interest are plasma process diagnostics and process control for thin film silicon solar cells.

Kiichiro Uchino is currently Professor at the Department of Applied Science for Electronics and Materials at Kyushu University in Fukuoka, Japan. He is working on laser-aided diagnostics of plasmas and various industrial plasma applications.

Masayuki Kuzuya is currently the Dean of College of Pharmaceutical Sciences, and Professor of Pharmaceutical Physical Chemistry at Matsuyama University, Japan. His research interest is in the field of fundamental studies of organic plasma chemistry and its bio- and pharmaceutical applications.

Akira Mizuno is Professor of the Department of Ecological Engineering at Toyohashi University of Technology. His research focus is on applied electrostatics and high-voltage engineering, which includes electrostatic precipitation, environmental application of plasma, sterilization and other environmental friendly technologies as well as manipulation and measurement of single DNA molecules. He is editor of the International Journal of Plasma Environmental Science and Technology.

List of Contributors

Tetsuya Akitsu
University of Yamanashi
Interdisciplinary Graduate School
of Medicine and Engineering
Takeda 4-3-11, Kofu
Yamanashi 400-8511
Japan

Hidenori Akiyama
Kumamoto University Graduate
School of Science and Technology
Kurokami 2-39-1
Kumamoto 860-8555
Japan

Masahiro Akiyama
Kumamoto University Graduate
School of Science and Technology
Kurokami 2-39-1
Kumamoto 860-8555
Japan

Eleftherios Amanatides
University of Patras
Department of Chemical
Engineering
Plasma Technology Laboratory
P.O. Box 1407, Patras, 26504
Greece

José-Luis Andújar
Universitat de Barcelona
FEMAN Group, Dep. Física
Aplicada i Òptica
IN2UB, Martí i Franqués
1, 08028
Barcelona

Angel Barranco
Instituto de Ciencia de Materiales
de Sevilla (CSIC-Univ. Sevilla)
Avda. Américo Vespucio 49
41092, Sevilla
Spain

Enric Bertran
Universitat de Barcelona
FEMAN Group, Dep. Física
Aplicada i Òptica
IN2UB, Martí i Franqués, 1
08028
Barcelona

Veronika Biskupičová
Comenius University in
Bratislava
Faculty of Mathematics
Physics and Informatics
Mlynská dolina F2
SK 84248 Bratislava
Slovakia

Industrial Plasma Technology. Edited by Yoshinobu Kawai, Hideo Ikegami, Noriyoshi Sato, Akihisa Matsuda,
Kiichiro Uchino, Masayuki Kuzuya, and Akira Mizuno

ISBN: 978-3-527-32544-3

Laïfa Boufendi
Orléans University
GREMI Laboratory, PB 6744
45067 Orléans Cedex 2
France

Marjory Cavarroc
Orléans University
GREMI Laboratory, PB 6744
45067 Orléans Cedex 2
France

Jen-Shih Chang
McMaster University
McIARS and Department of
Engineering Physics
Hamilton
Ontario, L8S 4M1
Canada

Victor Ciupina
Ovidius University
Department of Plasma Physics
Faculty of Physics, Chemistry
Electronics and Oil Technology
Mamaia 124, Constanta 900527
Romania

Mirelsa Contulov
Ovidius University
Department of Plasma Physics
Faculty of Physics, Chemistry
Electronics and Oil Technology
Mamaia 124, Constanta 900527
Romania

Carles Corbella
Universitat de Barcelona
FEMAN Group, Dep. Física
Aplicada i Òptica
IN2UB, Martí i Franqués, 1
08028
Barcelona

José Cotrino
Instituto de Ciencia de Materiales
de Sevilla (CSIC-Univ. Sevilla)
Avda. Américo Vespucio 49
41092, Sevilla
Spain

Ferencz S. Denes
University of Wisconsin
Center for Plasma-Aided
Manufacturing
1410 Engineering Drive
Madison, WI 53706
USA

Benjamin Denis
European Commission
Joint Research Centre
Institute for Health and
Consumer Protection
Via E. Fermi 2749
I-21027 Ispra
Italy

and

Ruhr–Universität Bochum
Institute for Electrical
Engineering and Plasma
Technology
Universitätsstrasse 150
44780 Bochum
Germany

Virginia Dinca
Ovidius University
Department of Plasma Physics
Faculty of Physics, Chemistry
Electronics and Oil Technology
Mamaia 124, Constanta 900527
Romania

Christophe Donnet
Université Jean Monnet
LHC (UMR CNRS 5516)
Saint-Etienne, 42000
France

Mirosław Dors
Robert Szewalski Institute of
Fluid Flow Machinery
Centre for Plasma and Laser
Engineering
Polish Academy of Sciences
Fiszera 14, Gdańsk, 80–952
Poland

Juan P. Espinós
Instituto de Ciencia de Materiales
de Sevilla (CSIC-Univ. Sevilla)
Avda. Américo Vespucio 49
41092, Sevilla
Spain

Zsuzsanna Fabry
University of Wisconsin
Department of Pathology and
Laboratory Sciences
Madison, WI 53705
USA

Julien Fontaine
Ecole Centrale de Lyon
LTDS (UMR CNRS 5513)
Ecully Cedex, 69134
France

Syohei Fujita
Nagoya University
Graduate School of Engineering
Department of Materials, Physics
Energy Engineering
464-8603 Nagoya
Japan

Agustín R. González-Elipe
Instituto de Ciencia de Materiales
de Sevilla (CSIC-Univ. Sevilla)
Avda. Américo Vespucio 49
41092, Sevilla
Spain

Nobuya Hayashi
Saga University
Faculty of Science and
Engineering
1 Honjo-machi
Saga-shi, Saga 840-8502
Japan

Junko Hieda
Nagoya University
Graduate School of Engineering
Department of Materials
Physics and Energy Engineering
464-8603 Nagoya
Japan

Masaru Hori
Nagoya University
Department of Electrical
Engineering and Computer
Science
Furu-cho, Chikusa-ku
Nagoya 464-8603
Japan

Hideo Ikegami
Professor Emeritus
Nagoya University and National
Institute for Fusion Science
Shiga 24-601
Nagoya 462-0037
Japan

Yuichiro Imanishi
NGK Insulator, Ltd.
2-56 Suda-cho
Mizuho
Nagoya 467-8530
Japan

Takahiro Ishizaki
National Institute of Advanced
Industrial Science and
Technology (AIST)
Materials Research Institute for
Sustainable Development
463-8560 Nagoya
Japan

Shinya Iwashita
Kyushu University
Department of Electronics
744 Motooka
Fukuoka 819-0395
Japan

Mário Janda
Comenius University in
Bratislava
Faculty of Mathematics
Physics and Informatics
Mlynská dolina F2
SK 84248 Bratislava
Slovakia

Hongquan Jiang
University of Texas at Arlington
Department of Chemistry &
Biochemistry
and Materials Science &
Engineering
700 Planetarium Place
130 CPB, Arlington, TX 76019
USA

Marie Christine Jouanny
Orléans University
GREMI Laboratory, PB 6744
45067 Orléans Cedex 2
France

Keiko Katayama-Hirayama
University of Yamanashi
Interdisciplinary Graduate School
of Medicine and Engineering
Takeda 4-3-11, Kofu
Yamanashi 400-8511
Japan

Sunao Katsuki
Kumamoto University
Bioelectrics Research Center
Kurokami 2-39-1
Kumamoto 860-8555
Japan

Yoshinobu Kawai
Kyushu University
Interdisciplinary Graduate School
of Engineering Sciences
Kasugakoen 6-1
Kasuga, 816-8580
Japan

Siti Khadijah Za aba
University of Yamanashi
Interdisciplinary Graduate School
of Medicine and Engineering
Takeda 4-3-11, Kofu
Yamanashi 400-8511
Japan

Kazunori Koga
Kyushu University
Department of Electronics
744 Motooka
Fukuoka 819-0395
Japan

Nina Kolesárová
Comenius University in
Bratislava
Faculty of Mathematics
Physics and Informatics
Mlynská dolina F2
SK 84248 Bratislava
Slovakia

Shin-ichi Kondo
Gifu Pharmaceutical University
Laboratory of Pharmaceutical
Physical Chemistry
5-6-1, Mitahora-Higashi
Gifu 502-8585
Japan

Ivan Košinár
Comenius University in
Bratislava
Faculty of Mathematics
Physics and Informatics
Mlynská dolina F2
SK 84248 Bratislava
Slovakia

Daniela Kunecová
Comenius University in
Bratislava
Faculty of Mathematics
Physics and Informatics
Mlynská dolina F2
SK 84248 Bratislava
Slovakia

Masayuki Kuzuya
Matsuyama University
Department of Pharmaceutical
Physical Chemistry
Faculty of Pharmaceutical
Sciences
4-2 Bunkyo-cho, Matsuyama,
Ehime 790-8578
Japan

Ondřej Kylián
Charles University Faculty of
Mathematics and Physics
V Holešovickách 2
Prague 8, 180 00
Czech Republic

and

European Commission
Joint Research Centre
Institute for Health and
Consumer Protection
Via E. Fermi 2749
I-21027 Ispra
Italy

Aurelia Mandes
Ovidius University
Department of Plasma Physics
Faculty of Physics, Chemistry
Electronics and Oil Technology
Mamaia 124, Constanta 900527
Romania

Viktor Martišovitš
Comenius University in
Bratislava
Faculty of Mathematics
Physics and Informatics
Mlynská dolina F2
SK 84248 Bratislava
Slovakia

Ester Marotta
Università di Padova
Department of Chemical Sciences
Via Marzolo 1, 35131
Padova
Italy

Dimitrios Mataras
University of Patras
Department of Chemical Engineering
Plasma Technology Laboratory
P.O. Box 1407, Patras, 26504
Greece

Akihisa Matsuda
Osaka University
Graduate School of Engineering Science
Toyonaka
Osaka, 560-8531
Japan

Abdelaziz Mezeghrane
Orléans University
GREMI Laboratory, PB 6744
45067 Orléans Cedex 2
France

Maxime Mikikian
Orléans University
GREMI Laboratory, PB 6744
45067 Orléans Cedex 2
France

Hiroshi Miyata
Kyushu University
Department of Electronics
744 Motooka
Fukuoka 819-0395
Japan

Jerzy Mizeraczyk
Robert Szewalski Institute of Fluid Flow Machinery
Centre for Plasma and Laser Engineering
Polish Academy of Sciences
Fiszera 14
Gdańsk, 80–952
Poland

and

Gdynia Maritime University
Department of Marine Electronics
Morska 83
Gdynia, 81–225
Poland

Akira Mizuno
Toyohashi University of Technology
Department of Ecological Engineering
Hibarigaoka 1-1, Tempaku-cho
Toyohashi
Aichi, 441–8580
Japan

Imrich Morva
Comenius University in Bratislava
Faculty of Mathematics
Physics and Informatics
Mlynská dolina F2
SK 84248 Bratislava
Slovakia

Marcela Morvová
Comenius University in Bratislava
Faculty of Mathematics
Physics and Informatics
Mlynská dolina F2
SK 84248 Bratislava
Slovakia

Geavit Musa
Ovidius University
Department of Plasma Physics
Faculty of Physics, Chemistry
Electronics and Oil Technology
Mamaia 124, Constanta 900527
Romania

Tatsuyuki Nakatani
TOYO Advanced Technologies
Co., Ltd.
5-3-38 Ujina-higashi Minami-ku
Hiroshima, 734-8501
Japan

Anna Niewulis
Robert Szewalski Institute of
Fluid Flow Machinery
Centre for Plasma and Laser
Engineering
Polish Academy of Sciences
Fiszera 14
Gdańsk, 80–952
Poland

Yuki Nitta
TOYO Advanced Technologies
Co., Ltd.
5-3-38 Ujina-higashi Minami-ku
Hiroshima, 734-8501
Japan

Kenji Nose
University of Tokyo
Institute of Industrial Science
4-6-1 Komaba, Meguro-ku
Tokyo, 153-8505
Japan

Tetsuji Oda
University of Tokyo
Department of Electrical
Engineering and Information
Systems
3-1 Hongo 7-Chome Bunkyo-chu
Tokyo, 113–8656
Japan

Hiroshi Ohkawa
University of Yamanashi
Interdisciplinary Graduate School
of Medicine and Engineering
Takeda 4-3-11, Kofu
Yamanashi 400-8511
Japan

Ryo Ono
University of Tokyo
Department of Electrical
Engineering and Information
Systems
3-1 Hongo 7-Chome Bunkyo-chu
Tokyo, 113–8656
Japan

Cristina Paradisi
Università di Padova
Department of Chemical Sciences
Via Marzolo 1, 35131
Padova
Italy

Cristian Petric Lungu
National Institute for Laser
Plasma and Radiation Physics
PO Box MG-36
Bucharest, 077125
Romania

Janusz Podlinski
Robert Szewalski Institute of
Fluid Flow Machinery
Centre for Plasma and Laser
Engineering
Polish Academy of Sciences
Fiszera 14, Gdańsk, 80–952
Poland

Jozsef Prechl
University of Wisconsin
Department of Pathology and
Laboratory Sciences
Madison, WI 53705
USA

and

University of Texas at Arlington
Department of Chemistry &
Biochemistry
and Materials Science &
Engineering
700 Planetarium Place
130 CPB, Arlington
TX 76019
USA

Hubert Rauscher
European Commission
Joint Research Centre
Institute for Health and
Consumer Protection
Via E. Fermi 2749
I-21027 Ispra
Italy

Massimo Rea
Università di Padova
Department of Electrical
Engineering
Via Gradenigo 6
35131, Padova
Italy

John Robertson
Cambridge University
Department of Engineering
Trumpington Street
Cambridge, CB2 1PZ
UK

Pablo Romero-Gómez
Instituto de Ciencia de Materiales
de Sevilla (CSIC-Univ. Sevilla)
Avda. Américo Vespucio 49
41092, Sevilla
Spain

François Rossi
European Commission
Joint Research Centre
Institute for Health and
Consumer Protection
Via E. Fermi 2749
I-21027 Ispra
Italy

Miguel Rubio-Roy
Universitat de Barcelona
FEMAN Group, Dep. Física
Aplicada i Òptica
IN2UB, Martí i Franqués, 1
08028
Barcelona

Ana Ruiz
European Commission
Joint Research Centre
Institute for Health and
Consumer Protection
Via E. Fermi 2749
I-21027 Ispra
Italy

Nagahiro Saito
Nagoya University
Graduate School of Engineering
Department of Molecular Design
and Engineering
464-8603 Nagoya
Japan

Matyas Sandor
University of Wisconsin
Department of Pathology and
Laboratory Sciences
Madison, WI 53705
USA

Yasushi Sasai
Gifu Pharmaceutical University
Laboratory of Pharmaceutical
Physical Chemistry
5-6-1, Mitahora-Higashi
Gifu 502-8585
Japan

Noriyoshi Sato
Tohoku University
Kadan 4-17-113
Sendai 980-0815
Japan

Milko Schiorlin
Università di Padova
Department of Chemical Sciences
Via Marzolo 1, 35131
Padova
Italy

Heidi A. Schreiber
University of Wisconsin
Department of Pathology and
Laboratory Sciences
Madison, WI 53705
USA

Naohiro Shimizu
NGK Insulator, Ltd.
2-56 Suda-cho
Mizuho
Nagoya 467-8530
Japan

Masaharu Shiratani
Kyushu University
Department of Electronics
744 Motooka
Fukuoka 819-0395
Japan

Robert D. Short
University of South Australia
Mawson Institute
Mawson Lakes
5095, South Australia
Australia

Katharina Stapelmann
European Commission
Joint Research Centre
Institute for Health and
Consumer Protection
Via E. Fermi 2749
I-21027 Ispra
Italy

and

Ruhr–Universität Bochum
Institute for Electrical
Engineering and Plasma
Technology
Universitätsstrasse 150
44780 Bochum
Germany

David A. Steele
University of South Australia
Mawson Institute
Mawson Lakes
5095, South Australia
Australia

Kunihide Tachibana
Kyoto University
Department of Electronic Science and Engineering
Nishikyo-ku Kyoto 615-8510
Japan

Osamu Takai
Nagoya University
EcoTopia Science Institute
464-8603 Nagoya
Japan

Kazunori Takashima
Toyohashi University of Technology
Department of Ecological Engineering
Hibarigaoka 1-1, Tempaku-cho
Toyohashi, Aichi, 441-8580
Japan

Hiromu Takatsuka
Mitsubishi Heavy Industries Ltd.
Solar Power System Business Unit
Renewable Energy Business
Tsukuba-machi 6-53
Isahaya, 854-0065
Japan

Yoshiaki Takeuchi
Mitsubishi Heavy Industries Ltd.,
Nagasaki R. & D. Center
Fukahori-machi 5-717-1,
Nagasaki, 851–0392
Japan

Yves Tessier
Orléans University
GREMI Laboratory, PB 6744
45067 Orléans Cedex 2
France

Kentaro Tomita
Kyushu University
Interdisciplinary Graduate School of Engineering Sciences
Kasugakoen 6-1
Kasuga, Fukuoka 816-8580
Japan

Gerard Touchard
Université de Poitiers
Laboratoire d'Etudes Aérodynamiques
CNRS, Bd Marie et Pierre Curie
BP 30179, 86962, Futuroscope Cedex
France

Masao Tsuji
Yamanashi Prefecture Industrial Technology Center
Ohtsu-cho 2094
Yamanashi 400-0055
Japan

Kiichiro Uchino
Kyushu University
Interdisciplinary Graduate School of Engineering Sciences
Kasugakoen 6-1
Kasuga, Fukuoka 816-8580
Japan

Rodica Vladoiu
Ovidius University
Department of Plasma Physics
Faculty of Physics, Chemistry
Electronics and Oil Technology
Mamaia 124, Constanta 900527
Romania

Takayuki Watanabe
Tokyo Institute of Technology
Department of Environmental
Chemistry and Engineering
4259-G1-22, Nagatsuta
Midori-ku, Yokohama, 226-8502
Japan

Yukio Watanabe
Kyushu University
Kyushu Electric College
4-4-5 Sumiyoshi
Fukuoka 812-0018
Japan

Yukinori Yamauchi
Matsuyama University
Department of Pharmaceutical
Physical Chemistry
Faculty of Pharmaceutical
Sciences
4-2 Bunkyo-cho, Matsuyama,
Ehime 790-8578
Japan

Yasuhiro Yamauchi
Mitsubishi Heavy Industries Ltd.
Solar Power System Business
Unit
Renewable Energy Business
Tsukuba-machi 6-53
Isahaya, 854-0065
Japan

Akira Yonesu
Saga University
Faculty of Science and
Engineering
1 Honjo-machi, Saga-shi
Saga, 840-8502
Japan

Toyonobu Yoshida
University of Tokyo
School of Engineering
Department of Materials
Engineering
7-3-1 Hongo
Bunkyo-ku, Tokyo, 113-8656
Japan

Francisco Yubero
Instituto de Ciencia de Materiales
de Sevilla (CSIC-Univ. Sevilla)
Avda. Américo Vespucio 49
41092, Sevilla
Spain

Noureddine Zouzou
Université de Poitiers
Laboratoire d'Etudes
Aérodynamiques
CNRS, Bd Marie et Pierre Curie
BP 30179, 86962, Futuroscope
Cedex
France

Alla Zozulya
University of Wisconsin
Department of Pathology and
Laboratory Sciences
Madison, WI 53705
USA

and

University of Geneva
Medical Faculty Geneva
Rue Gabrielle- Perret-Gentil 4
CH-1211 Geneva 14
Switzerland

1
Introduction to Plasmas

Hideo Ikegami

1.1
Plasmas

In physics and engineering, the word "plasma" means electrically conductive ionized gas media composed of neutral gases, ions, and electrons. Words like solid-state plasmas can be used instead of plasmas, because they show certain semiconductor phenomena analogous to known gaseous plasma phenomena, such as current-driven instabilities, a traveling-wave amplification, plasmon excitations, and so on.

It will not be exaggerating to say that the space in the universe is filled mostly with plasmas. An impressive example of plasmas in our daily life is lightning, which discharges a current from 30 kA up to 100 MA emitting light and radio waves. In such lightning plasmas, the electron temperature can approach as high as 28 000 K (2.4 eV) and electron densities may exceed 10^{24} m^{-3}(10^{18} cm^{-3}). One can also estimate that lightning plasmas are almost fully ionized by considering that there are 3.3×10^{22} m^{-3}(3.3×10^{16} cm^{-3}) neutral atoms in the atmosphere at room temperature.

1.1.1
Plasma Characteristics

Plasmas can be loosely described as an electrically neutral medium of electrons, positive ions, and neutrals in the gas phase which have the Maxwellian velocity distribution for the electron as follows [1]:

$$f_e(\vec{v}) = n_e A_e \exp\left(\frac{-m_e v^2}{2\kappa T_e}\right) \tag{1.1}$$

where $f_e(\vec{v})\,d\vec{v}$ is the electron number density, κ is Boltzmann constant, and T_e is the electron temperature. Each of the plasma components, the electrons, ions, and neutrals, has different density, mass and temperature, but $n_e = n_i = n_0$ which is the plasma density.

Industrial Plasma Technology. Edited by Yoshinobu Kawai, Hideo Ikegami, Noriyoshi Sato, Akihisa Matsuda, Kiichiro Uchino, Masayuki Kuzuya, and Akira Mizuno

ISBN: 978-3-527-32544-3

There are two important parameters that characterize plasmas: one is the electron plasma frequency and the other is the Debye length. For simplicity, consider a one-dimensional electron slab on a uniform ion background. If the electron slab is displaced from its neutral position by an infinitesimal distance δx, the electric field generated along the slab is $E_x = -en_0\delta x/\varepsilon_0$, where ε_0 is the permittivity of the free space, and from the equation of motion,

$$m_e d^2\delta x/dt^2 = -eE_x = -e^2 n_0 \delta x/\varepsilon_0 \tag{1.2}$$

and the electron plasma frequency is given by

$$\omega_p^2 = n_0 e^2/\varepsilon_0 m_e \tag{1.3}$$

$$f_p = 9 \times 10^3 n_0^{1/2} \text{ Hz} \quad \text{for } n_0 \text{ in cm}^{-3}$$

The electron plasma frequency implies that electric potential fluctuations in plasmas with their frequencies below the electron plasma frequency must be suppressed. For fluctuations with higher frequencies, the electrons cannot respond and waves are generated in the plasma.

Now let us consider the potential distribution around ions in the plasma. If the electrons have no thermal energy, a cloud of electrons would be trapped around each ion, and there would be no electric field present in the body of the plasma outside the cloud.

However, if the electrons have thermal energies, the electron cloud stays at the radius where the potential energy is approximately equal to the thermal energy, and the shielding is not complete anymore.

For simplicity, the Poisson's equation in one-dimension is given as

$$\varepsilon_0 d^2\phi/dx^2 = e(n_e - n_i) \tag{1.4}$$

$$= en_0\left\{[\exp(e\phi/\kappa T_e)] - 1\right\} \tag{1.5}$$

$$= n_0 e^2\phi/\kappa T_e \quad \text{for } e\phi/\kappa T_e \ll 1 \tag{1.6}$$

The Debye length λ_D is thus given by

$$\lambda_D = \left(\varepsilon_0 \kappa T_e/n_0 e^2\right)^{1/2} \tag{1.7}$$

$$\lambda_D = 743 \times T_e^{1/2} n_0^{1/2} \text{ cm} \ \left(\text{for } T_e \text{ in eV and } n_0 \text{ in cm}^{-3}\right)$$

and the solution of Eq. (1.6) is given by

$$\phi = \phi_0 \exp(-|x|/\lambda_D) \tag{1.8}$$

For the plasma to be quasineutral, charged particles must be close enough so that each particle influences many nearby charged particles, rather than just interacting with the closest particle, which is a distinguishing feature of plasmas. The quasineutral plasma approximation is valid when the number of charged particles within the Debye sphere is much higher than unity, $n_0\lambda_D^3 \gg 1$.

From Eqs. (1.8) and (1.3), one also finds that the Debye length is the distance that the thermal electron can travel during one period of the plasma frequency. Obviously, the Debye length must be short compared to the physical size of the plasma.

1.2 Discharge Plasmas

When high power is injected into a relatively high-pressure (such as 10^5 Pa) gas, "thermal plasmas" are produced with the temperature as high as 15,000 K ($T_e = T_{ion} = T_{gas}$). "Although the plasma depends very much on the input power, there is no temperature difference among electrons, ions and neutrals for the pressure above 10^4 Pa, because of heavy collisions; this kind of plasma is called "*thermal plasma*"." On the contrary, in the lower pressure range below 10^4 Pa, the temperature difference between the electrons and ions increases ($T_e \gg T_{ion} = T_{gas}$), and the kind of plasma is called "*glow discharge plasma*". According to a rough estimation, in the glow plasma the ion temperature stays at about 300 K, but the electron temperature tends to increase from 5,000 K (0.5 eV) at 10^4 Pa to 20,000 K (2 eV) at 1 Pa.

Plasmas in laboratories for researches and industrial uses are mostly generated by electrical discharges in vacuum chambers of various sizes equipped with a gas feeder and a pumping system. Inside the chamber, various gadgets are implemented around a substrate on a position controllable holder with a heater and monitors.

1.2.1 Glow Plasma

A glow discharge plasma generally operates in the pressure range from 100 to 0.1 Pa and the degree of ionization remains a few percent. Quite often inert gases like argon or helium are used to generate the main plasma into which various seed gases like CCl_4, CF_4, $SiCl_4$, SiH_4, and so on, are introduced depending upon specific subjects of plasma processing.

What is the role of the glow plasma in the system? Firstly, the electrons in the inert gas plasma interacting with the seed gas molecules create useful radicals for synthesizing materials, which may be called "*plasma chemistry*." Secondly, the electrons can also produce etchant atoms, usually Cl or F, which combine with substrate atoms and fly off as volatile molecules. The processing is known as "*dry etching*" to fabricate nanoscale structures. Thirdly, the electric field across the plasma sheath straightens the orbit of etching ions, and flat etching surfaces can be structured. Fourthly, the potential drop across the plasma sheath can be controlled by controlling the electron temperature of the main plasma facing the substrate.

When it comes to plasma processing, key issues are accumulated at the plasma–substrate boundary, and one will find controlling of the plasma electron temperature is the crucial issue, because it is proportional to the ion bombardment energy.

The plasmas generated by the application of DC or low-frequency-RF ($<$100 kHz) electric field in the gap between two metal electrodes are probably the most common glow discharge plasmas.

Capacitively coupled plasma (CCP) and inductively coupled plasma (ICP) are both glow discharge plasmas produced with high-frequency (typically 13.56 MHz)

RF sources. These are widely used in the field of plasma processing such as microfabrication and integrated-circuit manufacturing with dry etching and plasma-enhanced chemical vapor deposition. These are not only employed for producing CCP and ICP but are also widely employed to generate glow discharge plasmas with various kinds of electrodes to meet specific requirements. A large ladder-shaped electrode, for example, a kind of self-discharging RF antenna, is devised for fabricating a large-scale solar panel. Many innovative ideas have been proposed to realize low-cost/high-efficiency thin-film silicon solar cells.

Helicon wave plasma (HWP) [2] can also generate glow plasmas with an order of magnitude higher density than CCP and ICP with the same power, but the mechanism is not fully understood yet. HWP typically requires a complicated electrode structure with the coaxial magnetic field for the wave propagation. Concerning the effect of the magnetic field, in most cases the ions have little effect, but the electrons are anisotropic with their properties in the direction parallel to the magnetic field being different from those in the direction perpendicular to it.

1.2.2
Atmospheric Plasmas

It may be interesting to note that in the history of plasma application only a few products carried or used plasmas in their final stage, for example, surface modifier, electrosurgical instrument, arc welder, plasma spray, plasma display (TV), fabric synthesizer, and so on. These are all devised to use atmospheric, or nearly atmospheric, plasmas. On the other hand, glow plasmas are used as a tool for manufacturing.

- **Plasma spray** is the typical thermal plasma generated using high power supplies. The development of the plasma spray processing started with the invention of the DC plasma torch in the 1960s. In the thermal spraying process, raw materials are sprayed in either powder, solution, or vapor form, and metallurgical processes are studied. The fundamental role of the plasma is to provide the extremely high temperature environment and its flow, which are not achievable by any combustion flames [3].
- **Streamer corona** was once extensively studied in relation to electrosurgical instruments [4, 5]. When high RF voltages (300 kHz, 5 kV) are applied to a sharp electrode tip mounted on a narrow piping and at the same time gases like He or Ar are sent into the same piping from a side arm, a long streamer corona plasma of He or Ar extends steadily out of the piping tip. The plasma stream gives rise to coagulation without excessive drying, sterilization, and destruction of the spore's genetic activity.
- **Dielectric barrier discharge** is widely used in the web treatment of fabrics. The application of the discharge plasma to synthetic fabrics and plastics modifies their microstructure, and functionalizes the surface to allow for paints, glues, and similar materials to adhere.

References

1. Chen, F.F. (1984) *Introduction to Plasma Physics and Controlled Fusion*, 2nd edn, Plenum Press, New York.
2. Chen, F.F. (2008) *Advanced Plasma Technology*, Wiley-VCH Verlag, KgaA, Weinheim, Chapter 6.
3. Kambara, M., Huang, H., and Yoshida, T. (2008) *Advanced Plasma Technology*, Wiley-VCH Verlag, KgaA, Weinheim, Chapter 23.
4. Gordon, M.G. and Beuchat, C.E. (1994) Technique for incorporating an electrode within a nozzle. US Patent 5,320,621, Birtcher Medical Systems, Inc., Irvine, June 14, 1994.
5. Fleenor, R.P. (1994) Apparatus for supporting an electrosurgical generator and interfacing such with an electrosurgical pencil and an inert gas supply. US Patent 5,330,469, Beacon Laboratories, Inc., Broomfield, July 19, 1994.

2
Environmental Application of Nonthermal Plasma

Akira Mizuno

2.1
Introduction

Nonthermal plasma (NTP) at atmospheric pressure produces highly reactive radicals (O, OH, etc.) that promote chemical reactions to treat gaseous pollutants [1–5]. To improve the energy efficiency of plasma chemical processes, a combination of NTP and catalysts has been proved to be effective, and applied in several commercialized products such as indoor air cleaning and VOC decomposition [6, 7]. The surface of the TiO_2 catalyst is possibly activated by ionic oxygen radicals, and the catalyst shows increased oxidation ability for gaseous pollutants when the surface is exposed to discharge plasma. This combination was used in the first mass production apparatus for indoor air cleaning. There are several ways to generate NTP. Pulsed streamer discharge, dielectric barrier discharge (DBD), and packed bed are well known. In addition, honeycomb discharge has recently been developed [8]. Selection of the type of discharge and catalyst is very important to improve the efficiency. NTP can be applied to liquids as well [9, 10]. For example, nitric acid can be reduced to ammonia. Wet-type plasma reactor is very effective in cleaning exhaust gases, and the conversion of NO_3 to NH_4 enhances the absorption of NO_2 as well as reduces water consumption.

2.2
Generation of Atmospheric Nonthermal Plasma

There are several types of NTP that can be used for environmental remediation. Figure 2.1 shows some of the typical nonthermal discharge plasmas. (a) Pulsed streamer discharge can be generated using fast rising pulse voltages applied to a nonuniform electrode geometry [11]. (b) Packed bed is a kind of DBD using dielectric pellets between the electrodes stressed by pulsed or AC voltage. Ionization takes place at the contact points between the dielectric pellets. To improve the energy efficiency and selectivity, a combination of plasma and catalyst is deemed suitable and packed bed can easily realize this combination [12]. (c) Surface discharge can

Industrial Plasma Technology. Edited by Yoshinobu Kawai, Hideo Ikegami, Noriyoshi Sato, Akihisa Matsuda, Kiichiro Uchino, Masayuki Kuzuya, and Akira Mizuno

ISBN: 978-3-527-32544-3

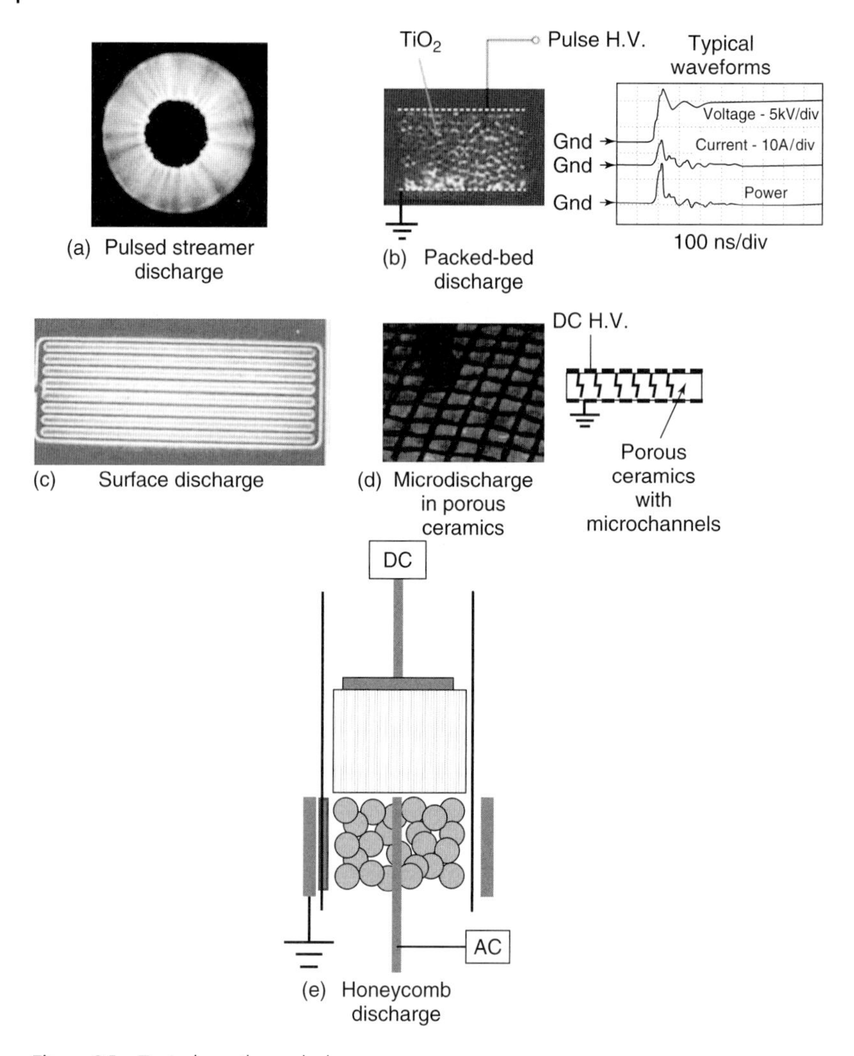

Figure 2.1 Typical nonthermal plasma.

be generated around a metal electrode strip attached to the surface of an insulating plate. The strip electrode is stressed by AC high voltage with respect to an induction electrode embedded in the insulating plate. The discharge appears as a short pulse due to the presence of the insulating plate [13]. (d) When a DC voltage is applied to a point-to-plane gap with a porous ceramic plate placed on the ground plane, corona discharge occurs at the point and breakdown takes place in each channel of the porous plate. Using a mesh electrode on a porous ceramic, channel discharge can also be generated. This channel discharge, designated as "microdischarge," is similar to back corona and can be used to generate NTP inside porous plate or honeycomb [14]. (e) In order to ionize honeycombs consisting of fine channels, a packed-bed discharge is used in front, and DC electric field is applied across the honeycomb. This electrode configuration enables the ionization of the fine channels (1 mm^2) of a honeycomb made of cordierite [8].

Silent discharge (or DBD) has been used for a long time for ozone generation. Corona discharge has also been used widely in many applications. Corona radical shower uses spray of reactive gas received from a corona point.

These discharge plasmas produce radicals for chemical reactions. Quantitative measurement of radicals, however, is difficult. A relatively reliable method is to use selective chemical reactions. For instance, CO is oxidized faster by OH than by O radicals [15]. Using this chemical reaction, OH production in pulsed streamer corona discharge in Argon gas has been measured. By selecting the appropriate concentration of CO, the concentration of generated OH can be controlled. Recently, optical measurements of OH production and other radicals have been reported [16]. Advancement of these measurements is necessary to improve the efficiency of chemical reactions promoted by NTP.

2.3 Indoor Air Cleaning System Using Plasma and Catalyst Combination

Contamination of indoor environments is a great concern as a cause of many health problems. Plasma/catalyst combination has been applied to indoor air cleaners [6, 12]. Figure 2.2 shows a first mass-produced plasma reactor attached to a room air conditioner. The reactor consists of wire-plate electrode system stressed by pulses superimposed on DC voltage as shown in Figure 2.2b. The DC voltage contributes to the collection of suspended particles. To enhance the removal performance of gaseous pollutants, TiO_2 catalyst was placed just after the electrode system to be exposed to the electric field, as depicted in Figure 2.2a. The catalyst was prepared by coating an aluminum mesh with the catalyst powders.

Results of the one-pass test are given in Table 2.1. The measurement was carried out at room temperature and atmospheric pressure. The velocity of the air flow was 1.0 m s^{-1}, the gas residence time was 10 ms, and the pressure drop in the reactor was less than 1 mm H_2O. Acetaldehyde (CH_3CHO) diluted

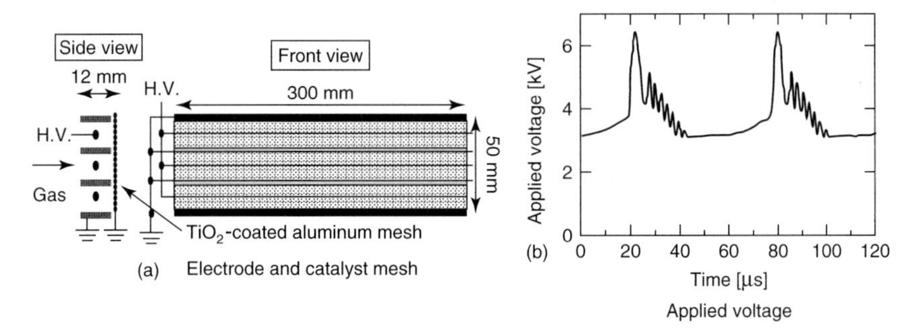

Figure 2.2 Indoor air cleaning system for air conditioner using plasma/catalyst combination.

Table 2.1 Performance of the indoor air cleaning system with plasma/catalyst combination.

Voltage (kV)	Particle collection efficiency (%) (for 0.5-μm diameter)	Acetaldehyde removal efficiency (%) (Inlet concentration: 1.0 ppm)
0	12	0.5
3.0 (DC) + 2.6 (pulse)	70	27 15 (without the catalyst mesh)

with dry air (initial concentration: 1.0 ppm) was used as the sample gas. Without stressing the reactor, absorption produced acetaldehyde removal efficiency of about 0.5% and particle collection efficiency of 12%. With stress to the reactor, 70% particle removal (0.5 μm size) and 27% acetaldehyde removal were achieved. The ozone generation during this operating condition was less than 0.1 ppm and the energy efficiency was 55.4 eV per molecule-CH_3CHO. With large air-circulation rate of the air conditioner, this one-pass efficiency would be practically effective.

Synergetic effect of the combination of pulsed streamer corona and TiO_2 was studied. TiO_2 is known as a *photocatalyst*, and the oxidation ability of TiO_2-coated mesh stained by indigo-carmine dye was investigated. When oxidized, the dye becomes clear. Figure 2.3 shows the TiO_2- coated mesh and the comparison of decolorization before and after oxidation.

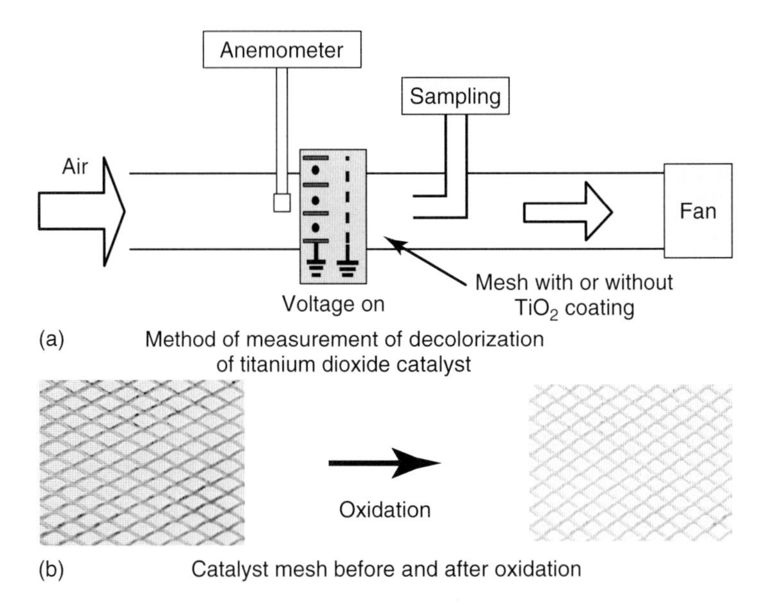

Figure 2.3 Oxidation of indigo-carmine dye on TiO_2 catalyst surface.

Table 2.2 Decolorization of indigo carmine on the surface of TiO_2 catalyst combined with the plasma.

			TiO_2 coating on the mesh	Color density (arbitrary unit, before the experiment, the density was set at 1.00)
Plasma on (ozone: less than 0.05 ppm)			+	0.6
Air flow 1.0 m s^{-1}, direction: plasma to mesh			−	0.99
Plasma on, air flow 2.5 m s^{-1}	Direction	Electrode system to mesh	+	0.76
		Mesh to electrode system	+	0.88
Plasma off, exposed to air with ozone of 0.2 ppm			+	0.84
			−	1
UV ray (quartz glass was inserted between the electrode and the catalyst)			+	0.99

A color meter was used for quantitative measurement of the dye's color. Table 2.2 compares the color change under different conditions. Combination of plasma and the catalyst was effective to oxidize the dye with air flow from plasma electrode system to mesh. At an air-flow velocity of 1.0 m s^{-1}, ozone concentration was less than 0.05 ppm. Ozone might affect the oxidation; therefore, the direction of the air flow was reversed. Thus, the air passes at first through the catalyst mesh, and then arrives in the plasma zone ensuring that ozone does not reach the surface of the TiO_2 mesh. This is in addition to the increase of flow velocity to 2.5 m s^{-1}. The dye on the mesh was still oxidized. For comparison purpose, the effect of 2.5-ppm ozone supplied by an ozonizer was tested. The dye was oxidized at the same level as that with the plasma. Also, the effect of UV from the plasma was checked. The UV did not activate the catalyst. These results suggest that O_3 ions (in this case, positive ions) may activate TiO_2 surface exposed to plasma.

Ogata *et al.* reported on the mechanism of activation of TiO_2 surface being exposed to plasma in argon [17]. Oxygen atom on the lattice of TiO_2 and H_2O trapped on the surface could be activated to become oxygen radicals. Generation of oxygen radicals from O_3 on various surfaces has been studied. Further understanding of surface chemistry coupled with plasma will contribute to improving the energy efficiency of surface plasma chemical reactions.

Indoor air cleaners for homes and offices have been widely used, and the combination of plasma and catalyst has been adopted in these applications.

2.4
Diesel Exhaust Cleaning

Cleaning diesel engine exhaust is potentially an important application of NTP technology, and various research works have been carried out [18, 19]. Reduction of NO_x" to their individual elements (N_2 and O_2) is the most attractive way, especially for exhaust from vehicles. In the two-stage plasma process as shown in Figure 2.4, NO is oxidized to NO_2 and fed to the catalyst, where low-temperature reduction of NO_2 is achieved by adding a reductive agent such as ammonia [20]. With Co–ZSM-5 catalyst at 150 °C, NO_x can be removed by catalytic reaction. For example, the following reactions were observed: NO at 200 ppm concentration was completely oxidized to NO_2 with SIE of 27 kJ Nm^{-3} in the plasma reactor, and introduced to the catalyst section. To assist the oxidation of NO to NO_2 in the discharge plasma reactor, 1200 ppm of ethylene was added. At the inlet of the catalyst section, 200 or 400 ppm of ammonia was added. At 150 °C, continuous reduction of NO_2 was observed without the need for input energy. The NO concentration was higher at the outlet of the catalyst bed than at the outlet of the plasma reactor. This

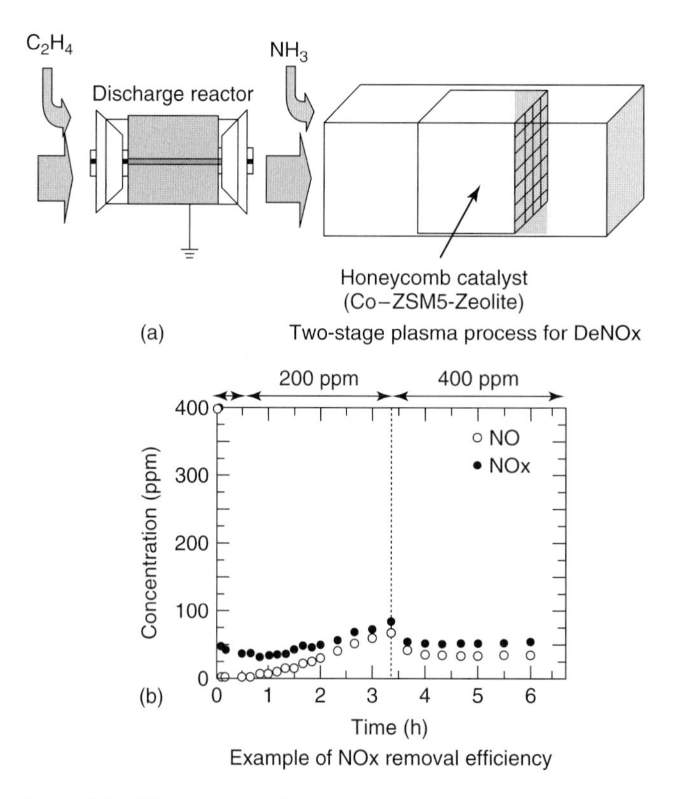

Figure 2.4 NO_x removal in the two-stage process with Co–ZSM-5 catalyst (SIE: 27 kJ N^{-1} m^{-3}, GHSV: 1820 h^{-1}, temperature: 150 °C, C_2H_4: 1200 ppm, NH_3: 100–400 ppm).

indicates that part of NO_2 is converted to NO over the catalyst bed. This observation supports the reductive removal of NO_2 over the Co–ZSM-5 catalyst.

It should be noted that there is another NO_x reduction process in which ratio of NO and NO_2 should be 1 : 1. This reaction is suitable for plasma process since the ratio can be controlled rapidly by adjusting the intensity of the plasma.

Carbon soot could be used as a reductive agent for NO_x. Using a packed bed with Pt-catalyst as shown in Figure 2.5, the reduction of NO was confirmed [21]. The plasma reactor consisted of a stainless steel rod (outer diameter 7 mm) extending along the axis of the ceramic tube and serving as the high-voltage electrode. Inner diameter and length of the ceramic tube were 16 and 150 mm, respectively. Ground electrode was wrapped around the outer surface of the ceramic tube. Catalyst pellets, $Pt-Al_2O_3$ (2.5–4.0 mm in diameter) were packed between the stainless steel rod and the ceramic tube. Pulsed square high voltage of 15 kV peak with 20-ns rising time and 120-Hz frequency was used. A diesel engine exhaust with a flow rate of 6.0 L min^{-1} was used as the test gas with space velocity of 36 000 h^{-1} in the plasma reactor. The pressure drop across the reactor was about 380–540 Pa. The measurement was made under rich oxygen (13–15%) and low temperature (423 K) conditions. The diesel exhaust particulate (DEP) in the exhaust was collected on the catalyst pellets with more than 95% collection efficiency and estimated to be 5.0–7.0 mg m^{-3}. By exposing the catalysts to the pulsed plasma, the collected DEP was continuously oxidized to CO_2.

As shown in Figure 2.6, concentrations of NO, NO_x, and C_2H_4 were rapidly decreased with stress to the reactor, and a large amount of CO_2 was formed at the same time. Under this test condition, 73% rate of removal of NO_x was obtained at energy consumption of SIE 43 kJ N^{-1} m^{-3} (44 eV per molecule-NO). This energy requirement is about 5% of the engine output power. Further reduction to less than 30 kJ N^{-1} m^{-3} is a target value for exhaust gas cleaning.

NO_x and hydrocarbons (mainly C_2H_6 and CH_4) were converted to N_2, CO_2, H_2O, and N_2O as shown in Table 2.3. CO_2 was generated from ethylene and carbon soot. Color of the pellets became brown to black when carbon soot was attached. It was confirmed that the surface recovered its original color after being exposed to the plasma. Generation of N_2O should be suppressed by further improvement

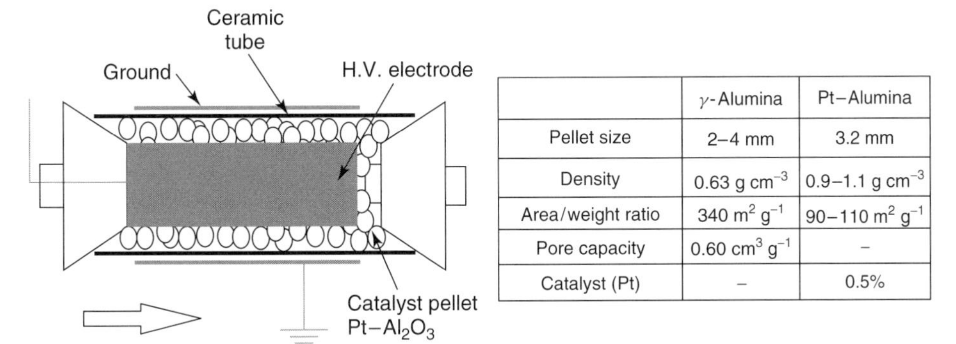

	γ-Alumina	Pt–Alumina
Pellet size	2–4 mm	3.2 mm
Density	0.63 g cm^{-3}	0.9–1.1 g cm^{-3}
Area/weight ratio	340 m^2 g^{-1}	90–110 m^2 g^{-1}
Pore capacity	0.60 cm^3 g^{-1}	–
Catalyst (Pt)	–	0.5%

Figure 2.5 Packed-bed reactor with catalyst for diesel exhaust cleaning.

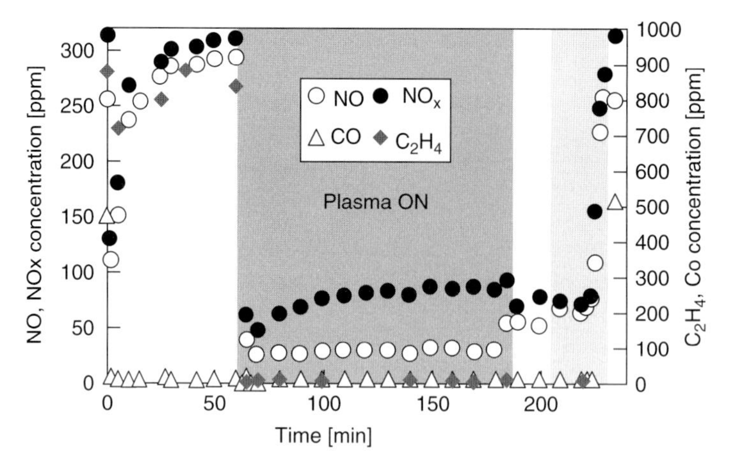

Figure 2.6 Concentration of NO_x and other gaseous components using Pt-catalyst. Diesel engine exhaust 6.0 L min^{-1}, additive: C_2H_4 900 ppm, SV: 36 000 h^{-1}, SIE: 43–46 kJ N^{-1} m^{-3}.

Table 2.3 By-products of the NO_x treatment using the Pt-catalyst packed bed.

	Inlet gas	Plasma off	Plasma on
NO (ppm)	255	294	30
NO_x (ppm)	313	310	85
C_2H_2 (ppm)	881	837	5
CO (ppm)	514	12	11
N_2O (ppm)	0	0	98
CO_2 (%)	3.85	3.9	4.3

of reaction conditions. Packed bed can also be used to generate ammonia at temperatures less than 150 °C for catalytic reduction of NO in diesel exhaust.

As noted, the combination of NTP and catalyst is effective for improving the selectivity and energy efficiency of plasma chemical reactions. Inside the honeycomb catalyst, NTP can be generated as depicted in Figure 2.1e.

2.5
Interaction of Atmospheric Plasma with Liquid

Using a wet-type plasma reactor shown in Figure 2.7, oxidation of SO_2 was studied [22]. Water was introduced to the reactor to form a water film on the inner surface. Methanol solution (2.5%) was also tested as an absorbent. The experiments were carried out at 20 °C and 1 atm. Typical gas composition in this study was

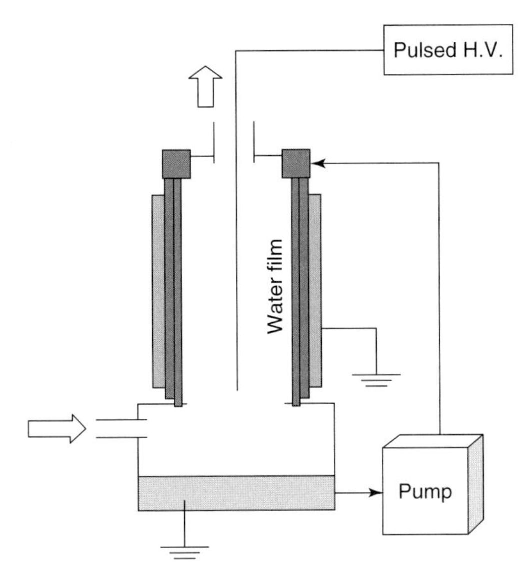

Figure 2.7 Wet-type plasma reactor to promote liquid-phase chemical reactions.

oxygen (10%), SO_2 (500 ppm), and the remaining N_2. Most of SO_2 was absorbed into the water film, and the extent of oxidation of gaseous SO_2 by the discharge plasma was small. The by-product in the liquid absorbent was analyzed with the ion chromatography (IC). The results obtained are summarized in Table 2.4. Ionic species in water resulting from the absorption of SO_2 is highly pH dependent. In this experiment, the pH range was $2 \leq \text{pH} \leq 7$, and most of the S(IV) in the water absorbent existed as HSO_3^- in the absence of plasma application. When the plasma was applied, however, this trend disappeared and the concentration of SO_4^{2-} ion was much higher than that of HSO_3^- ion. These results indicate that the oxidation of SO_2 (i.e., oxidation of HSO_3^- ion to SO_4^{2-} ion) proceeds in the liquid phase with relatively large rate. Even the discharge plasma exists only in the gas phase where the radical production takes place, and radical reaction in liquid is promoted. In comparison with water absorbent, liquid-phase oxidation of HSO_3^- could not be observed when 2.5% of methanol solution was used. The ratio between

Table 2.4 By-product distribution under different reaction conditions (specific input energy: 28.5 kJ N^{-1} m^{-3}).

Liquid	Plasma	ΔC (ppm)	S(IV) : SO_4^{2-}
H_2O	OFF	350	90 : 10
	ON	420	30 : 70
MeOH (2.5%)	OFF	220	100 : 0
	ON	300	72 : 28

S (IV) indicates bisulfite (HSO_3^-) in this case.

SO_2 removal by plasma and by absorption, and the percentages of ionic species were almost the same. This means that the formation of SO_4^{2-} was negligible in the methanol solution. One reason may be the large rate constant of the reaction of methanol with OH radical in the liquid phase (or methanol is a radical scavenger).

The experimental observation in the wet-type plasma reactor strongly suggests that the gas-phase pulsed corona directed to a liquid surface can be used to promote radical reactions in liquids. This conforms to previous findings where plasma chemical processes promoted by electrical discharge above or inside liquid could possibly be applied in decolorization, and sterilization of water and other liquids [23].

2.6 Concluding Remarks

NTP at atmospheric pressure has been utilized for cleaning indoor air, VOCs, sterilization of surface [24], and so on. NTP can possibly be applied to diesel exhaust cleaning, water treatment, fuel reforming [25–27], chemical synthesis, surface modification, thin-film deposition, and other important industrial applications. In order to improve the selectivity and energy efficiency, the combination of NTP and appropriate catalyst or absorbing surface has been found effective. The mechanism, however, has not yet been clarified. Further understanding of the interaction of NTP and solid/liquid surface will lead to various applications.

References

1. Mizuno, A. (2007) Industrial applications of atmospheric non-thermal plasma in environmental remediation. *Plasma Phys. Control. Fusion*, **49**, A1–A15.
2. Hackam, R. and Akiyama, H. (2000) Air pollution control by electrical discharges. *IEEE Trans. Dielectr. Electr. Insul.*, **7**, 654–683.
3. Penetrante, B.M. and Schultheis, S.E. (1993) *Non-thermal Plasma Techniques for Pollution Control*, NATO ASI Series, Springer-Verlag, Vol. 34.
4. Prieto, G., Prieto, O., Gas, C.R., Mizuno, A., Takashima, K., and Yamamoto, T. (2008) Nonthermal plasma reactors and plasma chemistry. *Int. J. Environ. Waste Manage.*, **2**, 349–398.
5. Kim, H.H. (2004) Nonthermal plasma processing for air-pollution control: a historical review, current issues, and future prospects. *Plasma Processes Polym.*, **1**, 91–110.
6. Mizuno, A., Kisanuki, Y., Masanobu Noguchi, S.K., Lee, S.H., Hong, Y.K., Shin, S.Y., and Kang, J.H. (1999) Indoor air cleaning using a pulsed discharge plasma. *IEEE Trans. Ind. Appl.*, **35**, 1284–1288.
7. Hyun-Ha, K., Atsushi, O., and Shigeru, F. (2006) Application of plasma-catalyst hybrid processes for the control of NO_x and volatile organic compounds, in *Trends in Catalysis Research* (ed. L.P. Bevy), Nova Science Publishers, Inc., New York, pp. 1–50.
8. Hensel, K., Sato, S., and Mizuno, A. (2008) Sliding discharge inside glass capillaries. *Plasma Sci., IEEE Trans.*, **36**, 1282–1283.
9. Sun, B., Sato, M., and Clements, J.S. (1999) Use of a pulsed high-voltage discharge for removal of organic compounds in aqueous solution. *J. Phys. D: Appl. Phys.*, **32**, 1908–1915.

10. Mededovic, S. and Locke, B.R. (2007) The role of platinum as the high voltage electrode in the enhancement of Fenton's reaction in liquid phase electrical discharge. *Appl. Catal. B: Environ.*, **72**, 342–350.
11. Masuda, S., Nakatani, H., Yamada, K., Arikawa, M., and Mizuno, A. (1984) Production of monopolar ions by traveling wave corona discharge. *IEEE Trans. Ind. Appl.*, **IA-20**, 694–702.
12. Mizuno, A., Yamazaki, Y., Ito, H., and Yoshida, H. (1992) AC energized ferroelectric pellet bed gas cleaner. *IEEE Trans. Ind. Appl.*, **28**, 535–540.
13. Masuda, S., Hosokawa, S., Tu, X., and Wang, Z. (1995) Novel plasma chemical technologies – PPCP and SPCP for control of gaseous pollutants and air toxics. *J. Electrostatics*, **34**, 415–438.
14. Hensel, K., Katsura, S., and Mizuno, A. (2005) DC microdischarges inside porous ceramics. *IEEE Trans. Plasma Sci.*, **33**, 574–575.
15. Su, Z.Z., Ito, K., Takashima, K., Katsura, S., Onda, K., and Mizuno, A. (2002) OH radical generation by atmospheric pressure pulsed discharge plasma and its quantitative analysis by monitoring CO oxidation. *J. Phys. D: Appl. Phys.*, **35**, 3192–3198.
16. Ono, R. and Oda, T. (2008) Optical diagnosis of pulsed streamer discharge under atmospheric pressure. *Int. J. Plasma Environ. Sci. Technol.*, **1**, 123–129.
17. Ogata, A., Einaga, H., Kabashima, H., Futamura, S., Kushiyama, S., and Kim, H.H. (2003) Effective combination of nonthermal plasma and catalysts for decomposition of benzene in air. *Appl. Catal. B: Environ.*, **46**, 87–95.
18. Okubo, M., Arita, N., Kuroki, T., and Yamamoto, T. (2007) Carbon particulate matter incineration in diesel engine emissions using indirect nonthermal plasma processing. *Thin Solid Films*, **515**, 4289–4295.
19. Yoshioka, Y. (2008) Recent developments in plasma De-NO_x and PM (particulate matter) removal technologies from diesel exhaust gases. *Int. J. Plasma Environ. Sci. Technol.*, **1**, 110–122.
20. Kim, H.H., Takashima, K., Katsura, S., and Mizuno, A. (2001) Low-temperature NO_x reduction processes using combined systems of pulsed corona discharge and catalysts. *J. Phys. D: Appl. Phys.*, **34**, 604–613.
21. Matsui, Y., Sawada, J., Koyamoto, I., Takashima, K., Katsura, S., and Mizuno, A. (2005) After-treatment of NO_x using combination of non-thermal plasma and oxidative catalyst prepared by novel impregnation. *J. Adv. Oxid. Technol.*, **8**, 255–261.
22. Kinoshita, Y., Saito, K., Takashima, K., Katsura, S., and Mizuno, A. (2001) Removal of SO_2 using a wet-type non-thermal plasma reactor. Conference Record – IAS Annual Meeting (IEEE Industry Applications Society), pp. 693–697.
23. Li, J., Sato, M., and Ohshima, T. (2007) Degradation of phenol in water using a gas-liquid phase pulsed discharge plasma reactor. *Thin Solid Films*, **515**, 4283–4288.
24. Tanino, M. and Mizuno, A. (2007) Sterilization using dielectric barrier discharge at atmospheric pressure. *Int. J. Plasma Environ. Sci. Technol.*, **1**, 108–113.
25. Okumoto, M., Hyun Ha, K., Takashima, K., Katsura, S., and Mizuno, A. (2001) Reactivity of methane in nonthermal plasma in the presence of oxygen and inert gases at atmospheric pressure. *IEEE Trans. Ind. Appl.*, **37**, 1618–1624.
26. Prieto, G., Okumoto, M., Shimano, K.I., Takashima, K., Katsura, S., and Mizuno, A. (2001) Reforming of heavy oil using nonthermal plasma. *IEEE Trans. Ind. Appl.*, **37**, 1464–1467.
27. Nozaki, T. and Okazaki, K. (2005) Non-equilibrium reaction fields created by atmospheric pressure plasmas and application to advanced utilization of natural gas resource. *Nihon Enerugi Gakkaishi/J. Jpn. Inst. Energy*, **84**, 462–467.

3
Atmospheric Plasma Air Pollution Control, Solid Waste, and Water Treatment Technologies: Fundamental and Overview

Jen-Shih Chang

3.1
Introduction

It has been suggested that air pollution has a large impact not only on human health but also on the economy, where approximately 0.06–0.09% of the population experiences premature death caused by air pollution [1, 2] and 2.5–3 times of these numbers are admitted to hospital [2]. The average cost of mortality was US$4M per person and adult chronic bronchitis costs an average of $279 000 per person for treatment [2].

3.1.1
Nonthermal Plasma Air Pollution Control

Nonthermal plasma pollution control technology may assist to reduce these health and economic impacts in the near future. Air pollutants can be categorized as follows [5, 6, 13]:

1) Particulate matter
2) Acid gases such as SO_x, NO_x, HCl, and so on.
3) Greenhouse gases such as CO_x, N_xO_y, PFCs (para-fluorocarbons), and so on.
4) Ozone depletion substances (ODSs) such as Freons, Halon, and so on.
5) Volatile organic compounds (VOCs) such as toluene, xylene, trichloroethylene (TCE), trichloroethane (TCA), ethylene glycol monoethyl ether (EGM), and so on.
6) Toxic gases such as Hg, dioxins, and so on.
7) Radioactive gases such as isotopes of C, Rn, I, Kr, and so on.

The three options for their treatment at the sources are as follows:

1) Particulates collection by electrostatic precipitators and filters, and adsorption of gaseous contaminants by water or adsorbents, followed by landfill or incineration.

Industrial Plasma Technology. Edited by Yoshinobu Kawai, Hideo Ikegami, Noriyoshi Sato, Akihisa Matsuda, Kiichiro Uchino, Masayuki Kuzuya, and Akira Mizuno

ISBN: 978-3-527-32544-3

2) Decomposition or conversion of gaseous pollutants to innocuous gases such as water, oxygen, and nitrogen.
3) Conversion of gaseous pollutants to recyclable materials such as fertilizers and dry ice.

Option (1) has been the traditional approach but owing to the secondary pollution and misuse of resources, it is desirable to use option (3) for highly concentrated pollutants for resource recovery, and option (2) for low-concentration pollutants for detoxification.

Gaseous pollution control technologies for acid gases (NO_x, SO_x, etc.) [1, 6], VOCs [3], greenhouse gases [4, 6], ODS [5, 6], and so on, have been commercialized based on catalysis, incineration, and adsorption methods. However, the nonthermal plasma techniques based on electron beams and corona discharges become significant due to the advantages such as lower costs, higher removal efficiency, smaller space volume, and so on. In order to commercialize this new technology, the pollution gas removal rate, energy efficiency of removal, pressure drop of reactors, and usable by-products production rates must be improved and identification of major fundamental processes, optimizations of reactor, and power supply for an integrated system must be done.

3.1.2
Thermal Plasma Solid Waste Treatment

For solid waste treatment such as incinerator ash and low radiation level nuclear waste treatment, the technology was already commercialized by oxidation principle. However, the potential for the reduction approaches is also analyzed. The liquid wastes considered are (i) waste oils from automobile and chemical industries; (ii) toxic fluids such as PCBs from electrical industries; (iii) ozone depletion fluids such as CFCs, Halon, and so on; and (iv) solvents from chemical and electronic industries. The solid wastes considered are (i) industrial wastes such as automobile tires, plastics, petroleum process by-products, and so on; (ii) medical wastes from hospitals and pharmaceutical industries; (iii) municipal wastes; and (iv) electronic wastes and semiconductor process by-products. Solid-waste disposal has become a major concern since there are only a limited number of landfill disposal sites. In the past, incineration of solid wastes was considered a solution to the landfill site limitation since solid-waste volume can be reduced to 1/6 by conversion to incinerator ash and heat and electricity recovery from the incineration processes is possible. The landfill of fly and bottom ashes has been considered a potential environmental hazard due to leaching of heavy metals from the ashes after landfill. Hence recycling them as construction material is beneficial not only economically but is also less impacting on the environment in the long term. The thermal plasma municipal incinerator ash volume reduction system is a recycle method that converts bottom and fly ashes from municipal incinerators to usable construction materials with simultaneous recovery of metals [6].

3.1.3 Electrohydraulic Water Treatment

The application of plasma technologies for the treatment of drinking and wastewater is not new. Four categories of plasma treatment technologies exist, including plasma injection, remote, indirect, and direct. Plasma injection method generates plasma on top of water or at the bottom of reactor [7]. Remote plasma technologies involve plasma generation in a location away from the medium to be treated (e.g., ozone). Indirect plasma technologies generate plasma near to, but not directly within, the medium to be treated (e.g., UV, electron beam). More recently, direct plasma technologies (i.e., electrohydraulic discharge) have been developed that generate plasma directly within the medium to be treated, thereby increasing treatment efficiency. Three types of electrohydraulic discharge systems (pulsed corona electrohydraulic discharge (PCED), pulsed arc electrohydraulic discharge (PAED), and pulsed power electrohydraulic discharge (PPED)) have been employed in numerous environmental applications, including the removal of foreign objects (e.g., rust, zebra mussels), disinfection, chemical oxidation, and the decontamination of sludge [6, 7].

In this work, all these recent developments are introduced and the fundamental physics, chemistry, engineering, and economics are discussed.

3.2 Type of Plasmas

Different types of plasmas are normally categorized as in Table 3.1 based on the gas temperature and operating gas pressure. For solid waste treatment, due to the melting and evaporation requirements, high gas temperature and high gas pressure plasma are used. For water treatment, depending on the target pollutants, low temperature and high gas pressure plasma are used. However, the key effect will be the nature of high electron temperature in plasma status, where the plasma is the mixture of electron, negative and positive ions, excited species, and neutral atoms and molecules as shown in Figure 3.1.

Similar to the latent heat for phase changes between solid and liquid and liquid and gas, gas-to-plasma phase change required ionization or/and attachment potentials as shown in Figure 3.1. The effect of gas temperature is in the form of rotational and vibrational excitation of molecules until onset of thermal ionization. In plasmas, atoms can be electronically excited and some of the metastable state atoms can have a lifetime of the order of milliseconds where molecules can be rotationally, vibrationally, or electronically excited and some of the metastable molecules also have lifetimes of a few seconds. Plasma is also chemically active, and reactions such as ionization, excitation, dissociation, attachment, detachment, atom and molecule transfer, and so on, can coexist. Reaction rate will be higher for reactions of radicals but with ions and electrons involved in reactions, the reaction rates become few orders faster than most neutral species reactions as shown in Table 3.2.

Table 3.1 Types of industrial plasma [5].

Type	Pressure		Gas temperature		Electron temperature		Typical applications
	High	Low	High	Low	High	Low	
Low pressure plasma		O		O	O		Semiconductor Lump and lasers Display
Non thermal plasma	O			O	O		Air pollution control Waste treatment Polymer coating Polymer Treatments
Thermal plasma	O		O		O		Solid waste treatments Coating Ceramic processing Water treatments Cutting and welding
Nuclear fusion plasma		O	O		O		Energy Military

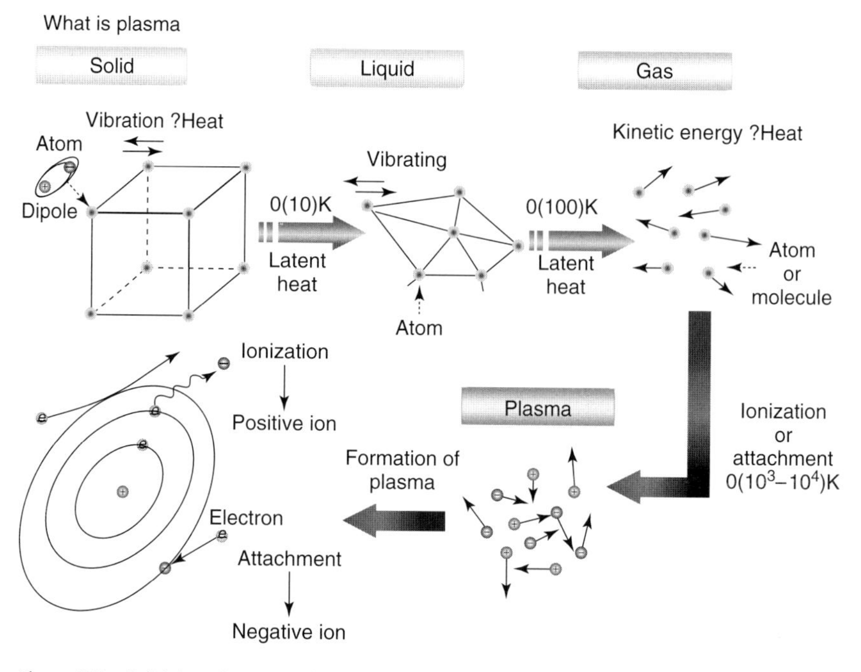

Figure 3.1 Solid, liquid, gas, and plasma status [5].

Table 3.2 Chemical reaction and rates.

Reaction	Two-body reaction rate ($cm^3 s^{-1}$)	Three-body reaction rate ($cm^6 s^{-1}$)
Molecule–molecule	10^{-14}–10^{-31}	10^{-30}–10^{-40}
Atom/radical–molecule	10^{-11}–10^{-24}	10^{-30}–10^{-36}
Ion-atom/molecule	10^{-9}–10^{-13}	10^{-28}–10^{-32}
Electron-molecule/molecule	10^{-7}–10^{-11}	10^{-27}–10^{-35}
Positive–negative ion	10^{-6}–10^{-8}	10^{-25}–10^{-26}
Electron–ion	10^{-6}–10^{-7}	10^{-26}–10^{-28}
Molecule/radical-aerosol	(10^{-5}–10^{-10}) Rp (nm) Rp: diameter of aerosol	
Comparison of dominant reaction (reaction rate) × (molecule density) × (reactant density) (× (third body molecule density))		

Modified from [16].

3.3 Plasma Chemistry

In classical plasma chemistry, thermal dissociation energy was defined from the ground state of molecules. However, under electron impact dissociation, dissociation energy can be less than a few times lower since the electron impact dissociation depends on the vibrational level of molecules and obeys Franck–Condon principle. Hence quantum chemistry model is applied [8]. In classical plasma chemistry, all the chemical reactions are assumed to be initiated by radical reactions and these radicals are generated by the direct electron impact molecule dissociation and ionization reactions. However, computer modeling [9] and recent LIF direct measurements [10, 11] show that radicals in a nonthermal plasma have a much longer life than classical recombination-based lifetime, since these radicals can be generated and their afterglow through ion–molecule reactions, dissociate recombination of ions and electrons, attachment and detachment reactions [8–13].

The gas–temperature dependence of reaction rate is more complex depending on the vibrational excitation especially when it involved electrons and this affects the role of radicals differently as compared with room temperatures. For example, the reaction rate or cross section of dissociative recombination decreases with increasing gas and hence vibrational temperature, while reaction rate and threshold energy of dissociative attachment and electron impact ionization increases and decreases, respectively, with increasing vibrational temperature [8]. Hence, the vibrational excitation that causes a major energy loss for the nonthermal plasma pollution control processes, as proposed by Penetrante *et al.* [15], should be reconsidered. On the other hand, the recent kinetic modeling of thermal plasma shows a significantly different degree of ionization than classical Saha-type ionization model and even the existence of negative ions in thermal plasmas [14]. Ion-induced aerosol formation is

Table 3.3 Reaction kinetics rate equations and relative importance between ionic and radical reaction [13].

$$\frac{dN}{dt} = k_1[A][B] + k_2[C^+][B] - \cdots$$

where

$$A + B \xrightarrow{k_1} N + M$$

and

$$C^+ + B \xrightarrow{k_2} N + L$$

However,

$$[A] \gg [C^+];\ k_1 \ll k_2$$

Thus,

$$k_1[A] \approx k_2[C^+]$$

also an important process in gaseous pollution control since aerosol surface reaction rate is a few orders of magnitude faster than the electronic, ionic, and radical reactions, and the relative order of these reaction rates are summarized in Table 3.2 [16]. Table 3.2 shows that both the two- and three-body reaction rate increase when plasma converted atoms or molecules to radicals, ions, and electrons, where reaction rate is faster for molecule $<$ radical $<$ ion $<$ electron $<$ aerosol surface reactions. Although radical density is much higher than ion density, the effectiveness to react with pollutant molecules is of a similar order as shown in reaction kinetics equation in Table 3.3 [13], where the product of radical reaction k_1 [A] is of the same order of magnitude as that of the product of ionic reaction $k_2[C^+]$ for production of [N]. Here, Table 3.3 shows the rate equations of species [N] formed by radical reaction of species [A] and [B] with reaction rate k_1 and ionic reaction $[C^+]$ and [B] species with reaction rate k_2. Table 3.2 also shows that the effect of aerosol or dust particle surface reactions is a significant process in nonthermal plasma pollution control as observed experimentally. The models that do not take into consideration ionic and/or the formation of aerosol particles should be revisited [9, 12, 13, 15, 17, 18]. The mechanism of nonthermal plasma gaseous pollution hence controls more complex processes as summarized in Figure 3.2 [16]. Next generation numerical modeling and analyses of experimental results should be based on these considerations. Chemical by-products should be measured in greater detail [10, 11, 31].

3.4 Plasma Reactor

3.4.1 Nonthermal Plasma Reactor

In nonthermal plasma gaseous pollution control, electron beam, barrier discharge, superimposed DC/AC streamer corona discharge, flow-stabilized corona discharge, and pulsed corona and pulse power techniques are used for the plasma sources [1–6]. However, in order to design more efficient plasma reactor and power supply,

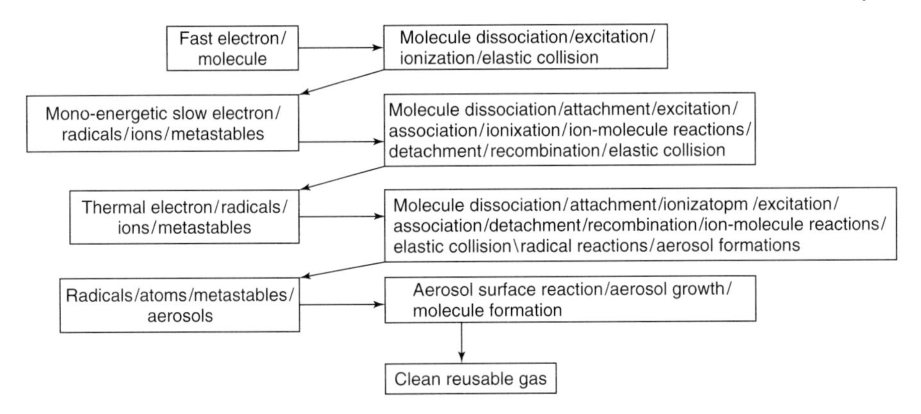

Figure 3.2 Mechanism of plasma pollution control (modified from [16]).

a physics of these nonthermal plasmas must be understood. For example, the secondary gas flow due to the electrohydrodynamic origin must be actively used to reduce the pressure drop of the reactor where the pumping power is the highest energy used components in an industrial process [19, 20]. A simple model to estimate plasma parameters from discharge current, applied voltage, and power is also presented.

DC streamer corona discharge is unstable and easily transitions to spark discharges; hence, the following approaches are used to stabilize discharge [1, 16]:

1) No sharp-edge discharge electrodes [19]
2) Flow stabilized corona [21]
3) Superimposed AC/DC discharge [22]
4) Wall stabilized (microgap) discharge [23]
5) External magnetic field [24]
6) Barrier discharge [25] and backed bed [26]
7) Pulsed corona discharges [27]
8) Plasma-catalyst and plasma adsorbent system based on the above reactors (Figure 3.3).

Typical plasma reactors are shown in Figure 3.4 and the corresponding characteristics are shown in Table 3.4.

3.4.2 Thermal Plasma Reactors

Two different types of thermal plasma processes do exist; one based on DC or AC submerged arc discharge inside materials and the other based on self-standing plasma torch [5, 8] above the material being treated. Thermal plasmas [8] considered are DC and AC arc reactors and DC, RF-inductive, RF-capacitive, and microwave plasma torches for solid waste treatment and water treatment. Typical thermal plasma reactor is shown in Figure 3.5 and their gas temperature environments are 7500–25 000 K.

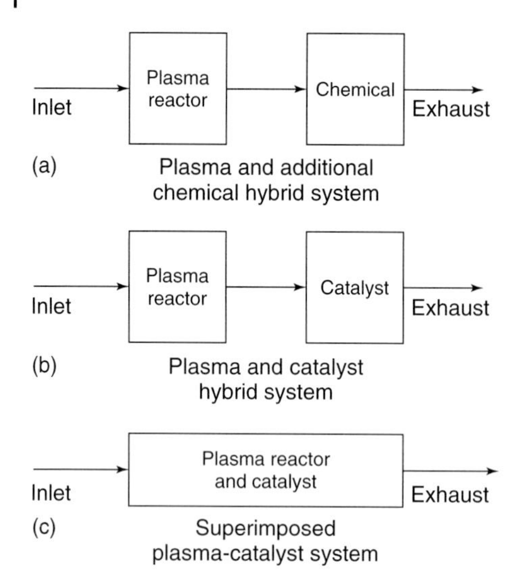

Figure 3.3 Plasma-catalysis and adsorbent systems [3].

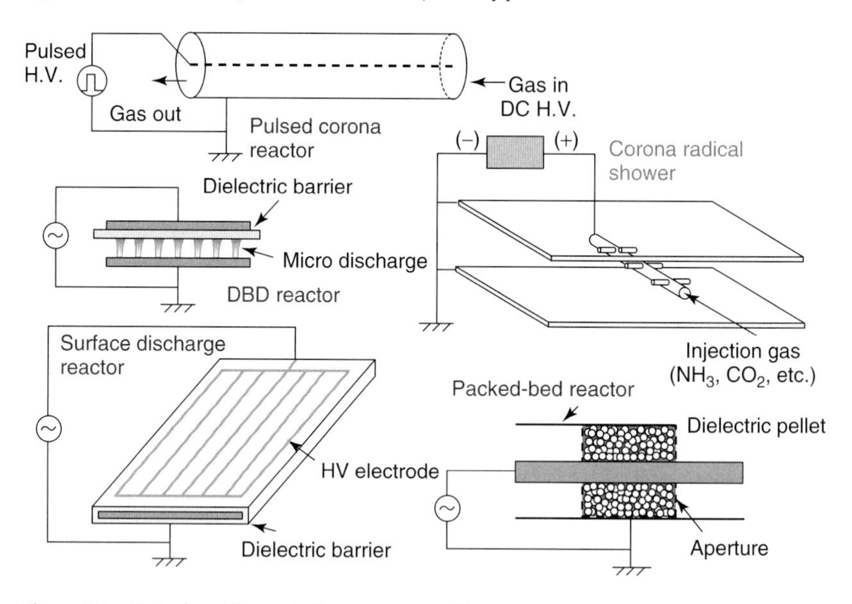

Figure 3.4 Typical nonthermal plasma reactors [1].

3.4.3 Electrohydraulic Discharge

The electrohydraulic discharge systems differ in several operational characteristics, as summarized in Table 3.5, due to their different configurations as well as the different amounts of energy injected into each type of system. The PCED system employs discharges in the range of 1 J per pulse, while the PAED and PPED

Table 3.4 Characteristics of various plasma systems.

	Voltage (V)	Current (A)	Frequency (Hz)	Power supply efficiency (%)	Pressure drop	Flowrate/flow channel ($Nm^3 h^{-1}$)
Electron beam	100 k to 200 M	1 m to 10 (DC)	DC-10	80–95	Small	10^3–10^5
Barrier discharge	5–20 k	1 m to 10	10–100 k	30–80	Large	10^{-2}–1
Pulsed corona	30–200 k	10 m to 1 k	10–1 k	20–75	Middle	1 – 10^2
Flow stabilized corona	10–100 k	10–100 m	DC	90–95	Middle	1 – 10^2
Arc discharge	0.1–0.5 k	10–100	DC	70–90	Middle	10^{-1}–10
High-frequency discharge	0.1–0.5 k	1 m to 1 k	1–100 k	50–70	Small	10^{-2}–10
Microwave discharge	0.1–0.5 k	1 m to 1 k	>1 G	30–60	Small	10^{-2}–1

Modified from [16].

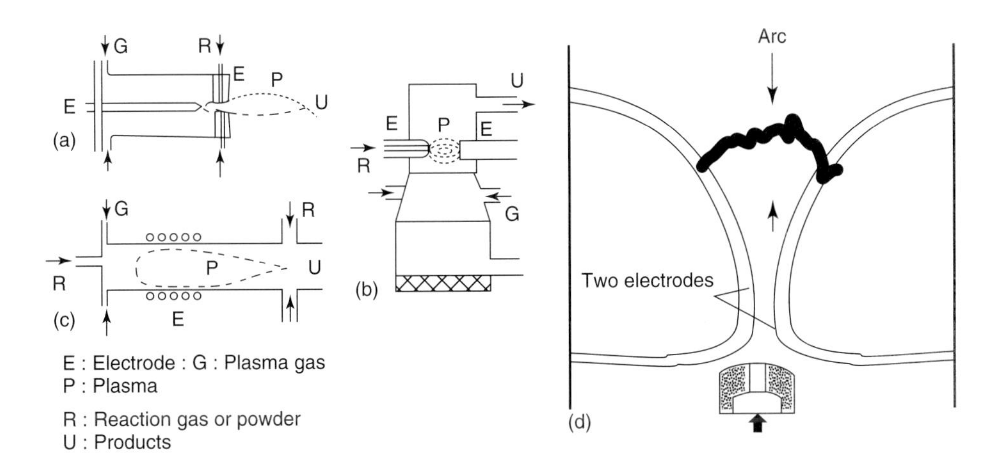

Figure 3.5 Thermal plasma reactors. (a) DC plasma torch; (b) transferred arc plasma; (c) inductively coupled RF plasma torch; and (d) gliding arc plasma [5, 8].

systems use discharges in the range of 1 kJ per pulse and larger. The pulsed corona system operates at a frequency of 10^2–10^3 Hz with the peak current below 100 A and the voltage rise occurring on the order of nanoseconds. A streamer-like corona is generated within the liquid to be treated, weak shock waves are formed, and a moderate number of bubbles are observed [37]. This system also generates weak ultraviolet (UV) radiation [36] and forms radicals and reactive species in the narrow region near the discharge electrodes.

PAED employs the rapid discharge of stored electrical charge across a pair of submerged electrodes to generate electrohydraulic discharges forming a local plasma region. The PAED system operates at a frequency of $10^{-2}-10^{2}$ Hz with the peak current above 10^{3} A and the voltage rise occurring on the order of microseconds. An arc channel generates strong shock waves with a cavitation zone containing plasma bubbles and transient supercritical water conditions. This system generates strong UV radiation and high radical densities, which have been observed to be short-lived in the cavitation zone. Pulsed spark electrohydraulic discharge (PSED) system characteristics are similar to those of PCED, with a few characteristics falling between those of PCED and PAED. More recently, RF-bipolar PSED system was developed for the inactivation of *Escherichia coli* [36, 37]. The PPED system operates at a frequency of $10^{-3}-10^{1}$ Hz with the peak current in the range of $10^{2}-10^{5}$ A. The voltage rise occurs on the order of nanoseconds. This type of system generates strong shock waves and some moderate UV radiation.

Traditional water and wastewater treatment technologies can be broadly classified into three categories: biological, chemical, and physical processes. Biological methods are typically used to treat both municipal and industrial wastewaters, and are generally not successful at degrading many toxic organic compounds. Chemical processes typically involve the addition of a chemical to the system to initiate a transformation; some chemicals commonly added, such as chlorine, are often associated with negative effects on both human and environmental health. Regulations governing the use of many of these chemicals are becoming increasingly stringent. Physical processes do not involve chemical transformation, and therefore generally serve to remove contaminants from the bulk fluid phase through concentrating them in a liquid or sludge phase. Many drinking water systems are challenged with difficult-to-treat target compounds such as organics (e.g., NDMA) and pathogens (e.g., *Cryptosporidium*, viruses, etc.). To be effectively removed from drinking water supplies, many of these recalcitrant compounds require specialized, and often target-specific, treatment technologies. The necessity of multiple treatment technologies adds considerably to the cost, especially for small systems. In order to meet the challenges presented by continuously emerging contaminants and increasingly stringent regulations, new and promising treatment technologies, such as electrohydraulic discharge, must be developed.

Depending on the technology, the treatment mechanisms generated by plasma technologies include [7, 36, 37] (i) high electric fields; (ii) radical reactions (e.g., ozone, hydrogen peroxide); (iii) UV irradiation; (iv) thermal reactions; (v) pressure waves; (vi) electronic and ionic reactions; and (vii) electromagnetic pulses (EMPs). In general, both electron and ion densities are proportional to the discharge current, while UV intensity, radical densities, and the strength of the pressure waves generated are proportional to the discharge power. Direct plasma technologies (i.e., electrohydraulic discharge) have the potential to be more efficient than either indirect or remote plasma technologies as they capitalize, to some degree, on all of these mechanisms due to the direct application. Figure 3.6 shows the treatment mechanisms initiated by PAED.

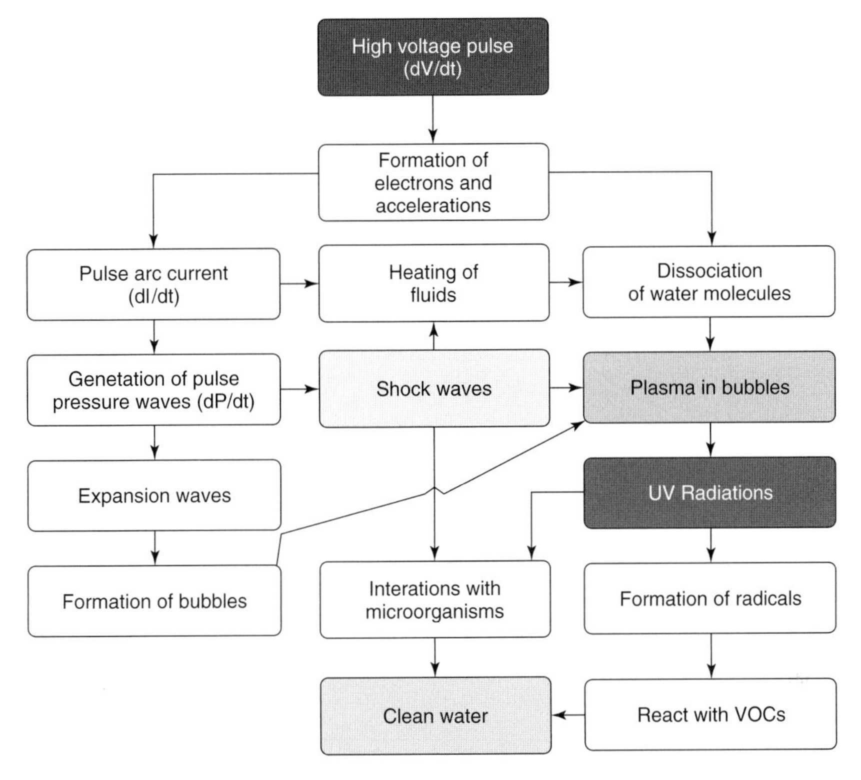

Figure 3.6 Water treatment mechanisms initiated by PAED [36].

3.5
Determination of Plasma Parameters and the Other Phenomena

The order of magnitude of time and discharge channel averaged plasma parameters such as electric field E, electron temperature $<T_e>$, and plasma density $<N_e>$, can be estimated by discharge current–voltage characteristics [28]. Time and discharge channel averaged discharge current consists of conduction current I_c and displacement current I_d component; $I_t = I_c + I_d$. For corona and spark discharges, the conduction current can be approximated by [28]

$$I_c \approx eN_e u_e EA$$

where e is the elementary electronic charge, u_e is the mobility of electrons, and A is the electrode surface area. Hence, plasma density can be expressed by

$$N_e \approx \frac{I_c}{eu_e EA} = \frac{I_c}{eu_{eo} AE^{1+b}}$$

where $u_e = u_{eo}E^b$, w is the power law constant determined from swarm experiment [30], and u_{eo} is the electron mobility at small electric field (constant) [12, 30].

Electric field distribution can be determined from the Poisson's equation as follows:

$$\Delta E = -eN_e/\gamma$$

where γ is the dielectric constant. The first-order approximation for electric field assumed no space charge effect (Laplace field), normally determined from the gap distances between two electrodes d as follows:

$$E_r = \frac{V_t}{d}$$

where E_r is the so-called reduced electric field and V_t is the applied voltage. If Poisson field can be approximated by $< E > \approx < E_r^w >$, the plasma density can be determined by

$$< N_e > \approx \frac{I_c}{e\mu_{eo}A(\frac{V_t}{d})^{w(1+b)}} = c_2 I^\alpha = \frac{c_1 I_c}{V_t^a}$$

$$a = w(1+b), \quad \alpha = \frac{w}{(b+2w)}$$

and mean electron temperature based on Einstein correlation can be determined by swarm parameters [26] as follows:

$$< kT_e/e > \approx c_o < E^f > = c_o(V_e/d)^{fw} = u_e/D_e$$

where c_0, c_1, and c_2 are constant depending on environmental gases and D_e is the diffusion coefficient.

For dry air, f and b are -0.43 and 0.501, respectively, and $0 < w < 1$ for $4 \ll (L/D_D)^2 \ll 0$, where L is the characteristic length and D_D is the Debye length [26].

The other associated physical phenomena in nonthermal plasma induced by streamer coronas are as follows [1–6]:

1) electrohydrodynamically induced gas flow which also assists transport of radicals and ions [19, 32];
2) photon emission especially UV light [26]; and
3) ion-induced aerosol particle formations [9, 12, 13].

3.6 Engineering and Economics

3.6.1 Nonthermal Plasma Pollution Control

For engineering of nonthermal plasmas, three components, (i) power supply developments, (ii) reactor development, and (iii) overall process developments, are required equally. In order to develop scaling of systems, the specific energy density (SED)

$$\text{SED} = \frac{\text{input electric power}}{\text{gas flow rate}} (\text{kWh m}^{-3}) \text{ or } (\text{J m}^{-3})$$

and the energy efficiency of pollutant removal

$$E_y = \frac{\text{g(mass of pollutant removed)}}{\text{kWh(electrical energy consumed)}}$$

were used, where E_y-[SED] characteristics can be used effectively to make economic evaluation [29] with scale-up based on the pilot scale test. Hence, experimental results are not only expressed in terms of applied voltage, current, or power but also both SED and E_y are shown to be present. Typical energy efficiency observed for NO_x and SO_x removal in pilot scale tests are $0.1 - 2$ kg kWh^{-1} for flow stabilized corona discharge radical shower system, $50 - 100$ g kWh^{-1} for electron beam, and $10 - 50$ g kWh^{-1} for pulsed corona and barrier discharges [1–6, 29, 33]. Typical economic evaluation for a 500-MW coal-fired power station SO_x and NO_x control results are shown in Figure 3.7 [29], where the optimum value for each system also changes depending on the treated gas flow rates and pollutant concentrations.

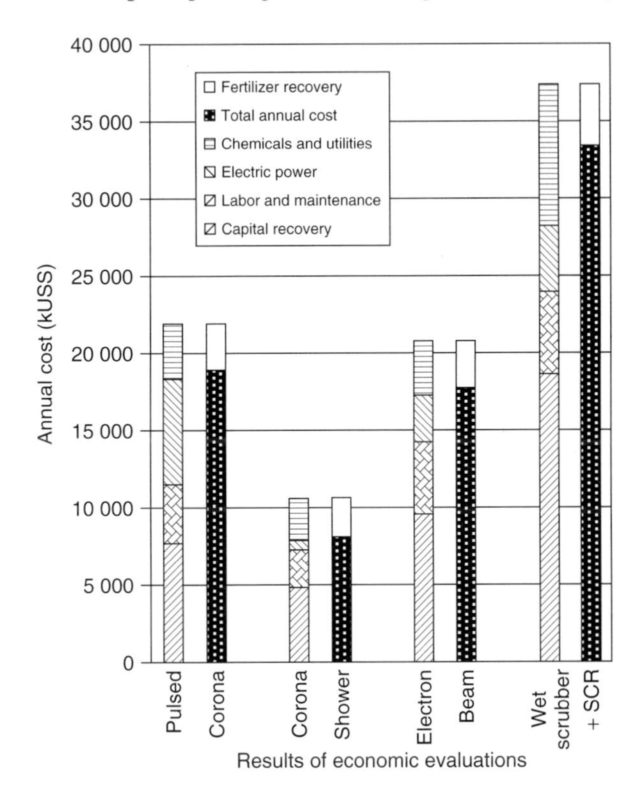

Figure 3.7 Economic evaluation of plasma pollution control systems [29].

Figure 3.7 shows that the plasma-based pollution control system has a potential to minimize the cost of pollution control if the right engineering can be applied. For the treatment of VOCs and PFCs, similar economic evaluation results are also obtained [3, 4].

3.6.2 Thermal Plasma Solid Waste and Water Treatment

The energy efficiency of solid waste treatment can be evaluated as follows: waste (tons) / kWh of electrical power consumed as a function of waste treated per day. For example, from the results of the observation of an incineration ash treatment system by Chang *et al.* [35], it was concluded that for an optimum of 0.8 kWh kg^{-1} an ash treatment reactor with a capacity of more than 25 tons per day is required.

For water treatment, the treatment time factor is important for the removal of chemical pollutants or for disinfection. Hence, these time factors will be lumped as accumulated energy input per volume kWh/m^3 versus mol, g or ppm/m^3, or #cfu/m^3 for contaminants or bacteria in water.

3.7 Concluding Remarks

It has been suggested that air pollution has a large impact not only on human health but also on the economy, where the population experiences premature death caused by air pollution and admission to hospital. The average cost of mortality was US$4M per person and adult chronic bronchitis costs an average of $279 000 per person for treatment [2]. Nonthermal plasma pollution control technology may assist to reduce these health and economic impacts in the near future. The fundamental study of physics and chemistry of this plasma technology gives us an engineering baseline data for an economically feasible pollution control technology [6, 29]. On the basis of research in the past 30 years [1–6, 34], the types of nonthermal plasma technologies that are feasible for each gaseous pollutant treatment are summarized in Table 3.6. For a solid waste treatment, a similar comparison is not well established. Hence, in order to understand the mechanisms, and hence the optimized process, more fundamental research on its physics and chemistry is required. Also engineering and economic evaluation must be conducted.

Because electrohydraulic discharge systems do exploit all of the treatment mechanisms generated by the plasma reaction, both chemical and physical, these technologies have the ability to effectively treat a range of contaminants broader than that of other conventional and emerging technologies [6]. Preliminary research has indicated that PAED offers advantages over indirect plasma methods in that it can provide comparable or superior treatment of microorganisms, algae, volatile organics, nitrogenous municipal waste compounds, and some in organics [36, 37]; these observations are qualitatively summarized in Table 3.7. Moreover,

Table 3.5 Characteristics of electrohydraulic discharge systems [37].

Property	Pulsed corona (PCED)	Pulsed arc (PAED)	Pulsed power (PPED)	Pulsed spark (PSED)
Operating frequency (Hz)	10^2-10^3	$10^{-2}-10^2$	$10^{-3}-10^1$	10^3-10^4
Current (A)	10^1-10^2	10^3-10^4	10^2-10^5	10^2-10^3
Voltage (V)	10^4-10^6	10^3-10^4	10^5-10^7	10^3-10^4
Voltage rise (s)	$10^{-7}-10^{-9}$	$10^{-5}-10^{-6}$	$10^{-7}-10^{-9}$	$10^{-6}-10^{-8}$
Pressure wave generation	Weak	Strong	Strong	Moderate
UV generation	Weak	Strong	Moderate	Weak

Table 3.6 Plasma parameters for various nonthermal plasmas and the most feasible applications [2, 38].

	Plasma density	Electron temperature	Gas temperature	Electric field	Treatment flue gas
Electron beam	Very high	Extremely high	Low	Very low	Acid gases, VOCs
Barrier discharge (silent/surface)	High	Medium	Low	Medium	Oxidation of VOCs or acid gases
Barrier discharge (ferro-electric)	Low	High	Low	Very high	PFCs, oxidation of VOCs
Pulsed corona	High	Medium	Low	High	VOCs
Pulsed power	Very high	High	Medium	High	Acid gases
Capillary	High	Low	Medium	Low	VOCs
Flow stabilized corona	Locally high	Locally high	Low	High	Acid gases, VOCs, toxic gases
Arc/plasma torch	Extremely high	Locally high	Extremely high	Low	ODS/VOCs toxic gases
RF discharge	High	Medium	High	Low	ODS/VOCs
Microwave discharge	High	Medium	Medium	Medium	ODS/VOCs

these benefits are available concurrently from one technology as opposed to a series of treatment technologies. In addition, preliminary investigations conducted with a limited number of target compounds have indicated that effective water treatment with PAED utilizes less than 50% of the kilowatt-hours required by other plasma technologies (e.g., UV) for equivalent levels of treatment [37]. The plasma techniques based on plasma injection methods should also be evaluated and compared [39].

Table 3.7 Comparison of plasma and conventional water treatment processes [36].

Target compounds	Cl/ClO_2	Ozone	Electron beam	PCED	PAED	UV-C
Microorganisms	Adequate	Good	Adequate	Good	Good	Good
Algae	None	Partial	None	Partial	Good	Adequate
Urine components	Adequate	Good	Good	Good	Good	None
VOCs	None	Adequate	Good	Good	Adequate	None
Inorganics	None	Partial	Partial	Adequate	Adequate	None

References

1. Chang, J.S. (2008) Physics and chemistry of plasma pollution control technology. *Plasma Sources Sci. Technol*, **17**, 450–462.
2. Ontario Ministry of the Environments (2002) Report on Smog Plan for Ontario, Canada.
3. Urashima, K. and Chang, J.S. (2001) Removal of volatile organic compounds from air streams and industrial flue gases by non-thermal plasma technology. *IEEE Trans. Dielectr. Electr. Insul.*, **7**, 602–614.
4. Chang, M.B. and Chang, J.S. (2006) Abatement of PFCs from semiconductor manufacturing processes by non-thermal plasma technologies: A critical review. *Ind. Chem. Eng. Res.*, **45**, 4101–4109.
5. Chang, J.S. (2006) Atmospheric plasmas. *J. Plasma Nucl. Fusion*, **82**, 682–692.
6. Chang, J.S. (2001) Recent development of plasma pollution control technology: A critical review. *Sci. Technol. Adv. Mater.*, **2**, 571–576.
7. Locke, B.R., Sato, M., Sunka, P., Hoffmann, M.R., and Chang, J.S. (2006) Electrohydraulic discharge and nonthermal plasma for water treatment. *Ind. Chem. Res.*, **45**, 882–905.
8. Beuthe, T. and Chang, J.S. (1995) Gas discharge phenomena, in *Handbook of Electrostatic Processes*, Chapter 9 (eds J.S.Chang, A.J. Kelly, and J. Crowley), Marcel Dekker Inc., New York.
9. Mazig, H. (1991) Chemical processes in electron beam flue gas control. *Adv. Chem. Phys.*, **130**, 315–386.
10. Ono, R. and Oda, T. (2002) Optical diagnostics of pulsed streamer discharge under atmospheric pressure. *J. Electrost.*, **55**, 333–340.
11. Kanazawa, S., Tanaka, H., Kajiwara, A., Ohkubo, T., Nomoto, Y., Kocik, M., Mizeraczyk, J., and Chang, J.S. (2007) LIF imaging of OH radicals in DC positive streamer coronas. *Thin Solid Films*, **515**, 4266–4271.
12. Chang, J.S., Ichikawa, Y., Hobson, R., and Kaneda, T. (1983) *Atomic and Molecule Processes in an Ionized Gases*, Tokyo Denki University Press, Tokyo.
13. Chang, J.S., Lawless, P.A., and Yamamoto, T. (1991) Corona discharge processes. *IEEE Trans. Plasma Sci.*, **19**, 1152–1166.
14. Beuthe, T.G. and Chang, J.S. (1999) Chemical kinetic modelling of nonequilibrium Ar-H_2 thermal plasma processes. *Jpn. J. Appl. Phys.*, **38**, 98060–98065.
15. Penetrante, B.M., Haiao, M.C., and Bersley, J.N. (1997) Chemical kinetics of plasma pollution control. *Jpn. J. Appl. Phys.*, **36**, 5007–5014.
16. Chang, J.S. (2000) Recent development of gaseous pollution control technologies based on non-thermal plasma. *Oyoubutsuri (Appl. Phys.)*, **69**, 268–277 (in Japanese).
17. Morrow, R. and Lowke, J.J. (1997) Modelling of plasma NO_x removal. *J. Phys. D.: Appl. Phys.*, **30**, 614–620.

18. Li, R. and Xin, I. (2000) Modelling of radical injection processes. *Chem. Eng. Sci.*, **55**, 2481–2489.

19. Mizeraczyk, J., Kocik, M., Dekowski, J., Dors, M., Podinski, J., Ohkubo, T., Kanazawa, S., and Kawasaki, T. (2001) PIV measurement of electrohydrodynamic flow in model electrostatic precipitator. *J. Electrost.*, **51**, 272–277.

20. Chang, J.S. (1989) The role of H_2O on the formation of NH_4NO_3 aerosol particles and De–NO_x under the corona discharge treatment of combustion flue gases. *J. Aerosol Sci.*, **8**, 1087–1090.

21. Maezono, I. and Chang, J.S. (1988) Flow enhanced corona discharge: corona torch. *J. Appl. Phys.*, **59**, 2322–2324.

22. Yan, K., Yamamoto, T., Kanazawa, S., Ohkubo, T., Nomoto, Y., and Chang, J.S. (2001) NO removal characteristics of a corona radical shower system under DC and AC/DC superimposed operations. *IEEE Trans. Ind. Appl.*, **37**, 1499–1504.

23. Kohno, H., Berezin, A.A., Shibuya, A., and Chang, J.S. (1998) VOC removal by capillary tube non-thermal plasma reactors. *IEEE Trans. Ind. Appl.*, **34**, 953–996.

24. Park, J.Y., Kim, G.H., Kim, J.D., Koh, H.S., and Lee, D.C. (1998) NO_x removal using dc corona discharge with magnetic field. *Combust. Sci. Technol.*, **133**, 65–78.

25. Eliasson, B. and Kogelschatz, U. (1991) Barrier discharge ozone generations. *IEEE Trans. Plasma Sci.*, **19**, 1063–1077.

26. Takagi, K., Chang, J.S., and Kostov, K.G. (2000) Atmospheric pressure nitrogen plasmas in ferro-electric packed bed barrier discharge reactor. *IEEE Trans. Dielectr. Electr. Insul.*, **15**, 481–489.

27. Masuda, S. and Nakao, H. (1990) Control of NO_x by positive and negative pulsed corona discharges. *IEEE Trans. Ind. Appl. Soc.*, **26**, 374–383.

28. Loeb, L. (1965) *Electrical Corona*, University of California Berkeley Press.

29. Urashima, K., Kim, S.J., and Chang, J.S. (2003) The scale-up and evaluation of non-thermal plasmas for power plants. *J. Adv. Oxid. Technol.*, **6**, 123–131.

30. Huxley, L.G.H. and Compton, R.W. (1974) *The Diffusion and Drift of Electrons in Gases*, John Wiley & Sons, Inc., New York.

31. Marotta, E., Callea, A., Ren, X., Rea, M., and Paradisi, C. (2007) A mechanism study of pulsed corona processing of hydrocarbons in air at ambient temperature and pressure. *Int. J. Plasma Environ. Sci. Technol.*, **1**, 39–45.

32. Dekowski, J., Mizeraczyk, J., Kocik, M., Dors, M., Podlinski, S., Kanazawa, S., Ohkubo, T., and Chang, J.S. (2004) Electrohydrodynamic flow and its effect on Ozone rransport in corona radical shower reactor. *IEEE Trans. Plasma Sci.*, **32**, 370–279.

33. Chang, J.S., Urashima, K., Tong, X.Y., Liu, W.P., Wei, H.Y., Yang, F.M., and Lin, X.J. (2003) Simultaneous removal of NO_x and SO_2 from coal boiler flue gases by DC corona discharge ammonia radical shower systems: pilot plant tests. *J. Electrost.*, **57**, 313–323.

34. Kim, H.H. (2004) Plasma VOC pollution control. *Plasma Processes Polym.*, **1**, 91–110.

35. Inaba, T. and Iwao, T. (2001) Treatment of waste by DC arc discharge plasmas. *IEEE Trans. Dielectr. Electr. Insul.*, **7**, 684–692.

36. Chang, J.S., Dickson, S., Guo, Y., Urashima, K., and Emelko, M.B. (2008) Electrohydraulic discharge direct plasma water treatment processes, plasma technology, in *Advanced Plasma Technology*, Chapter 24 (eds R. d'Agostino, P. Favia, Y. Kawai, H. Ikegami, N. Sato, and F. Arefi-Khonsari), Wiley-VCH Verlag GmbH, Wienheim, pp. 421–434.

37. Chang, J.S. and Urashima, K. (2008) Concurrent treatment of chemical and biological contaminants in water by pulsed arc electrohydraulic discharge plasmas, in *Plasma Assisted Decontamination of Biological and Chemical Agents* (eds S. Guceri and A. Fridman), Springer Science, pp. 87–97.

38. Chang, J.S. (2003) Next generation integrated electrostatic gas cleaning system. *J. Electrost.*, **57**, 273–219.

39. Yasuoka, Y., Maebara, T., Katsuki, J., Namihira, T., Kaneko, T., and Hatakeyama, R. (2008) Generation of under water discharge plasma and its property. *J. Plasma Fusion Res.*, **84**, 666–673.

4
Optical Diagnostics for High-Pressure Nonthermal Plasma Analysis

Tetsuji Oda and Ryo Ono

4.1
Introduction

High-pressure plasma is a very effective technique in decomposing and processing various gaseous materials. However, plasma reaction mechanism, especially nonthermal plasma mechanism, at high pressure, is not yet known well compared to the low-pressure plasma, and so on. If we insert some measuring instruments in the plasma region, the insertion-perturbation is very large. Optical diagnosis is very useful for plasma analysis with low interference with the plasma. There are many optical measuring methods developed for the discharge analysis. Among them, the most typical method is optical emission spectroscopy which needs an optical spectrograph, photo-detector, and some optical accessories such as lens and photo-arraignment apparatus. Optical spectral analysis is strongly dependent on the optical emission by the plasma and other information cannot be obtained. Optical imaging is also a useful technique for the high pressure plasma analysis. Optical absorption is effective in detecting various species with good sensitivity. Atomic-photo absorption and infrared absorption spectroscopy are typical examples. Absorption sensitivity is strongly dependent on the target atom, ion, molecule, and so on. It is not easy to detect low-concentration materials if the target sensitivity is not large. Currently, we have very strong light sources, and an optical measuring method is much improved by using such strong laser beam interaction with other species, or photons. One such source is laser-induced fluorescence (LIF). When we use pulse laser excitation, we can check many species at any moment. For electron excitation, the laser beam wavelength must be adjusted very precisely, which needs an expensive tunable narrow-band laser system. If some atoms, molecules, and other species are irradiated with a laser beam, there are some inelastic scattering lights observed together with elastic scattering photons (wavelength is not changed). The inelastic-scattered photon energy change is termed as *optical wave-number shift*. If that value is dependent on the species and independent of the beam wavelength, it is termed as *Raman shift*. The method to detect this shift is usually called *Raman spectroscopy*, which is also useful to detect various species. Thomson scattering method is also famous

Industrial Plasma Technology. Edited by Yoshinobu Kawai, Hideo Ikegami, Noriyoshi Sato, Akihisa Matsuda, Kiichiro Uchino, Masayuki Kuzuya, and Akira Mizuno

ISBN: 978-3-527-32544-3

for the detection of electron density and energy in gas phase but not yet succeeded by the authors (please refer to Chapter 32). In this chapter, we like to introduce some optical diagnosis methods and their results for nonthermal plasma analysis.

4.2 Experimental Examples

4.2.1 Optical Emission Photograph

For nonthermal plasma, streamer development is very important, especially in the case of pulse discharge. As streamer development is very fast, we need a very high speed frame camera or streak camera to observe this [1]. The authors used several high-speed cameras (shutter speed: from 200 ps to several ms with image intensifiers) for this observation. One electric circuit (pulse power supply) to detect such pulse streamer development is shown in Figure 4.1. Many primary streamers (fragmentlike, appears till 12 ns after the discharge) and secondary streamers

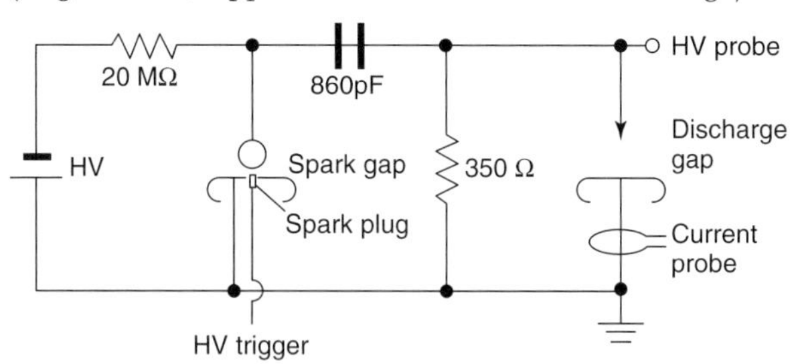

Figure 4.1 High-voltage pulse generating circuit.

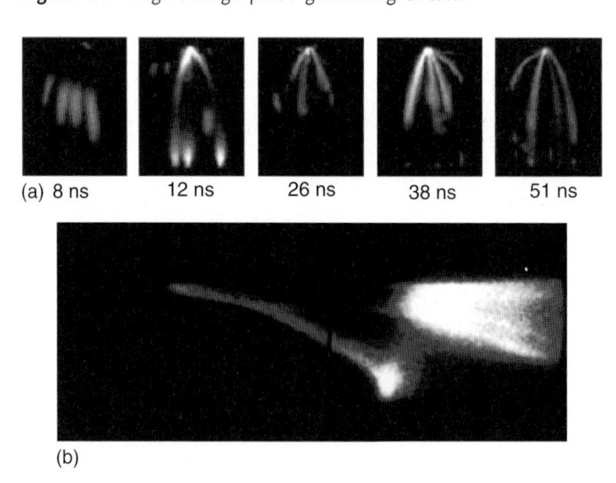

Figure 4.2 Pulse streamer development of positive corona. (a) A 5-ns interval (exposure time is 5 ns). (b) Streak image.

(bright zone) are detected from the needle after 20 ns as shown in Figure 4.2 where (a) shows 5-ns exposure picture and (b) shows streak mode picture (slit-frame moves from left to right with time: left trace is primary streamer). If the ambient gas changes, those streamers show different patterns.

4.2.2 Optical Emission Spectra

Many optical emission spectra of such high-pressure plasmas have been reported. For example, Sato group at Gunma University reported the spectra of water discharge plasma.

We have used a special spectrograph (Hamamatsu C5095), which can provide one-dimensional position-information spectra. One example is shown in Figure 4.3 where the high-pressure plasma shown is just below the jet nozzle [2]. In this case, the gas is compressed dry air (CDA) and the spectrograph measurement area is from just 1 mm below the plasma jet nozzle to 21 mm below the nozzle. Line intensity of OH is not uniform at the vertical position because the discharge is completed inside the nozzle.

4.2.3 Laser-Induced Fluorescence (LIF)

4.2.3.1 OH Radical

In the oxidation process of high-pressure plasma, OH radical and atomic oxygen should be the most effective radicals. For a better understanding of those radicals, LIF is a typically useful technique to observe them at any moment. For example, there are many wavelengths of light to detect OH radicals. We used 248-nm photons generated by the KrF tunable excimer laser (Lambda Physics Complex 150T). The related electron band level in this LIF observation is shown in Figure 4.4. The excitation is 248 nm and fluorescence is 297 nm. In this case, the predissociation rate is very large and quenching is negligibly small compared to the dissociation, indicating that the LIF signal intensity is not affected by environmental molecular compositions. However, the sensitivity is very small and we need a high-power narrow-band laser source. Our experimental setup is shown in Figure 4.5. It is possible to take OH radical images where most OH can be detected in the streamer channel as shown in Figure 4.6. Since these images are of different time periods, they do not appear the same, though the trend is similar these photos are not of the same timing, they are not exactly the same but the tendency is the same. OH density profiles from the needle (positive) to plate for three different applied peak voltages are shown in Figure 4.7 [3]. With a slight increase of the applied voltage, OH density increases greatly. OH relaxation time constant is also examined by this method. Water concentration in the gas affects the OH density and lifetime. For OH radical generation, 282-nm excitation LIF is also very useful because of the high sensitivity (sensitivity of OH radicals is about 100 times larger than that of 248 nm). Kanazawa *et al.* reported OH radicals at 4 cm from the plasma nozzle. The authors' group also observed this.

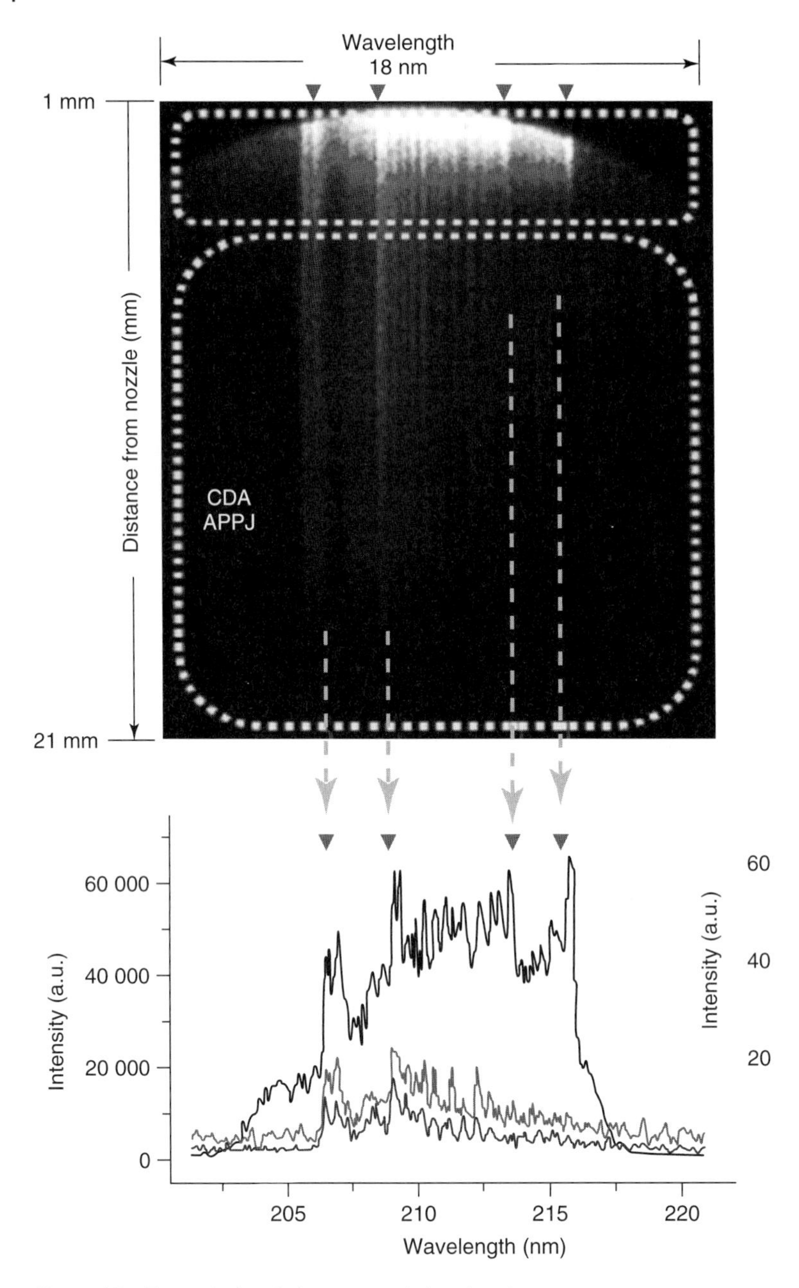

Figure 4.3 The optical emission spectra below the plasma jet nozzle from 1 to 21 mm. The upper image shows wavelength shift around 309 nm in horizontal axis and the position from the nozzle in vertical axis. Brightness is signal intensity. The lower lines are spectra intensity for 2.5, 5, and 10 mm below the nozzle.

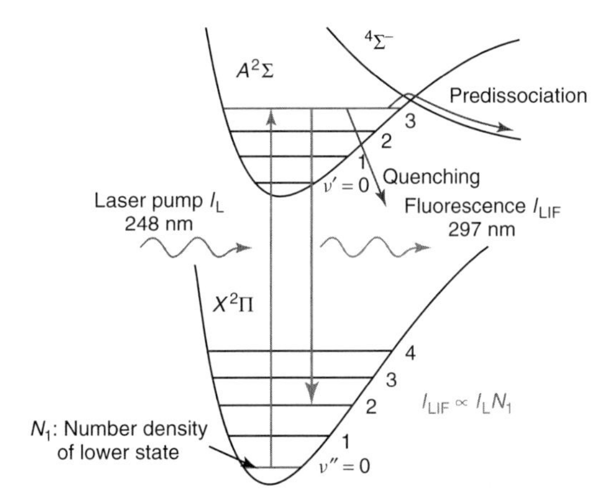

Figure 4.4 The electron energy level related with LIF of OH (248 nm).

4.2.4 LIF of Atomic Oxygen Radical

LIF signal of the atomic oxygen is also observed at 226-nm photo excitation (two-photon excitation), which is called as two-photon-assisted laser-induced fluorescence (TALIF). Figure 4.8 shows electron energy level of atomic oxygen where two photons near 226 nm excite electron to $3p^3P$ level and emission is infrared, which is not easy to detect. One experimental data is shown in Figure 4.9 [5]. The life time of the atomic oxygen is dependent on the environmental gas composition and about submillisecond order. The decrease of atomic oxygen is in good agreement with the production of the ozone in dry condition (collision with the oxygen molecule is important) and may generate OH radicals in humid air.

4.2.5 LIF of NO Molecule

Fundamental state NO can be detected by 226-nm laser excitation LIF which is pretty easy. NO decomposition by the plasma is often observed as shown in Figure 4.10 [6]. The photo on the left top shows the optical emission of the streamer and the other photos show NO LIF. At the plasma generation, NO in the streamer region is decomposed immediately after plasma generation. After about several milliseconds, NO is mostly decomposed, indicating that some long life radicals decompose NO and new NO appears with NO flow from the left side [6].

4.2.6 N_2 (A) LIF in the Plasma

Excited nitrogen A state can be directly detected by infrared excitation LIF techniques [7]. There are many LIF signals from N_2 (A) state. We use the following parameters.

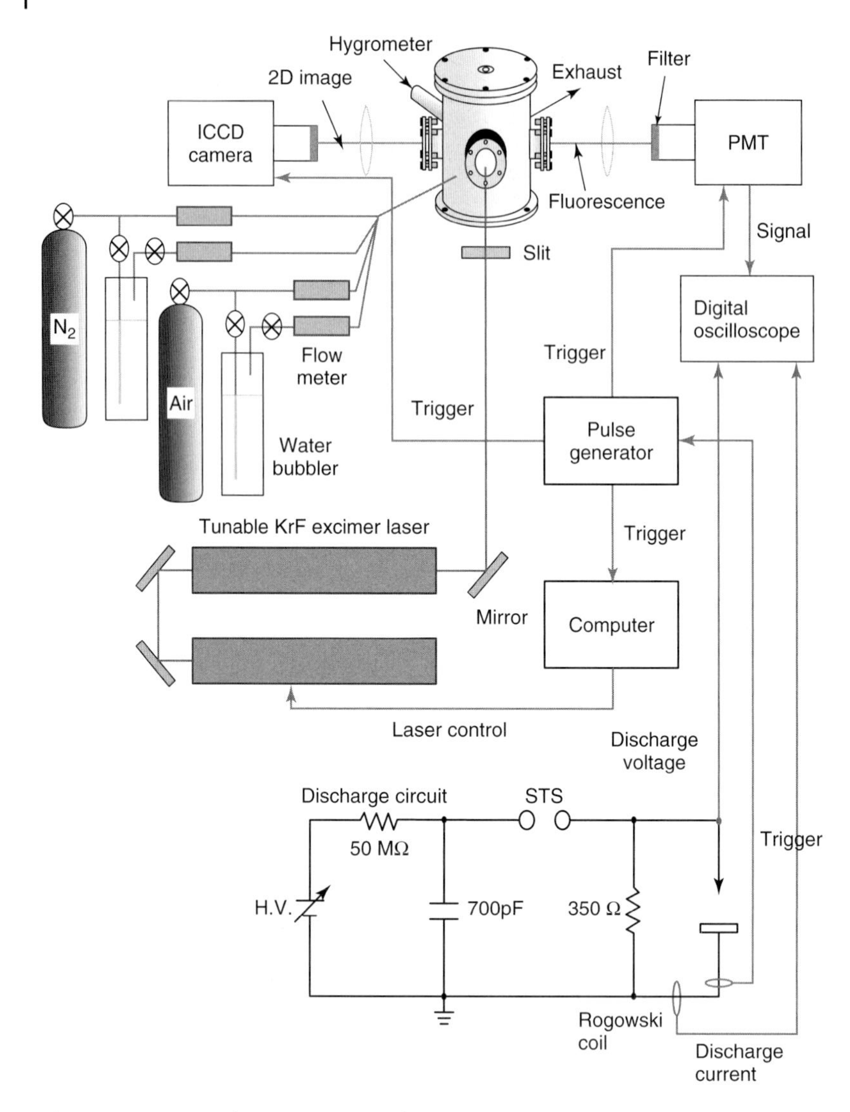

Figure 4.5 KrF LIF observation system for OH (248 nm).

The vibrational ground state $N_2(A^3\sum_u^+, v''=0)$ is excited to $N_2(B^3\Pi_g, v''=4)$ by a 618-nm dye laser irradiation, and then the fluorescence from $N_2(B^3\Pi_g, v''=4)$ $\rightarrow N_2(A^3\sum_u^+, v''=1)$ is measured with a photomultiplier tube (PMT) through an optical band-pass filter (676 ± 5 nm) [7]. One example of LIF signal intensity profile from the discharge needle to the plate at 4 μs after the negative pulse discharge where the laser energy is 1.5 mJ per pulse is shown in Figure 4.11. The rather flat distribution suggests that $N_2(A)$ was produced mainly in the primary streamer.

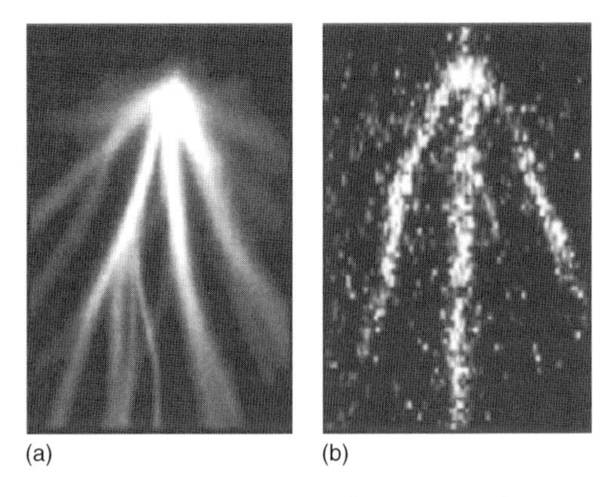

Figure 4.6 Streamer images of (a) an optical emission photo and (b) an LIF image [3].

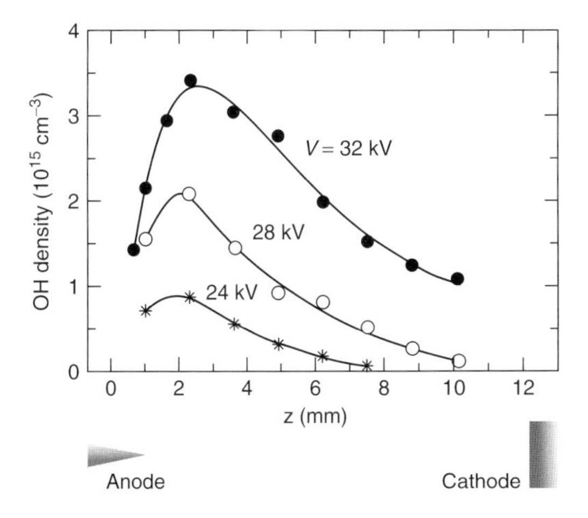

Figure 4.7 OH density distributions from the needle (anode) to the plate (cathode) for different discharge peak voltages [4].

If the gas contains NO, the excited nitrogen excites NO and NO $- \gamma$ irradiation can be detected. This correlation was also examined experimentally and this flat distribution of NO $- \gamma$ irradiation was also confirmed.

4.2.7 O_2 LIF

The oxygen molecule is also excited by 248-nm photon and the LIF of oxygen molecules is also observed by the same equipment for OH radical measurement that is used in excimer laser system [8]. With the LIF data, some attempts can be made to measure the molecular or radical temperatures if they are in equilibrium. One example of $O_2(\nu'' = 6)$ density (excitation laser is 248 nm) and calculated temperature change versus the time after the pulse discharge is shown

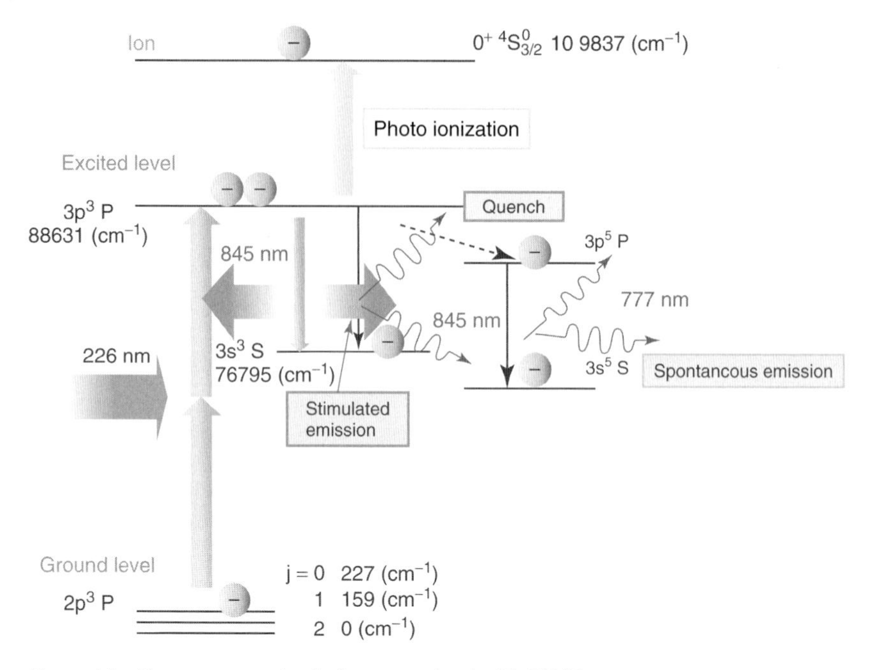

Figure 4.8 Electron energy level of oxygen related with TALIF.

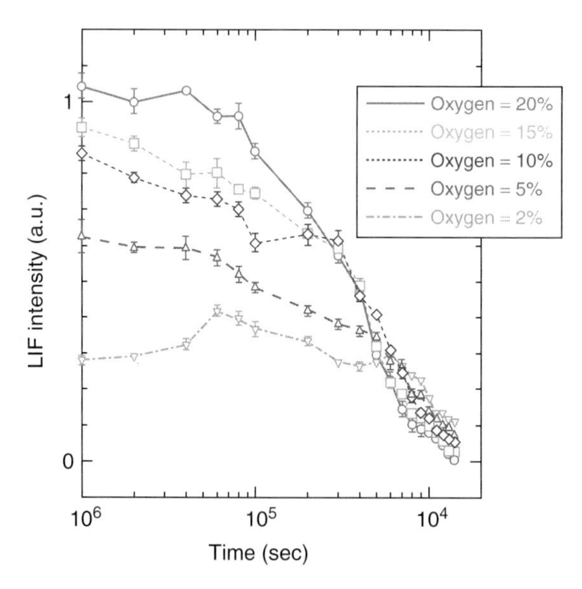

Figure 4.9 Atomic oxygen LIF signal decay with time for different concentrations of O_2 to N_2.

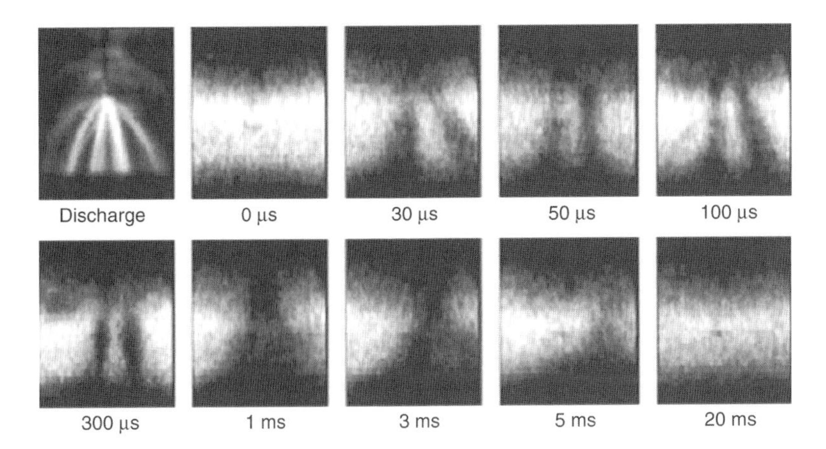

Figure 4.10 NO LIF after the plasma discharge.

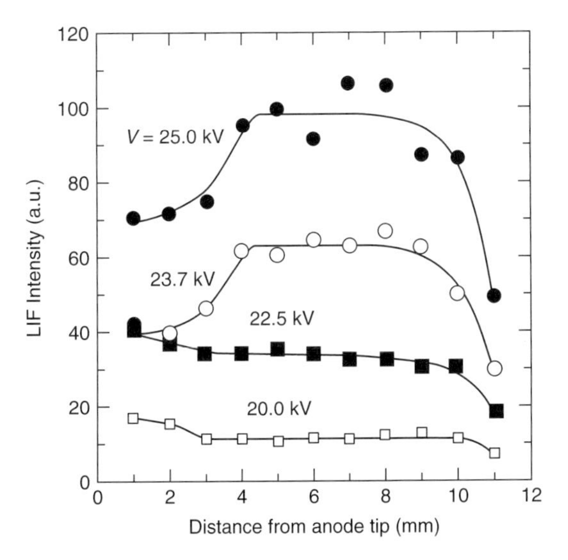

Figure 4.11 One-dimensional distribution of N_2(A) density for different pulse voltage (N_2 atmosphere).

in Figure 4.12 where the excited oxygen molecule immediately decays with a time constant of the order of 10 μs. Just after the pulse discharge, the oxygen gas temperature is rather high (about 1500 °C) and the concentration of excited O_2 decreases by one-tenth within 10 μs.

4.2.8
Optical Absorption

Not only the optical emission spectroscopy but also the optical absorption is a very useful technique to identify various species including some radicals or basis.

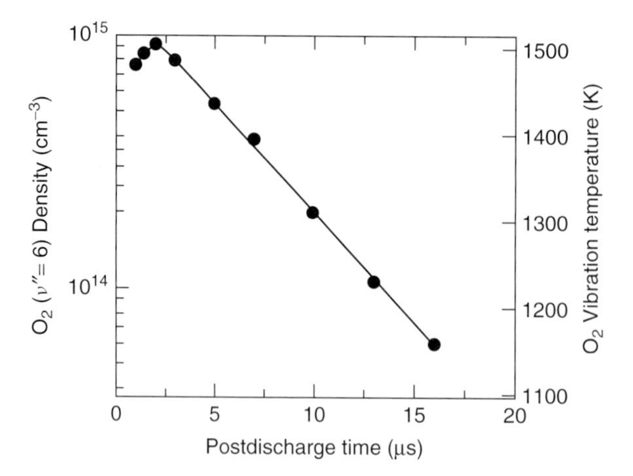

Figure 4.12 Time decay of $O_2(\nu'' = 6)$ density after the plasma discharge.

FFTIR (fast-Fourier transform infrared spectrograph) is one good example to analyze various gas components but very quick phenomena are not detected by this method. For the fast phenomena, the laser absorption method is very effective if the wavelength is adjusted to a high absorption level. One example is the generated ozone that absorbs 248 nm of light very well as shown in Figure 4.13 [9]. It was

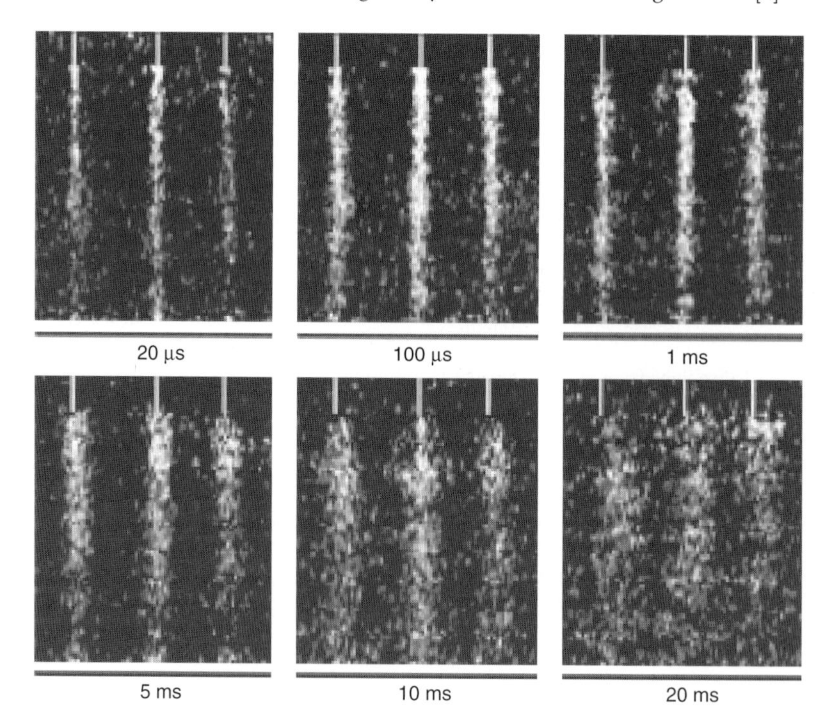

Figure 4.13 Ozone profile after the discharge under the discharge needle.

found that the ozone is produced much more in the second streamer than in the primary streamer as shown in Figure 4.2 [9].

4.3 Conclusions

Because of space constraint, other topics, such as Raman spectroscopy, which will give details of other factors including temperature, have not been discussed in this chapter, . Such optical measuring methods for high-pressure plasma are just beginning to be formulated and we already have many new results. At some points, nonthermal plasma is rather active as simulated and new plasma processes will be devised. High-pressure glow discharge analysis will be done in the near future.

References

1. Ono, R. and Oda, T. (2003) Formation and structure of primary and secondary streamers in positive pulsed corona discharge – effect of oxygen concentration and applied voltage. *J. Phys. D*, **36**, 1952–1958.
2. Chiang, C.C., Ono, R., and Oda, T. (2008) Optical emission spectra of atmospheric pressure non-equilibrium plasma jet. Proceedings of the International Workshop on Electrostatics Ishigaki, 23_I.
3. Ono, R. and Oda, T. (2002) Dynamics and density estimation of hydroxyl radicals in a pulsed corona discharge. *J. Phys. D*, **35**, 2133–2138.
4. Ono, R. and Oda, T. (2008) Measurement of gas temperature and OH density in the afterglow of pulsed positive corona discharge. *J. Phys. D*, **41**, 035204.
5. oda, T., Yamashita, Y., Takezawa, K., and Ono, R. (2006) Oxygen atom behaviour in the nonthermal plasma. *Thin Solid Films*, **506-507**, 669–673.
6. Oda, T. and Ono, R. (2002) Application of LIF-techniques to atmospheric pressure nonthermal plasma. *Proc. SPIE*, **4460**, 263–273.
7. Ono, R., Tobaru, C., Teramoto, Y., and Oda, T. (2008) Observation of N2(A) metastable in pulsed positive corona discharge using laser-induced fluorescence. Annual Conference of IEEE/IAS, 12P7.
8. Ono, R. and Oda, T. (2008) Measurement of O2 (v = 6) density inpulsed corona discharge using laser-induced fluorescence. International Conference on Electrical Engineering 2008, to be presented (in Japanese).
9. Ono, R. and Oda, T. (2004) Spatial distribution of ozone density in pulsed corona discharges observed by two-dimensional laser absorption method. *J. Phys. D*, **37**, 730–735.

5
Laser Investigations of Flow Patterns in Electrostatic Precipitators and Nonthermal Plasma Reactors

Jerzy Mizeraczyk, Janusz Podlinski, Anna Niewulis, and Mirosław Dors

5.1
Introduction

Recent rapid progress in laser technologies has resulted in the development of measurement methods useful in investigations of gas, liquid, and particle flow structures. These methods are laser visualization and particle image velocimetry (PIV) [1].

PIV is used to determine the flow velocity vectors in a selected cross section of the flow, called the *observation plane*. The observation plane is set in the flow by introducing a laser beam into it in the form of a laser sheet (Figure 5.1). Then two approaches of PIV are possible: two- or three-dimensional. When using two-dimensional particle image velocimetry (2D PIV) method, the laser sheet is assumed to be infinitely thin, and only one charge coupled device (CCD) camera is used for monitoring the observation plane. This enables measuring two components of the velocity vectors in the observation plane. The velocity vector measurement is based on observation of the movement of seeding particles that cross the observation plane. The flow seeding particle movement is monitored by the laser sheet light scattered by the particles. The scattered laser light forms the flow image that is recorded by the CCD camera. From two successive images, the velocity vectors of the seeding particles can be found. When the seeding particles follow the gas flow, the gas flow velocity in the observation plane can be determined. In the other case, PIV method delivers the particle velocity field in the gas.

Three-dimensional particle image velocimetry (3D PIV) method is based on the fact that a real laser sheet has a definite thickness. As a result, three velocity vector components in the observation plane can be determined when two CCD cameras observing the same measurement area under different angles as shown in Figure 5.1.

The PIV techniques have been proved to be very useful for measuring the flow velocity fields in electrostatic precipitators [2–5] and nonthermal plasma reactors [6, 7]. The electrostatic precipitators have been used for dust particle collection for decades [8]. Recently, narrow electrostatic precipitators (ESPs) have become a subject of interest because of their possible application in diesel engines.

Industrial Plasma Technology. Edited by Yoshinobu Kawai, Hideo Ikegami, Noriyoshi Sato, Akihisa Matsuda, Kiichiro Uchino, Masayuki Kuzuya, and Akira Mizuno

ISBN: 978-3-527-32544-3

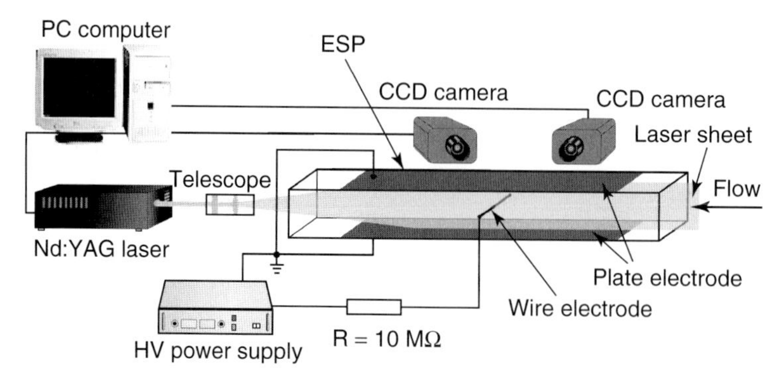

Figure 5.1 Schematic setup for 3D PIV measurement of flow velocity field in an electrostatic precipitator.

Diesel engines emit fine particles, in the size range of $7.5 \times 10^{-3} - 1.0\,\mu m$ [9], that are harmful for human and animal health. Measures such as diesel engine modification, special fuel additives, alternative fuels, after-treatment systems or diesel particulate filters (DPFs) decrease particulate emission from diesel engines [9–13]. However, the first three measures are not effective for reduction of diesel soot. On the other hand, the DPFs removal efficiency is very high (90–95% [11]), but DPFs are very expensive because of their high energy consumption and high maintenance costs. Recently, several authors have proposed electrostatic precipitation as an alternative control of diesel particulate emission [14–19].

The nonthermal plasma reactors are of interest because of their usefulness for cleaning air and water from pollutants [20–22].

In this chapter, we present results of the PIV measurements of flow velocity fields in selected electrostatic precipitators and in a nonthermal plasma reactor for water treatment.

5.2
PIV Experimental Setups

2D and 3D PIV measurements of flow velocity fields were carried out by the authors in electrostatic precipitators [2–5] and nonthermal plasma reactors [6, 7].

The apparatus used in these experiments consisted of an ESP or nonthermal plasma reactors, high voltage supply, and standard PIV equipment for the measurement of velocity fields. An experimental setup for 3D PIV measurement in an electrostatic precipitator with a single high-voltage electrode (wire) inserted between two collecting electrodes (steel plates) is presented in Figure 5.1.

The standard PIV equipment consisted of a twin second harmonic Nd-YAG laser system ($\lambda = 532$ nm), imaging optics (cylindrical telescope), one or two CCD cameras, and a PC. A laser sheet of 1-mm thickness, formed from the Nd-YAG laser beam by the cylindrical telescope, was introduced into the ESP to form an observation plane. The images of the particles following the flow in the laser sheet were recorded by two FlowSense M2 cameras. The CCD camera active element size was 1186 × 1600 pixels. The captured images were transmitted to the PC for digital analysis.

All the velocity fields presented in this chapter resulted from the averaging of 100 measurements, which means that each velocity map was time-averaged. On the basis of the measured velocity fields, the flow streamlines were calculated.

5.3
Results – PIV Measurements

In this chapter, several examples of PIV measurements in electrostatic precipitators are presented. These selected results concern the electrohydrodynamic (EHD) flow in the following:

1) wide ESP with (transverse) wire-plate electrode system;
2) wide ESP with (transverse) spike-plate electrode system;
3) narrow ESP with transverse wire-plate electrode system; and
4) narrow ESP with longitudinal wire-plate electrode system.

Also, selected results of PIV measurements *in a nonthermal plasma reactor for water treatment* are shown.

5.3.1
3D EHD Flow in Wide ESP with (Transverse) Wire-Plate Electrode System

In this section, results of 3D PIV measurements of the flow velocity fields in a relatively wide wire-plate-type ESP (width-to-height ratio was 2) are presented. The ESP was an acrylic box (height 10 cm, width 20 cm, length 100 cm) with a wire discharge electrode (transversely placed in respect to the primary flow direction) and two plate collecting electrodes. Air flow seeded with a cigarette smoke was blown along the ESP duct with an average velocity of 0.6 $m\,s^{-1}$. Either positive or negative DC voltage (up to 28 kV) was applied to the wire electrode through a 10 MΩ resistor. The 3D PIV velocity field measurements (instantaneous images of the flow taken with an exposure time of 6 ns, instantaneous particle flow velocity fields, and the time-averaged particle flow velocity fields resulted from averaging of 100 instantaneous PIV images) were carried out in four parallel observation planes stretched along the ESP duct, perpendicularly to the wire electrode and plate electrodes. The first plane (plane A) was placed in the ESP mid-plane, that is, 10 cm from the side wall, the others (planes B, C, D) were placed 60, 20, 10 mm from the side wall.

Figures 5.2 and 5.3 show the instantaneous images of the particle flow in planes A, B, C, and D in the ESP for positive and negative voltage of 19.5 kV. The corresponding 3D flow velocity fields are shown in [5].

The obtained results show that the flow in the ESP mid-plane is almost two-dimensional when its time-averaged behavior is considered. However, in a short

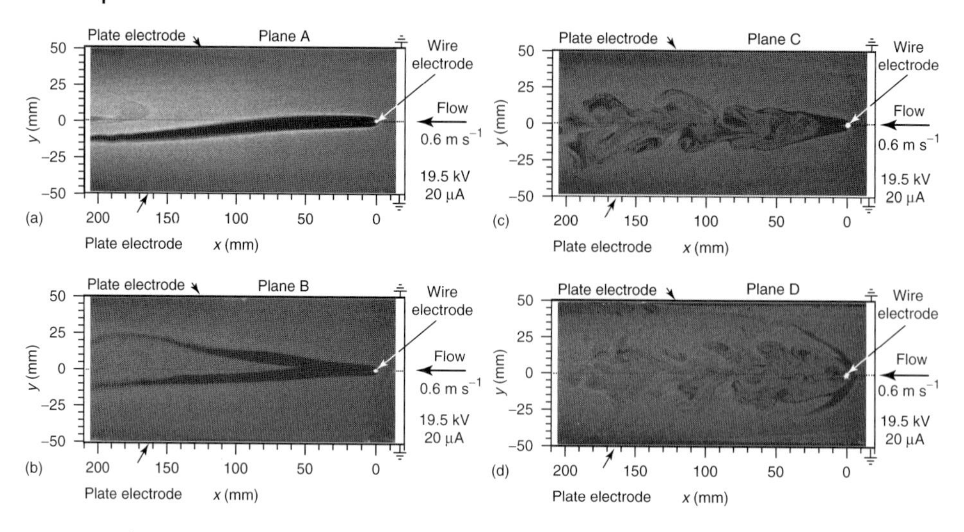

Figure 5.2 Instantaneous images of the flow in planes A, B, C, and D in the wide ESP for positive voltage of 19.5 kV. Ehd number was 8.4×10^6 and Ehd/Re^2 ratio was 0.57. Exposure time: 6 ns.

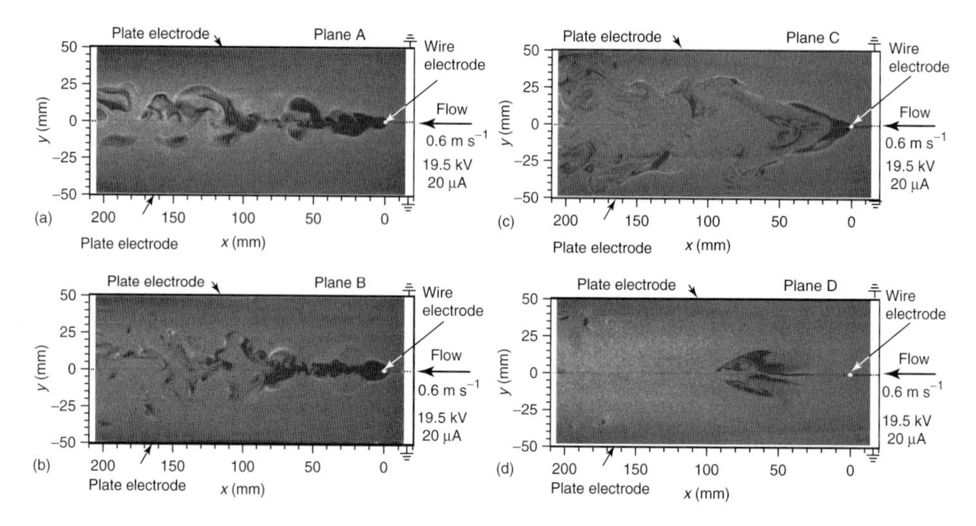

Figure 5.3 Instantaneous images of the flow in planes A, B, C, and D in the wide ESP for negative voltage of −19.5 kV. Ehd number was 6.2×10^6 and Ehd/Re^2 ratio was 0.43. Exposure time: 6 ns.

time period, the flow in all the longitudinal cross sections of the ESP is turbulent (as it is also seen in Figures 5.2 and 5.3) with a relatively high instantaneous velocity z-component, and, therefore, the flow must be considered as three-dimensional. The analysis of results for all four measurement planes showed strong influence of the side wall on the flow patterns in the ESP. This influence is stronger for the negative voltage polarity. These facts have to be considered when analyzing and modeling the flow in ESPs, also with a relatively high width : height ratio.

5.3.2 3D EHD Flow in Wide ESP with (Transverse) Spike-Plate Electrode System

This ESP, similar to the wide ESP with wire-plate electrodes, had a spike electrode (200-mm long, 1-mm thick, and 30-mm tip-to-tip wide) mounted parallel to the plate electrodes in the ESP middle, transversely to the primary flow direction. The spike tips were directed primary flow upstream on one side of the electrode, and downstream on the other (Figure 5.4). The distance from the spike electrode to the plate electrodes was 50 mm.

The ambient air flow seeded with cigarette smoke was blown along the ESP duct with an average velocity of 0.6 m s^{-1}. The positive voltage applied to the spike electrode was 27.9 kV and the discharge current was 210 μA.

2D and 3D PIV measurements were carried out in 14 planes placed perpendicularly to the plate electrodes. Three different parallel planes (A, B, and C) passed along the ESP by the upstream- and downstream-directed spike tips as well as in

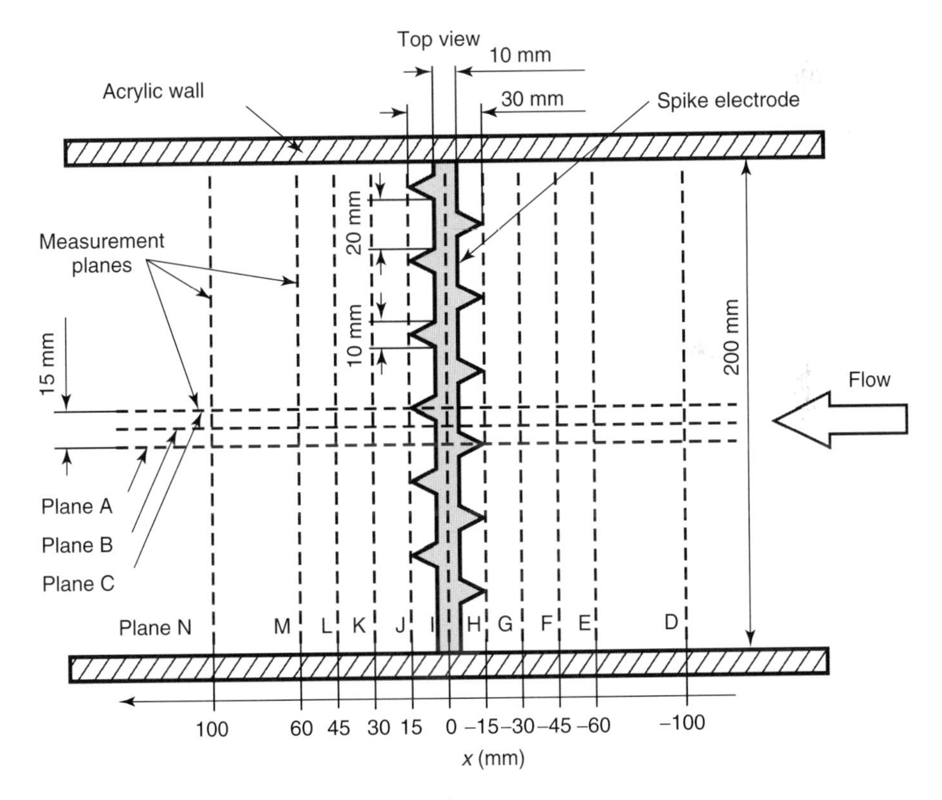

Figure 5.4 Top-view schematic of the spike-plate ESP.

between them. This allowed us to study the influence of the spike tips and their position in respect to the primary flow on the velocity field in the ESP.

The averaged flow streamlines (obtained from the corresponding velocity fields) in planes A, B, and C are shown in Figure 5.5. In plane A, (passing through the upstream spike) the pairs of strong vortices, covering the whole ESP height in the upstream and in the downstream regions, appear. The upstream vortices block the main flow and force it to move toward the collecting plate electrodes. On the other hand, the downstream vortices do not allow the main flow to move near the plate electrodes. As a result, after passing the spike electrode, the main flow turns toward the ESP center and passes between the downstream vortices. In plane B

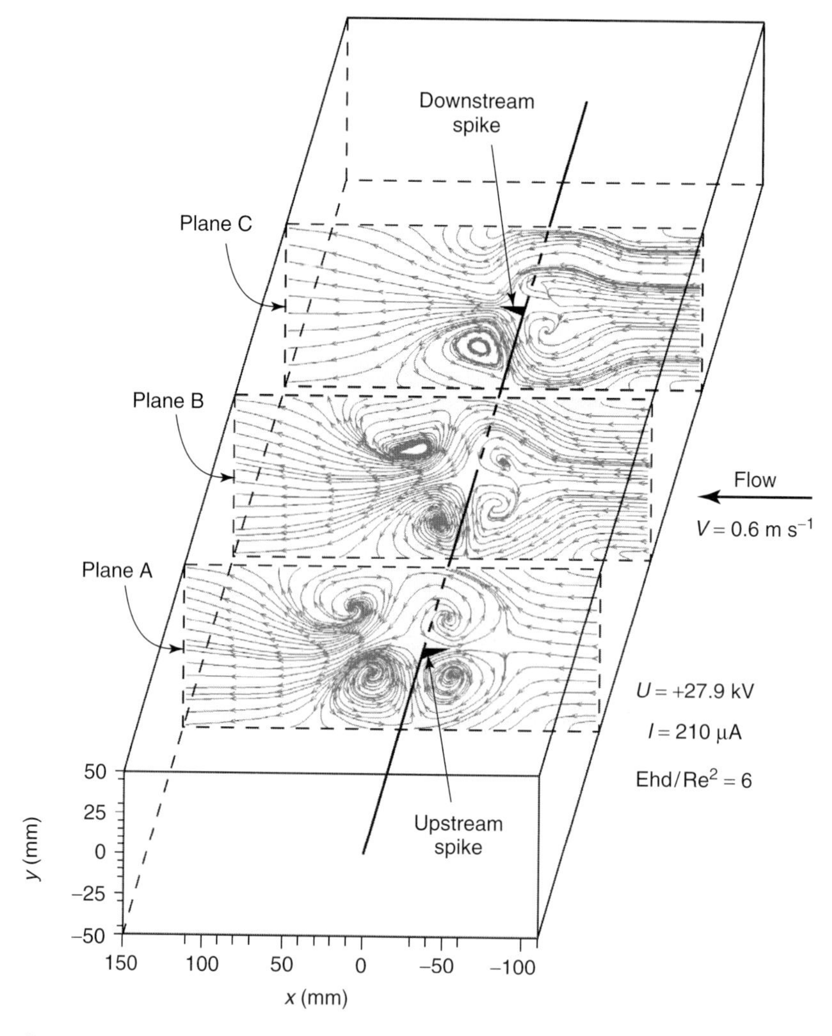

Figure 5.5 Outline of the spike-plate ESP with the averaged flow streamlines measured in planes A, B, and C.

(the mid-plane between upstream and downstream spikes), two pairs of vortices (upstream and downstream) also develop but the vortices in this plane are smaller and less active than those in plane A. The movement of the flow in plane B is similar to that in plane A; however, the flow in plane B is less toward the collecting plate electrodes. The influence of the EHD force on the flow pattern in plane C on the upstream side of ESP is even lower than that in plane B. Notice that plane C passes though the downstream spike, and thus there is no corona discharge on the upstream side. Relatively small vortices are formed before the spike electrode and a considerable part of the primary flow on the upstream side moves not much disturbed by the EHD forces, never reaching the collecting electrodes.

After passing the spike electrode, the flow in plane C enters the discharge region produced by the downstream spike. The strong EHD forces, existing in this region, form two vortices, which deflect a part of the flow toward the collecting plate electrodes.

The PIV measurements in planes A, B, and C showed that the flow pattern depends on whether the measuring plane contains a discharge source (i.e., the spike tip) or not, and on the direction of the discharge in respect to the primary flow. The results of 3D PIV measurements in the y–z planes, presented in Figure 5.6, confirmed this observation.

A relation between the patterns of particle deposits and the collecting plate electrodes in the ESP and flow patterns was found, revealed by the PIV measurements. Figure 5.7 shows a schematic view of the spike-plate ESP with one collecting plate electrode covered by the particle deposits (a similar particle deposit pattern is on the other collecting plate). Also, the average flow streamlines in plane A are shown. It is seen that nail-like deposits were formed on the collecting plate around plane

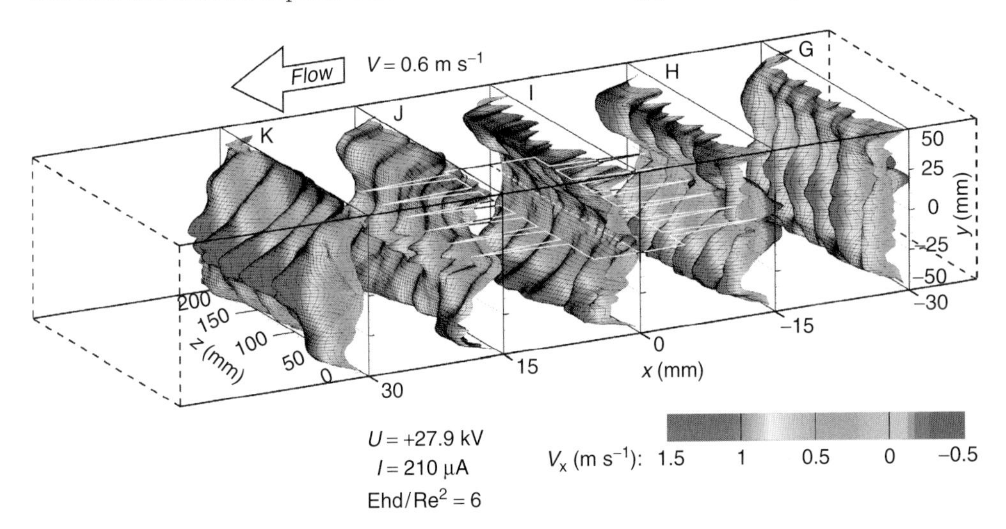

Figure 5.6 Averaged flow velocity x-component measured in planes G–K in the spike-plate ESP. Positive voltage of 27.9 kV was applied, discharge current was 210 μA.

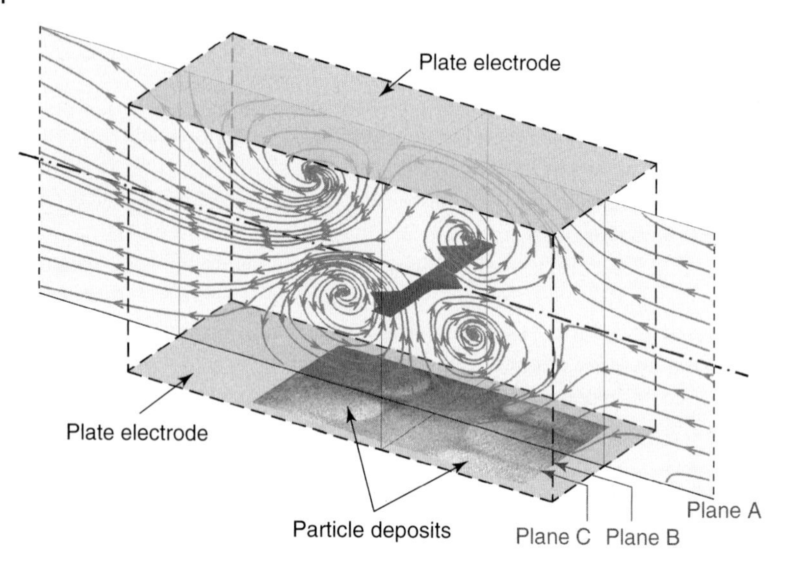

Figure 5.7 Schematic view of the spike-plate ESP with photograph of the particle precipitated on the bottom plate electrode and with the averaged flow streamlines measured in plane A for a positive voltage of 27.9 kV and a discharge current of 210 μA.

A on the upstream side of the ESP. In contrast, on the downstream side of the ESP there are no particles deposits on the collecting electrode, along plane A. Such a behavior of the particle collection can be explained by the flow pattern in plane A. As described above (see also Figure 5.5), the strong EHD forces formed by the discharge from the upstream spike (in plane A) force the majority of incoming particles to move toward the collecting plate where they deposit. On the other hand, the rest of the incoming particles, in plane A, not deposited before the spike electrode, move toward the ESP center, and leave the ESP with a little chance to deposit. As a result, the collecting electrode in plane A on the downstream side of ESP is free of deposits. In a similar way, one can explain the deposit patterns formed in planes B and C, taking into account the corresponding flow patterns shown in Figure 5.5 (planes B and C). This shows a close correlation between the electrohydrodynamically induced flow and the particle deposits on the collecting electrodes. According to that there should be a relation between the EHD flow patterns and the particle collection efficiency. However, a clear conclusion in this matter has not been presented yet.

5.3.3
3D EHD Flow in Narrow ESP with Transverse Wire-Plate Electrode System

In this section, results of 2D PIV measurements of the flow velocity fields in a narrow ESP are presented. The narrow ESP, with a wire electrode that is placed

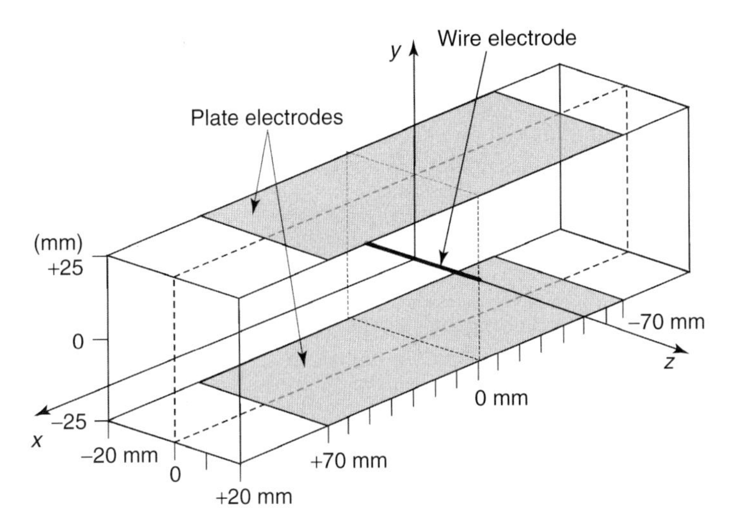

Figure 5.8 The narrow ESP with transverse wire electrode.

traverse to the flow, was a transparent acrylic box 900-mm long, 40-mm wide, and 50-mm high (Figure 5.8). The primary flow average velocity was 0.6 m s^{-1}. The positive voltage was 18 kV. The total discharge current was 50 μA.

2D PIV measurements were carried out in the y–z plane across the ESP duct at x = +30 mm (Figure 5.9). The flow in this plane is slightly disturbed by the

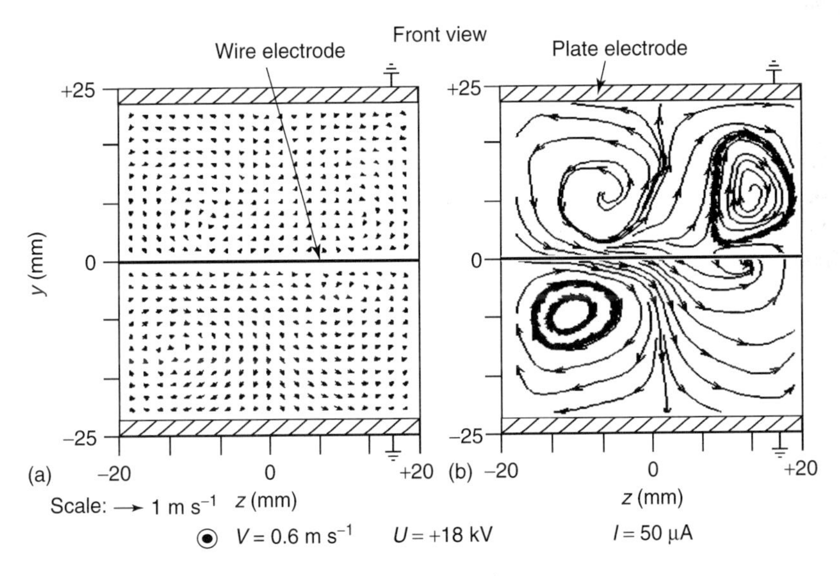

Figure 5.9 Flow velocity field (a) and the corresponding flow streamlines (b) in the narrow ESP (with transverse wire electrode) in the transverse plane (x = +30 mm) for a positive voltage of 18 kV.

EHD forces, taking into account that the transversely induced flow moves with a maximum velocity of 0.3 m s^{-1}. However, the highest vortex velocity (about 0.45 m s^{-1}) was observed in the transverse plane at +70 mm (70 mm behind the wire electrode). Four relatively weak vortex structures are formed in this plane. The particles in the vortices move toward the plate electrodes, then, if not collected, along the plate electrodes and side walls. The four vortex structures are mainly caused by the narrow cross-section of the ESP duct.

5.3.4
3D EHD Flow in Narrow ESP with Longitudinal Wire-Plate Electrode System

This narrow ESP, (transparent acrylic box, 90-mm long, 30-mm wide, and 30-mm high) had a wire electrode, called a *longitudinal wire electrode,* mounted in the middle of the ESP, along the flow direction (Figure 5.10). The primary flow average velocity was 0.6 and 0.9 m s^{-1}. The positive voltage was 10 kV. The total discharge current was 65 μA.

The flow pattern measured using PIV in the y–z plane at $x = 0$ mm (Figure 5.11) is heavily influenced by the EHD forces. Four vortices are formed in this plane. The flow is strongly directed toward the plate electrodes. Then, near the plate electrodes, the flow turns toward the side walls, and move along them. Finally, the flow turns again toward the wire electrode. The maximum flow velocity in the vortices is 1.1 m s^{-1}, that is, much higher than that in the transverse wire electrode case (Section 5.3.3).

The four vortices start to form at the beginning of the discharge region, that is, at x = −40 mm (Figure 5.12). Starting from this point, the vortices developed and became strongest in the central mid-plane (x = 0 mm). At x = +60 mm the

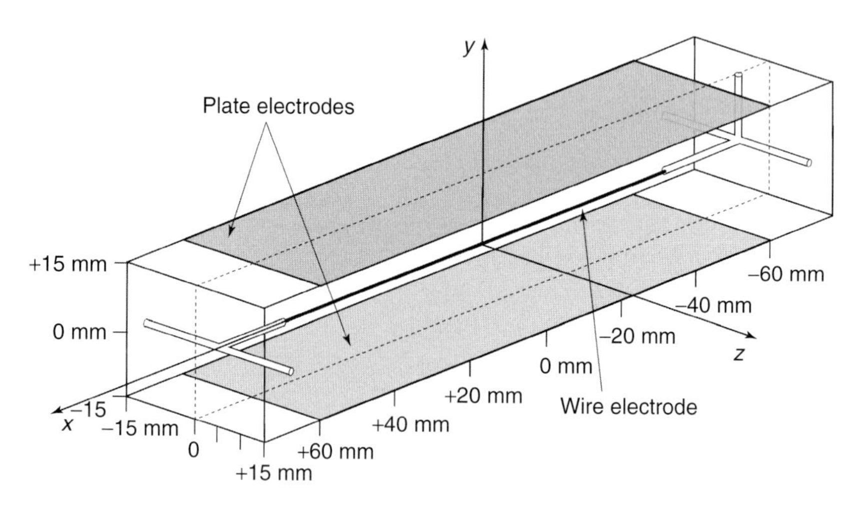

Figure 5.10 The narrow ESP with longitudinal wire electrode.

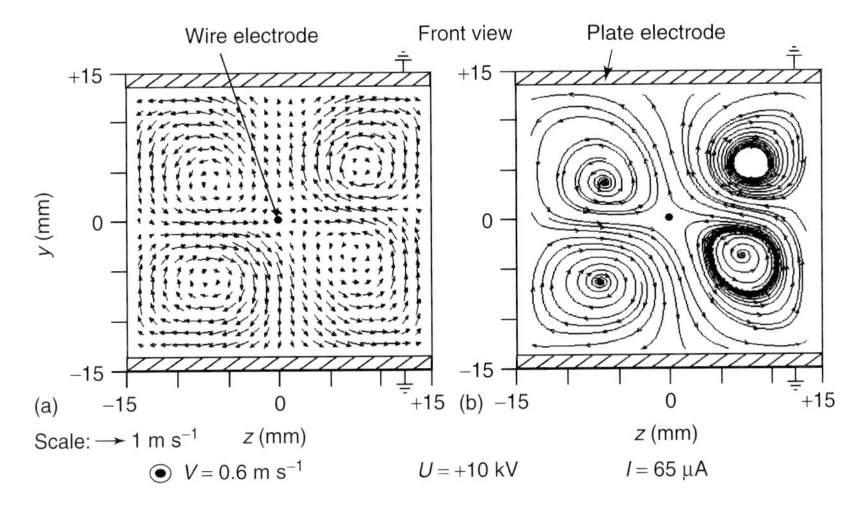

Figure 5.11 Flow velocity field (a) and the corresponding flow streamlines (b) in the narrow ESP (with longitudinal wire electrode) in the transverse plane (x = 0 mm) for a positive voltage of 10 kV.

vortices weaken. In general, in the whole discharge region the vortex structures are regular and symmetric.

On the basis of the flow patterns shown in Figures 5.11 and 5.12, one can conclude that the flow in the ESP with longitudinal wire electrode has the form of four spiral vortices moving along the flow (Figure 5.13).

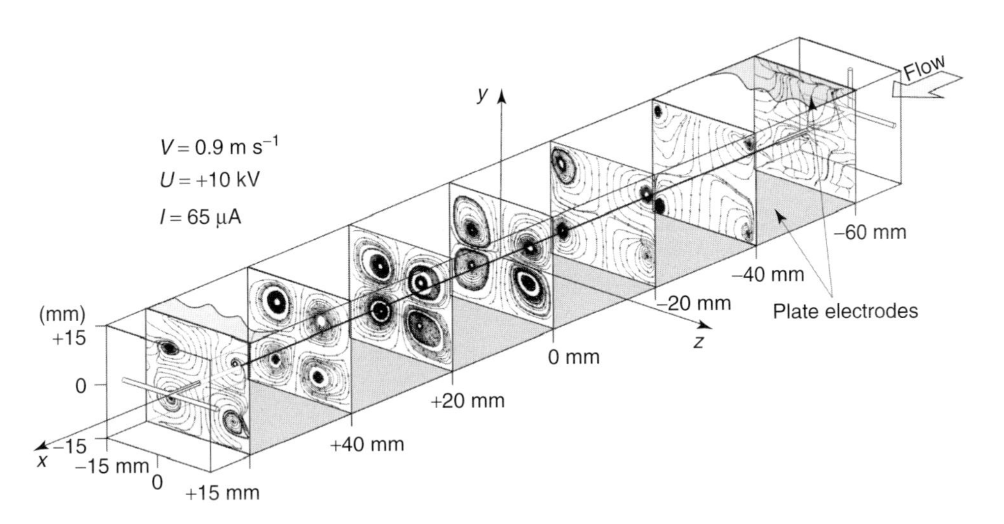

Figure 5.12 The flow streamlines in several transverse planes in the narrow ESP with longitudinal wire electrode.

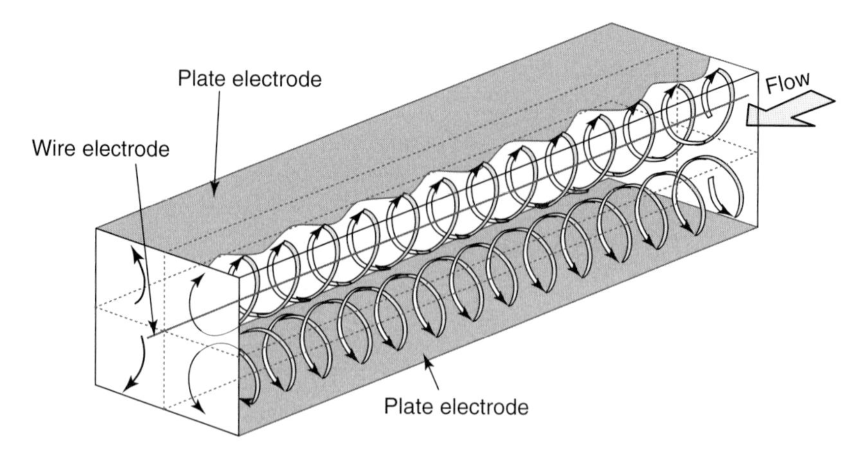

Figure 5.13 Illustration of the spiral flows in the narrow ESP with longitudinal wire electrode.

5.3.5
EHD Flow in Nonthermal Plasma Reactor for Water Treatment

An application of high-voltage pulsed discharge for water purification from organic compounds has been investigated for many years [23–26]. The pulsed discharge forms plasma channels in water. The plasma channels emit ultraviolet radiation and generate shock waves and gas bubbles. The parameters of the bubbles were investigated in [27–29].

In our study of the EHD flow in water reactor, the water reactor was a glass box (35-mm wide, 150-mm long, and 120-mm high) filled with distilled water. Three stainless steel needles (grounded) and a brass cylinder (stressed) were mounted in the reactor. The distance between the needles was 50 mm, whereas the needle–cylinder spacing was 55 mm. The measurements were carried out with one or three needles. The pulsed positive discharge was generated between the grounded needle electrode(s) and the negatively polarized, high-voltage cylinder. The negative high-voltage pulses of 25 kV were applied to the cylinder electrode from discharge capacitors having a capacitance of 0.7, 2, and 2.7 nF. The current pulses, of duration about 100 μs, had an amplitude up to 0.5 A. The pulse repetition rate of 50 Hz was determined by the angular velocity of the rotating element in a spark gap switch.

2D PIV equipment, as described above, was employed for monitoring the flow in the reactor. The bubbles produced during the discharge were used as a seed. The PIV measurements were carried out in several measurement planes in the reactor. Examples of the bubble flow patterns are shown in Figures 5.14 and 5.15.

Figure 5.14 shows the bubbles flow patterns in the plane along the middle of the reactor when a single-needle electrode was used. It is seen that the bubble flow from the needle tip toward the cylindrical electrode was relatively strong (maximum velocity around 54 mm s^{-1}). Bubble vortices were formed under the

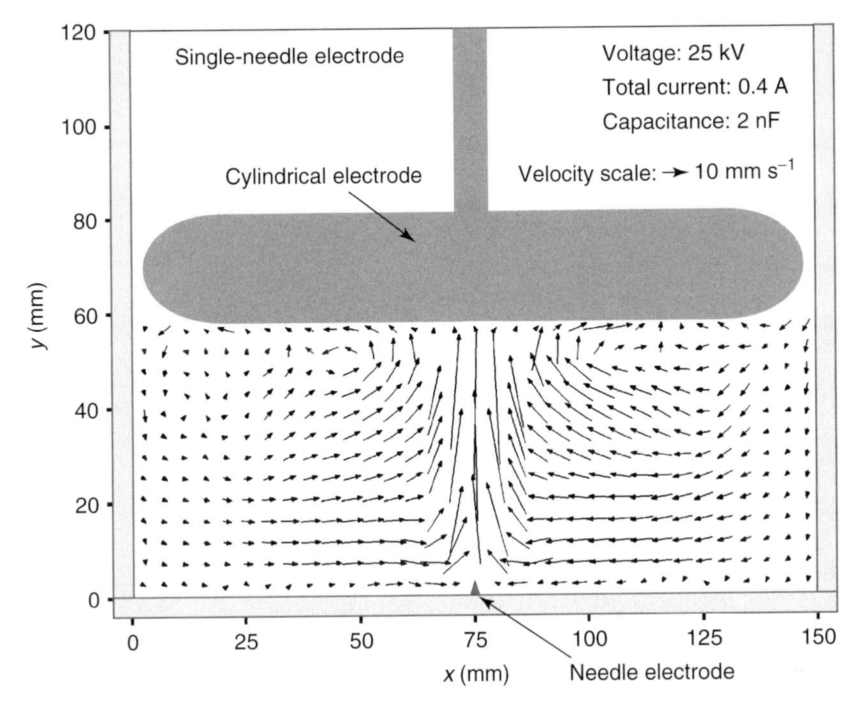

Figure 5.14 Bubble flow velocity field along the reactor with single-needle electrode. The discharge capacitor was of 2 nF capacitance.

cylindrical electrode on both sides of the discharge region. They transported the bubbles back toward the discharge region. Owing to it, the water was stirred in the whole reactor volume and the whole reactor was filled with bubbles.

Figure 5.15 shows the bubble flow patterns in the reactor with three-needle electrode measured in the same plane as in Figure 5.14. It can be seen that three bubble flow streams, corresponding to each discharge region, were formed. On both sides of each discharge region, vortices were formed. They transported bubbles downward in the direction of needles. The maximum velocity of bubbles was 38.0 mm s^{-1} (in the needle-cylinder direction).

The analysis of the PIV results showed that the bubble flow velocity in the reactor with single-needle electrode increased with capacitance of the supplying capacitor (0.7–2.0 nF). The supplying capacitor also influenced the size of vortices and the place of their formation. Higher capacitance resulted in larger vortices. Their centers were shifted from the vicinity of the sidewalls toward the discharge region. The influence of supplying capacitor on the bubble flow velocity and the bubble flow patterns was insignificant in the reactor with three-needle electrodes. However, it must be noticed that, in this case, the range of examined capacitance was relatively low (2.0–2.7 nF).

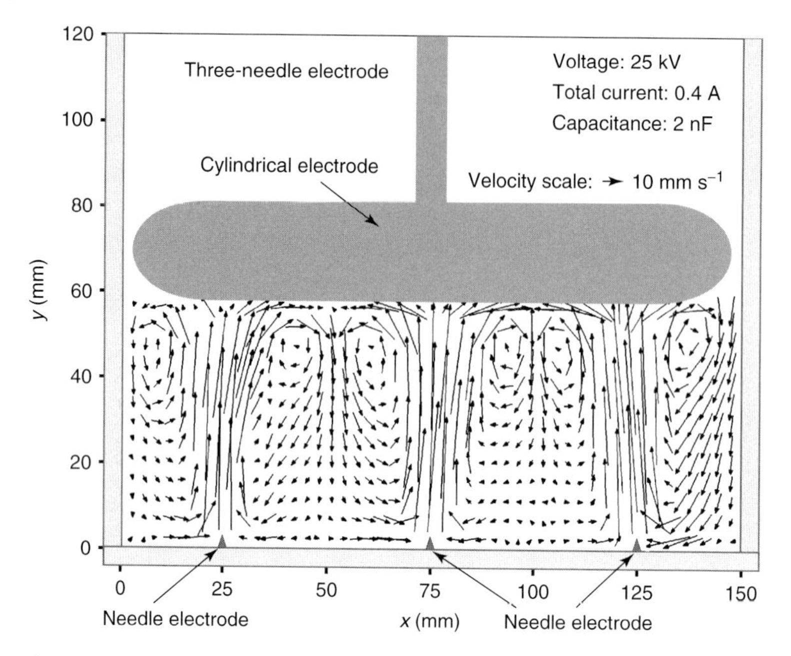

Figure 5.15 Bubble flow velocity field along the reactor with three-needle electrode. The discharge capacitor was of 2.7 nF capacitance.

The results of PIV measurements showed that the pulsed positive streamer discharge in water caused good mixing of the water, due to the generation of bubbles, which were electrohydrodynamically distributed in the whole water volume.

5.4 Conclusions

Our measurements of the velocity fields showed that 2D and 3D PIV are very useful for investigations of flow in ESPs and nonthermal plasma reactors. The flow in all tested ESPs was strongly affected by the EHD forces. The complicated flow patterns in ESPs are supposed to influence the particle collection. However, at present, there is no common opinion about the relation between the EHD flow and particle collection efficiency in ESPs. The results showed that the PIV technique can be used for studying transport of species, for example, electrically active species such as CH, H, O, H_2O, and O_3, in nonthermal plasma reactors.

PIV techniques are continuously developing. PIV method for high temporal resolution (10^4 PIV images per second) measurement of turbulent flow has been recently developed. PIV techniques have also been adopted to measure velocity fields in flow objects of submillimeter dimensions with micron scale spatial resolution. The microscale PIV includes studying the pumping transport and mixing of microflows, and heat transfer in microchannels.

Acknowledgment

This work was supported by the Ministry of Science and Higher Education (grants PB 1857/B/T02/2007/33 and 3892/B/T02/2008/35).

References

1. Raffel, M., Willert, Ch., and Kompenhans, J. (2007) *Particle Image Velocimetry, A Practical Guide*, Springer-Verlag, Berlin , Heidelberg.
2. Podliński, J., Dekowski, J., Mizeraczyk, J., Brocilo, D., Urashima, K., and Chang, J.S. (2006) EHD flow in a wide electrode spacing spike-plate ESP under positive polarity. *J. Electrost.*, **64**, 498–505.
3. Niewulis, A., Podliński, J., Kocik, M., Barbucha, R., Mizeraczyk, J., and Mizuno, A. (2007) EHD flow measured by 3D PIV in a narrow electrostatic precipitator with longitudinal-to-flow wire electrode and smooth or flocking grounded plane electrode. *J. Electrost.*, **65**, 728–734.
4. Podliński, J., Niewulis, A., Mizeraczyk, J., and Atten, P. (2008) ESP performance for various dust densities. *J. Electrost.*, **66**, 246–253.
5. Podliński, J., Niewulis, A., and Mizeraczyk, J. (2008) Electrohydrodynamic turbulent flow in a wide wire-plate electrostatic precipitator measured by 3D PIV method. XI International Conference on Electrostatic Precipitation, Hanghzou, pp. 134–139.
6. Podliński, J., Metel, E., Dors, M., and Mizeraczyk, J. (2008) Bubble flow measurements in pulsed streamer discharge in water using particle image velocimetry. *Inst. Phys. Conf. Ser.*, **142**, 012–036.
7. Podliński, J., Dekowski, J., Jasiński, M., Zakrzewski, Z., and Mizeraczyk, J. (2002) Influence of the gas flow on the microwave torch plasma flame structure. *Czech. J. Phys.*, **52** (Suppl. D), D736–D742.
8. White, H.J. (1963) *Industrial Electrostatic Precipitation*, Wesley Publishing Company, Inc.
9. Baumgard, K.J. and Johnson, J.H. (1992) The effect of low sulfur fuel and a ceramic particle filter on diesel exhaust particle size distribution. *SAE Trans.*, **101**, 691–699.
10. Neeft, J.P.A., Makkee, M., and Moulijn, J.A. (1996) Diesel particulate emission control. *Fuel Process. Technol.*, **47**, 1–69.
11. Walker, A.P. (2004) Controlling particulate emissions from diesel vehicles. *Top. Catal.*, **28**, 1–4.
12. Fino, D. (2007) Diesel emission control: catalytic filters for particulate removal. *Sci. Technol. Adv. Mater.*, **8**, 93–100.
13. Knecht, W. (2008) Diesel engine development in view of reduced emission standards. *Energy*, **33**, 264–271.
14. Thimsen, D.P., Baumgard, K.J., and Kotz, T.J. (1990) The Performance of an Electrostatic Agglomerator as a Diesel Soot Emission Control Device, SAE Paper, 900330, pp. 173–182.
15. Kittelson, D.B., Reinersen, J., and Michalski, J. (1991) Further Studies of Electrostatic Collection and Agglomeration of Diesel Particles, SAE Paper, 910329, pp. 145–163.
16. Wadenpohl, C. and Loeffler, F. (1994) Electrostatic agglomeration and centrifugal separation of diesel soot particles. *Chem. Eng. Process.*, **33** (5), 371–377.
17. Farzaneh, M., Marceu, A.K., and Lachance, P. (1994) Electrostatic capture and agglomeration of particles emitted by diesel engines. *IEEE*, 1534–1537.

18. Saiyasitpanich, Ph., Keenera, T.C., Khangb, S.-J., and Lua, M. (2007) Removal of diesel particulate matter (DPM) in a tubular wet electrostatic precipitator. *J. Electrost.*, **65**, 618–624.
19. Boichot, R., Bernis, A., and Gonze, E. (2008) Agglomeration of diesel particles by an electrostatic agglomerator under positive DC voltage: experimental study. *J. Electrost.*, **66**, 235–245.
20. Chang, J.S. (2001) Recent development of plasma pollution control technology: a critical review. *Sci. Technol. Adv. Mater.*, **2**, 571–576.
21. Chang, J.S. (2008) Physics and chemistry of plasma pollution control technology. *Plasma Sources Sci. Technol.*, **17** (4), 045004 (6pp) (online).
22. Urashima, K. and Chang, J.S. (2000) Removal of volatile organic compounds from air streams and industrial flue gases by non-thermal plasma technology. *IEEE Trans. Dielectr. Electr. Insul.*, **7** (5), 602–614.
23. Sugiarto, A.T., Ito, S., Ohshima, T., Sato, M., and Skalny, J.D. (2003) Oxidative decoloration of dyes by pulsed discharge plasma in water. *J. Electrost.*, **58**, 135–145.
24. Grymonpre, D.R., Sharma, A.K., Finney, W.C., and Locke, B.R. (2001) The role of Fenton's reaction in aqueous phase pulsed streamer corona reactors. *Chem. Eng. J.*, **82**, 189–207.
25. Sunka, P., Babicky, V., Clupek, M., Lukes, P., Simek, M., Schmidt, J., and Cernak, M. (1999) Generation of chemically active species by electrical discharges in water. *Plasma Sources Sci. Technol.*, **8**, 258–265.
26. Locke, B.R., Sato, M., Sunka, P., Hoffmann, M.R., and Chang, J.S. (2006) Electrohydraulic discharge and non-thermal plasma for water treatment. *Ind. Eng. Chem. Res.*, **45**, 882–905.
27. Tsouris, C., Shin, W., Yiacoumi, S., and DePaoli, D.W. (2000) Electrohydrodynamic velocity and pumping measurements in water and alcohols. *J. Colloid Interface Sci.*, **229**, 335–345.
28. Sato, M., Hatori, T., and Saito, M. (1997) Experimental investigation of droplet formation mechanisms by liquid-liquid system. *IEEE Trans. Ind. Appl.*, **33**, 1527–1534.
29. Jomni, F., Aitken, F., and Denat, A. (1999) Dynamics of microscopic bubbles generated by a corona discharge in insulating liquids: influence of pressure. *J. Electrost.*, **47**, 49–59.

6
Water Plasmas for Environmental Application

Takayuki Watanabe

6.1
Introduction

Wastes of 450 million tons a year are discharged by the economic activities of mass production and mass disposal in Japan. Thermal plasmas provide the advanced tool for waste treatment. For the environmental problems, thermal plasmas have received much attention due to their high chemical reactivity, easy and rapid generation of high temperature, high enthalpy to enhance the reaction kinetics, oxidation and reduction atmosphere in accordance with required chemical reaction, and rapid quenching capability (10^5–10^6 K s^{-1}) to produce chemical nonequilibrium compositions [1–9]. Thermal plasmas have been widely applied to many fields because of these advantages. Moreover, engineering advantages such as smaller reactor, lower capital cost, portability allowing on-site destruction, and rapid start-up and shut-down offer efficient decomposition of hazardous and waste materials.

Waste decomposition using thermal plasmas has been reported for various kinds of wastes, such as halogenated hydrocarbon [10–23], polychlorinated biphenyl (PCB) [24, 25], hydrocarbon [26–28], polymer [29–31], organic waste [28, 32–37], used tires [38–40], and medical waste [41, 42]. Waste materials can be efficiently degraded by thermal plasmas under reducing or oxidizing conditions. However, thermal plasmas have been mainly used as a high-temperature source. This indicates that thermal plasmas may have more capability for waste treatment, if they are utilized effectively as chemically reactive gas.

Thermal plasmas are usually generated in electric arcs, in inductively coupled discharge, or in microwave discharge. The gases commonly used to form thermal plasmas are argon, air, nitrogen, and hydrogen. In certain applications, thermal plasmas with addition of reactive gas are desirable to enhance the chemical reactivity of the plasma. Water plasmas are especially suitable for halogenated hydrocarbon decomposition because hydrogen and oxygen combine with the liberated halogen and carbon atoms to form HCl, HF, and CO_2 to prevent recombination reactions that result in the reformation of halogenated hydrocarbons after the decomposition.

Industrial Plasma Technology. Edited by Yoshinobu Kawai, Hideo Ikegami, Noriyoshi Sato, Akihisa Matsuda, Kiichiro Uchino, Masayuki Kuzuya, and Akira Mizuno

ISBN: 978-3-527-32544-3

The water plasma process for hazardous waste decomposition is more effective than air plasmas and the conventional incineration process. Another advantage of water plasmas is the high enthalpy. The mole enthalpy of steam is 1.7 MJ mol^{-1} at 10 000 K where the electrical conductivity is high enough to maintain stable plasmas. This molar enthalpy is higher than that of nitrogen (1.5 MJ mol^{-1}) or argon (0.24 MJ mol^{-1}).

When water plasmas are applied to waste treatments, the use of an additional steam generator is unsuitable because the steam generator requires a complicated system including the heating-up of the steam feeding line for preventing condensation. In this study, we have focused on the decomposition of organic compounds including perfluorocarbons (PFCs) and hydrofluorocarbons (HFCs) by water plasmas generated under atmospheric pressure. Thermodynamic equilibrium calculations were performed between 300 and 6000 K to predict the products after the decomposition. In the experiments, the decomposition was performed by pure-water plasma generated by direct current (DC) discharge without a commercially available steam generator.

6.2
Water Plasma Generation and Its Characteristics

6.2.1
Water Plasma Generation

Thermal plasmas with steam addition have been widely studied for industrial application of waste treatments. Applications of DC argon plasma for halogenated hydrocarbon decomposition have been operated with injection of oxidizing gas such as steam and oxygen to the plasma jet [11–15]. A 150-kW plant for decomposition of Australian's stockpile of halons has been opened in Melbourne. The use of steam as the oxidizing gas improves the decomposition performance, reducing the residual levels of halogenated hydrocarbon, and eliminating CF_4 production as the by-product, with suppression of soot formation.

DC plasmas can be generated efficiently with a configuration that is simpler than those for radio frequency (RF) and microwave plasmas, although steam cannot pass through the electrode region because of the electrode erosion by oxidation. If oxidizing gas such as steam is used in conventional DC plasma, severe erosion of the electrodes occurs resulting in the limitation of the lifetime and contamination in the prepared substances with the electrode material. Kim *et al.* [25] used nitrogen or argon to protect tungsten cathode. Glocker *et al.* [16] used hydrogen to protect tungsten cathode. These protection gases are not favorable for industrial applications owing to economical reason. Considering the cost reduction, pure-water plasmas without injection of other gases are desired for waste treatments of DC plasma process.

DC plasma systems such as water-stabilized arc have been developed for high-temperature source [43–48]. The water-stabilized arc is ignited in a center of vortex of water, which is created in a cylindrical arc chamber with tangential injection of water. The plasma is generated by heating and ionization of steam that is produced by evaporation of water from the inner surface of the vortex. The flow rate of plasma is thus controlled by a balance of heat transfer in the arc column and cannot be adjusted independently like in ordinary gas-stabilized arc.

Plasma generation system generally requires complex sub-equipment such as gas supply unit and cooling system. Especially water plasma system needs a heating equipment to prevent condensation of steam through the reactor. Thus, efficient water plasma generation system is required without complicated systems. We have successfully developed the pure-water DC plasmas as shown in Figure 6.1. The experimental apparatus consists of a plasma torch, a reaction tube, a pump for control of water feeding, and a DC power supply. The water plasma torch is a DC nontransferred plasma arc generator of coaxial design with a cathode of hafnium embedded into a copper rod and a nozzle-type copper anode [17, 21, 28].

The developed torch enables the generation of pure-water plasma without the need for a commercially available steam generator. This feature of the torch results from the simple steam generation process: liquid water from the reservoir is heated up and evaporates at the anode region to form the plasma supporting gas. Simultaneously, the anode is cooled by water evaporation; therefore, the electrodes

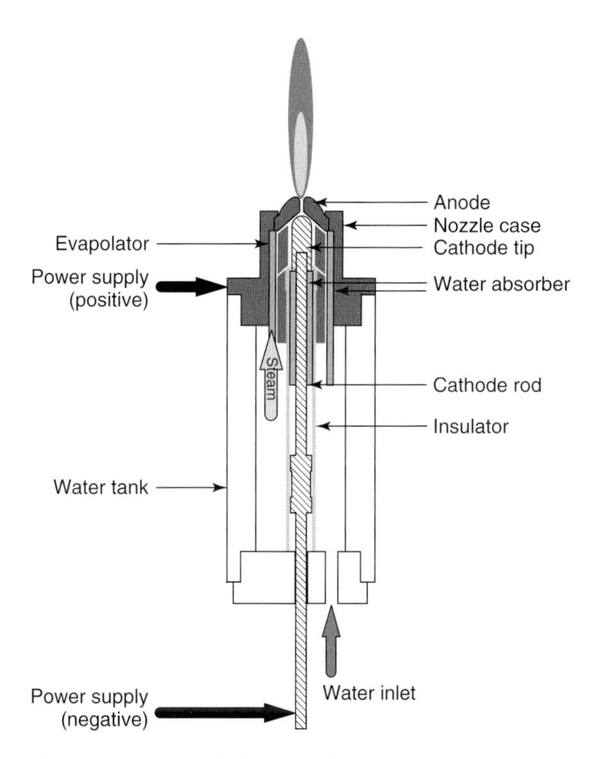

Figure 6.1 Schematic diagram of water plasma torch.

require no additional water cooling. The distinctive steam generation method offers a portable lightweight plasma generation system that does not require the gas supply unit, and thus the high energy efficiency results from the lack of requirement for additional water cooling. Furthermore, hydrogen and oxygen in the produced gas suppress the recombination of by-products. These features of the proposed plasma generation method, which are not readily achievable by other methods, allow for a simple and effective water plasma generation system.

The measured feed rate of the plasma gas produced from water was 40 mg s^{-1} at the arc current of 7.0 A. The plasma gas flow rate was 70 mg s^{-1} in the case of methanol or ethanol solution. Plasma gas flow rate increases with increasing alcohol concentration because of higher vapor pressure. The flow rate of plasma gas is controlled by the balance between the heat transfer from the arc and the evaporation of alcohol solution. Higher vapor pressure of the methanol solution causes larger feed rate of the plasma supporting gas evaporated from the methanol solution.

6.2.2 Temperature Measurements

Temperature measurement is indispensable to generate high-temperature source for radicals with high reactivity to completely decompose the injected wastes in water plasma processing. The excitation temperature of the water plasma in the region just downstream of the nozzle exit was measured by emission spectroscopy (iHR550, Horiba Jobin Yvon). The emission of H_α and H_β atom lines from hydrogen were received by a charge coupled device (CCD) detector. The excitation temperatures were determined from the Boltzmann plot from hydrogen atoms.

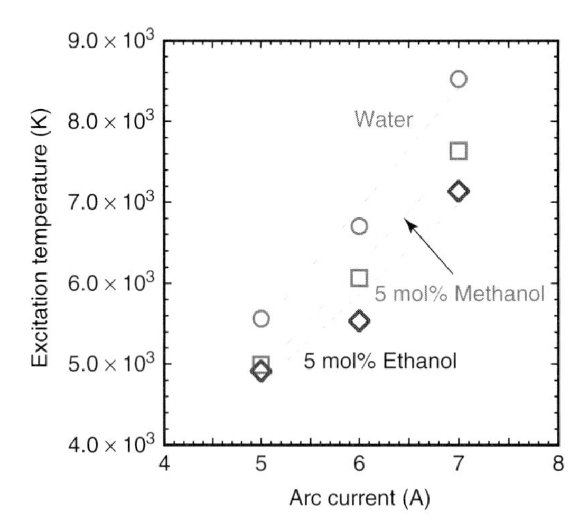

Figure 6.2 Excitation temperature at the nozzle exit of the plasma jet for water, 5 mol% methanol, and 5 mol% ethanol.

The effect of alcohol concentration on the excitation temperature at the nozzle exit for water, methanol–water, and ethanol–water plasmas is presented in Figure 6.2. The excitation temperature of water was about 8500 K at the arc current of 7.0 A. The increasing of alcohol concentration from 1.0 to 10 mol% resulted in decreasing excitation temperature of plasma because of larger decomposition energy at higher concentration.

6.3
Decomposition of Halogenated Hydrocarbons

HFCs are being developed to replace chlorofluorocarbons (CFCs) and hydrochlorofluorocarbons (HCFCs) for use primarily in refrigeration and air conditioning equipment. HFCs do not significantly deplete the stratospheric ozone layer, but they are powerful greenhouse gases with global warming potentials ranging from 140 (HFC-152a) to 11 700 (HFC-23). Thus, in 1997, HFCs are considered as one of the six target greenhouse gases under the Kyoto Protocol, which requires reduction of greenhouse gases by 7% below the 1990 levels in the period 2008–2010.

PFCs such as CF_4, C_2F_6, and C_3F_8 have been widely used for plasma etching as well as chamber cleaning, following chemical vapor deposition in semiconductor manufacturing processes. However, their usage is being regulated internationally because their gases cause greenhouse effect and their global warming potential is several thousand times higher compared with that of CO_2.

Decomposition of PFCs and HFCs is conventionally accomplished by process optimization, recycle and recovery, and abatement processes. There are few alternative chemical data for replacing PFCs because semiconductor manufacturing processes are complex and diverse. Recycle and recovery are appropriate for the reduction of large quantities of PFCs. Abatement techniques include combustion, catalytic decomposition, and plasma abatement. Combustion is the most developed technology, but the complete decomposition is difficult to achieve because the process temperature is not high enough for the decomposition. Catalytic decomposition system is located downstream of the dry pump and designed to treat emissions from multiple chambers with continued operation, thereby providing economy of scale. Since the early 1990s, nonthermal plasma, as an effective technology for controlling PFCs emissions, has been evaluated and studied by many researchers. However, there are problems with these methods, such as the formation of by-products and high operation cost.

Thermal plasmas have been employed to decompose halogenated hydrocarbons through direct injection into the high-temperature region of thermal plasmas. In comparison to other decomposition systems for halogenated hydrocarbons, thermal plasma decomposition system has unique features, such as rapid decomposition with large throughput, quick start-up and shut-down, and greater flexibility of the operating parameters. These potential advantages are ideally suited to provide for industrial applications for halogenated hydrocarbon decomposition.

In the industrial applications, RF steam plasmas generated at 200 torr have performed high-level decomposition with parts per million range of destructed CFCs or halon in the off-gas [10]. A second commercial system of CFC decomposition with RF plasmas has been operated in Kitakyushu to decompose CFCs with steam plasmas at atmospheric pressure. These commercial systems have been used to decompose halogenated hydrocarbons by thermal plasmas with steam as the main plasma gas.

The PLASCON process, developed by CSIRO in Australia, uses a DC plasma torch [12–15]. The process has decomposed a wide range of halogenated hydrocarbons such as CFCs, HFCs, and PCB. The halogenated hydrocarbons were injected with an oxidizing gas (steam or oxygen) at the end of the argon plasma jet. Decomposition efficiency of PCB and CFCs is more than 99% for 1.3–37 kg h^{-1} and 100 kg h^{-1} input flow rate, respectively. However, CF_4 was emitted as the by-product from the decomposition processes. As an example for HFCs decomposition, Ohno *et al.* [22] studied the decomposition of HFC-134a using steam and air plasma. When the steam and air flow rate were set at 12 and 90 l min^{-1}, respectively, the decomposition of HFC-134a to the 99.99% level was demonstrated at 30 A.

Kim *et al.* [25] succeeded in decomposing a mixture of 27% PCB and 73% CCl_4 by DC steam plasma torch at 100 kW. The decomposition rate and removal efficiency were 99.9999% at 0.98 kg kWh^{-1} feeding rate. The mixture wastes were completely decomposed and converted to gases, such as CO, CO_2, HCl, Cl_2, and CH_4, and carbon particles.

Watanabe *et al.* [23] decomposed HFC-134a by the water plasma under atmospheric pressure. The decomposition efficiency of 99.9% can be obtained up to 0.43 mmol kJ^{-1} of the ratio of HFC-134a feed rate to the arc power, and the maximum feed rate was estimated to be 160 g h^{-1} at 1.0 kW of the arc power.

6.3.1
Thermodynamic Consideration

Thermodynamic equilibrium consideration is useful for comprehending the behavior of materials and development of predictive models because it can be used to optimize the composition of the off-gas after the decomposition. In this study, the thermodynamic equilibrium compositions were calculated by FACT (Centre for Research in Computational Thermochemistry, Canada), which is a database software for searching chemical equilibrium in which Gibbs free energy is the lowest. The purpose of the thermodynamic consideration is to estimate the effect of process parameters on halogenated hydrocarbon decomposition and to predict the optimal quenching temperature for the off-gas after the decomposition. The chemical composition obtained from the calculation is probably not practical due to the limited reaction rates; however, the thermodynamic equilibrium is useful to assess the important species in halogenated hydrocarbon decomposition with water plasmas. All computations were conducted with the assumption that the complete thermodynamic equilibrium was achieved at a constant pressure of 101 kPa.

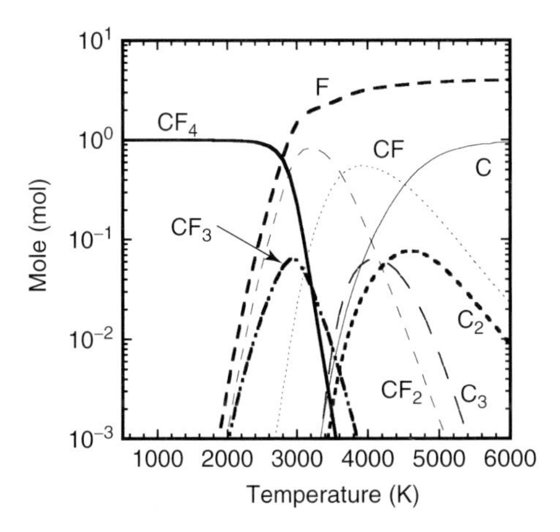

Figure 6.3 Equilibrium composition of pyrolyzed 1.0 mol CF_4.

Thermal equilibrium composition of pyrolyzed CF_4 is shown in Figure 6.3. The stability of CF_4 up to 2800 K indicates that high temperature is required for CF_4 decomposition. At 4600 K, decomposed species such as F, C, C_2, and CF are found in the system. More complex species such as CF_2, CF_3, and C_3 exist on cooling of the off-gas in the temperature range of 2000–5000 K. It indicates that CF_4 is decomposed at high temperatures, but the by-products such as CF_4 and/or other PFCs and HFCs are formed at quenching step. From the equilibrium diagram, we can see that the largest quantities of CF, CF_2, and CF_3 radicals are generated while thermal decomposition of CF_4 takes place at 4000, 3200, and 2900 K, respectively. Suppression of CF, CF_2, and CF_3 radicals is important for by-product formation in CF_4 decomposition processes.

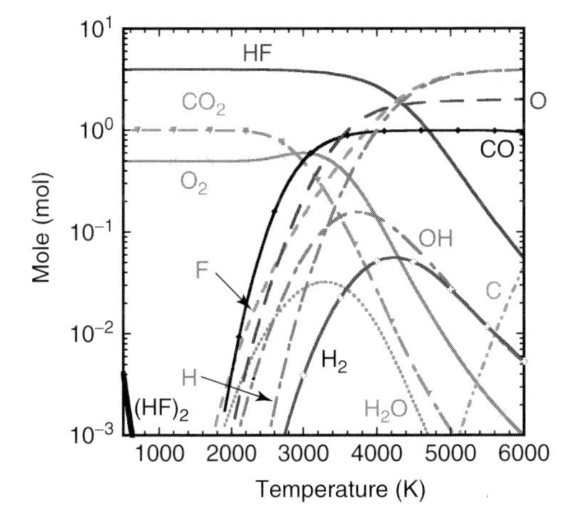

Figure 6.4 Equilibrium composition of 2.0 mol H_2O_2 and 1.0 mol CF_4.

Thermal equilibrium composition of CF_4 decomposition with water plasmas is shown in Figure 6.4. In this calculation, water is added to CF_4 decomposition to suppress CF_4 formation after the decomposition. As shown in this figure, CF_4 formation as well as $CF_{x,}(x: 1-4)$, radical is not observed at any temperature after the decomposition when water is added. The equilibrium diagram indicates that water dissociates into oxygen and hydrogen atoms above 3500 K, and these oxygen and hydrogen atoms play important roles in the reformation control of undesirable by-products; hydrogen atom removes fluorine from the system to produce hydrofluorine (HF). HF is the most stable fluoride under 4400 K without formation of undesirable by-products of halogenated hydrocarbons. Correspondingly, use of water plasmas for decomposition of PFCs can be considered as an environmentally clean process. In comparison with the equilibrium composition of pyrolyzed CF_4 without water as shown in Figure 6.3, carbon disappears to produce CO and CO_2. The water plasma reduces carbon clusters, such as C_2 and C_3, as well as carbon fluoride, such as CF, CF_2, and CF_3.

6.3.2
Experimental Setups

The decomposition system for halogenated hydrocarbons is presented in Figure 6.5. The system includes a DC water plasma torch, a reaction tube for the decomposition of halogenated hydrocarbons, and a neutralization vessel for the fluorine absorption in the off-gas. In this study, CF_4 or HFC-134a is injected through the injection hole toward the water plasma jet. The presence of an oxidizing gas improves the decomposition efficiently and prevents the formation of soot. The hot mixture of gases is formed from the mixing of the plasma gas, and the injected substances are

Figure 6.5 Decomposition system for halogenated hydrocarbons using water plasmas.

rapidly quenched in 1.0 M NaOH solution after consequent complicated chemical reactions in the reaction tube. The quenching step is important to suppress by-product formation in the plasma waste treatment.

The decomposition mechanism of halogenated hydrocarbons by the water plasma was investigated using several diagnostic methods. The produced gas was analyzed by gas chromatography (GC) equipped with a thermal conductivity detector (Shimadzu, GC-8A) and a quadrupole mass spectrometer (QMS, Ametek, Dycor Proline). The produced liquid was analyzed by GC and a total organic carbon (TOC) analyzer (Shimadzu, TOC-V CSN).

6.3.3
Experimental Results

The CF_4 injected into the water plasma was decomposed immediately in the high-temperature region, and then the carbon from CF_4 was converted to CO and CO_2. Besides, fluorine was converted to HF in the reactor due to the high temperature for production of reactive radicals and electrons. The effect of DC arc current on the recovery of fluorine which was produced from CF_4 decomposition is shown in Figure 6.6. The flow rates of both CF_4 and the supplied O_2 were set at $0.215\,l\,min^{-1}$.

The recovery of fluorine was estimated from Eq. (6.1).

$$\text{Recovery of fluorine} = (\text{absorbed fluorine NaOH solution})/(\text{total fluorine feed}) \tag{6.1}$$

The recovery of fluorine was estimated from the pH measurements by neutralization titration of the solution in the neutralization vessel. From the recovery of fluorine, the decomposition efficiency of CF_4 can be evaluated. Fluorine in the off-gas after the decomposition easily dissolved in water; therefore, the recovery of

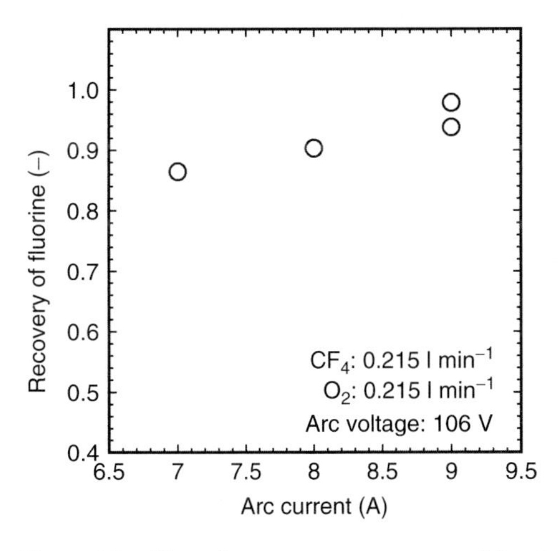

Figure 6.6 Effect of arc current on recovery of fluorine for CF_4 decomposition.

fluorine is a good indication of the CF_4 decomposition. The off-gas composition also provides the decomposition rate, which is the same as the recovery of fluorine. The negligible difference between the recovery of fluorine and the decomposition rate indicates that there is no by-product formation in the off-gas after the CF_4 decomposition. As the arc current increases, the recovery of fluorine increases because the plasma enthalpy increases with the increasing of arc current. When the arc current is 9.0 A, CF_4 is completely decomposed and the recovery of fluorine reaches 99%.

The compositions of the off-gas as a function of the arc current are presented in Figure 6.7. The compositions were analyzed by GC. The conversion to CO_2 increases with increase in the arc current. When the arc current is 9.0 A, the products in the off-gas were CO_2(84%), H_2(10%), O_2(5%), and CO(1%). This result is close to the results obtained from the thermodynamic analysis. Therefore, the water plasma for the treatment of PFCs is considered as an environmental and economical process.

The concentrations of CO_2 and CO in the off-gas also provide important information of the decomposition mechanism, because these concentrations are related to the reaction temperature. Higher arc current leads to increase in CO_2 and CO production. Besides, the selectivity of CO increased with increasing arc current, because CO is more stable than CO_2 at higher temperature. For the determination of the decomposition performance in detail, comprehensive kinetic simulation is required.

The QMS spectrum of the gas component after decomposition of 1.0 l min^{-1} CF_4 using the water plasma without O_2 addition is shown in Figure 6.8a. The spectra of 2(H_2^{2+}), 12(C^+), 16(O^+), 18(H_2O^+), 19(F^+), 25(C_2H^+), 28(CO^+), 31(CF^+), 32(O_2^+), 44(CO_2^+), 50(CF_2^+), and 69(CF_3^+) were observed. The QMS spectrum of CF_4 decomposition with O_2 addition is presented in Figure 6.8b. The decomposition of CF_4 was performed with the water plasma at the CF_4 flow rate of 1.0 l min^{-1} while keeping the O_2 feed at 0.215 l min^{-1}. Comparing Figure 6.8a and b, when the O_2 was added into the plasma, the peak intensities of C_2H^+, CF^+, CF_2^+, and CF_3^+,

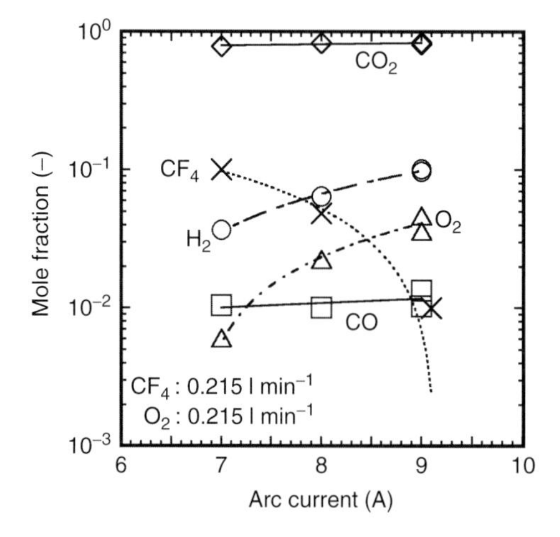

Figure 6.7 Effect of arc current on gaseous product after decomposition of CF_4.

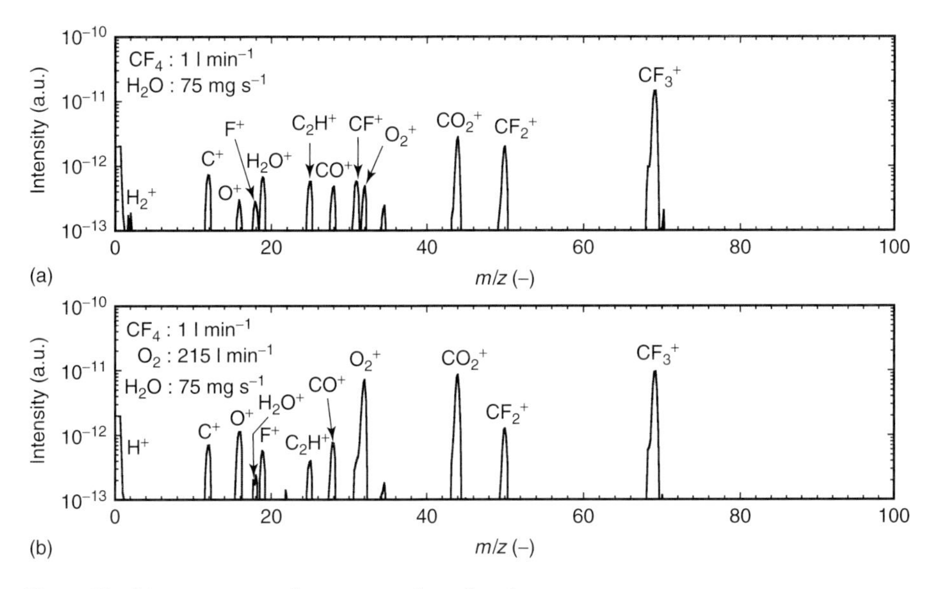

Figure 6.8 Mass spectrum of gaseous product after decomposition of 1.0 l min^{-1} CF_4 at 7 A: (a) without O_2 and (b) with O_2: 0.215 l min^{-1}.

correspondingly to $m/z = 25$, 31, 50, and 69, respectively, decreased substantially, and the peak intensities of CO and CO_2 increased. The by-products are suppressed by adding O_2 in the process of the decomposition of CF_4 by the water plasma.

In the case of HFC-134a decomposition, HFC-134a (0.8 l min^{-1}) was completely decomposed at 7.0 A and the fluorine recovery is more than 99% while the HFC-134a flow rate is 0.215 l min^{-1}. Arc current at 7.0 A is enough for 99% decomposition of HFC-134a at 1.0 l min^{-1} injection. However, the by-products of CH_4, C_2H_2, CH_2F_2, and CHF_3 were generated in the off-gas.

The chemical analysis of the solution in the neutralization vessel was carried out to check the by-product from the HFC-134a decomposition. TOC in the solution for the off-gas treatment was measured after the decomposition. There was a tendency of a decrease in TOC with increase in the feed rate of HFC-134a, but the absolute value of TOC (3×10^{-4} mol l^{-1}) can be negligible compared with the solubility of HFC-134a in water (1.5×10^{-2} mol l^{-1}). The measured organic carbon (OC) was also negligible in the solution. Therefore, the main component in the solution is HF after the HFC-134a decomposition.

6.4 Conclusion

The decomposition of HFC-134a and CF_4 was performed by the water plasma under atmospheric pressure.

1) For decomposition of halogenated hydrocarbons by the water plasma, oxygen radical is important for the suppression of by-product formation. The suitable quenching temperature after CF_4 decomposition is 3600 K.
2) CF_4 was completely decomposed by the water plasma at 0.215 l min^{-1} feed rate at 9.0 A as the arc current. The CF_4 decomposition rate increases with increasing amount of O_2. The stoichiometric mixture ratio of O_2 to CF_4 should be 0.6 to achieve well CF_4 decomposition. In this condition, the by-products of COF_2, HFC, and the recombination of CF_4 are not observed in the off-gas.
3) The supply of 7 A for the water plasma generation was enough for 99% decomposition of 1.0 l min^{-1} HFC-134a. However, the by-products of CH_4, C_2H_2, CH_2F_2, and CHF_3 were generated in the off-gas.

References

1. Mostaghimi, J. and Boulos, M.I. (1990) *J. Appl. Phys.*, **68**, 2643–2648.
2. Sakano, M., Watanabe, T., and Tanaka, M. (1999) *J. Chem. Eng. Jpn.*, **32**, 619–625.
3. Tanaka, Y.and Sakuta, T. (2002) *J. Phys. D*, **35**, 468–476.
4. Watanabe, T.and Sugimoto, N. (2004) *Thin Solid Films*, **457**, 201–208.
5. Tanaka, Y. (2004) *J. Phys. D*, **37**, 1190–1205.
6. Atsuchi, N., Shigeta, M., and Watanabe, T. (2006) *Int. J. Heat Mass Transf.*, **49**, 1073–1082.
7. Watanabe, T., Atsuchi, N., and Shigeta, M. (2006) *Int. J. Heat Mass Transf.*, **49**, 4867–4876.
8. Shigeta, M., Atsuchi, N., and Watanabe, T. (2006) *J. Chem. Eng. Jpn.*, **39**, 1255–1264.
9. Watanabe, T., Atsuchi, N., and Shigeta, M. (2007) *Thin Solid Films*, **515**, 4209–4216.
10. Takeuchi, S., Takeda, K., Uematsu, N., Komaki, H., Mizuno, K., and Yoshida, T. (1995) Proceedings of the 12th International Symposium on Plasma Chemistry, pp. 1021–1026.
11. Snyder, H.R.and Fleddermann, C.B. (1997) *IEEE Trans. Plasma Sci.*, **25**, 1017.
12. Murphy, A.B.and McAllister, T. (1998) *J. Appl. Phys. Lett.*, **73**, 459.
13. Murphy, A.B. (1999) *Ann. N. Y. Acad. Sci.*, **891**, 106–123.
14. Murphy, A.B.and McAllister, T. (2001) *Phys. Plasma*, **8**, 2565.
15. Murphy, A.B., Farmer, A.J.D., Horrigan, E.C., and McAllister, T. (2002) *Plasma Chem. Plasma Process.*, **22**, 371.
16. Glocker, B., Nentwig, G., and Messerschmid, E. (2000) *Vacuum*, **59**, 35.
17. Watanabe, T.and Shimbara, S. (2003) *High Temp. Mater. Process.*, **7**, 455–474.
18. Föglein, K.A., Szabó, P.T., Dombi, A., and Szépvölgyi, J. (2003) *Plasma Chem. Plasma Process.*, **23**, 651–664.
19. Föglein, K.A., Szépvölgyi, J., Szabó, P.T., Mészáros, E., Pekker-Jakab, E., Babievskaya, I.Z., Mohai, I., and Károly, Z. (2005) *Plasma Chem. Plasma Process.*, **25**, 275–288.
20. Föglein, K.A., Szabó, P.T., Babievskaya, I.Z., and Szépvölgyi, J. (2005) *Plasma Chem. Plasma Process.*, **25**, 289–302.
21. Watanabe, T. (2005) *ASEAN J. Chem. Eng.*, **5**, 30–34.
22. Ohno, M., Ozawa, Y., and Ono, T. (2007) *Int. J. Plasma Environ. Sci. Technol.*, **1**, 2.
23. Watanabe, T.and Tsuru, T. (2008) *Thin Solid Films*, **516**, 4391–4396.
24. Tock, R.W.and Ethindton, D. (1988) *Chem. Eng. Commun.*, **71**, 177–187.
25. Kim, S.W., Park, S.H., and Kim, J.H. (2003) *Vacuum*, **70**, 59–66.
26. Kim, K.S., Seo, J.H., Ju, W.T., and Hong, S.H. (2005) *IEEE Trans. Plasma Sci.*, **33**, 813–823.

27. Kim, S.C.and Chun, Y.N. (2008) *Renew. Energy*, **33**, 1564–1569.
28. Nishioka, H., Saito, H., and Watanabe, T. (2008) *Trans. Mater. Res. Soc. Jpn.*, **33**, 691–694.
29. Tang, L., Huang, H., Zhao, Z., Wu, C.Z., and Chen, Y. (2003) *Ind. Eng. Chem. Res.*, **42**, 1245–1150.
30. Park, H.S., Kim, C.G., and Kim, S.J. (2006) *J. Ind. Eng. Chem.*, **12**, 216–223.
31. Tang, L.and Huang, H. (2007) *Fuel Process. Technol.*, **88**, 549–556.
32. Dlugogorski, B.Z., Berk, D., and Munz, R.J. (1992) *Ind. Eng. Chem. Res.*, **31**, 818–827.
33. Nishikawa, H., Ibe, M., Tanaka, M., Ushio, M., Takemoto, T., Tanaka, K., Takahashi, N., and Ito, T. (2004) *Vacuum*, **73**, 589–593.
34. Nishikawa, H., Ibe, M., Tanaka, M., Takemoto, T., and Ushio, M. (2006) *Vacuum*, **80**, 1311–1315.
35. Vaidyanathan, A., Mulholland, J., Ryu, J., Smith, M.S., and Circeo, L.J.Jr. (2007) *J. Environ. Manage.*, **82**, 77–82.
36. Lemmens, B., Elslander, H., Vanderreydt, I., Peys, K., Diels, L., Oosterlinck, M., and Joos, M. (2007) *Waste Manage.*, **27**, 1562–1569.
37. Huang, H.and Tang, L. (2007) *Energy Convers. Manage.*, **48**, 1331–1337.
38. Chang, J.S., Gu, B.W., and Looy, P.C. (1996) *J. Environ. Sci. Health*, **A31**, 1781–1799.
39. Huang, H., Tang, L., and Wu, C.Z. (2003) *Environ. Sci. Technol.*, **37**, 4463–4467.
40. Rutberg, Ph.G., Bratsev, A.N., and Ufimtsev, A.A. (2004) *High Technol. Plasma Process.*, **8**, 433–445.
41. Nema, S.K.and Ganeshprasad, K.S. (2002) *Curr. Sci.*, **83**, 271–278.
42. Park, H.S., Lee, B.J., and Kim, S.J. (2005) *J. Ind. Eng. Chem.*, **11**, 353–360.
43. Hrabovsky, M., Konrad, M., Kopecky, V., and Sember, V. (1997) *IEEE Trans. Plasma Sci.*, **25**, 833–839.
44. Hrabovsky, M. (2002) *Pure Appl. Chem.*, **74**, 429–433.
45. Djakov, E.B., Hrabovsky, M., Kopecky, V., and Jones, G.R. (2004) *High Temp. Mater. Process.*, **8**, 185–193.
46. Hrabovsky, M., Kopecky, V., Chumak, O., Kavka, T., and Konrad, M. (2004) *High Temp. Mater. Process.*, **8**, 575–583.
47. Hrabovsky, M., Konrad, M., Kopecky, V., Hlina, M., Kavka, T., Chumak, O., Oost, G.V., Beeckman, E., and Defoort, B. (2006) *High Temp. Mater. Process.*, **10**, 557–570.
48. Oost, G.V., Hrabovsky, M., Kopecky, V., Konrad, M., Hlina, M., Kavka, T., Chumak, A., Beeckman, E., and Verstraeten, J. (2006) *Vacuum*, **80**, 1132–1137.

7
Chemistry of Organic Pollutants in Atmospheric Plasmas

Ester Marotta, Milko Schiorlin, Massimo Rea, and Cristina Paradisi

7.1
Introduction

Atmospheric air plasmas are very complex oxidizing environments which are easily and conveniently produced by means of various types of electric discharges. Such plasmas can be exploited for many environmental applications [1, 2], including air pollution control especially for the removal of volatile organic compounds (VOCs) in low concentration [1–6]. Major research efforts in this field focused initially on improving the process energy efficiency and, more recently, on gaining a better control on the emissions produced. Ideally, atmospheric air plasma-processing of VOCs should lead to their exhaustive oxidation to CO_2, a goal seldom reached. Only a partial characterization is usually available of the other products of VOC plasma treatment, which often include CO, volatile organic oxidation intermediates but also nonvolatile organics and particles. In addition, electric discharges in air produce undesired side products such as NO_x and ozone. Recent developments in VOC plasma processing concern various combinations of atmospheric plasma with heterogeneous catalysts to achieve better energy efficiency and product selectivity [1, 2, 7, 8].

Product control and process optimization require a good understanding of the chemical reactions involved in the VOC degradation process. The strong oxidizing power of atmospheric air plasmas makes them applicable to the processing of all kinds of VOCs, which come in huge numbers and in a large variety of chemical composition, structure, and reactivity. Despite their abovementioned strong oxidizing power, such plasmas display some selectivity so that it has been found that different VOCs are oxidized with different efficiencies [9]. As organic chemists we intend to exploit such differences to arrive at a better characterization and understanding of the chemistry of atmospheric plasmas. Essential for this analysis is the wealth of kinetic data available in the literature for the reactions of many VOCs with species which are also present in atmospheric air plasmas, notably electrons, O atoms, OH radicals, and many important atmospheric ions: the cations $O_2^{+\bullet}$, $N_2^{+\bullet}$, O^+, N^+, NO^+, the anions $O^{-\bullet}$, $O_2^{-\bullet}$, $O_3^{-\bullet}$, and their hydrates.

Depending on the attacking species, different initiation steps will occur in the oxidation of any specific VOC: excitation, bond dissociation, atom abstraction,

Industrial Plasma Technology. Edited by Yoshinobu Kawai, Hideo Ikegami, Noriyoshi Sato, Akihisa Matsuda, Kiichiro Uchino, Masayuki Kuzuya, and Akira Mizuno

ISBN: 978-3-527-32544-3

radical addition, electron attachment, charge transfer, and ion–molecule reactions [2]. Anyone of such processes activates the oxidation of VOC molecules by converting them into some VOC-derived organic radicals, $R^{\bullet}$. The subsequent oxidation of such radicals initiates with their trapping by molecular oxygen and is known from studies of the natural oxidative degradation pathways of VOCs in the troposphere [10].

It is well known that different corona regimes are characterized by different distributions in density, energy, and space of electrons, and other reactive species [2]. It is, therefore, to be expected that for a given input energy different power supplies should produce different chemical outcomes. However, despite the wealth of literature data on VOC processing, the results produced by different corona regimes with a single VOC could not usually be compared, because the experiments were performed with different reactors and under different experimental conditions with regard to the composition of the gas being treated (relative amounts of oxygen and humidity, VOC concentration). Thus, it is well known that the corona phenomenon depends strongly on the configuration and size of the electrodes and on the interelectrode distance. On the other hand, it is also well known that the efficiency of VOC processing within a given corona reactor, under well-defined experimental conditions, changes greatly with the initial concentration of VOC $[VOC]_0$, usually decreasing significantly with increasing $[VOC]_0$. [1, 2, 11, 12]. It was therefore of interest to develop a single reactor capable of sustaining different corona regimes and to compare their performance in the oxidation processing of selected VOCs. A brief account is offered here of our comparative studies made with the developed apparatus, a large wire/cylinder bench-top corona reactor which can be powered by DC and pulsed voltages of either polarity. It will be shown that through these comparisons important mechanistic insight can be gained and that the oxidation of VOCs is initiated by reaction with different species depending on the specific corona regime used. A powerful diagnostic tool in our work is the integrated analysis of DC corona current/voltage characteristics coupled with that of the ionized components of the plasma and of their reactions. To this end we are using APCI-MS (atmospheric pressure chemical ionization–Mass Spectrometry) which interfaces a corona chamber at atmospheric pressure with a quadrupole mass analyzer. With our APCI-MS apparatus, which has a time window of about 500 μs, we can thus detect and monitor the ion population within the plasma formed from any gas mixture of desired composition (air, VOC, humidity).

7.2 Experimental

7.2.1 Chemicals

Pure air used in the experiments was a synthetic mixture (80% nitrogen–20% oxygen) from Air Liquide with specified impurities of H_2O (<3 ppm) and of C_nH_m

(<0.5 ppm). Air mixtures of *n*-hexane (500 ppm ± 3%), *i*-octane (500 ppm, ±2%), CO_2 (520 ppm, ±2%), and CO (499 ppm, ±1%) were purchased from Air Liquid in cylinders loaded to a pressure lower than their respective condensation limits. Liquide samples of *n*-hexane, *i*-octane, benzene, and toluene were the products of Aldrich (purity ≥ 99%).

7.2.2
Experiments with the Corona Reactor

Corona reactor (wire/cylinder configuration; cylinder size: 38.5 mm i.d. × 600 mm; wire size 1 mm × 620 mm), power supplies, gas line, and the apparatus used for electric and chemical diagnostics were described in detail in previous publications [13–15]. As shown in the schematic drawing of Figure 7.1a, the reactor is used in a flow-through mode of operation, which allows changing of the input energy by either of the two ways: the applied voltage is changed while maintaining a constant gas flow rate; alternatively, the gas flow rate is changed while maintaining a constant applied voltage. All experiments reported in this work were conducted by changing the applied voltage at a constant gas flow rate of 450 mL min^{-1}. The experimental procedure and the analytical protocols used were as described previously [13–15]. Briefly, for each applied voltage, a 1-mL aliquot of the treated gas mixture was withdrawn with a gastight volumetric syringe from the sampling port just downstream of the reactor. GC-FID quantitative analysis of the aliquots yielded VOC conversion data, $[VOC]/[VOC]_0$. Plots of $[VOC]/[VOC]_0$ against E,

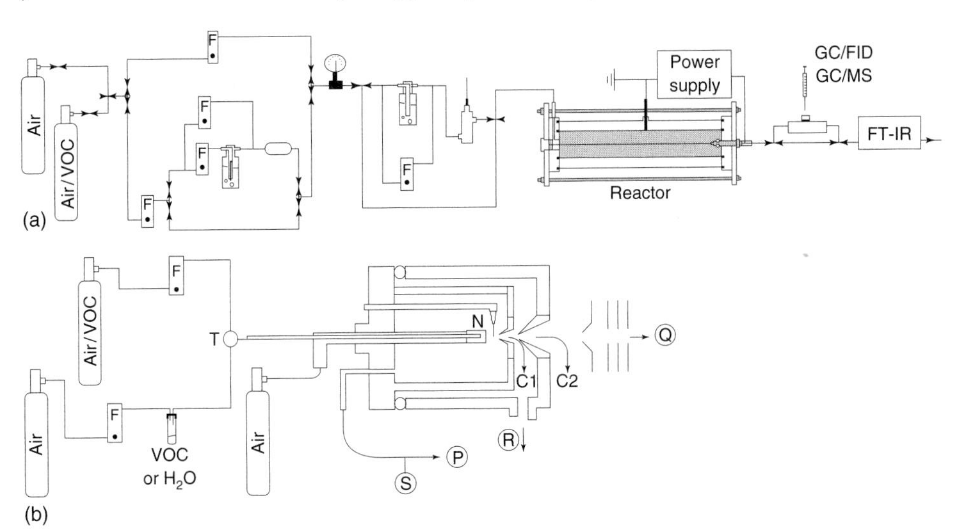

Figure 7.1 Schematics of the experimental apparatus: (a) corona reactor, gas line, and diagnostics system; (b) APCI-mass spectrometer. Legend for symbols: F = flow meter; T = T-junction; P = diaphragm pump; S = septum-capped sampling port; N = needle electrode at high voltage; C1 = sampling cone; C2 = skimmer cone; R = rotary pump; Q = quadrupolar mass analyzer.

the corresponding energy per unit volume or *specific input energy* ($kJ \cdot L^{-1}$), were interpolated with the exponential function of Eq. (7.1) to determine the energy constant k_E, which is a measure of the process energy efficiency.

$$[VOC]/[VOC]_0 = e^{-k_E E} \tag{7.1}$$

The specific input energy for DC [14] and pulsed [15] corona was determined as described previously. Since k_E depends on $[VOC]_0$ [1, 2, 11, 12], in order to compare the efficiency of the different corona regimes, all experiments were conducted at the same initial concentration of VOC (500 ppm).

Current/voltage characteristics of DC corona were monitored both in the presence and in the absence of the selected VOC, always in 500 ppm concentration. For each applied voltage the mean current intensity was measured with a multimeter after a stabilization time of 5 min.

7.2.3
Analysis of Ions

APCI spectra in air plasma were acquired with a TRIO 1000 II instrument (Fisons Instruments, Manchester, U.K.), equipped with a Fisons APCI source as described in earlier publications [16, 17]. The ion source, schematically drawn in Figure 7.1b, is kept at atmospheric pressure by a flow of synthetic air ($4000 - 5000\ mL \cdot min^{-1}$) introduced through the nebulizer line, a capillary of about 2 mm i.d.. Vapors of the desired VOC, stripped from a liquid sample by an auxiliary flow of synthetic air (typically $5{-}50\ mL \cdot min^{-1}$), enter the APCI source through a capillary (i.d. = 0.3 mm) running coaxially inside the nebulizer line. A second line allows for the introduction of water vapors as desired. The needle electrode for corona discharge was kept at 3000 V. Ions leave the source through an orifice, about 50 μm in diameter, in the counter electrode (the "sampling cone", held at 0–150 V relative to ground), cross a region pumped down to about 10^{-2} torr, and, through the orifice in a second conical electrode (the "skimmer cone", kept at ground potential), reach the low-pressure region hosting the focusing lenses and the quadrupole analyzer. Prior to the introduction of the organic compound, a preliminary analysis is routinely conducted to monitor the "background" spectra with only air (or humid air) introduced into the APCI source.

7.3
Results and Discussion

In Figure 7.2, we report the decay of *n*-hexane in dry air as a function of specific input energy in our corona reactor under identical experimental conditions (dry air, initial concentration of VOC: 500 ppm) except for the power supply: +DC (•), −DC(▲), and +pulsed (♦) [13, 15, 18]. Not only is a much greater efficiency observed when energy is provided by pulsed corona than by DC corona, but also significant differences are found between +DC and −DC corona. Similar trends were also

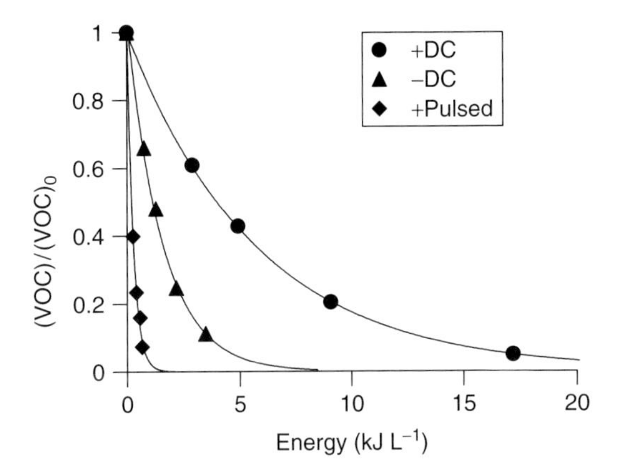

Figure 7.2 Efficiency of removal of *n*-hexane (500 ppm initial concentration in dry air) by different types of corona at room temperature and pressure.

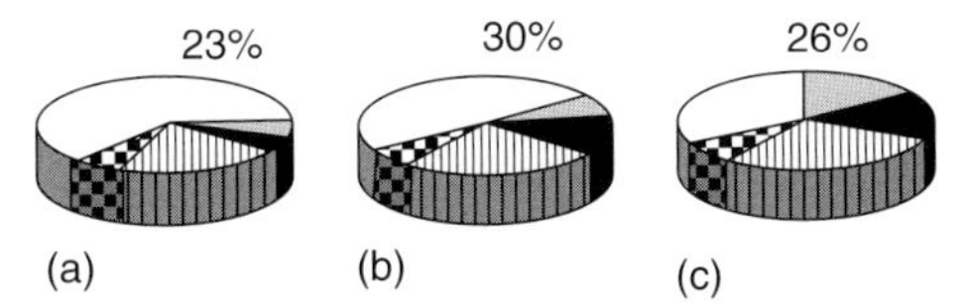

Figure 7.3 Comparison of product distributions in *n*-hexane processing with (a) +pulsed, (b) −DC, and (c) +DC corona at similar conversion extents. Color code: lines = residual *n*-hexane, checks = volatile organic byproducts, white = unaccounted carbon, gray = CO_2, black = CO.

observed with *i*-octane [15] and toluene [19] thus indicating that the efficiency of hydrocarbon processing decreases in the order +pulsed > −DC > +DC. Interestingly, the observed product distributions are also significantly different for the three energization modes tested. An example is shown in Figure 7.3, which compares product data for three experiments in which *n*-hexane was decomposed by treatment with +pulsed, −DC, and +DC corona to achieve similar conversions (the residual hexane was 23, 30, and 26%, respectively). It is evident that different product distributions are obtained under the different energization modes. Notably, the extent of unaccounted carbon, due to nonvolatile and/or particulate byproducts, is the largest with +pulsed corona (60%) and decreases in the order −DC (50%) > +DC (32%). In contrast, the selectivity for CO_2 production follows an opposite trend, namely +DC (16%) is better than −DC and + pulsed (6% for both).

The process response to the presence of humidity in the air is also very characteristic of the mode of corona energization. Figure 7.4 shows decomposition data for *i*-octane induced by +DC (circle symbols) and −DC (triangle symbols) in dry (open symbols) and humid air (40%RH, RH = relative humidity) (closed

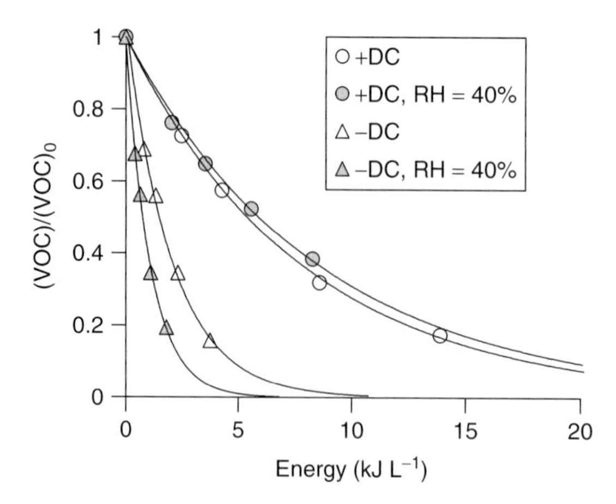

Figure 7.4 Effect of humidity on the efficiency of removal of *i*-octane (initial concentration 500 ppm in air) by DC corona.

symbols). It is seen that the efficiency of the process induced by −DC corona increases in humid air whereas a slight decrease is found with +DC corona. Similar observations were also made with *n*-hexane [14] and toluene [19], as shown by the energy constants summarized in Table 7.1. As for +pulsed corona, the presence of humidity in the air improves the efficiency of hydrocarbon processing, but the effect is much less pronounced than for −DC corona. This can be appreciated by the data reported in Table 7.1 under the heading "Δ%" which represents the percent efficiency change in going from dry to humid (40% RH) air. In the case of toluene, for example, with −DC corona the efficiency of hydrocarbon processing in humid air doubles with respect to dry air, whereas with +pulsed corona a modest increment of 15% is experienced under the same conditions.

Note that for *n*-hexane two sets of data are reported in Table 7.1 for +DC and −DC experiments. The data in parentheses were recorded with the reactor in an earlier configuration [14], which was later slightly modified to allow for the acquisition of optical emission spectra [18]. The modification to house the optical lens required increasing slightly the length of the emitting wire while maintaining its original diameter (1 mm) and composition (stainless steel). Moreover, the wire morphology also changed from the original entwined threads, not anymore available, to a smooth cylindrical shape. This new electrode was used in the present work as well as in that reported in [13, 18, 19].

The data in Table 7.1 also serve to establish a relative reactivity scale for the investigated hydrocarbons and thus to address the question of structure/reactivity relationship. Relative reactivity data are shown in Table 7.2, taking *n*-hexane as the reference compound and using consistent sets of data. Thus, the relative reactivity of *i*-octane versus *n*-hexane was calculated using, for the latter, the data in parentheses which were obtained at the same time and with the same reactor configuration as the data for *i*-octane [14]. Table 7.2 shows that

Table 7.1 Efficiency of VOC processing by +DC, −DC, and +pulsed corona in dry and humid (40% RH) air.

	$k_E(\text{kJ} \cdot \text{L}^{-1})$								
	+DC			−DC			+pulsed		
VOC[a]	Dry	Humid	Δ%[b]	Dry	Humid	Δ%[b]	Dry	Humid	Δ%[b]
n-Hexane[c]	0.20	0.18	−10%	0.77	1.1	+43%	2.1	2.2	+30%
	(0.17[d])	(0.15[d])		(0.58[d])	(1.1[d])	(+90%)			
i-Octane[d]	0.13	0.12	−8%	0.42	0.75	+79%	–	–	–
Toluene[e]	0.14	0.13	−7%	0.41	0.81	+98%	3.1	3.5	+15%

[a] Initial concentration of VOC was 500 ppm.
[b] Percentage increment/decrement of k_E in humid air relative to dry air.
[c] Ref. [18].
[d] Ref. [14].
[e] Ref. [19].

Table 7.2 VOCs relative reactivity data in processing by +DC, −DC, and +pulsed corona in dry and humid (40% RH) air.

	+DC		−DC		+pulsed	
VOC[a]	Dry	Humid	Dry	Humid	Dry	Humid
n-Hexane	(1.0)	(1.0)	(1.0)	(1.0)	(1.0)	(1.0)
i-Octane	0.76	0.80	0.72	0.68	–	–
Toluene	0.70	0.72	0.53	0.74	1	1.6

[a] Initial concentration of VOC was 500 ppm.

n-hexane is more reactive than both *i*-octane and toluene under DC corona conditions, whereas an inversion in the relative order of reactivity is found under +pulsed corona conditions, with toluene being about 1.5 times more reactive than *n*-hexane.

The effect of humidity is most interesting not only for obvious practical implication but also because it provides important clues on the mechanisms of VOC oxidation. One major outcome due to the presence of water within the air plasma is the formation of OH radicals which are among the strongest known oxidants of VOCs. There are many ways by which OH radicals can be formed by corona discharges in humid air. These include H_2O reactions with electrons, neutrals and ions as shown in Eqs. (7.2)–(7.6b).

$$H_2O + e \rightarrow {}^{\bullet}OH + {}^{\bullet}H + e \tag{7.2}$$

$$O\,(^1D) + H_2O \rightarrow 2^{\bullet}OH \tag{7.3}$$

$$N_2^{+\bullet} + H_2O \rightarrow N_2H^+ + {}^{\bullet}OH \tag{7.4a}$$

$$\rightarrow H_2O^{+\bullet} + N_2 \tag{7.4b}$$

$$H_2O^{+\bullet} + H_2O \rightarrow H_3O^+ + {}^{\bullet}OH \tag{7.5}$$

$$O_2^{+\bullet} + H_2O + M \rightarrow O_2^{+\bullet}\,(H_2O) + M \tag{7.6a}$$

$$O_2{}^{+\bullet}\,(H_2O) + H_2O \rightarrow H_3O^+ + {}^{\bullet}OH + O_2 \tag{7.6b}$$

In "dry" air, the origin of OH radicals is to be attributed to any residual water present and to the reaction of VOC molecules with atomic oxygen, as shown for toluene, as an example, in Eq. (7.7)

$$O\,(^1D) + C_6H_5CH_3 \rightarrow {}^{\bullet}OH + C_6H_5CH_2^{\bullet} \tag{7.7}$$

How important are these reactions for each of the different corona regimes compared in this work? How many OH radicals are present in each case? To compare the relative $^{\bullet}OH$ density produced by +DC, −DC, and +pulsed corona in our reactor, we have used a chemical reactivity probe which was developed for atmospheric chemistry determinations [20] and later used also for determinations in atmospheric plasmas [13, 15, 21, 22]. The probe is based on the known reaction of OH radicals with CO for which a rate constant of 2.28×10^{-13} $cm^{-3} \cdot molecule^{-1} \cdot s^{-1}$ was reported at 296 K [23].

$$CO + {}^{\bullet}OH \rightarrow CO_2 + {}^{\bullet}H$$
$$k = 2.28 \cdot 10^{-13} cm^3 \cdot molecule^{-1} \cdot s^{-1}\ (T = 296\ K) \tag{7.8}$$

The experiments were conducted by subjecting CO (500 ppm in air) to +DC, −DC, and +pulsed treatment in our reactor, both in dry and in humid (40% RH) air. The results [13, 15] are summarized in Figure 7.5. It is seen that the concentration of CO does not change in dry air even under conditions of high input energies, indicating that the reactive species present in dry air plasmas are not capable of oxidizing carbon monoxide. In contrast, a decay of CO concentration is observed in all experiments conducted in humid air, with the efficiency of reaction (7.8) being, however, greatly different depending on the mode of energization: +pulsed > −DC > +DC.

In the light of these results one can now interpret the reactivity data of Table 7.1 and conclude that the oxidation of hydrocarbons in humid air is greatly favored by attack of OH radicals in −DC, while in +DC this reaction route is not important since in the presence of OH radicals a slight decrease in efficiency is observed. Considering that OH radicals are much more reactive than O atoms toward hydrocarbons (see, for example, Eqs. (7.9) and (7.10) concerning *i*-octane [24, 25]), the fact that OH radicals are not involved in hydrocarbon oxidation in +DC processing reasonably suggests that reaction with O atom is also not very likely and that alternative routes should be considered.

$$i\text{–}C_8H_{18} + O(^3P) \rightarrow {}^{\bullet}OH + \text{products}$$
$$k = 9.13 \times 10^{-14} cm^3 \cdot molecule^{-1} \cdot s^{-1}\ (T = 307\ K) \tag{7.9}$$

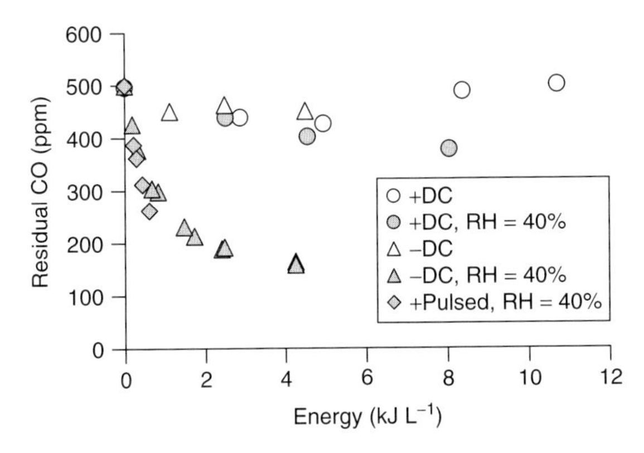

Figure 7.5 Efficiency of CO (500 ppm initial concentration) oxidation under different corona types in dry and humid air.

$$i\text{–}C_8H_{18} + {}^{\bullet}OH \rightarrow H_2O + \text{products}$$
$$k = 3.34 \times 10^{-12}\,\text{cm}^3 \cdot \text{molecule}^{-1} \cdot \text{s}^{-1}\ (T = 298\ \text{K}) \quad (7.10)$$

We believe that ion–molecule reactions are important in these systems. Such conclusion is not simply based on the exclusion of a significant role by reactions with neutrals ($^{\bullet}$OH and O) and on the consideration that ion–molecule reactions are intrinsically much faster than reactions with neutrals (compare the rate constants of Eqs. (7.9) and (7.10) with those of Eqs. (7.11–7.14) [26, 27]) but is also supported by direct experimental observations made in the study of current/voltage characteristics of DC coronas and of the corresponding ionized species.

$$i\text{–}C_8H_{18} + O_2^{+\bullet} \rightarrow \text{products} \qquad k = 1.8 \times 10^{-9}\,\text{cm}^3 \cdot \text{s}^{-1}\ (T = 300\ \text{K}) \quad (7.11)$$

$$i\text{–}C_8H_{18} + N_2^{+\bullet} \rightarrow \text{products} \qquad k = 2.0 \times 10^{-9}\,\text{cm}^3 \cdot \text{s}^{-1}\ (T = 300\ \text{K}) \quad (7.12)$$

$$i\text{–}C_8H_{18} + NO^{+} \rightarrow \text{products} \qquad k = 1.8 \times 10^{-9}\,\text{cm}^3 \cdot \text{s}^{-1}\ (T = 300\ \text{K}) \quad (7.13)$$

$$i\text{–}C_8H_{18} + H_3O^{+} \rightarrow \text{products} \qquad k = 4.9 \times 10^{-10}\,\text{cm}^3 \cdot \text{s}^{-1}\ (T = 300\ \text{K}) \quad (7.14)$$

Figure 7.6 reports current/voltage characteristics measured for −DC and +DC corona in pure air and in air containing 500 ppm of benzene. The data show that while −DC corona current/voltage profiles show no response to the presence of benzene in the air (Figure 7.6b), with +DC corona the current detected in benzene-containing air is significantly lower than in pure air (Figure 7.6a). Since corona current is due to ion transport across the drift region of the interelectrode gap and depends on the ion mobility, our data indicate that different ions are

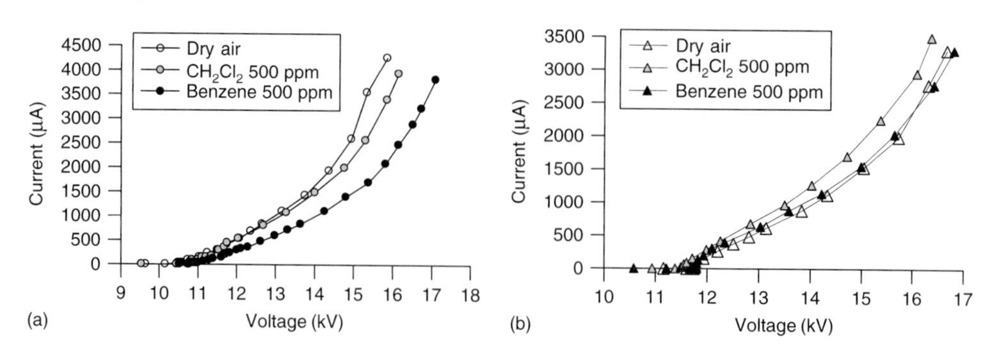

Figure 7.6 Current/voltage characteristics for (a) +DC and (b) −DC corona in pure air and in air containing dichloromethane and benzene.

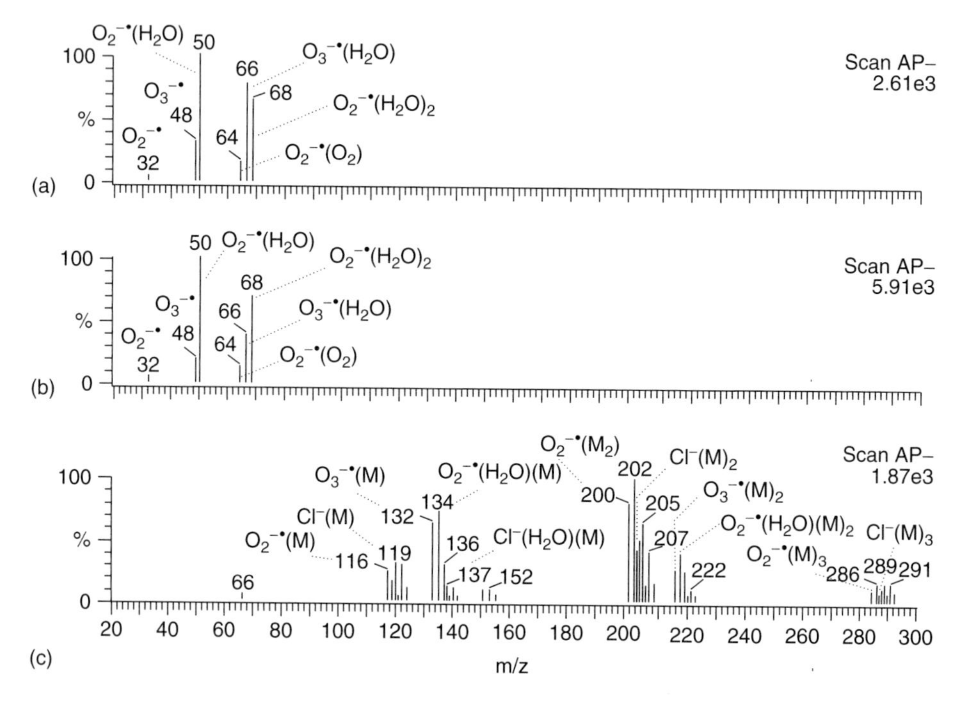

Figure 7.7 Negative-ion APCI mass spectra obtained at ambient temperature and pressure with (a) pure air, (b) benzene in air, and (c) CH_2Cl_2 in air.

involved in pure air and in hydrocarbon-containing air. Specifically, the average ion mobility is lower in hydrocarbon-containing air than in pure air. These observations are supported and rationalized by the results of ion studies. The APCI-MS mass spectra reported in Figure 7.7 show that with −DC corona the same negatively charged ions are observed in pure air (Figure 7.7a) and in benzene-containing

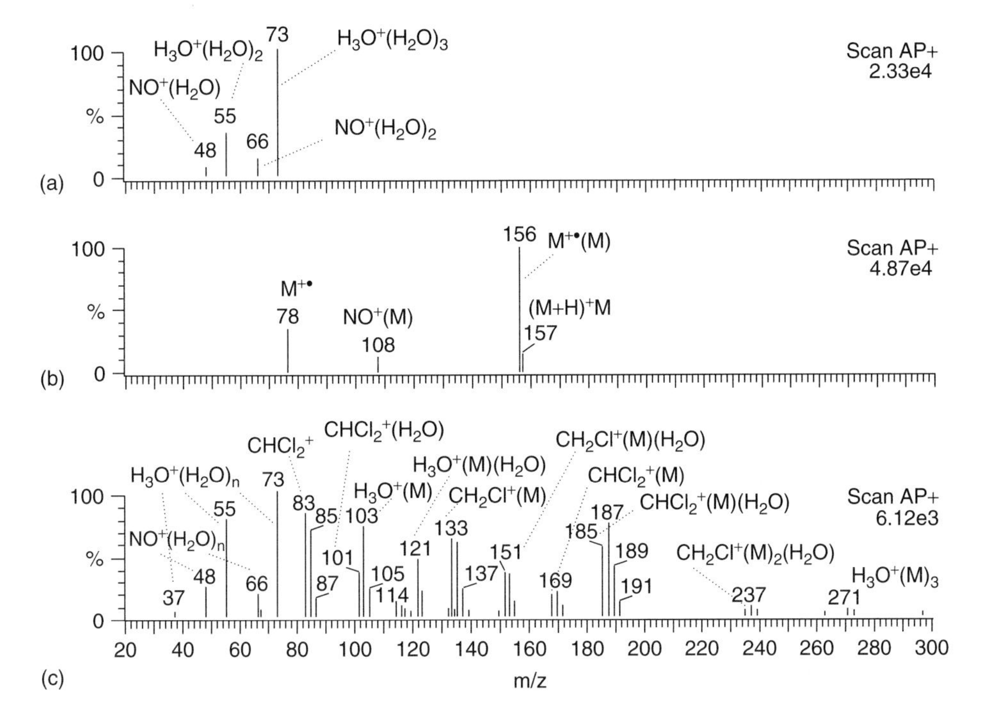

Figure 7.8 Positive-ion APCI mass spectra obtained at ambient temperature and pressure with (a) pure air, (b) benzene in air, and (c) CH_2Cl_2 in air.

air (Figure 7.7b). These are $O_2^{-\bullet}(H_2O)_n$ ($n = 0\text{–}2$: m/z 32, 50, 68), $O_3^{-\bullet}(H_2O)_n$($n = 0\text{–}1$: m/z 48, 66), $O_2^{-\bullet}(O_2)$(m/z 64).

In contrast, with +DC corona the spectrum of pure air (Figure 7.8a) differs markedly from that of air containing benzene (Figure 7.8b). Thus, while in pure air major ions are $H_3O^+(H_2O)_n$ ($n = 2\text{–}3$: m/z 55, 73) and $NO^+(H_2O)_n$ ($n = 0\text{–}1$; m/z 30, 48), in the presence of benzene the most abundant ions are evidently the benzene radical cation, $C_6H_6^{+\bullet}$ (m/z 78), its cluster with a neutral benzene $(C_6H_6^{+\bullet})(C_6H_6)$ (m/z 156) and the complex $(C_6H_6)NO^+$ (m/z 108). Such ions have different compositions, sizes, and shapes from those detected in pure air and are therefore reasonably expected to have different mobilities. Thus, despite its low concentration (500 ppm), benzene produces major effects on the ion population formed by +DC in air and, consequently, on the corona current/voltage characteristic. In contrast, no effects are seen for −DC corona. The behavior of benzene is very similar to that of other hydrocarbons studied previously, notably *n*-hexane [13], *i*-octane [13], and toluene [19].

Such observations, combined with the evidence presented above that OH radicals do not take part in the rate-limiting initiation of hydrocarbon oxidation by +DC corona, suggest that ion–molecule reactions play an important role in these processes.

While these conclusions appear to be of general validity for hydrocarbons (the same behavior was observed with benzene (this work), *n*-hexane [13, 28], *i*-octane [13, 28], and toluene [19]), the picture changes drastically with other VOCs, notably halogen-containing organic compounds. Negative-ion APCI mass spectra of chloro- and bromo-containing VOCs [16, 17, 29, 30] are generally rich in signals due to VOC-derived ions including Cl^-, Br^-, and clusters thereof. This behavior is consistent with the known high electron affinity of alkyl halides, which are readily ionized via electron transfer and charge transfer reactions. Dichloromethane is no exception and produces a wealth of negatively charged ions within the air plasma produced by −DC corona (Figure 7.7c), consisting mainly of ion–molecule complexes of Cl^-, $O_2^{-\bullet}$, and $O_3^{-\bullet}$ with H_2O and dichloromethane molecules. Consistently, significant differences are seen in the current/voltage profiles measured for −DC corona in pure air and in air containing dichloromethane in 500-ppm concentration (Figure 7.6b). As for the effect of dichloromethane on +DC corona, this is similar to that of the hydrocarbons. Thus, the +DC corona current is modified with respect to pure air, being somewhat lower at any applied voltage (Figure 7.6a). The APCI positive mass spectrum of dichloromethane (Figure 7.8c) is very complex and comprises many ions derived from VOC ionization/fragmentation or from ion–molecule complex formation in which VOC molecules are neutral ligands. Tentative attributions are shown on the spectrum (Figure 7.8c), always with reference, in the case of Cl-containing species, to the isotopic peak of the lowest mass, that is, the signal which is due, within each cluster, to ions containing only the isotope ^{35}Cl.

7.4
Conclusions

The results presented and discussed in this paper support the view that different corona regimes (DC or pulsed, positive or negative polarity) applied within the same reactor under the same experimental conditions can support different mechanisms of VOC oxidation. We propose that, in spite of the overall rather similar general features of hydrocarbon processing with +DC, −DC, and +pulsed corona, the crucial initiation steps are different. Specifically, with +DC corona ionic reactions, which are intrinsically much faster than radical reactions, prevail not only in dry air but also in humid air. In contrast, with −DC and +pulsed corona radical initiation steps occur, involving mainly O atoms in dry air and $^\bullet OH$ in humid air. Electron-induced bond dissociation is also probably an important decay channel, especially for toluene [12]. This might be the reason for the inversion in the relative reactivity order observed for toluene in +pulsed corona with respect to DC processing. The elemental composition and chemical structure of the decomposing VOC have expectedly a major effect on its reactivity under the specific corona regime considered. We have shown here that, in contrast to the behavior of hydrocarbons, the currents measured for −DC corona in air are significantly increased by the addition of trace amounts of dichloromethane (500 ppm). Consistently, the negative-ion APCI mass spectra are also greatly affected due to the prevalence of VOC-derived Cl-containing anions.

Acknowledgments

Financial support by the University of Padova (Progetto Interarea 2005) is gratefully acknowledged.

References

1. Van Veldhuizen, E.M. (2000) *Electrical Discharges for Environmental Purposes: Fundamentals and Applications*, Nova Science Publishers, New York.
2. Fridman, A. (2008) *Plasma Chemistry*, Cambridge University Press, Cambridge.
3. Kim, H.-H. (2004) *Plasma Processes Polym.*, **1**, 91–110.
4. Penetrante, B.M. (1993) *Nonthermal Plasma Techniques for Pollution Control*, Springer-Verlag, New York.
5. Odic, E., Paradisi, C., Rea, M., Parissi, L., Goldman, A., and Goldman, M. (1999) Treatment of organic pollutants by corona discharge plasma, in *The Modern Problems of Electrostatics with Applications in Environment Protection*, Vol. 63 (eds I.I. Inculet, F.T. Tanasescu, and R. Cramariuc), Springer, Bucharest pp 143–160.
6. Urashima, K. and Chang, J.-S. (2000) *IEEE Trans. Dielectr. Electr. Insul.*, **7**, 602–614.
7. Van Durme, J., Dewulf, J., Leys, C., and Van Langenhove, H. (2008) *Appl. Catal. B: Environ.*, **78**, 324–333.
8. Harling, A.M., Demidyuk, V., Fischer, S.J., and Whitehead, J.C. (2008) *Appl. Catal. B: Environ.*, **82**, 180–189.
9. Harling, A.M., Glover, D.J., Whitehead, J.C., and Zhang, K. (2008) *Environ. Sci. Technol.*, **42**, 4546–4550.
10. Atkinson, R. (2000) *Atmos. Environ.*, **34**, 2063–2101.
11. Rudolph, R., Francke, K.-P., and Miessner, H. (2002) *Plasma Chem. Plasma Process.*, **22**, 401–412.
12. Blin-Simiand, N., Jorand, F., Magne, L., Pasquiers, S., Postel, C., and Vacher, J.-R. (2008) *Plasma Chem. Plasma Process.*, **28**, 429–466.
13. Marotta, E., Callea, A., Ren, X., Rea, M., and Paradisi, C. (2008) *Plasma Process. Polym.*, **5**, 146–154.
14. Marotta, E., Callea, A., Rea, M., and Paradisi, C. (2007) *Environ. Sci. Technol.*, **41**, 5862–5868.
15. Marotta, E., Callea, A., Ren, X., Rea, M., and Paradisi, C. (2007) *Int. J. Plasma Environ. Sci. Technol.*, **1**, 39–45.
16. Marotta, E., Scorrano, G., and Paradisi, C. (2005) *Plasma Processes Polym.*, **2**, 209–217.
17. Donò, A., Paradisi, C., and Scorrano, G. (1997) *Rapid Commun. Mass Spectrom.*, **11**, 1687–1694.
18. Zaniol, B., Schiorlin, M., Gazza, E., Marotta, E., Ren, X., Puiatti, M.E., Rea, M., Sonato, P., and Paradisi, C. (2008) *Int. J. Plasma Environ. Sci. Technol.*, **2**, 65–71.
19. Schiorlin, M., Marotta, E., Rea, M., and Paradisi, C. (2009) *Environ. Sci. Technol.*, **43**, 9386–9392.
20. Campbell, M.J., Farmer, J.C., Fitzner, C.A., Henry, M.N., Sheppard, J.C., Hardy, R.J., Hopper, J.F., and Muralidhar, V. (1986) *J. Atmos. Chem.*, **4**, 413–427.
21. Su, Z.-Z., Ito, K., Takashima, K., Katsura, S., Onda, K., and Mizuno, A. (2002) *J. Phys. D: Appl. Phys.*, **35**, 3192–3198.
22. Rudolph, R., Francke, K.-P., and Miessner, H. (2003) *Plasmas Polym.*, **8**, 153–161.
23. Lias, S.G. (2006) Ionization energy evaluation, in *NIST Standard Reference Database*, Vol. 69 (eds W.G. Maillard

and P.J. Linstrom), National Institute of Standard and Technology, Gaithersburg.

24. Herron, J.T. (1988) *J. Phys. Chem. Ref. Data*, **17**, 967–1026.
25. Atkinson, R. (1998) *Pure Appl. Chem.*, **70**, 1327–1334.
26. Arnold, S.T., Viggiano, A.A., and Morris, R.A. (1997) *J. Phys. Chem. A*, **101**, 9351–9358.
27. Arnold, S.T., Viggiano, A.A., and Morris, R.A. (1998) *J. Phys. Chem. A*, **102**, 8881–8887.
28. Marotta, E. and Paradisi, C. (2009) *J. Am. Soc. Mass Spectrom.*, **20**, 697–707.
29. Marotta, E., Bosa, E., Scorrano, G., and Paradisi, C. (2005) *Rapid Commun. Mass Spectrom.*, **19**, 391–396.
30. Nicoletti, A., Paradisi, C., and Scorrano, G. (2001) *Rapid Commun. Mass Spectrom.*, **15**, 1904–1911.

8
Generation and Application of Wide Area Plasma

Noureddine Zouzou, Kazunori Takashima, Akira Mizuno, and Gerard Touchard

8.1
Introduction

The research presented in this paper deals with electrical surface discharges, in other words with surface plasmas. We have performed, analyzed, and tried to optimize surface plasma for two different applications. The first one is in aeronautics and the second one is in the control of pollution. The results presented here with regard to these two different applications concern the creation of wide area surface plasma, as the efficiency of the plasma devices for these two applications seemed to be linked to the wideness of the plasma zone.

In the recent past, several researches have been carried out to use electrical discharges to generate and modify aerodynamic flows.

Discharges are used to precipitate smoke particles and can also be used to destroy the gathered particles.

The development of these two different applications was at the origin of the research of optimized performances of wide area plasma.

8.2
AirFlow Control

An airfoil, which can correspond for instance to the wings of an airplane, is generally submitted, during flight, to two different forces: one in the opposite direction of the motion called the *drag* and one from bottom to top (in the opposite direction of the weight) called the *lift* [1].

Thus, if we consider an airplane in motion at a constant velocity and altitude, the drag is equal and opposite to the force produced by the plane engines and the lift is equal and opposite to the weight of the plane. The drag is due to the friction of the air on the airplane body, the wings, and the tail. The lift is essentially due to the friction on the wings.

Industrial Plasma Technology. Edited by Yoshinobu Kawai, Hideo Ikegami, Noriyoshi Sato, Akihisa Matsuda, Kiichiro Uchino, Masayuki Kuzuya, and Akira Mizuno

ISBN: 978-3-527-32544-3

It is, thus, important to try to reduce the drag during takeoff and cruise but not for landing, and to control the lift, as a higher lift is needed during takeoff than for cruise or landing. Anyway, the shapes of different parts of a plane are designed in order to have an aerodynamic profile, which means a drag as small as possible and an appropriate lift for takeoff, cruise, and landing. An airplane is made of different parts submitted to airflows. Each of them constitutes an airfoil.

As the wings are mandatory parts of an airplane, it is important to examine the profile of these airfoils. The wings of an airplane are airfoils which must generate a drag as small as possible and a lift equal to the weight of the plane for a given inclination and velocity. After a lot of studies, some profiles of airfoils have been designed to fulfill this purpose. The most known are those proposed by the National Advisory Committee for Aeronautics (NACA).

We are now going to examine the effect of the airflow around an airfoil, such as airplane wings, to better understand the need for aerodynamic actuators.

A vehicle in motion is submitted to the friction of the air around its body. This is due to the attachment of the air molecules at the air/body interface, which induces a null relative velocity of the air at this interface, then, the viscosity of the air produces the so called "boundary layer" all around the body in motion. In this layer, the relative velocity of air around the body evolves from zero at the body wall, to a constant velocity far from the body. This relative external velocity is equal and opposite to the body velocity. We are first going to examine the boundary layer development on a flat plate.

We consider a flat plate and a uniform flow velocity U_∞ before the plate. Then, from the beginning of the plate (generally called the *leading edge*) a laminar boundary layer develops. The width of this layer increases with the distance to the leading edge. At the limit of the boundary layer the velocity is U_∞ (in practice, the limit is defined when the velocity is U_∞). The pressure is more important upstream that downstream, we say that the pressure gradient is negative. The velocity profile depends on the flow regime. Indeed two flow regimes exist: the laminar flow and the turbulent flow with a transition between laminar to turbulent.

At a given distance from the leading edge the transition from laminar to turbulent appears. This transition to turbulence is clearly discernible by a sudden and large increase of the boundary layer thickness.

Another important phenomenon that appears around an airfoil in motion is the so called "separation."

In the case of airfoils like plane wings, due to the curvature of the upper face of the wing, adverse pressure gradient appears. When the pressure gradient becomes positive, then separation takes place. In other words, the velocity gradient perpendicular to the plate decreases and when it becomes null the boundary layer detaches from the plate. This is the point of separation. Then instabilities appear and a counterflow exists in this region. The existence of such a zone on an airfoil induces a reduction in lift, which is important, and an increase in the drag. Generally the separation point is located close behind the leading edge of the wing.

In extreme conditions these consequences may be responsible for airplane stalling and must obviously be avoided, but, even in "normal" conditions, the separation is a worrying phenomenon, especially during takeoff, which markedly reduces the efficiency of the airplane. Thus, delaying separation on a wing plane has been one of the most important challenges in aeronautics for many years.

One way to delay the separation is to generate the transition from laminar to turbulent. Indeed, at the critical value of the Reynolds number (corresponding to the transition from laminar to turbulent), the drag coefficient suddenly decreases before increasing regularly again for larger Reynolds numbers. This "surprising" decrease of the drag at the transition is due to the shift downstream of the separation zone. Indeed, in the case of a turbulent boundary layer, the accelerating influence of the external flow extends further in the boundary layer due to turbulent mixing. Thus, the turbulent boundary layers are more negative-pressure resistant than the laminar ones. Momentum transfer by convection is much more efficient than by diffusion. Thus, momentum provided in areas close to the wall where the velocity is very small is much more important in turbulent boundary layers than in laminar ones. In consequence, the boundary layer detachment is delayed.

In order to realize this transition several devices have been used and tested. Among these different devices, plasma actuators have been proposed for the past 10 years [1, 2]. Their principle is rather simple: generation of plasma along the airfoil body in order to modify the boundary layer. Recently, researches focus on a special point, which is the generation or modification with plasma discharges of small turbulent structures before the transition zone in order to get an amplification of these structures by the flow and finally to induce the transition. The advantages of plasma actuators are their robustness and high frequency response, both due to the fact that plasma actuators do not have any moving parts.

8.3
Aerodynamic Plasma Actuators

During the last decade three kinds of plasma actuators have been used and tested. The first one is DC actuators. They use DC surface corona discharge, which takes place under nonuniform DC electric field. Application of a DC high voltage between wire and plane electrodes or two parallel wire electrodes with different diameters flush mounted on the surface of insulating plate generates airflow from thinner toward thicker electrodes. The maximum induced velocity is a few meters per second. The problem of surface DC corona discharge is that it can be significantly affected by the electrical properties of the ambient gas and dielectrics into which electrodes are inserted; thus it is difficult to generate stable discharge under some conditions such as high humidity, surface degradation, and dust deposition.

The second kind of actuator is dielectric barrier discharge (DBD) actuator. It is made of a dielectric barrier sandwiched between two electrodes. A high AC voltage is applied between the two electrodes [3]. Since sufficiently high voltage can be applied without arcing owing to the dielectric barrier, stable discharge is generated on the dielectric surface. DBD is stable and less sensitive to humidity compared with surface corona discharge. Plasma area was smaller than that of the surface DC corona discharge, and experiments conducted on different actuators seemed to show that the wider was the plasma zone the larger was the efficiency of the actuator. Thus, following this observation we have tried to generate wide area plasma. One way to get wide area plasma is to use a third kind of plasma actuator: the sliding discharge actuator. This actuator is in fact a combination of a DBD actuator and a DC one. Three main arrangements of such actuators are shown in Figure 8.1. The first arrangement used is shown in Figure 8.1a. In this arrangement the upstream electrode is connected to an AC + DC power supply, while the downstream electrode and the lower electrode are linked and grounded. In a similar arrangement (not shown in the figure) the upstream electrode is only connected to the AC power supply while the downstream electrode with the lower electrode is connected to a DC power supply; in this case, obviously, the results are the same as in the previous one. These arrangements generally give a wider plasma sheet than a simple DBD arrangement; however, the sheet is generally limited to a width of 4 cm [4]. In the second arrangement (shown in Figure 8.1b) the upstream electrode is grounded, while the downstream electrode is connected to a DC power supply and the lower one is connected to the AC power supply. With such an arrangement, increasing the DC voltage on the downstream electrode allows the formation of a plasma zone as large as the lower electrode. This is the reason why we decide to enlarge the lower electrode to the total distance between the upstream and downstream electrodes; this is the third arrangement (shown in Figure 8.1c). Such an arrangement can give rather wide plasma zone; however, in the case of plate configuration, as sown in Figure 8.1, the AC and DC potentials are limited by the appearance of sparks between the electrodes.

On the other hand we are very disappointed to notice that in spite of experiments on DBD, the electric wind generated is not proportional to the wideness of the plasma zone but seems to remain of comparable strength to that produced by a simple DBD. Thus, further investigations are necessary to know better the relation between the plasma and the ionic wind generated, in order to improve this device.

Nevertheless, the experience acquired in this study could be used in another domain: the destruction of precipitated particles at the exit of the diesel engine exhaust.

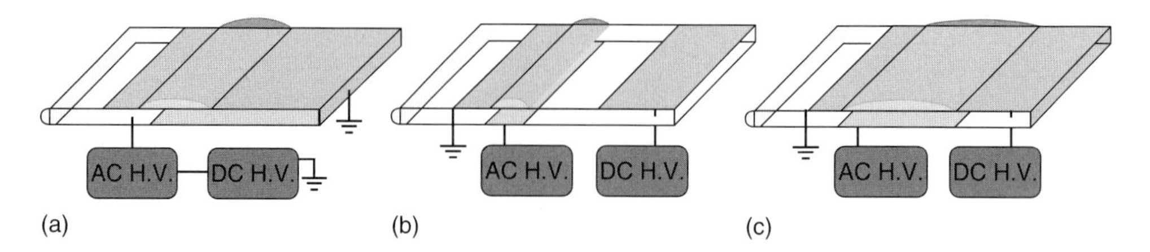

Figure 8.1 Different sliding discharge arrangements.

8.4
Particle Destruction

During the past years, several techniques have been developed for the collection of diesel soot and their treatment. The recent one is the ceramic diesel particulate filter (DPF) made of SiC or cordierite [5–7]. Diesel particles can also be collected by using dielectric barrier discharge-electrostatic precipitator (DBD-ESP) [8–11]. Main results show that particles are collected effectively on the dielectric surface. The collection efficiency is higher at higher applied voltages and lower flow rates. The effect of frequency on the collection efficiency depends on the frequency range, the electric power consumption, and the reactor configuration [11].

As with any filter, the DBD-ESP has to be emptied regularly to maintain performance. Indeed, after particle collection on the dielectric surface, the particles must be destroyed or swept from the surface to be collected elsewhere.

This section shows the ability of a surface dielectric barrier discharge (SDBD) to partially destroy the particles deposited on a dielectric surface. The used reactor is supplied with a sine high voltage. Particles generated from a diesel-oil fire are used to produce a thin sheet of deposit where the SDBD will occur. Then, the efficiency of the treatment is found out by analyzing the photographs of the dielectric surface and particle size distribution after treatment.

Figure 8.2 shows a schematic illustration of the experimental setup used for surface regeneration. The plasma reactor consists of three electrodes flush mounted on both sides of a Pyrex plate (2 mm-thick). The electrodes consist of thin aluminum strips 60 mm long and 20 mm wide. Two electrodes are grounded and the third one is connected to a high voltage power amplifier (Trek, 20/20C). Time-averaged photographs of the discharge are taken by a standard digital camera (exposure time = 10 seconds). All the experiments are carried out in air under atmospheric pressure and at room temperature (~300 K).

Figure 8.3a shows a photograph of the dielectric surface before treatment. The particles are deposited near the grounded electrodes. During the treatment (Figure 8.3b), a luminous plasma is observed on the dielectric surface, which is due

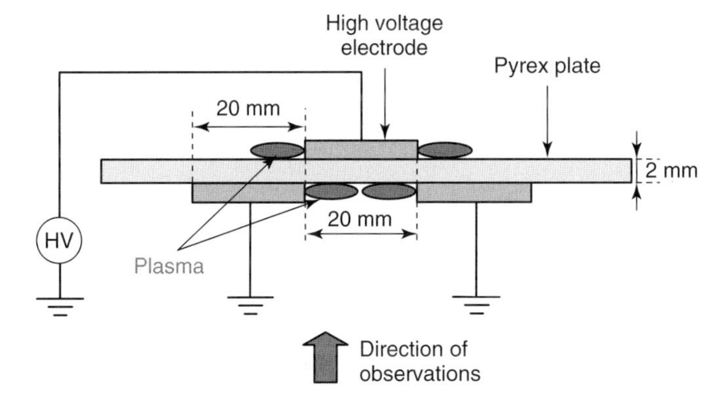

Figure 8.2 Schematic illustration of the experimental setup used for surface regeneration.

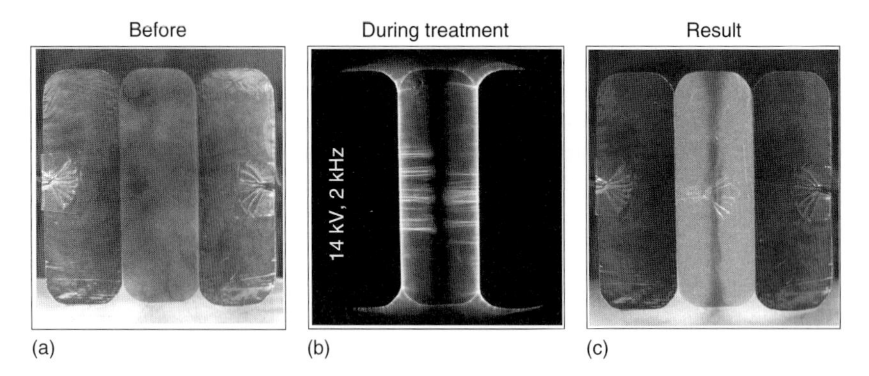

Figure 8.3 Photographs of the bottom view of the reactor: (a) before treatment, (b) during treatment, and (c) after treatment. Conditions: applied voltage = 14 kV_{peak}, frequency = 2 kHz.

to the establishment of the SDBD. The plasma is visually nonhomogeneous along the electrodes and microdischarges are intense locally. As shown in Figure 8.3c, the SDBD restores the dielectric surface in a better manner.

An aerosol spectrometer (Wellas-1000, sensor range of 0.18–40 μm, concentration up to 10^5 particles per cubic centimeter) has been used in order to measure the number and the mean size of the particles produced by the interaction between the plasma and the deposit. The results obtained show that the detected particles have the same initial mean particle size. The number of these particles increases at the beginning of the treatment, reaches a maximum, and decreases slowly with the treatment duration [12].

Particle removal mechanism is not fully understood but possible mechanism of the interaction between the surface discharge and the particle is as follows:

1) mechanical mechanism: the ionic wind generated by the discharge removes some of them;
2) electromechanical mechanism: the particles are charged electrically and transported away (out of the discharge area);
3) chemical mechanism: the discharge can generate atoms, radicals, and excited species with energetic electrons at moderate gas temperatures, which induces the oxidation of diesel soot particles.

In parallel to these experiments, we used a reactor capable of generating wide area plasma.

8.5
Large Sliding Discharge

Figure 8.4 shows a schematic illustration of the experimental setup. In this section, cylindrical reactors are employed in place of planar ones used previously for

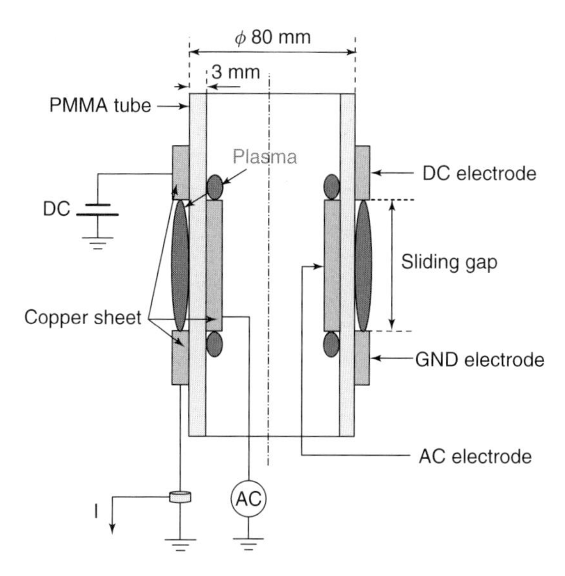

Figure 8.4 Schematic illustration of the experimental setup used for large sliding discharge.

aeronautic applications [2, 4, 13]. Indeed, as we have already seen, the planar configuration may sometimes generate an edge effect. Thus, in order to eliminate this effect and to create uniform plasma along the dielectric surface between the electrodes, a cylindrical configuration was used. This is important for the analysis of fundamental discharge phenomenon, because it is difficult to separate the edge effect from the experimental data. It is also easier to observe radial profile of the plasma in this configuration [14, 15].

We can see a 3D sketch of this reactor in Figure 8.5. The inner electrode has an adjustable size that helps it to vary the distance between the two outer electrodes. This property also helps it to fit exactly in the space between these two outer electrodes.

Plasma reactor used in this section consists of a poly(methyl methacrylate) (PMMA) tube and three copper electrodes. Outer diameter of the tube is 80 mm and the thickness is 3 mm. Two thin copper sheets are wrapped on the outer surface of the PMMA tube. One is connected to a DC high voltage power supply and the other is grounded. The inner electrode is connected to AC high voltage power amplifier (Trek 20/20C, ±20 kV, ±20 mA). Applied AC voltage is monitored with an oscilloscope (Lecroy 424, 200 MHz, 2GS s^{-1}) using internal probe of the high voltage power amplifier. The current is measured with a current transformer (Bergoz ACCT) having a bandwidth and accuracy of 0.16–160 kHz and some microamperes, respectively. Time-averaged photographs of the discharge were taken by a standard digital camera (Canon EOS 300D). All the experiments were carried out in air under atmospheric pressure and at room temperature. The gas flow inside and outside of the tube is not controlled.

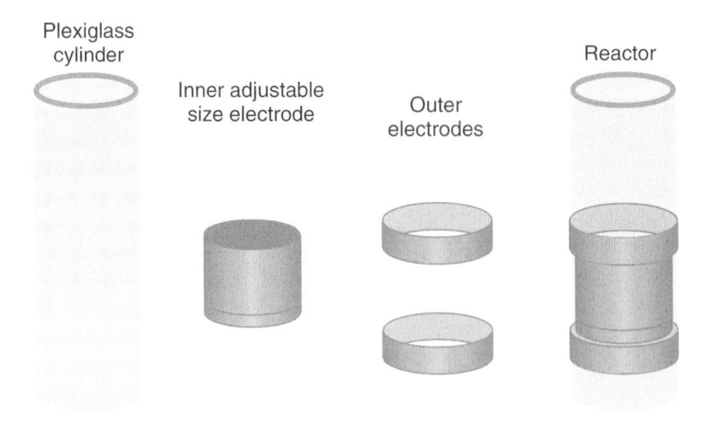

Figure 8.5 Sketch of the cylindrical reactor.

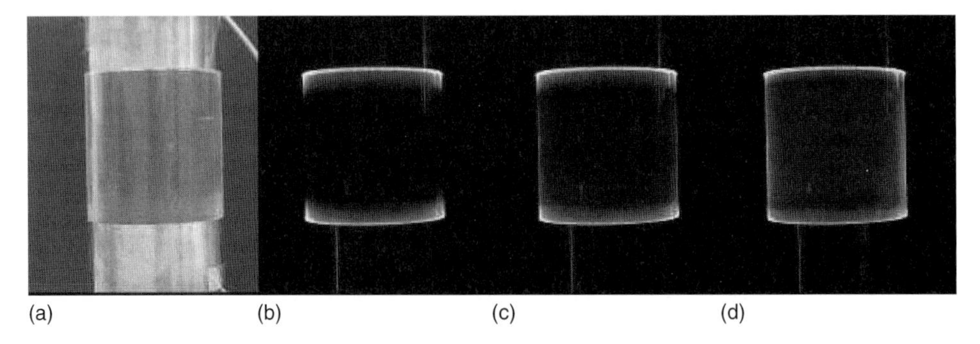

Figure 8.6 Time-averaged photographs of DBD and sliding discharge. Outer electrodes: DC high voltage electrode (top) and grounded one (bottom). Inner electrode: AC high voltage electrode ($V_{AC} = 18\,kV_{peak}$, $f = 1\,kHz$). (a) No plasma, (b) $V_{DC} = 0\,kV$, (c) $V_{DC} = -30\,kV$, and (d) $V_{DC} = -41\,kV$.

Figure 8.6 shows time-averaged photographs at different applied DC voltages in the case of the two outer electrodes separated by 80 mm. Without DC component (Figure 8.6b), luminous plasma is observed in the vicinity of the DC and GND electrodes show a typical SDBD.

Intense SDBD is also observed near the AC electrode inside the dielectric tube. When the applied DC voltage reaches $V_{DC} = -30\,kV$ (Figure 8.6c), a plasma sheet is observed between the DC and GND electrodes. The plasma is very stable and no arcing is observed although plasma sheet bridges the electrodes. These observations are related to the establishment of the sliding discharge. With higher applied DC voltages (Figure 8.6d), the plasma sheet becomes more intense and homogeneous, both along the electrode and along the axis of the tube.

Surface discharges are constituted of microdischarges distributed in time and space around the dielectric tube. Figure 8.7a shows typical measured current, discharge current, and applied AC voltage waveforms. The measured current is

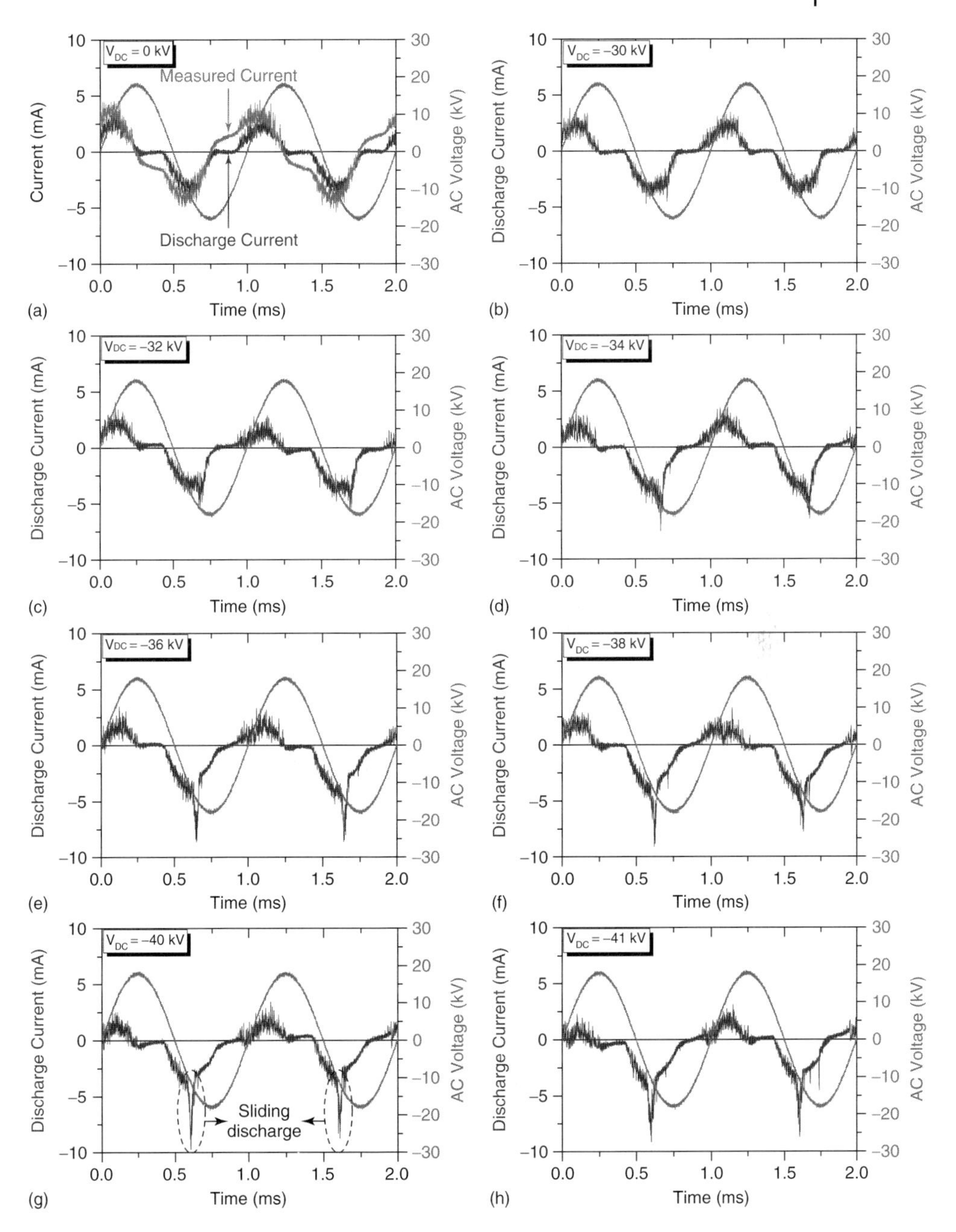

Figure 8.7 Time evolution of AC applied voltage and discharge current. Conditions: $V_{AC} = 18\,kV_{peak}$, $f = 1\,kHz$, (a) $V_{DC} = 0$, (b) $V_{DC} = -30\,kV$, (c) $V_{DC} = -32\,kV$, (d) $V_{DC} = -34\,kV$, (e) $V_{DC} = -36\,kV$, (f) $V_{DC} = -38\,kV$, (g) $V_{DC} = -40\,kV$, and (h) $V_{DC} = -41\,kV$.

composed of a capacitive current and the discharge current, which presents some current pulses.

The discharge current pulses, which correspond to one or several simultaneous microdischarges, can be recorded easily by using a shunt resistor. A current transformer, with a bandwidth of 0.16–160 kHz, reduces the number and the magnitude of these pulses, but gives a good illustration of the slow component associated with the discharge current. In the following discussion, only discharge current waveforms recorded by the transformer are presented.

Without DC high voltage (Figure 8.7a), only SDBD occurs and the discharge current is nearly symmetric between positive and negative half-cycles. Figure 8.7b–f show the discharge current when a DC high voltage is applied (between −30 and −41 kV). Additional component, associated with the sliding discharge, is observed during the negative half-cycles once the DBD current reaches the maximum. These results suggest that SDBD near the upper electrodes plays an important role in the generation of sliding discharge. During the negative half-cycle, the potential difference $V_{GND} - V_{AC}$ is higher than $V_{DC} - V_{AC}$. Thus, intense SDBD is generated near GND electrode. This SDBD plays the role of an ionizer, and then the space charge slides along the surface because of the electric field induced by the potential difference $V_{DC} - V_{GND}$.

These results agree well with the mechanism of establishing sliding discharge in which charge deposited by DBD slides toward the third electrode (grounded or linked to DC high voltage) [2].

Figure 8.8a shows the time-averaged photographs of the sliding discharge established on a bigger dielectric tube ($\phi_{ext} = 100$ mm). The plasma sheet remains intense and homogeneous, both along the electrode and along the axis of the tube. However, higher electric power is needed for larger diameters.

Figure 8.8 Time-averaged photographs of sliding discharge at different configurations. (a) $\phi_{ext} = 100$ mm, SD-gap = 80 mm, $V_{AC} = 18\,kV_{peak}$, $f = 1$ kHz, $V_{DC} = -41$ kV, and exposure time = 30 seconds; (b) $\phi_{ext} = 50$ mm, SD-gap = 130 mm, $V_{AC} = 30\,kV_{peak}$, $f = 200$ Hz, $V_{DC} = -43$ kV, and exposure time = 30 seconds.

Figure 8.8b shows the time-averaged photographs of sliding discharge created on large sliding gaps with an appropriate applied voltage value. In this case, sliding discharge covers a large gap between the two outer electrodes (about 130 mm) around a PMMA tube with an external diameter $\phi_{ext} = 50$ mm.

The minimum value of the high voltage ($V_{AC} - V_{DC}$) necessary for the establishment of sliding discharge depends on the sliding gap. In a recent publication, it has been shown that this value is between 5.5 and 6.5 kV cm^{-1} [14]. It was also shown that the voltage required for the onset of the sliding discharge increased with decreasing AC component and with increasing DC component. In addition, with a higher AC component, arcing was less likely to take place. These observations suggest that SDBD seeds and stabilizes the sliding discharge.

8.6
Conclusion

In this study, the goal was to perfect devices capable of providing wide area plasma zones to be used in two different applications: plasma actuators for airflow control, and a system for the destruction of particles that is combined with an electrostatic precipitator to make the diesel exhaust pollution free. In the case of airflow control, it is not yet clear that the sliding discharge system furnishing wide area plasma enhances the ability of the actuator to control airflow. Concerning the destruction of particles, the system is still under investigation. Nevertheless, this study gives an interesting behavior of the three-electrode system and shows that wide area plasma zones can be generated with this system. On the other hand, it should be noted that homogeneous surface discharge plasma under atmospheric pressure is interesting not only for electroaerodynamic actuators or destruction of particles but also for surface processes such as chemical vapor deposition (CVD), surface treatment, plasma-induced catalytic reactions, and so on.

References

1. Touchard, G. (2008) *Int. J. Plasma Environ. Sci. Technol.*, **2** (1), 1–25.
2. Moreau, E. (2007) *J. Phys. D: Appl. Phys.*, **40**, 605–636.
3. Pons, J., Moreau, E., and Touchard, G. (2005) *J. Phys. D: Appl. Phys.*, **38**, 3635–3642.
4. Moreau, E., Louste, C., and Touchard, G. (2008) *J. Electrost.*, **6-**, 107–114.
5. Durbin, T.D., Zhu, X., and Norbeck, J.M. (2003) *Atmos. Environ.*, **37** (15), 2105–2116.
6. Park, J.K., Park, J.H., Park, J.W., Kim, H.S., and Jeong, Y.I. (2007) *Sep. Purif. Technol.*, **55** (3), 321–326.
7. Christofides, P.D., Li, M., and Mädler, L. (2007) *Powder Technol.*, **175** (1), 1–7.
8. Kawada, Y., Kubo, T., Ehara, Y., Ito, T., Zukeran, A., Takahashi, T., Kawakami, H., and Takamatsu, T. (1999) Conference of the IEEE-IAS Annual Meeting, Vol. 2, Phoenix, pp. 1130–1135.
9. Byeon, J.H., Hwang, J., Park, J.H., Yoon, K.Y., Ko, B.J., Kang, S.H., and Ji, J.H. (2006) *J. Aerosol Sci.*, **37**, 1618–1628.
10. Dramane, B., Zouzou, N., Moreau, E., and Touchard, G. (2008) The 6th International Symposium on Non Thermal

Plasma Technology for Pollution Control and Sustainable Energy Development – ISNTPT'6, May 12–16, 2008, Taoyuan County.

11. Dramane, B., Zouzou, N., Moreau, E., and Touchard, G. (2009) *IEEE Trans. Dielec. Electr. Ins.*, **16**, 343–351.
12. Zouzou, N., Dramane, B., Moreau, E., Touchard, G., and Petit, J.M. (2008) SIA Conference: Diesel Engines After Treatment, November 27, 2008, Paris.
13. Louste, C., Artana, G., Moreau, E., and Touchard, G. (2005) *J. Electrost.*, **63**, 615–620.
14. Takashima, K., Zouzou, N., Moreau, E., Mizuno, A., and Touchard, G. (2007) *Int. J. Plasma Environ. Sci. Technol.*, **1**, 14–20.
15. Zouzou, N., Takashima, K., Moreau, E., Mizuno, A., and Touchard, G. (2007) The 28th International Conference on Phenomena in Ionized Gases – ICPIG'2007, July 15–20, 2007, Prague.

9
Nonthermal Plasma-based System for Exhaust Treatment under Reduced Atmosphere of Pyrolysis Gases

Marcela Morvová, Viktor Martišovitš, Imrich Morva, Ivan Košinár, Mário Janda, Daniela Kunecová, Nina Kolesárová, Veronika Biskupičová, and Marcela Morvová Jr

9.1
Introduction

The energy that we use (heat, electricity, fuel for motor vehicles) is usually obtained from fossil fuels such as coal, crude oil, or earth gas. Presumptive reserves, after deduction of 10% for requirement of chemistry, of crude oil will be for about 35 years, earth gas about 200 years, coal about 300 years. As long as our world depends on energy, we need sources that will last forever. Such sources, which are able to ensure sustainable development of society, are called renewable (sun, water, wind, geothermal energies, energy from biomass and waste, and energy carriers such as hydrogen).

Waste incinerators have been extensively used for more than 20 years with increasingly stringent emission standards in Japan, the EU, the US, and other countries. Mass burning is relatively expensive and, depending on the plant scale and flue-gas treatment, currently ranges from about 95 to 150 € per ton of waste. Waste-to-energy plants can also produce useful heat or electricity, which improves process economics.

Pyrolysis and gasification, considering the environmental aspect and low energy utilization efficiency of incineration (typically about 15%), appear to be a better option for thermal processing of waste. In pyrolysis and gasification processes, high temperatures are used to break down the waste containing mostly hydrocarbons with no (pyrolysis) or less oxygen than incineration (gasification). The pyrolysis process degrades the wastes to produce char, or ash, pyrolysis oil, and synthetic gas (i.e., syngas) at 500–1000 °C [1–3]. Despite many advantages over traditional incineration, pyrolysis and gasification of wastes still seem to have a long way to go in satisfying the energy profitability.

We have been studying this method for several years [4]. We have developed additional systems for the removal of all gas oxides including CO_2 [5]. The system itself does not produce metal oxides, furans, and dioxines and has an advantage that the final product, carbon char, is suitable for improving the soil properties [6].

Industrial Plasma Technology. Edited by Yoshinobu Kawai, Hideo Ikegami, Noriyoshi Sato, Akihisa Matsuda, Kiichiro Uchino, Masayuki Kuzuya, and Akira Mizuno

ISBN: 978-3-527-32544-3

An important aspect of this paper is that it shows an innovative access to the energetic and environmental problems (gas cleaning, waste processing) with a form of polygeneration of energy. Here we want to contribute to the emission-free process of liquid fuels (online condensation and distillation of tar) and hydrogen production with the possibility of online hydrogen storage. Up to now, most of the hydrogen produced from wastes isusing biologic processes [7, 8].

9.2 Experimental

Experimental pilot plant comprises pyrolysis oven (thermochemical degradation system), distillation and heat recuperation system, flying-particulate removal system, online analysis system, nonthermal plasma exhaust cleaning system, hydrogen absorber, odor removal system (biochemical cleaner) as well as necessary hydrodynamic components. The system works in the dynamic regime, where the total gas flow from pyrolysis oven is varied between 80 and 650 $m^3\ h^{-1}$ depending on the phase of the pyrolysis process and the location in pyrolysis system (in the distillation system, due to condensation the gas flow decreases down to about 300–500 $m^3\ h^{-1}$). All the systems involved work continuously.

Figure 9.1 shows a pilot experimental setup for the study of thermochemical processes, including the discharge system and other cleaning subsystems as well as the chamber for online hydrogen absorption.

The spontaneously regularly pulsing direct current (DC) electric discharge in streamers to spark transition regime was used for exhaust cleaning. The discharge operates in corona geometry; synergetic effect of electrode surface catalysis is present in the discharge gap. The strongly shining streamer channels migrate quickly along the stressed electrode of each discharge tube. The multifunctional discharge system used for the tests on pilot scale operates on gas flow volumes of 50–250 $m^3\ h^{-1}$ and comprises discharge system plurality of 24 discharge tubes connected in parallel to each other. One discharge chamber consists of the copper rod with an internal thread (stressed electrode) and a coaxial cylinder (nonstressed electrode), with an interelectrode distance of 6 mm. The length of one discharge tube is 50 cm. High voltage source with DC high voltage of both polarities up to 20 kV, maximum power 600 W, and maximum current 30 mA was applied for discharge generation [9]. The system works very efficiently on all oxides inside the exhaust (CO_2, CO, NO_X, and others) and also removes water vapor, and several hydrocarbons and VOCs present in the exhaust. The product is a solid particularly separated in by the discharge system and the second flying-particulate removal system situated after discharge system and before online hydrogen absorption system. The measurements as well as the working of all components of the pilot system usually go on continuously for 10 h.

For an online analysis of processes and product we use Testo 454 for gas flow parameter analysis (temperature, gas flow velocity, total gas flow, and static and dynamic pressure; measured every second at several places) and Maihak S715 from

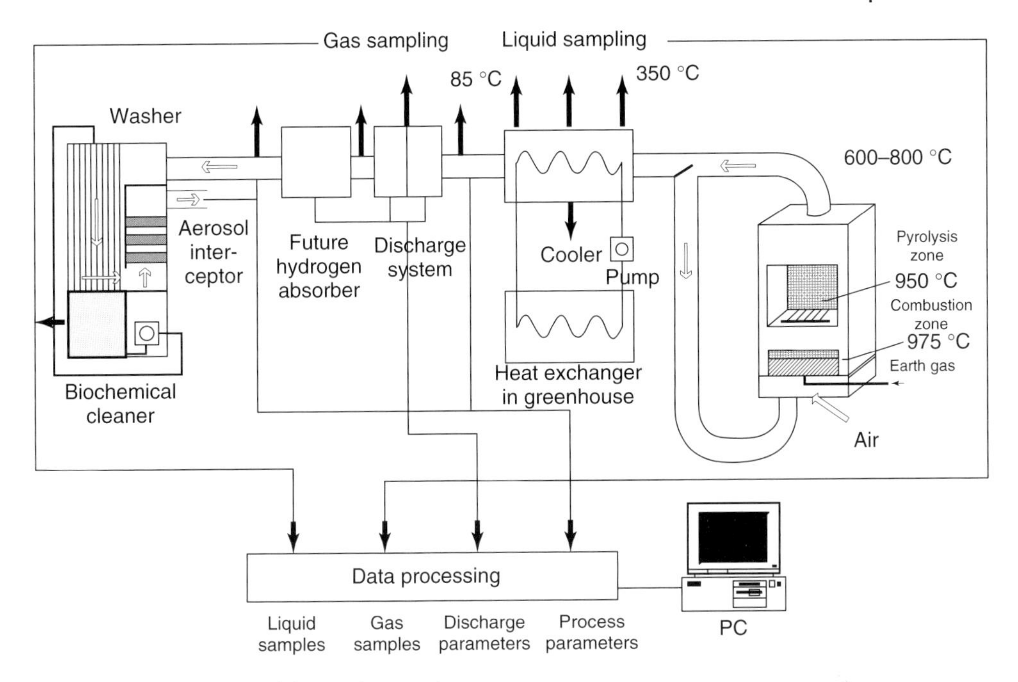

Figure 9.1 Apparatus used for pyrolysis studies.

Sick for online chemical analysis (CO_2, CO, NO_X, CH_4, O_2, and H_2; measured every 6 s at eight places). For ex-post analysis of products, we have used Fourier transform infrared (FTIR) absorption spectrometry, TGA-DTA, optical microscopy, and SEM. Inside the pilot system, nine extraction pipes for isokinetic sampling are installed (after oven, before and after distillation system, before and after low temperature distillation system for oil and paraffin separation, before and after discharge system, before and after online hydrogen absorber). We have analyzed relatively complicated products except of digital IR catalogs using [10].

9.3 Results

The thermochemical systems produce a typical exhaust containing high concentrations of various components. The gas flow from the oven varies from 80 to $650\,m^3\,h^{-1}$, depending on the phase of the pyrolysis process. At least two-thirds of it are condensable portions. That is why it is necessary to separate and condense this part of exhaust into liquid phase inside the recuperation/distillation unit. This results in energetic gas (CO, H_2, C_{1-4}), combustion exhaust (CO_2, CH_X, NO_X) and other, mostly unwanted, components.

As the outgoing gas flow from pyrolysis chamber changed very quickly, it was necessary to build an online absorption system for hydrogen storage.

After measurements, the absorbed hydrogen is removed together with the nanosized carbon absorber and inserted into the desorption system. The hydrogen in the desorption system is released and piped into the proton exchange membrane (PEM) fuel cell (Ballard Nexa^{RM}).

PEM fuel cell is an electrochemical system that uses hydrogen as the fuel and converts it into electricity, water, and heat. The system works at room temperature using proton exchange membranes and electrodes particularly containing platinum.

To use hydrogen effectively and to prevent destruction of platinum electrode inside the PEM fuel cell, it is necessary to separate the produced hydrogen from all oxides before absorption (discharge system) and later from residual hydrocarbons (selective desorption process, which is the subject of our further experiments).

It was possible to gain the basic information about the process from the ex-post FTIR spectra of solid condensation product produced from gas pyrolysis exhaust inside the discharge system and secondary tar condensed inside the distillation unit, which are given in Figure 9.2.

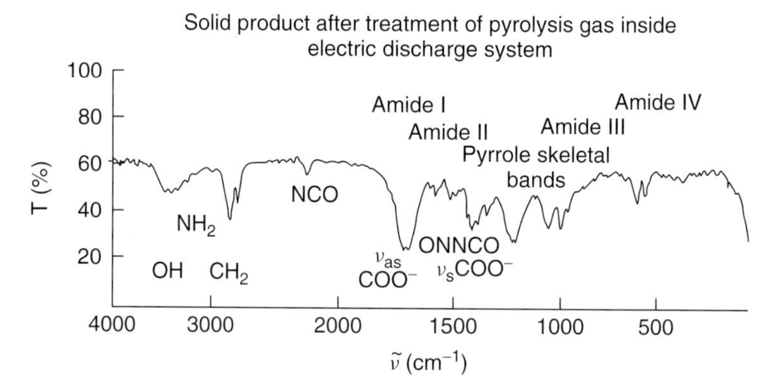

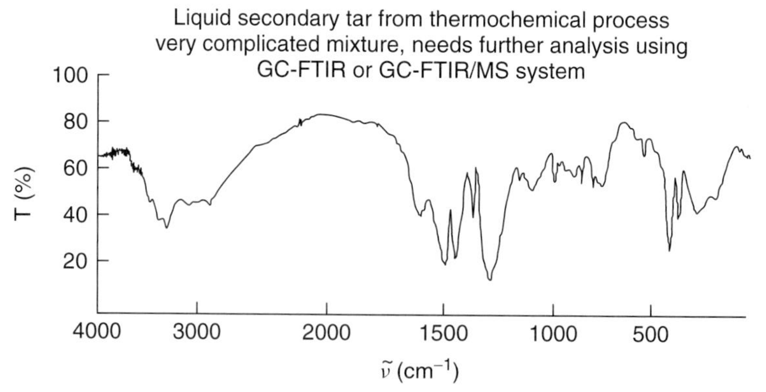

Figure 9.2 IR absorption spectra of solid condensation product produced from gas pyrolysis exhaust inside the discharge system and secondary tar condensed inside the distillation unit.

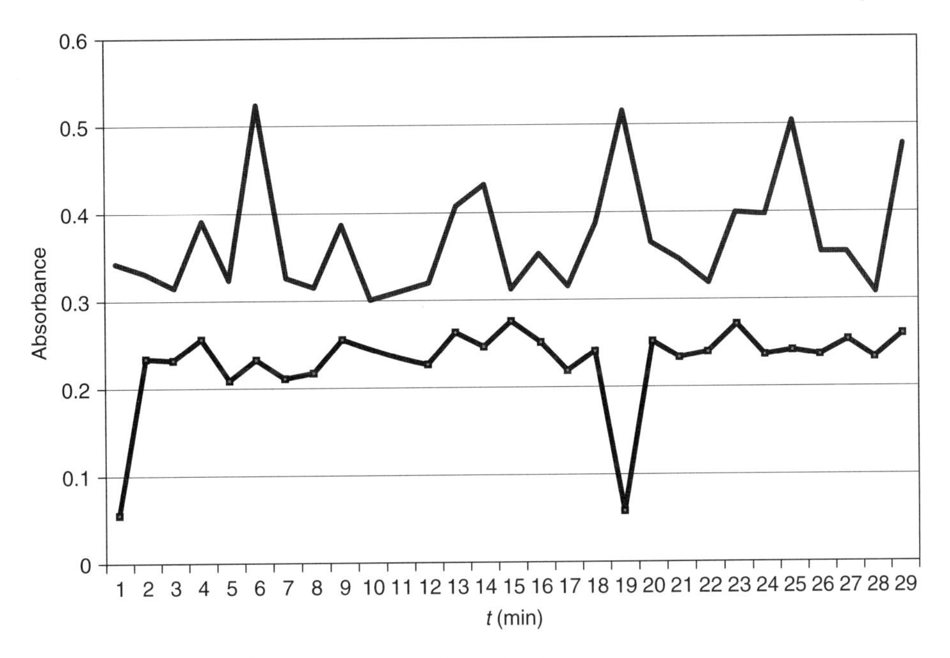

Figure 9.3 Changes in the concentration of CO_2 inside the pyrolysis chamber under the influence of electric discharge (time scale is measured according to the introduction of waste and last 30 min of pyrolysis), upper line – before discharge, lower line – after discharge.

To prepare the calibration and necessary data for online chemical analytic system Maihak S715, we have realized the measurement of whole FTIR spectra. The samples were introduced isokinetically from the pilot system into 10-cm long gas cells with KBr windows which were placed before and after the discharge system, while discharge was continually produced in 1-min intervals.

From these FTIR spectra, we have estimated the concentration of CO_2. The entire pyrolysis process lasted about 30 min. From the spectra before and after discharge every 1 min during a 30-min period, the curves of time development of CO_2 removal (using discharge action) were evaluated. The curves obtained in Figure 9.3 show CO_2 removal efficiency depending on time, from the input of raw material (waste, biomass) to the pyrolysis chamber.

As the solid product formed inside the discharge system was amino acid-based polymer of proteinoid nature, it was important to study such products microscopically. In the case of pyrolysis exhaust, additional components present in the exhaust (sulfur and others) lead to the formation of crossing nanosized (diameter about 3 nm) bridges inside the solid product, as can be seen from the SEM photograph in Figure 9.4.

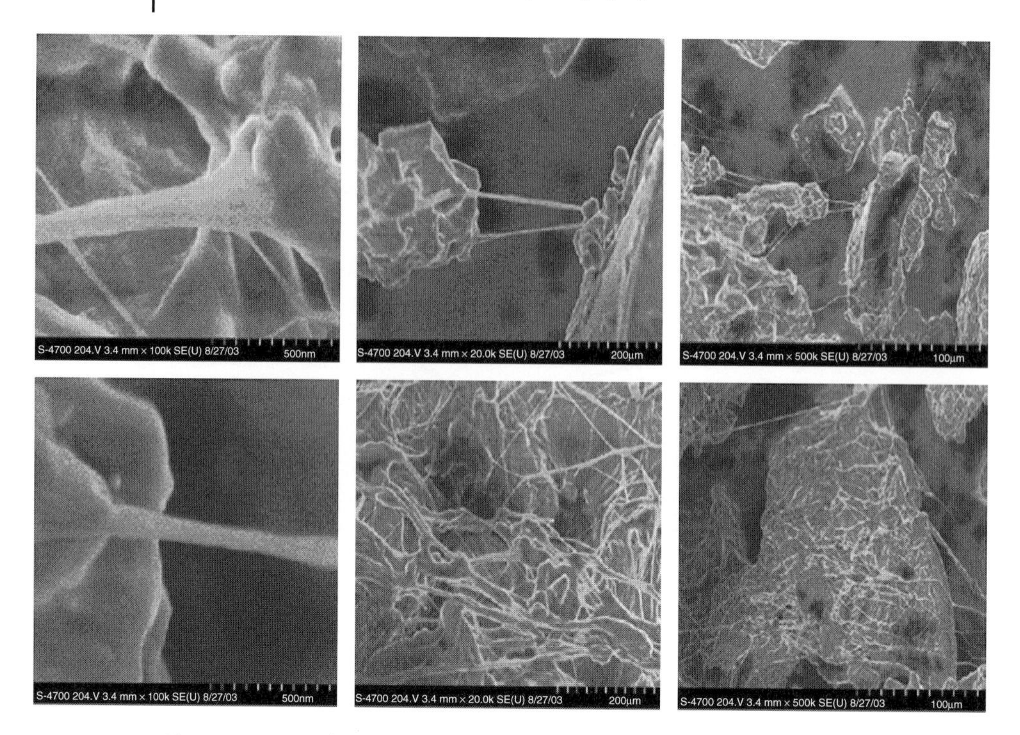

Figure 9.4 SEM photograph of proteinoid product obtained from pyrolysis and produced in the discharge system.

9.4 Conclusions

Inside the tested pyrolysis system, it is possible to recycle many types of wastes (separated municipal waste, agrowaste, gardening waste, excrements, residuum of water cleaning systems, biomass, and PET bottles), and produce energetic gas, liquid products containing valuable chemicals, and solid products similar to active carbon, which are suitable for many environmental purposes as well as for the improvement of soil properties. The amount of raw material that can be introduced into the pyrolysis chamber varies between 0.5 and 2 kg depending on the type of raw material used; it means that up to 4 kg of waste is generated per hour. The system gives relatively large amounts of energetic gas containing hydrogen (up to 60 $m^3\ h^{-1}$).

We have developed a method and equipment for the continuous solidification of CO_2 and exhaust cleaning using synergetic effect of electric discharge and catalysis induced by metal organic product formed on copper electrodes.

Online system for the absorption of hydrogen was installed and the preliminary tests were carried out during the pyrolysis process.

Acknowledgments

The authors wish to thank the Slovak Research and Development Agency APVV project No. 0267-06 and the Slovak Grant Agency VEGA grant No. 1/0193/09 for their support.

References

1. Digman, B. and Kim, D.-S. (2008) Review: alternative energy from food processing wastes. *Environ. Prog.*, **27** (4), 524–537.
2. Bridgwater, A.V. (1999) Principles and practice of biomass fast pyrolysis processes for liquids. *J. Anal. Appl. Pyrolysis*, **51**, 3–22.
3. Weerachanchai, P., Horio, M., and Tangsathitkulchai, C. (2009) Effects of gasifying conditions and bed materials on fluidized bed steam gasification of wood biomass. *Bioresour. Technol.*, **100**, 1419–1427.
4. Morvová, M., Morva, I., Janda, M., Hanic, F., and Lukáč, P. (2003) Combustion and carbonisation exhaust utilisation in electric discharge and its relation to prebiotic chemistry. *Int. J. Mass Spectrom.*, **223-224** (1-3), 613–625. ISSN 1387-3806.
5. Morvová, M., Hanic, F., and Morva, I. (2000) Plasma technologies for reducing CO2 emissions from combustion exhaust with toxic admixtures to utilisable products. *J. Therm. Anal. Calorim.*, **61** (1), 273–287. ISSN 1418-2874.
6. Svetková, K., Henselová, M., and Morvová, M. (2005) Effects of a carbonization product as additive on the germination, growth and yield parameters of agricultural crops. *Acta Agron. Hung.*, **53** (3), 241–250. ISSN 0238-0161.
7. Yang, P., Zhang, R., McGarvey, J.A., and Benemann, J.R. (2007) Biohydrogen production from cheese processing wastewater by anaerobic fermentation using mixed microbial communities. *Int. J. Hydrogen Energy*, **32**, 4761–4771.
8. Nandi, R. and Sengupta, S. (1998) Microbial production of hydrogen: an overview. *Crit. Rev. Microbiol.*, **24**, 61–84.
9. Morvová, M., Morva, I., and Machala, Z. (1998) Environmental applications of spontaneously pulsing discharge with corona geometry and developed ARC phase. 13th Symposium on Physics of Switching Arc, Vol. 2: Invited Papers – Brno: Department of Electrical Machines and Apparatus, FEECS TU, 1998. ISBN 80-214-1171-6, pp. 239–260.
10. Zhubanov, B.A., Agashkin, O.V., and Rychina, L.B. (1984) Atlas of IR Spectra of Heterocyclic Monomers and Polymers, Nauka of Kazachstan USR, Alma-Ata, p. 147.

10
Pharmaceutical and Biomedical Engineering by Plasma Techniques

Masayuki Kuzuya, Yasushi Sasai, Shin-ichi Kondo, and Yukinori Yamauchi

10.1
Introduction

Cold plasma is most characterized by a low gas temperature and a high electron temperature, and easily generated by electric discharges under reduced pressure. In addition, the methods for generating similar cold plasmas under atmospheric pressure, so-called dielectric barrier atmospheric pressure glow discharges (APGDs), have recently been developed [1] and the surface treatment with APGD plasma have actually attracted considerable interest in a wide variety of industrial applications [2]. One of the characteristics of surface treatment by cold plasma irradiation is the fact that it is surface limited (about 500–1000 Å) so only the surface properties can be changed without affecting the bulk properties.

In recent years, biomedical applications of cold plasma have rapidly grown due to the fact that the use of cold plasmas is very useful to treat heat-sensitive objects such as polymeric materials and biological samples. Considering biomedical applications of cold plasma, the targets of plasma treatment can be classified into two major categories: living and nonliving matter. The former applications include living tissue sterilization, tissue removal, blood coagulation, wound healing, and living cell treatment. In the applications, the interactions between plasma itself and the living matter are very important because the plasma is in direct contact with the living cells, tissues, and organs. Recently, demonstrations of plasma technology in treatment of living cells, tissues, and organs have created a new field at the intersection of plasma science and technology with biology and medicine, called *plasma medicine* [3], and the first International Conference on Plasma Medicine (ICPM) was also held in Texas, USA, from October 15 to 18, 2007 [4].

The latter applications include the fabrication and/or treatment of biomedical devices, the sterilization of biomedical devices, and so on. When cold plasma is irradiated onto polymeric materials, the plasma of inert gas emits intense UV and/or VUV ray to cause an effective energy transfer to solid surface and gives rise to a large number of stable free radicals on the polymer surface. In view of the fact that surface reactions of plasma treatment are initiated by such plasma-induced radicals, study of the resulting radicals is of utmost importance for

Industrial Plasma Technology. Edited by Yoshinobu Kawai, Hideo Ikegami, Noriyoshi Sato, Akihisa Matsuda, Kiichiro Uchino, Masayuki Kuzuya, and Akira Mizuno

ISBN: 978-3-527-32544-3

understanding the nature of plasma treatment. However, detailed studies of such plasma-induced surface radicals had not been carried out. We therefore conducted plasma irradiation of a wide variety of polymers, synthetic and natural, and the surface radicals formed were studied in detail by electron spin resonance (ESR) coupled with systematic computer simulations.

On the basis of the findings from a series of such studies, we were able to open up several novel applications of plasma surface treatment in the field of pharmaceutical and biomedical engineering. In the present chapter, the design and development of drug delivery system (DDS) and biomaterials on the basis of the nature of plasma-induced polymer radical formation and their reactivity are reviewed.

10.2
Nature of Plasma-Induced Polymer Radicals

Plasma-induced surface radicals permit reactions for surface modification in several different ways such as CASING (cross-linking by activated species of inert gas), surface graft and/or block copolymerization, and incorporation of functional groups [5–21]. All these techniques are referred to as *plasma surface treatment*.

Over the years, we have been working on the structural identification of plasma-induced surface radicals of various kinds of organic polymers as studied by ESR spectra coupled with the systematic computer simulations. One of the advantages of plasma irradiation over other types of radiations for the study of the polymer radicals is that the radical formation can be achieved with plasma irradiation for a brief duration using a simple experimental apparatus. The experimental setup for the plasma irradiation and ESR spectral measurement is schematically shown in Figure 10.1. This method makes it possible not only to study the polymer radicals without a significant change of polymer morphology but also to readily follow the ESR kinetics for the radical formation, so that systematic computer simulations can be carried out with a higher credibility.

Figure 10.2 shows the observed ESR spectra of plasma-induced surface radicals formed on several selected polymers relevant to the present study, together with the corresponding simulated spectra shown as dotted lines. On the basis of the systematic computer simulations, all the observed spectra in addition to those shown here were deconvoluted and the component radical structures have been identified.

On the basis of a series of such experiments, we were able to establish the general relationship between the structure of radicals formed and the polymer structural features. Cross-linkable polymers yield the midchain alkyl radical as a major component radical, while degradable polymers yield the end-chain alkyl radical as a major component radical, and if polymers are of branched structure or contain the aromatic ring, the cross-linking reactions occur preferentially on these moieties. One the common feature is that dangling-bond sites (DBSs) are more or less formed in all plasma-irradiated polymers as a result of CASING.

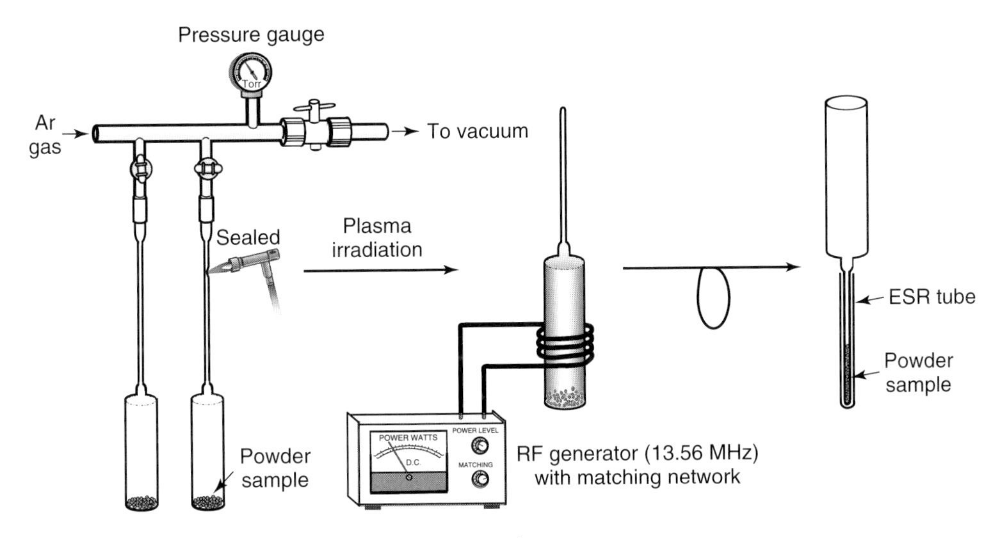

Figure 10.1 Schematic representation for plasma irradiation and ESR spectral measurement.

Since all plasma-irradiated polymers are eventually exposed to air for their practical use, we have to understand the nature of oxygenation, that is, the auto-oxidation.

Figure 10.3 shows a reaction scheme for the formation of peroxy radical and its ensuing process (hydroperoxide, alkoxyl radicals formation) demonstrating how auto-oxidation ends up with introduction of oxygen-containing functional groups such as hydroxyl groups, carboxyl groups, and so on, and dissipation of the surface radical formed. We have studied the nature of peroxy radical formation as an initial process of auto-oxidation.

Figure 10.4 shows several examples of ESR spectra of peroxy radicals formed immediately after exposure of the plasma-irradiated polymers to air, which correspond to those shown in Figure 10.2, as well as the simulated spectra shown as dotted lines. It can be seen that in some polymers the spectral pattern remained unchanged with only a lowering of the intensity, and in other polymers, the spectra have been completely converted to the ones exhibiting typical spectral patterns of peroxy radicals.

Note that, in most polymers, the intensity of peroxy radicals usually decreases to less than 30–40% of the original carbon-centered radicals immediately after exposure to air, except in polytetrafluoroethylene (PTFE), which can be best discussed on its comparison with high-density polyethylene (HDPE) to understand the nature of auto-oxidation in more detail.

As shown in Figure 10.5, exposure of plasma-irradiated HDPE to air at room temperature did not give the ESR spectra of peroxy radicals, but the ESR spectra did show the decrease in the spectral intensity. On the other hand, the peroxy radicals of PTFE are extremely stable for a long period of time at room temperature. The spectral intensity, therefore, is nearly the same as that of the original radicals [6].

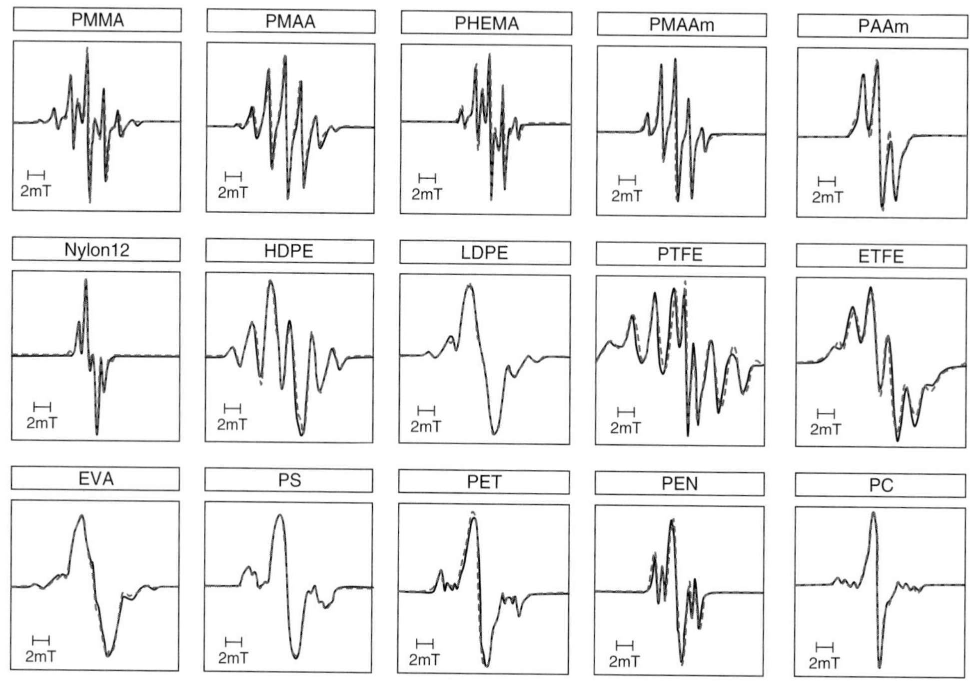

Figure 10.2 Room temperature ESR spectra of plasma-induced radicals in organic polymers, together with the simulated spectra shown as dotted lines. Plasma conditions: 40 W, Ar 0.5 torr, 3 minutes. PMMA, polymethylmethacrylate; PMAA, polymethacrylic acid; PHEMA, poly-(2-hydroxyethyl) methacrylate; PMAAm, polymethacrylamide; PAAm, polyacrylamide; HDPE, high-density polyethylenel LDPE, low-density polyethylene; PTFE, polytetrafluoroethylene; ETFE, (ethylene-tetrafluoroethylene) copolymer; EVA, (ethylene-vinylacetate) copolymer; PS, polystyrene; PET, polyethyleneterephtalate; PEN, polyethylenenaphthalate; PC, polycarbonate.

ROO· -COOOOR → 2 -CO·
$-O_2$
R· -COOR → 2 -CO·
-C· + O_2 ⇄ -C(O-O·)- → RH -COOH + R· —(-HO·)→ -CO·
→ -C(OH)- ; -C(=O)- ; -C(=O)-H
R-O-O· 33 kcal mol^{-1}
R-O-O-H 70 40 90 kcal mol^{-1}

Figure 10.3 Peroxy radical formation from carbon-centered radical with molecular oxygen and its reaction, resulted in introduction of oxygen functional groups on polymer surface.

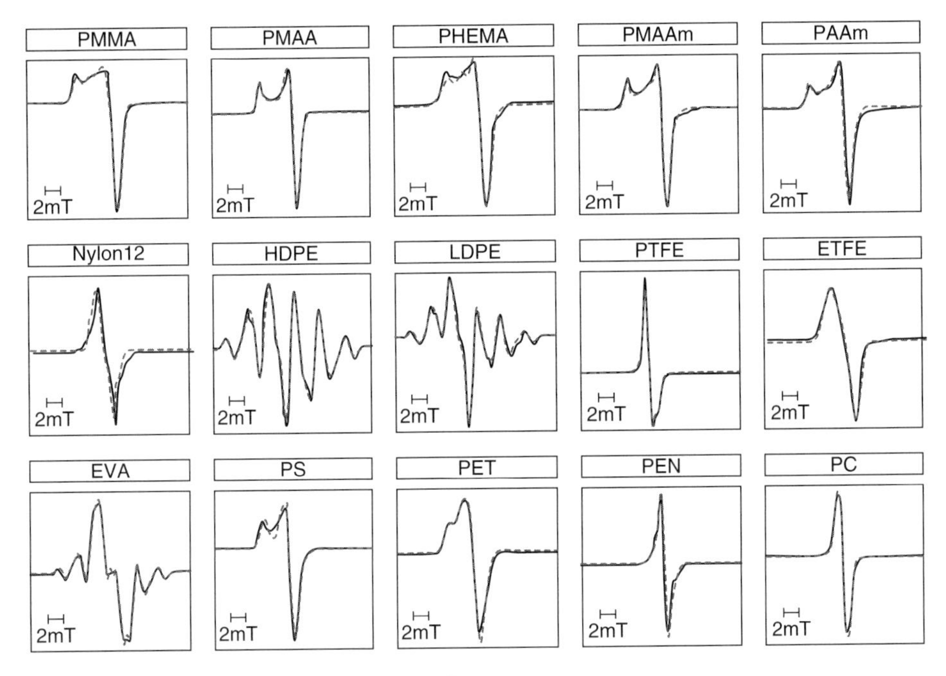

Figure 10.4 Room temperature ESR spectra of various plasma-irradiated polymers (corresponding to the spectrum in Figure 10.2, respectively) immediately after exposure to air together with the simulated spectra shown as dotted lines.

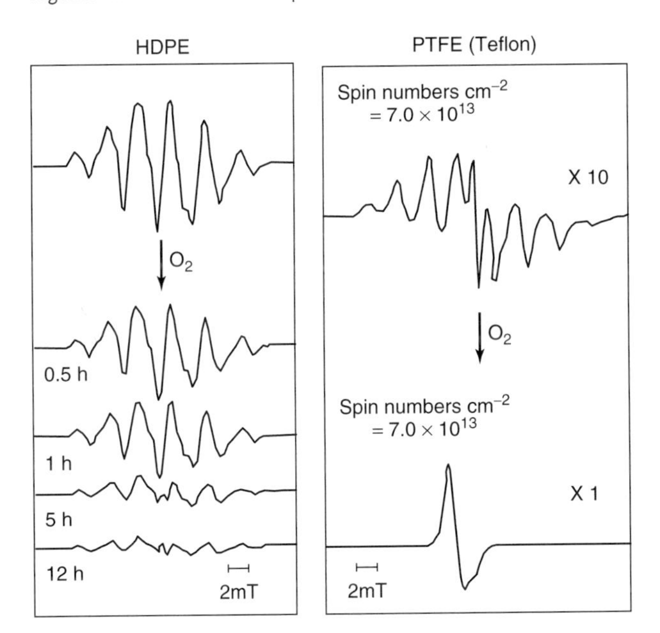

Figure 10.5 Difference in free radical reactivity with oxygen between HDPE and PTFE.

The extraordinary instability of HDPE peroxy radical can be ascribed to the rapid chain termination reaction through the hydroperoxide consuming several moles of molecular oxygen, due to the presence of abundant hydrogen atoms bonded to sp^3 carbons in HDPE, while the exceptional stability of PTFE peroxy radicals can be attributed to the absence of any abstractable hydrogen in PTFE to undergo the chain termination reactions. Because of occurrence of this type of oxygenation reaction, plasma treatment by inert gas plasmolysis has a tendency to result in the introduction of surface wettability in many polymers. It should be noted, however, that such a fact does not hamper our development of DDS preparation as demonstrated in drug release tests, which are described in the following sections.

10.3 Pharmaceutical Engineering for DDS Preparation by Plasma Techniques

For the most suitable therapy, development of controlled-release systems for drug delivery is one of the most active areas today in the field of drug research [22–41]. A wide variety of approaches of controlled-release DDS have been thus far investigated for oral application. Oral drug delivery is the most desirable and preferred method of administrating therapeutic agents for their systematic effects such as convenience in administration, cost-effective manufacturing, and high patient compliance compared with several other routes.

On the basis of findings from a series of studies on the nature of plasma-induced radical formation on a variety of organic polymers, we have developed plasma-assisted DDS preparation as schematically shown in Figure 10.6. Figure 10.7 illustrates the schematic representation for preparation of double-compressed (DC) tablets including the experimental setup for plasma irradiation on the tablets.

10.3.1 Preparation of Sustained-Release DDS from Plasma-Irradiated DC Tablets

The development of new active pharmaceutical ingredients (APIs) is often hampered or even blocked due to their side effects. [25–31]. Some of the severe side effects may be the early and high peak blood plasma concentration of APIs just after oral administration. This problem can be overcome by altering the blood plasma concentration profile so that a more gradual absorption rate is obtained. In that case, with a sustained-release DDS, so that the drug is slowly released over a prolonged period of time, is an ideal therapeutic system.

When oxygen plasma was irradiated on the outermost layer of the DC tablet, which consists of a drug as a core material and a mixture of plasma-cross-linkable and plasma-degradable polymer powders as wall materials, plasma-degradable polymers could be selectively eliminated and simultaneously the cross-linkable polymer undergoes a rapid cross-linking reaction resulting in the formation of a

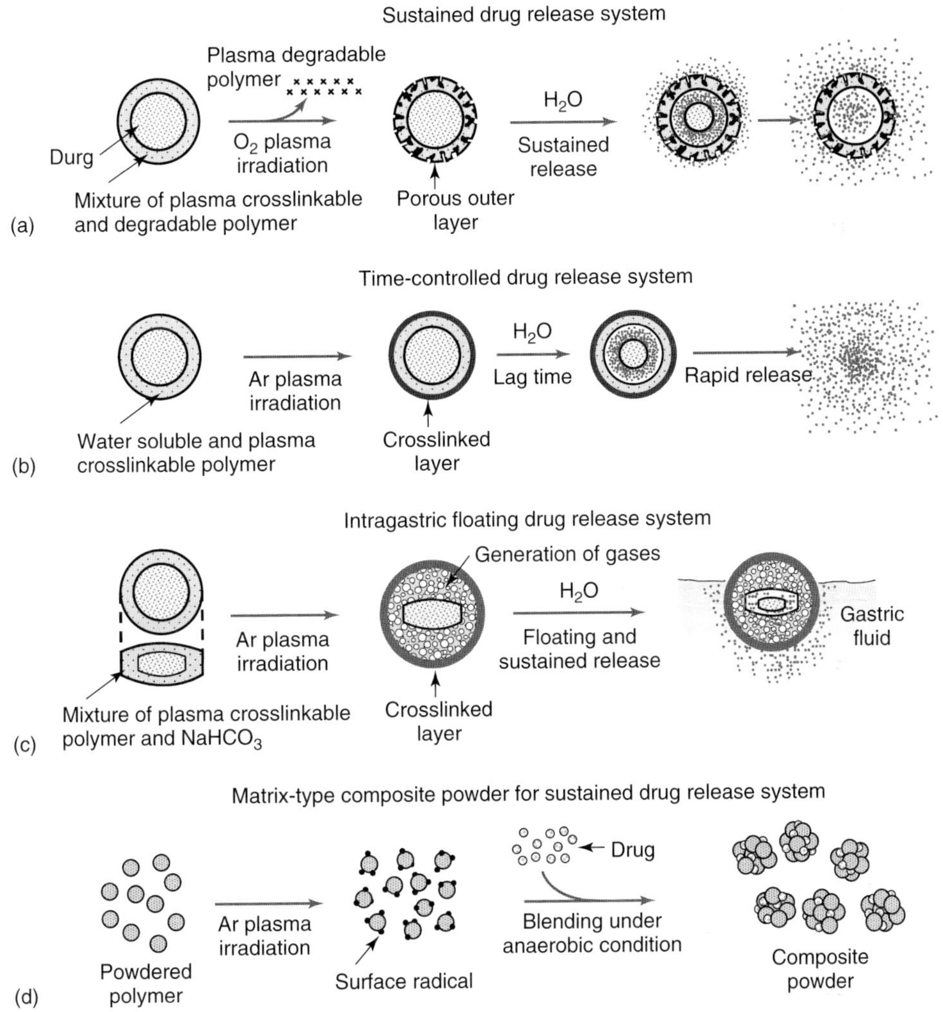

Figure 10.6 Conceptual illustration for preparation of DDS by plasma techniques.

porous outer layer on the tablet. As a result, the drugs could be released from the tablet through the resulting micropore [26–31].

Figure 10.8 shows the effect of oxygen plasma irradiation on theophylline release from the DC tablet as a representative example of the release test. As shown in Figure 10.8a, when a mixed powder of polystyrene (PS) and polyoxymethylene (POM) is used for the outer layer, it is seen that the release rate of theophylline increases as plasma irradiation duration increases, while the blank tablet did not exhibit any appreciable release of theophylline even with longer dissolution time [25]. Thus, the release profile of theophylline from DC tablet can readily be controlled by the selection of plasma operational tunings. On the basis of the

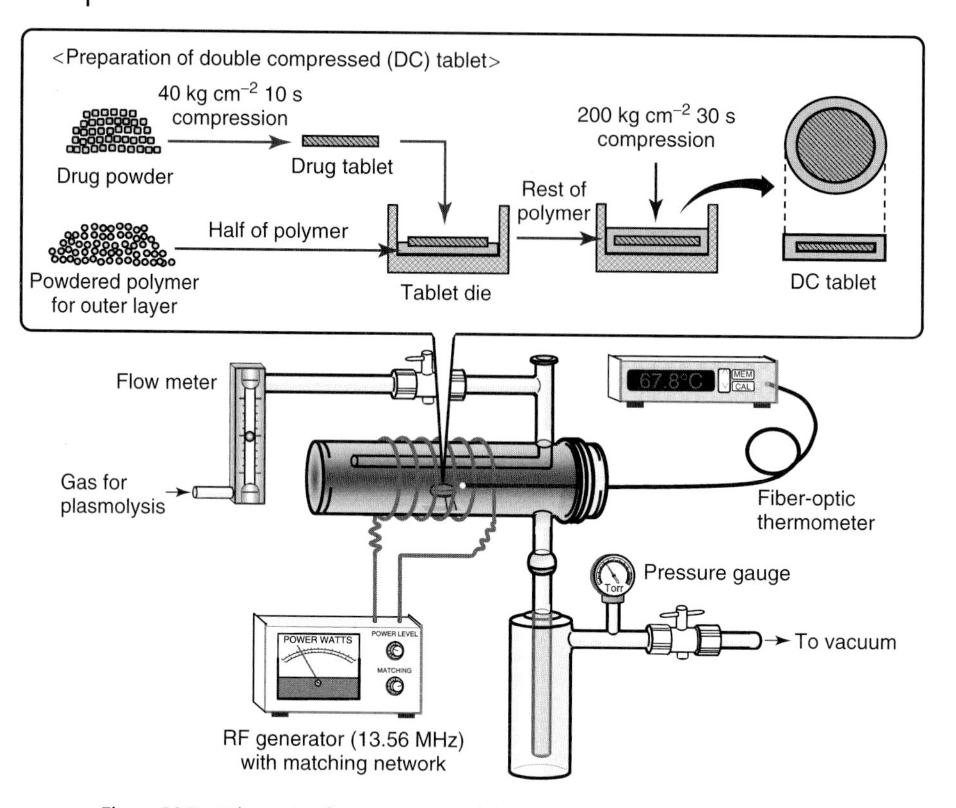

Figure 10.7 Schematics for preparation of double-compressed(DC) tablets including the experimental setup for plasma irradiation.

fact that the value of weight loss (shown in parentheses) increases as the plasma duration increases, it is apparent that the plasma-degradable polymer, POM, could be selectively eliminated by oxygen plasma irradiation, while plasma-cross-linkable PS undergoes the cross-linking reaction, to result in the formation of the porous outer layer on the tablet (Figure 10.9). Then, theophylline could be released from the tablet through the resulting micropore evidenced by the scanning electron micrographs (SEMs) pictures.

Similar work includes the preparation of the controlled-release tablet by using bioerodible polylactic acid (PLA) in place of PS. The sustained-release tablet was similarly obtained on the basis of the theophylline release test as shown in Figure 10.8b. Furthermore, DC tablet containing an insulin–PLA matrix tablet as core material was prepared and the changes in blood glucose levels after the subcutaneous implantation of the DC tablet in diabetic rats was examined (Figure 10.10) [27]. As the result, normal blood glucose levels were maintained for 10 days in the plasma-irradiated DC tablet and the release rate of insulin in the steady state from the plasma-irradiated DC tablet was 5 IU h^{-1} which was calculated from the data from 4 to 34 hours. These results indicated that DC tablet

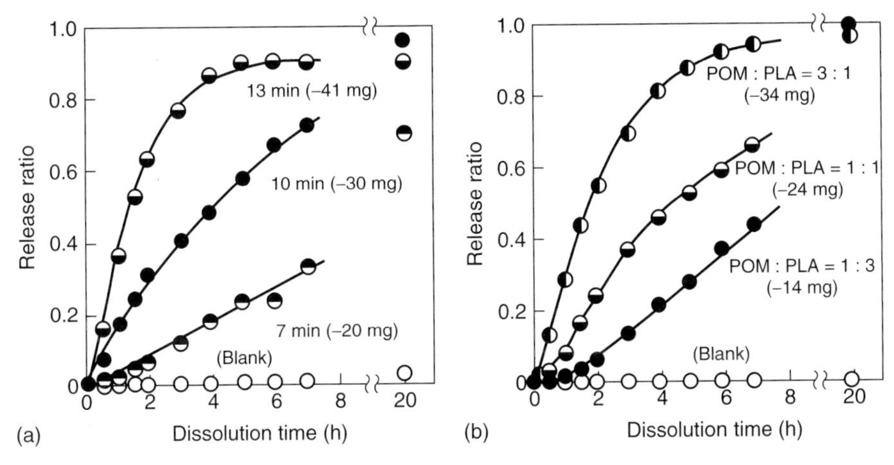

Figure 10.8 Effect of oxygen plasma irradiation on theophylline release from DC tablet. The values shown in parentheses denote the weight loss of the tablets after plasma irradiation. Core tablet: 100 mg (theophylline). For (a) outer layer: 80 mg (PS : POM = 1 : 1); plasma conditions: power: 50 W, pressure: 0.5 torr, O_2 50 ml min^{-1}. For (b) outer layer: 80 mg (PLA : POM = 1 : 3, 1 : 1, 3 : 1); plasma conditions : duration: 2 hours, power: 6 W, pressure: 0.5 torr, O_2 50 ml min^{-1}.

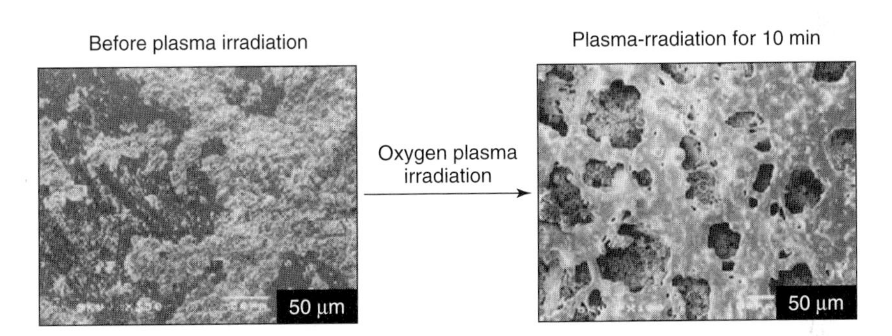

Figure 10.9 Scanning electron micrograph (SEM) of DC tablet using PS/POM (1 : 1) as outer layer before and after plasma irradiation. Plasma conditions: 50 W, O_2 0.5 torr, 50 ml min^{-1}.

consisting of PLA and POM as the outer layer can be applied to an implantable dosage form in the subcutaneous tissue.

When a water-soluble polymer is used as a wall material of the DC tablet, the drug release rate from the tablet would be dependent on the solubility of the polymer used. In fact, the rapid release from a DC tablet containing theophylline with the water-soluble polymer, polymethacrylic acid (PMAA) or polyacrylamide (PAAm), used as a wall material was suppressed by argon plasma-induced cross-linking reactions and changed into the slow release with a sigmoid release pattern due to decrease in the solubility of water-soluble polymers [30].

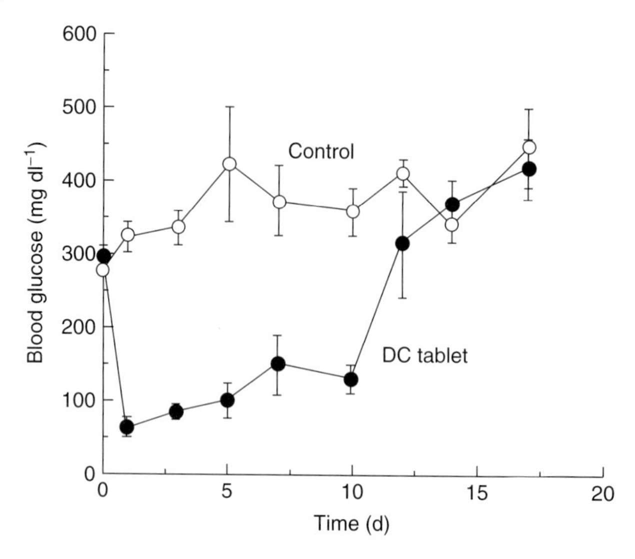

Figure 10.10 Change in blood glucose levels with time after the subcutaneous implantation of the DC tablet in rats. Outer layer: a mixed powder of PLA 10000 and POM (3/1), Core tablet: a mixed powder of PLA5000 and insulin (1/1). Plasma conditions: 6 W, O_2 0.5 torr, 50 ml min^{-1}, 3 hours.

10.3.2
Preparation of Time-Controlled Drug Release System by Plasma Techniques

Presently, therapy based on the factor of biorhythmic time is becoming more important in the progress toward an aging society in many countries, in addition to customary controlled-release systems [32–35]. Time-controlled release system has a function of a timer, so that the main technical point for the development of this system is how to control lag time and drug release after lag time.

It is well known that methacrylic–acrylic acid copolymers including their derivatives with various combinations and composition ratios of the monomers have been used as pharmaceutical aids for enteric coating agents commercially known as a *series of Eudragits.* These Eudragit polymers turn to be water-soluble in a certain specific pH solution, and they show different dissolution rates. The structures and the dissoluble pH values of several Eudragit polymers are shown in Figure 10.11.

Since plasma-crosslinkable acrylic monomers are one of the component polymers in Eudragits L100-55, argon plasma irradiation would lead to the suppression of Eudragit L100-55 solubility even in a dissoluble pH-value solution (pH $>$ 5.5) due to the occurrence of the surface cross-linking reactions. Thus, when Eudragit L100-55 is used as the wall material of a DC tablet, the initial drug release could be completely sustained for a certain period of time.

With this expectation in mind, we undertook argon plasma irradiation to examine the possibility of a rapid-release DC tablet of Eudragit L100-55, being converted into a delayed-release tablet, that is, the time-controlled DDS.

Eudragit L100

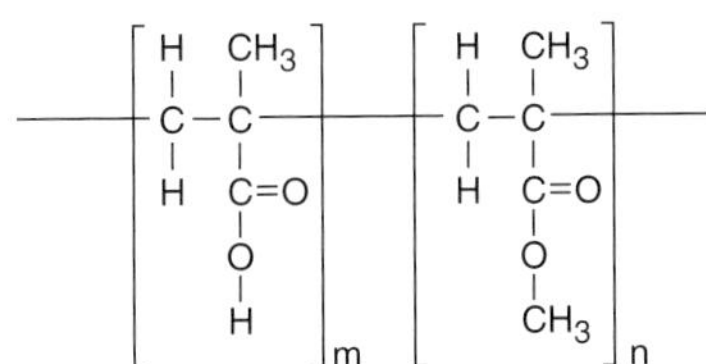

(m : n = 1 : 1, soluble at pH>6.0, T_g = ca. 160 ºC)

Eudragit S100

(m : n = 1 : 2, soluble at pH>7.0, T_g = ca. 160 ºC)

Eudragit L100

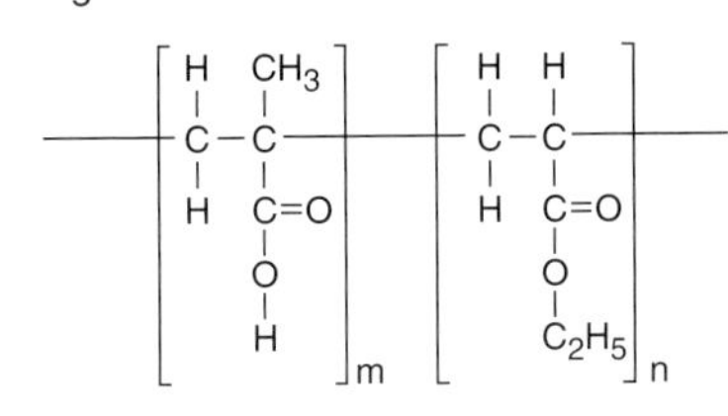

(m : n = 1 : 1, soluble at pH>4.5, T_g = ca. 110 ºC)

Eudragit RL (ERL)

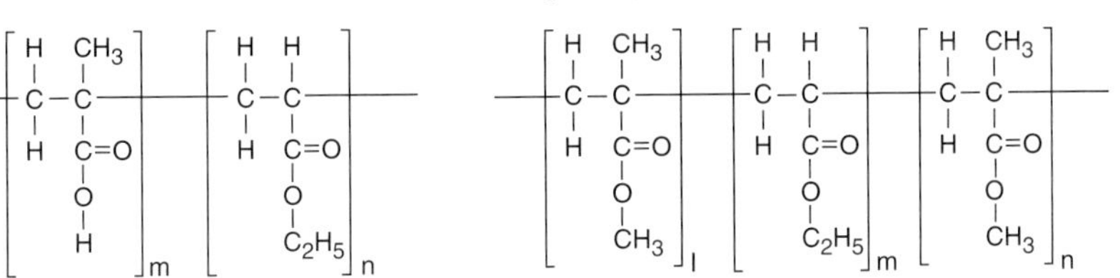

(l : m : n = 2 : 1 : 0.1, nonsoluble, T_g = ca. 60 ºC)

Figure 10.11 Structures and dissoluble pH values of several commercial Eudragit polymers used for enteric coating agents.

Figures 10.12 and 10.13 show the effect of argon plasma irradiation on theophylline release profiles in pH 6.5 buffer solution and the SEM pictures of the surface of Eudragit L100-55 tablet before and after argon plasma irradiation, respectively. It is seen that the Eudragit L100-55 tablets plasma-irradiated for 3 and 5 minutes have shown to produce prolongation of lag time for theophylline release.

The SEM pictures demonstrated that with a 5-minute irradiation the tablet surface had converted into a rather smooth layer with filling of the crack presenting at particle–particle interfaces by softening of Eudragit L100-55, and that with a 10-minute irradiation, the surface converted into a porous outer layer. It is considered that the porous layer was formed not only by the effect of plasma irradiation but also by physical actions such as evolved gas scattering accompanied by softening of the Eudragit L100-55 due to the plasma heat fusion.

10.3.3
Preparation of Intragastric FDDS by Plasma Techniques

Intragastric floating drug delivery system (FDDS) [36–38] has been noted as orally applicable to systems for the prolongation of the gastric-emptying time [39, 40]. Prolonged gastric retention improves bioavailability, reduces drug waste, and improves solubility for drugs that are less soluble in a high-pH environment. It also has applications for local drug delivery to the stomach and proximal small intestines. Gastro retention helps to provide better availability of new products with new therapeutic possibilities and substantial benefits for patients.

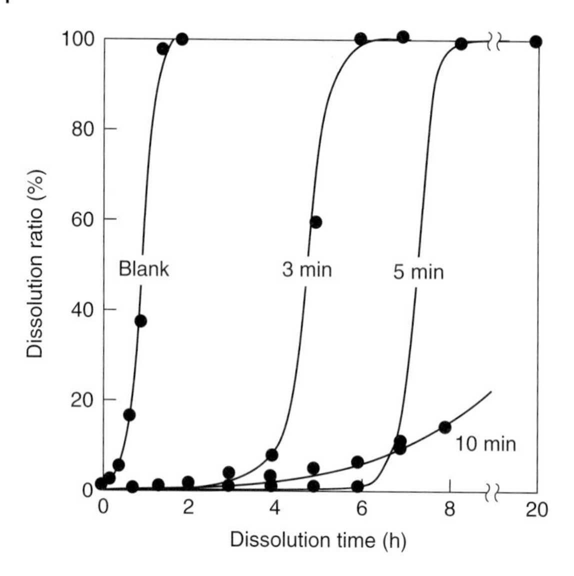

Figure 10.12 Effect of plasma duration on theophylline release from plasma-irradiated DC tablets of Eudragit L100-55 in pH 6.5 buffer solution. Plasma conditions: 50 W, Ar 0.5 torr, 50 ml min^{-1}.

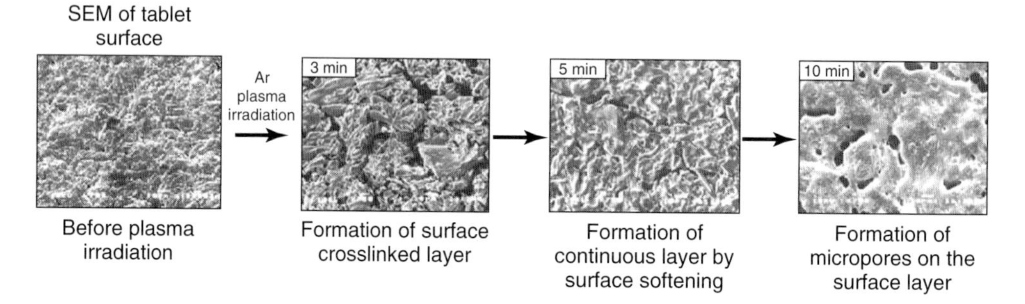

Figure 10.13 SEM pictures of Eudragit L100-55 tablet before and after plasma irradiation. Plasma conditions: 50 W, Ar 0.5 torr, 50 ml min^{-1}.

In the course of our study on plasma-assisted DDS preparation, we found that carbon dioxide was trapped in the tablet when argon plasma was irradiated onto the surface of a DC tablet composed of plasma-crosslinkable polymers possessing carboxyl group as an outer layer. It was considered that such tablets could be applicable to FDDS. On the basis of such findings, we have developed intragastric FDDS by plasma irradiation on DC tablets composed of a mixture of plasma-crosslinkable polymer and sodium acid carbonate ($NaHCO_3$) as the outer layer. Figure 10.14 shows the floating behavior of a plasma-irradiated DC tablet using a mixture composed of a 68/17/15 weight ratio of Povidone, Eudragit

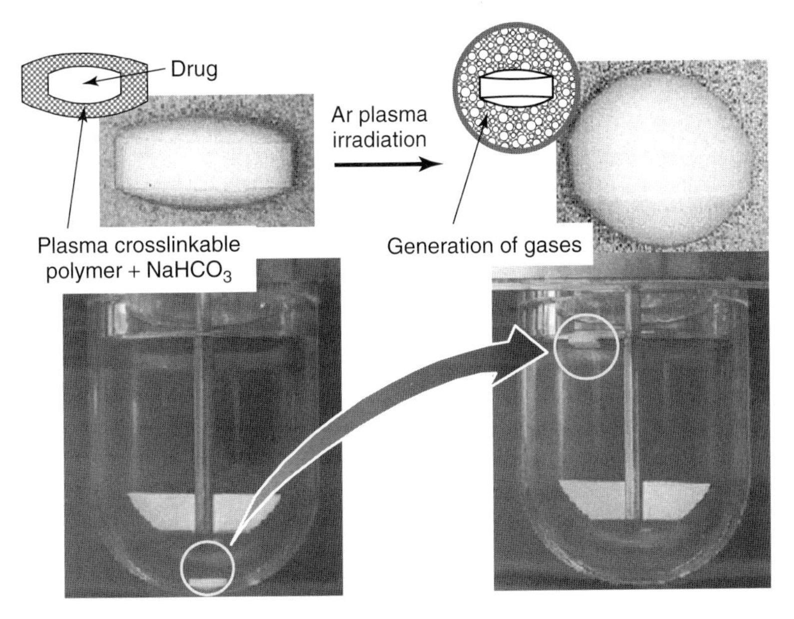

Figure 10.14 Photos of DC tablet for FDDS before and after plasma irradiation.

L100-55, and $NaHCO_3$ as an outer layer on the simulated gastric fluid. As shown in Figure 10.15, the plasma heat flux caused the thermal decomposition of $NaHCO_3$ to generate carbon dioxide and the resultant gases were trapped in the bulk phase of outer layer, so that the tablets turned to have a lower density than the gastric contents and remained buoyant in simulated gastric fluid for a prolonged period of time. In addition, the release of Fluorouracil (5-FU) from the tablet is sustained by occurrence of plasma-induced cross-linking reaction on the outer layer of tablet and the release rate of 5-FU can be well controlled by plasma operational conditions (Figure 10.16) [38].

10.3.4 Patient-Tailored DDS for Large Intestine Targeted-Release Preparations

With most of the currently available oral DDS devices, it is difficult for all patients to obtain the expected therapeutic effects of drugs administered, because of the individual difference in the environment such as pH value and the transit time in the gastrointestinal (GI) tract, which causes the slippage of time-related and positional timing of drug release [41]. From a viewpoint of the real optimization of drug therapy, in order to fulfill the specific requirements on drug release at the appropriate sites in GI tract, the "patient-tailored DDS" (tailor-made DDS) should be administered on the basis of the diagnosis of each patient's GI environment, which can be obtained by direct monitoring using a diagnostic system of the pH-sensitive radio telemetry capsule, so-called pH-chip.

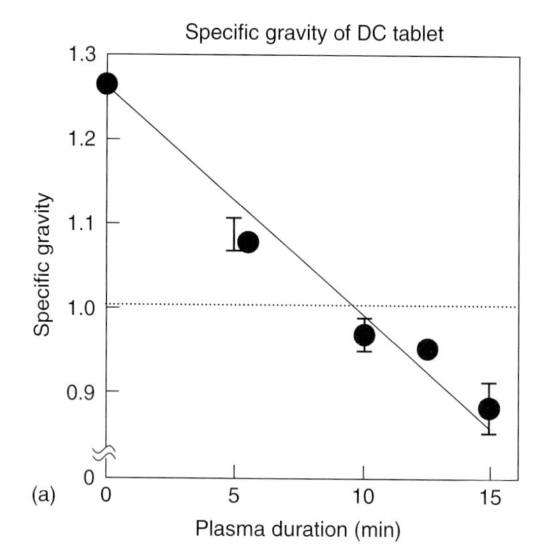

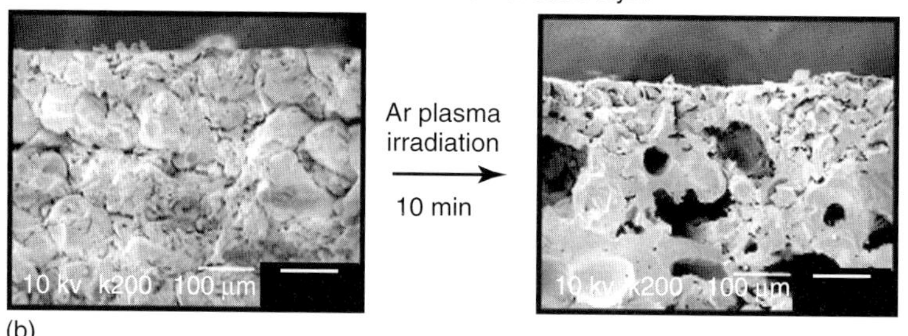

Figure 10.15 Effect of plasma irradiation on the specific gravity of DC tablet (a) and SEM pictures of cross section of DC tablet before and after plasma irradiation. Outer layer: a mixed powder of Povidone, Eudragit L100-55, and $NaHCO_3$ (68/17/15), Core tablet: 5-fluorouracil. Plasma condition: 20-Hz pulse frequency (on/off cycle = 35 ms/15 ms), 100 W, Ar 0.5 torr, 50 ml min^{-1}.

We have fabricated an experimental setup for the simulated GI tract for large intestine targeting, the dissolution test solution being changed in pH value corresponding to stomach (pH 1.2), small intestine (pH 7.4) and large intestine (pH 6.8), and examined the drug release test of plasma-irradiated DC tablet in the simulated GI tract.

Figure 10.17 shows the preliminary result of theophylline dissolution test in pH 6.8 test solution on the DC tablets using a mixture of Eudragits L100-55/RSPO (7 : 3) as outer layer. It is seen that the lag time has increased with the extension of plasma-irradiation time. The lag time has not been largely affected by treatment in pH 1.2 and pH 7.4 test solutions, which indicated the possibility for the development of the "patient-tailored DDS" targeting the large intestine such as the colon. We are now elaborating these initial studies aiming at more rapid drug

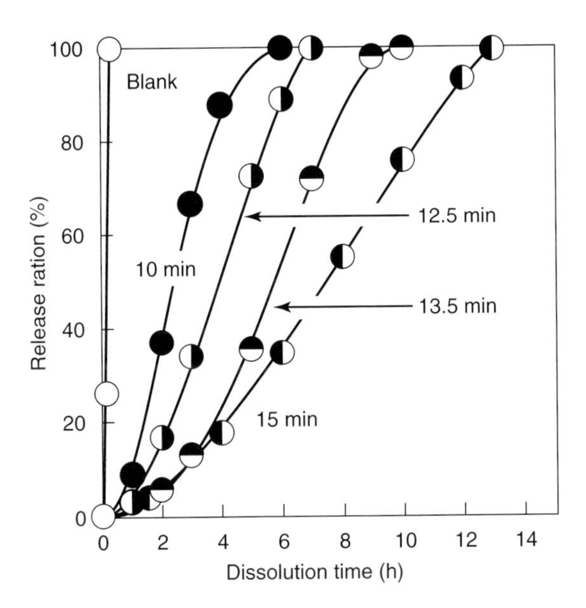

Figure 10.16 Effect of pulsed plasma duration on drug release from plasma-irradiated DC tablet. Outer layer: a mixed powder of Povidone, Eudragit L100-55 and $NaHCO_3$ (68/17/15), Core tablet: 5-fluorouracil. Plasma condition: 20-Hz pulse frequency (on/off cycle = 35 ms/15 ms), 100 W, Ar 0.5 torr, 50 ml min^{-1}.

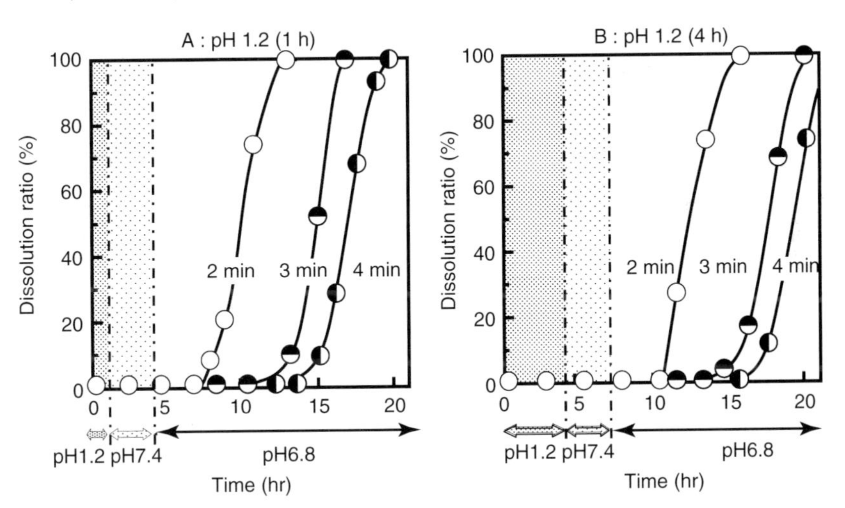

Figure 10.17 Release property of theophylline from helium plasma-irradiated DC tablet in the GI tract-simulated dissolution test: (a) for 1 hour in pH 1.2; (b) for 4 hours in pH 1.2. Outer layer: a mixed powder of Eudragit L100-55 and Eudragit RS (7/3), core tablet: theophylline. Plasma conditions: 30 W, He 0.5 torr, 50 ml min^{-1}.

release right after the drug preparations reached the prescribed pH value of the large intestine due to the contents of semisolid nature in the large intestine.

10.3.5
Preparation of Functionalized Composite Powders Applicable to Matrix-Type DDS

The recombination of solid-state radicals is significantly suppressed due to the restriction of their mobilities, unlike radicals in the liquid or gas phase [42]. Interactions between radicals at solid–solid interfaces do not occur under normal conditions [12]. On the other hand, we have reported the occurrence of mechanically induced surface radical recombination of plasma-irradiated polyethylene (PE) powder, low-density polyethylene (LDPE) and HDPE. Figure 10.18 shows the experimental setup for mechanical vibration of plasma-irradiated polymer powder to induce radical recombination and subsequent ESR spectral measurement.

As shown in Figure 10.19, the mechanical vibration caused effective decay in the ESR spectral intensity with change of the spectral pattern, especially in the case of plasma-irradiated LDPE, as the duration of mechanical vibration increased. This clearly indicated that on mechanical vibration, plasma-induced surface radicals successfully underwent solid-state radical recombination, since the spectral intensity did not appreciably decrease on standing at room temperature, so long as it was kept under anaerobic conditions.

For the preparation of matrix-type DDS, the mechanical vibration of plasma-irradiated PE powder was carried out in the presence of theophylline powder so as to immobilize the theophylline powder into PE matrix formed by interparticle linkage of PE powder. Figure 10.20 shows the theophylline release from the resulting composite powders of LDPE and HDPE. It is seen that the

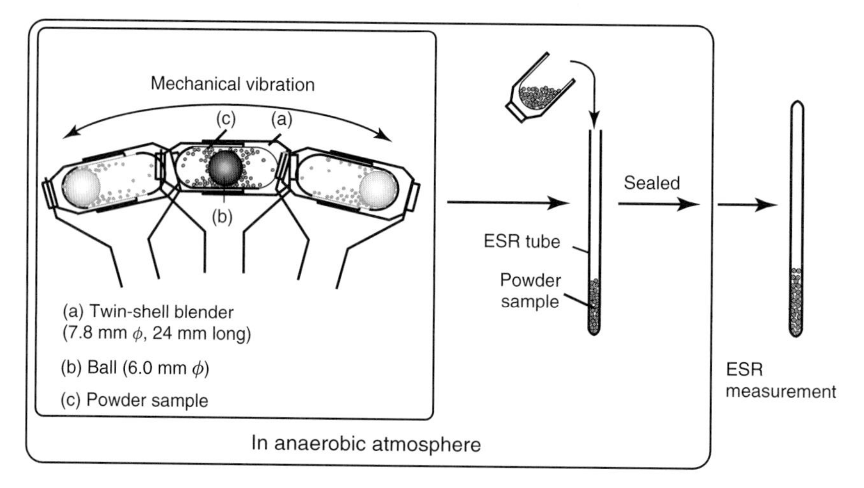

Figure 10.18 Schematic representation for mechanical vibration of plasma-irradiated polymer powder and subsequent ESR measurement.

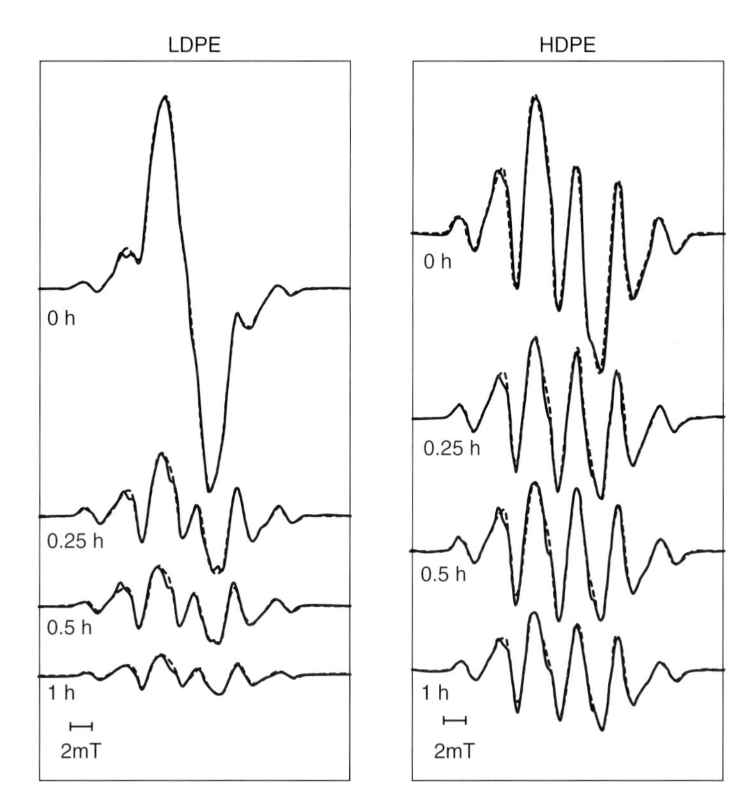

Figure 10.19 Progressive changes in observed ESR spectra of 10-minute plasma-irradiated LDPE and HDPE powders on mechanical vibration (60 Hz) in Teflon twin-shell blender, together with the simulated spectra shown as dotted lines. Plasma conditions: 40 W, Ar 0.5 torr, 10 minutes.

theophylline release is apparently suppressed from each of the plasma-irradiated PE powders, being proportional to the spin number of the surface radicals, due to trapping theophylline powder into the PE matrix [42]. It should be noted here that the theophylline release is further retarded from the tablet prepared by compressing the above composite PE powders.

10.4
Biomedical Engineering by Plasma Techniques

The wettability of polymer surface is an important characteristic relating to the biocompatibility for biomaterials. It is known, however, that the wettability introduced by plasma treatment decays with time after treatment. The mechanism has been ascribed to several reasons such as the overturn of hydrophilic groups into the bulk phase for crosslinkable polymers, and detachment of the hydrophilic

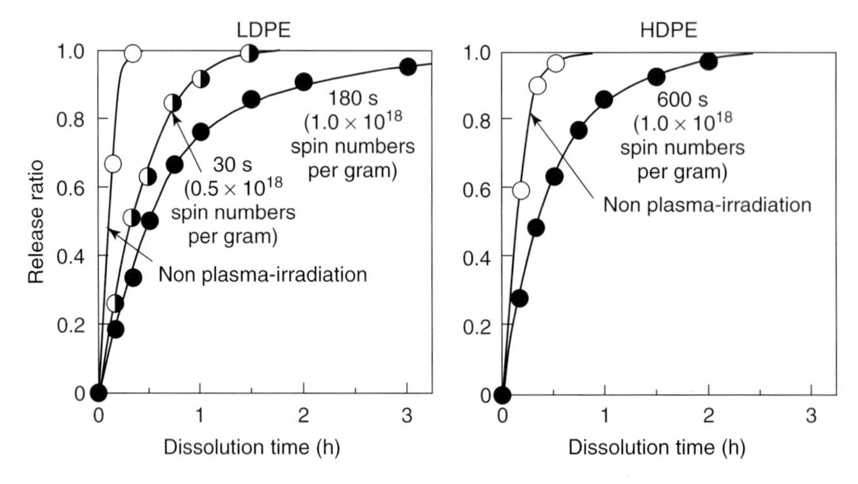

Figure 10.20 Theophylline release profiles from the composite powder composed of theophylline and Ar plasma- irradiated polyethylenes, LDPE and HDPE. LDPE plasma-irradiated for 60 seconds: 0.5×10^{18} spin/g, for 180 seconds: 1.0×10^{18} spin/g. HDPE plasma-irradiated for 60 seconds: 1.0×10^{18} spin/g. Plasma conditions : 40 W, Ar 0.5 torr, 1 minute.

lower molecular weight species from the surface for degradable polymers. We have reported a novel method to introduce a durable surface wettability and minimize its decay with time on several hydrophobic polymers – (polyethylene-naphthalate (PEN), LDPE, Nylon-12, and PS) [43–46]. The method involves a sorption of vinylmethylether-maleic anhydride copolymer (VEMA) into the surface layer and its immobilization by plasma-induced cross-linking reaction, followed by hydrolysis of maleic anhydride linkage in VEMA to generate durable hydrophilic carboxyl groups (VEMAC) on the surface (Figure 10.21). The present method was applied in the preparation of functionalized polyurethane-made catheter with durable surface lubricity.

10.4.1 Preparation of Clinical Catheter with Durable Surface Lubricity

One of the most important requirements of clinical catheters is the durability of the surface lubricity to diminish pain [43]. Figure 10.22 shows the representative data of measurement of surface slipperiness as a function of the number of repeated rubbings of the treated catheter against silicon rubber.

In can be seen that the resistance of the catheter containing VEMA without Ar plasma irradiation and that of the commercial catheter start to gradually increase after moving the catheter back and forth around 20–30 times in both cases, while that of the catheter containing VEMA Ar plasma-irradiated for 30 and 60 seconds remains low up to around 130–150 times. Prolonged plasma irradiation such as for 300 and 600 seconds duration, however, showed very poor durability of slipperiness, probably due to the formation of an extremely highly cross-linked

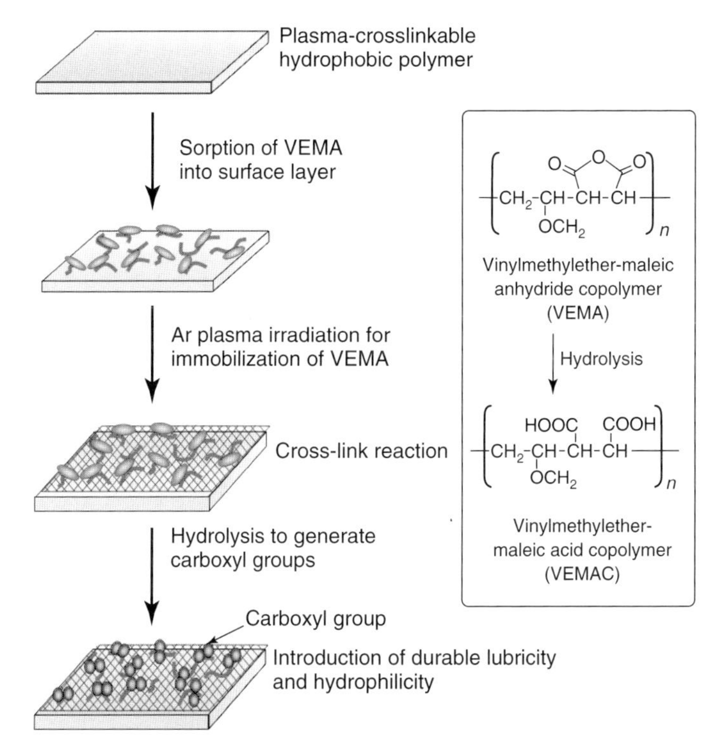

Figure 10.21 Conceptual illustration for introduction of durable hydrophilicity onto the polymer surface by plasma techniques.

surface. Thus, the result shows clearly a much higher functionality in terms of durability of surface lubricity.

10.4.2 Improvement of Cell Adhesion by Immobilizing VEMAC on Polymer Surface

In most types of cells, the adhesion to some substrates is a key primary process for cell development such as proliferation, survival, migration, and differentiation [46]. PS has been commonly used in a substrate for the *in vitro* cell culture owing to its excellent durability, low production cost, optical transparency in visible range, and nontoxicity. However, PS must be subjected to a surface treatment for biomedical use because it is a very hydrophobic polymer.

In order to improve the cell adhesion properties of a PS dish, VEMAC was immobilized on the surface using essentially the same method shown in Figure 10.21. Figure 10.23 shows the microscopic images of LNCap cells adhered on the VEMAC-immobilized PS dish (PS/VEMAC) after 6 hours in culture As shown in Figure 10.23, a distinct difference in cell attachment and spreading of LNCap cells between on PS/VEMAC and on nontreated PS dish was observed. The PS/VEMAC

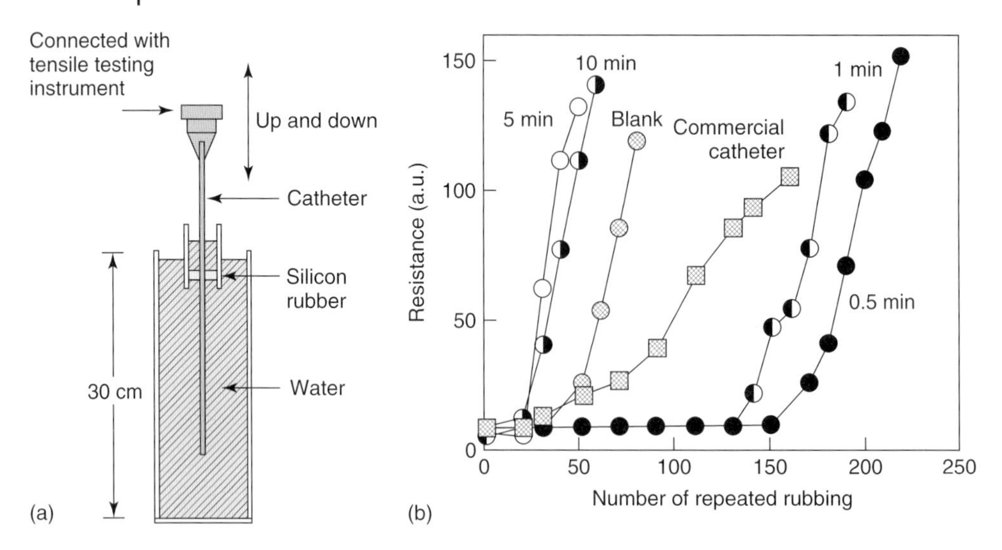

Figure 10.22 Experimental setup for measurement of surface lubricity of plasma-irradiated polyurethane-made catheter (a) and durability of the surface lubricity of plasma-assisted VEMAC-immobilized catheter in comparison with that of commercial catheter (b).

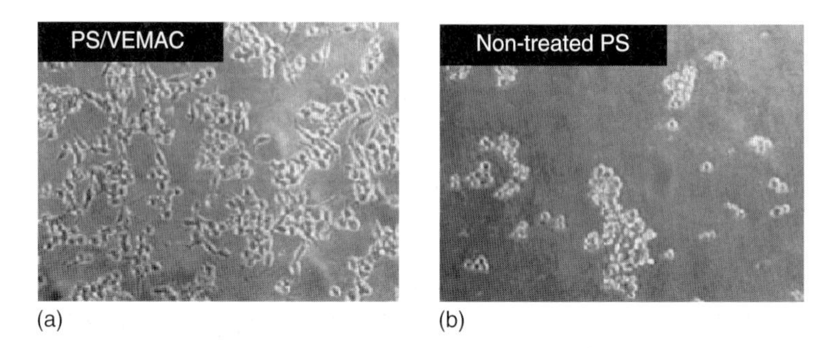

Figure 10.23 Microscopic images of LNCap cells on PS/VEMAC and nontreated PS dish after 6 hours in culture.

surface showed much better adhesion and spreading properties, while the adhered cells were not observed on nontreated PS surfaces. This result indicates that the PS/VEMAC surfaces prepared by the present method have preferential culturing properties of LNCap cells.

10.4.3
Plasma-Assisted Immobilization of Biomolecules onto Polymer Surfaces

Considerable interest has focused on the immobilization of several important classes of biomolecules such as DNA, enzyme, and protein, onto the water-insoluble supports. The development of DNA chips on which many kinds of *oligo*-DNA are immobilized, for example, has revolutionized the fields of genomics and bio-informatics [47]. However, all the current biochips are disposable and lack reusability, in part because the devices are not physically robust [48].

The method shown in Figure 10.20 has further been extended to application for the covalent immobilization of single-stranded *oligo*-DNA onto VEMAC-immobilized LDPE (LDPE/VEMAC) sheet by the reaction of 5′-aminolinker *oligo*-DNA with a condensation reagent. [49, 50] The 5′-aminolinker *oligo*-DNA, which possesses an aminohexyl group as a 5′-terminal group of DNA, is considered to be able to react with the carboxyl group on the surface of LDPE/VEMAC sheet. In fact, the resulting DNA-immobilized LDPE/VEMAC sheet was able to detect several complementary *oligo*-DNAs by effective hybridization.

To examine the reusability of DNA-immobilized LDPE/VEMAC sheet, we repeatedly conducted the hybridization and dehybridization of fluorescence-labeled complementary *oligo*-DNA on the same DNA-immobilized LDPE/VEMAC sheet, according to the general procedure to remove bounded target DNA from the chip (washing with hot water (90 °C) for 5 minutes). Figure 10.24 shows the result of the reusability test based on confocal laser microscope images of the DNA-immobilized

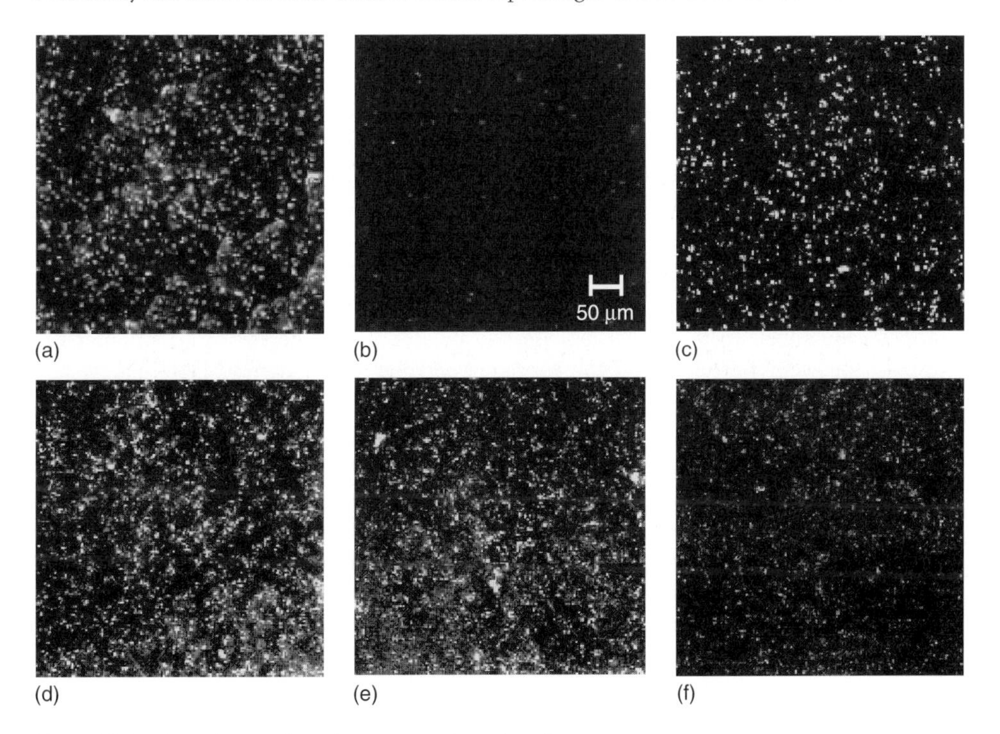

Figure 10.24 Scan image of the fluorescence intensity of LDPE–VEMAC–DNA sheet for reusability test. (a) Hybridization of complementary *oligo*-DNA, (b) after hot water rinse of sheet (a) for 5 minutes. Rehybridization of complementary *oligo*-DNA on the same sheet (c) two times, (d) five times, (e) seven times, and (f) eight times.

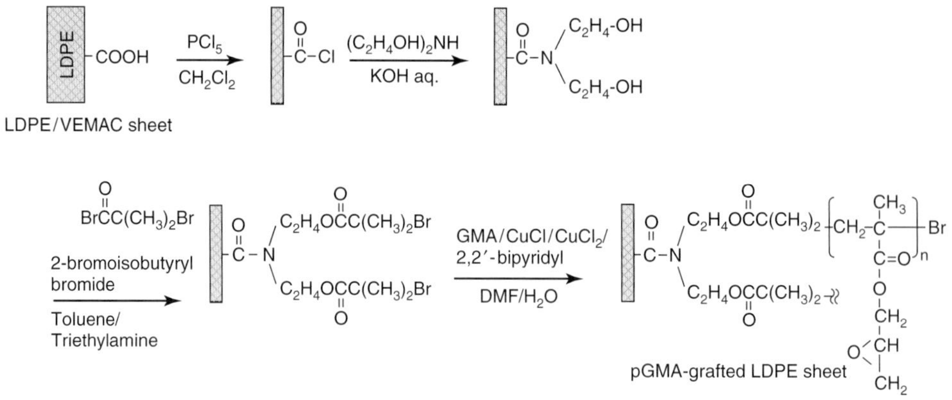

Figure 10.25 Reaction scheme for fabrication of pGMA brushes on LDPE sheet by ATRP.

LDPE/VEMAC sheet. It can be seen that fluorescence is observed at nearly the same level of intensity even after the repeated hybridization and dehybridization. The result indicated that the DNA-immobilized LDPE/VEMAC sheet obtained by the present method would be reusable.

Further, we used the LDPE/VEMAC surface for immobilization of enzyme [51, 52]. When the enzyme was immobilized covalently on solid surface, as is well known, a decrease in the enzyme activity was observed due to modifications in the tertiary structure of the catalytic sites. For the successful immobilization of enzymes on polymer substrate while retaining the activity, in this study, we prepared polyglycidylmethacrylate (pGMA) brushes on the LDPE/VEMAC sheet by atom transfer radical polymerization (ATRP) of GMA via carboxyl groups on the sheet. In the ATRP process, the polymerization degree of a monomer can be well controlled and the resultant polymer has a narrow molecular weight distribution [53]. Figure 10.25 shows the reaction scheme for the functionalization of LDPE/VEMAC surface. The epoxy group of pGMA can react readily and irreversibly with nucleophilic groups like $-NH_2$ under mild conditions. In fact, we succeeded in the covalent immobilization of fibrinolytic enzyme, urokinase, as a model enzyme through the direct coupling with epoxy groups of GMA on the surface thus prepared. Table 10.1 shows the relative surface concentration of immobilized urokinase and its activity. As can be seen in Table 10.1, the relative surface concentration of immobilized urokinase increased with the polymerization time for the fabrication of pGMA brushes. On the other hand, the activity of immobilized urokinase also increased in the pGMA-grafted LDPE sheet prepared by ATRP up to 2 hours, but it then leveled off under the present experimental conditions. Therefore, the ratio of active urokinase on pGMA-grafted LDPE sheet decreased with the increase in polymerization time. These results indicate that the LDPE surface with high enzymatic activity can be obtained by controlling the structure of interfaces between the enzyme and the substrate using the present method.

Table 10.1 The amount of immobilized urokinase and its activity on LDPE sheet.

Sample sheet	Immobilized UK ($\mu g\ cm^{-2}$)[a]	Activity ($IU\ cm^{-2}$)[b]	Ratio of active urokinase (%)
pGMA-grafted LDPE (ATRP for 2 h)	0.44 ± 0.08	35.66 ± 2.77	101.3
pGMA-grafted LDPE (ATRP for 4 h)	2.05 ± 0.08	31.34 ± 1.86	19.1
pGMA-grafted LDPE (ATRP for 6 h)	4.53 ± 0.15	32.96 ± 4.63	9.1

[a]The amount of immobilized urokinase on the pGMA-g-LDPE sheet was determined by Bradford dye binding assay using bovine gamma globulin as the standard.
[b]Activity of immobilized urokinase ($IU\ cm^{-2}$) was assayed using Glu–Gly–L-Arg–7-amido-4-methylcoumarin (GLU–GLY–L-ARG–MCA) as the substrate.

10.4.4 Basic Study on Preparation of Functionalized PVA–PAANa Hydrogel for Chemoembolic Agent

Vinyl alcohol–sodium acrylate copolymer (PVA–PAANa) is a well known *non-crosslinked hydrogel* (*water absorbent polymer*) due to the intense hydrogen-bonding network among the hydroxyl groups of polyvinyl alcohol moiety. The PVA–PAANa microsphere (about 100 μm) swells about 3.5 times in diameter its original size in human serum within a few minutes and can pass through a microcatheter. Recently, its microsphere has been applied to the chemoembolic agent used for transcatheter arterial embolization (TAE) in clinical trials on patients [54–57]. The PVA–PAANa microsphere is shape adjustable in nature according to the surrounding blood pressure because of the non-cross-linked structure, so that it has been shown to occlude the blood vessel much more effectively than any other conventional embolic agent such as gelatin sponge and lipiol.

In order to seek the possibility of further functionalization of PVA–PAANa such as a capability of controlling the ratio and rate of swelling by plasma processing, we have carried out argon plasma irradiation onto the PVA–PAANa microsphere and the surface radicals formed were studied by ESR by comparison with those of vinyl alcohol–acrylic acid copolymer (PVA–PAA) as well as its respective component homopolymers, PVA, PAA, and its sodium salt (PAANa) [58]. In fact, it was found that the ESR spectra have shown the vast difference in pattern between PVA–PAANa and PVA–PAA, demonstrating the strong sodium salt effect on the nature of plasma-induced surface radical formation (Figure 10.26). The systematic computer simulations of the ESR spectra revealed that the major spectral component was the radicals derived from PVA site for PVA–PAANa and the ones from PAA site for PVA–PAA. The SEM pictures indicated that the observed site selectivity for the surface radical formation has been derived from

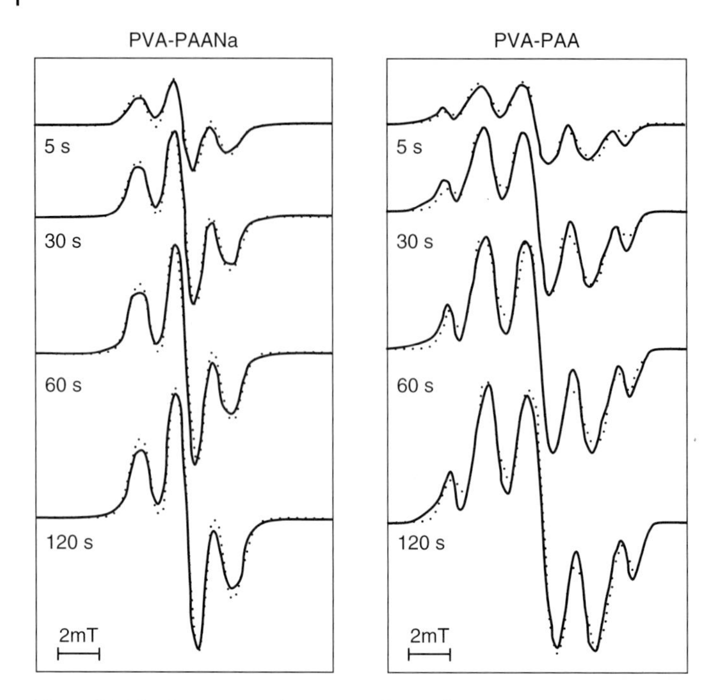

Figure 10.26 Progressive changes in observed ESR spectra of plasma-irradiated PVA–PAANa and PVA–PAA together with the simulated spectra shown as dotted lines.

the difference in the surface morphology between PVA–PAANa and PVA–PAA (Figure 10.27). PVA–PAANa forms the microphase separation structure with the condensed domain of PAANa site so as to reduce the effective surface area for the surface radical formation. The present result provides a basis for the future experimental design for giving an additional performance to PVA–PAANa, including the sustained-drug release function at the occluding site.

10.5 Conclusion

The present results have clearly shown that one can prepare a variety of desired DDS devices, if one selects the tailored polymers for wall materials of DC tablets as well as plasma operational conditions. The method of plasma-assisted DDS preparations has several advantages: (i) totally dry process, (ii) polymer surface modification without affecting the bulk properties, (iii) avoidance of direct plasma-exposure to drugs, and (iv) versatile control of drug release rates. Thus, it is hoped that more practical applications will be developed in the course of the attempt now in progress. It should be noted, however, that we have restricted the use of the organic polymers so as to manipulate the existing pharmaceutical aids licensed

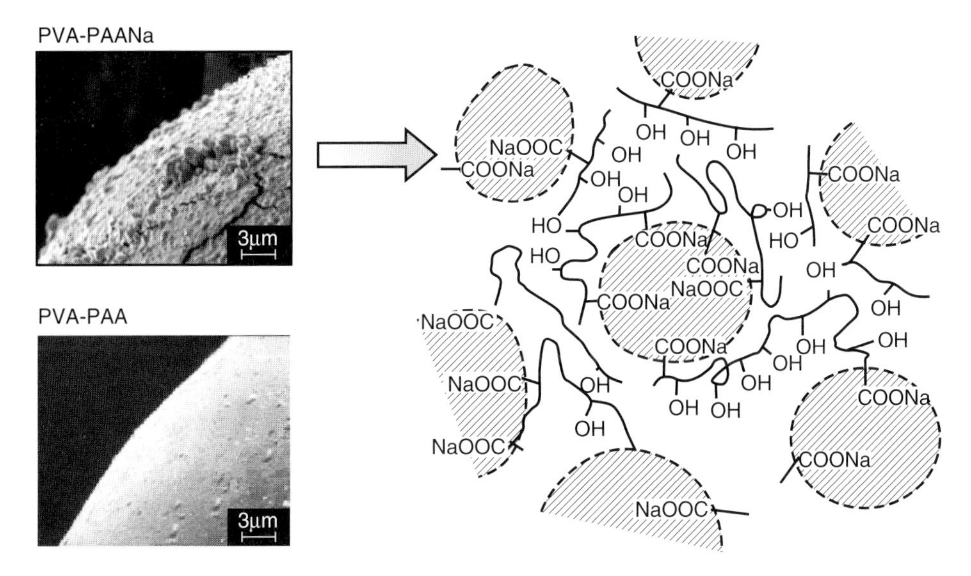

Figure 10.27 SEM pictures of PVA–PAANa and PVA–PAA, and the conceptual illustration of microphase separation structure of PVA–PAANa.

for practical use in patients, since pharmaceutical aids containing lower than 0.1% level impurity can be used without structural identification, in accordance with the harmonized tripartite guidelines in Japan, USA, and EU. Otherwise, unlike the process for obtaining the approval of industrial substances, obtaining the license for manufacturing new drug and quasi-drug substances is an extremely cost- and time-consuming process.

Acknowledgments

This work was supported in part by a scientific research grant from the Ministry of Education, Science, Sports and Culture of Japan (Grant No. 11672147, 14370730) and from the Japan Society for the Promotion of Science (Grant No. JSPS-RFTF 99R13101), which are gratefully acknowledged.

References

1. Kanazawa, S., Kogoma, M., Moriwaki, T., and Okazaki, S. (1988) *J. Phys. D: Appl. Phys.*, **21**, 838.
2. Stoffels, E., Sakiyama, Y., and Graves, D.B. (2008) *IEEE Trans. Plasma Sci.*, **36**, 1441.
3. Fridman, G., Friedman, G., Gutsol, A., Shekhter, A.B., Vasilets, V.N., and Fridman, A. (2008) *Plasma Processes Polym.*, **5**, 503.
4. International Conference on Plasma Medicine (ICPM)-I *http://plasma.mem.drexel.edu/icpm-1/*.
5. Kuzuya, M., Noguchi, A., Ishikawa, M., Koide, A., Sawada, K., Ito, A., and

Noda, N. (1991) *J. Phys. Chem.*, **95**, 2398.
6. Kuzuya, M., Ito, H., Kondo, S., Noda, N., and Noguchi, A. (1991) *Macromolecules*, **24**, 6612.
7. Kuzuya, M., Noguchi, A., Ito, H., Kondo, S., and Noda, N. (1991) *J. Polym. Sci., Part A: Polym. Chem.*, **29**, 1.
8. Kuzuya, M., Noda, N., Kondo, S., Washino, K., and Noguchi, A. (1992) *J. Am. Chem. Soc.*, **114**, 6505.
9. Kuzuya, M., Ishikawa, M., Noguchi, A., Sawada, K., and Kondo, S. (1992) *J. Polym. Sci., Part A: Polym. Chem.*, **30**, 379.
10. Kuzuya, M., Kondo, S., Ito, H., and Noguchi, A. (1992) *Appl. Surf. Sci.*, **60**, 416.
11. Kuzuya, M., Kamiya, K., Yanagihara, Y., and Matsuno, Y. (1993) *Plasma Sources Sci. Technol.*, **2**, 51.
12. Kuzuya, M., Niwa, J., and Ito, H. (1993) *Macromolecules*, **26**, 1990.
13. Kuzuya, M., Morisaki, K., Niwa, J., Yamauchi, Y., and Xu, K. (1994) *J. Phys. Chem.*, **98**, 11301.
14. Kuzuya, M., Yamauchi, Y., Niwa, J., Kondo, S., and Sakai, Y. (1995) *Chem. Pharm. Bull.*, **43**, 2037.
15. Kuzuya, M., Niwa, J., and Kondo, S. (1996) *Mol. Cryst. Liq. Cryst. Sci. Technol., Sect. A*, **277**, 703.
16. Kuzuya, M., Matsuno, Y., Yamashiro, T., and Tsuiki, M. (1997) *Plasmas Polym.*, **2**, 79.
17. Kuzuya, M., Yamashiro, T., Kondo, S., Sugito, M., and Mouri, M. (1998) *Macromolecules*, **31**, 3225.
18. Kuzuya, M., Kondo, S., Sugito, M., and Yamashiro, T. (1998) *Macromolecules*, **31**, 3230.
19. Kuzuya, M., Sasai, Y., and Kondo, S. (1999) *J. Photopolym. Sci. Technol.*, **12**, 75.
20. Kuzuya, M., Yamauchi, Y., and Kondo, S. (1999) *J. Phys. Chem. B*, **103**, 8051.
21. Yamauchi, Y., Sugito, M., and Kuzuya, M. (1999) *Chem. Pharm. Bull.*, **47**, 273.
22. Kuzuya, M. and Matsuno, Y. (1993) *Drug Deliv. Syst.*, **8**, 149.
23. Kuzuya, M., Kondo, S., and Sasai, Y. (2001) *Plasmas Polym.*, **6**, 145.
24. Kuzuya, M., Kondo, S., and Sasai, Y. (2005) *Pure Appl. Chem.*, **77**, 667.
25. Kuzuya, M., Noguchi, A., Ito, H., and Ishikawa, M. (1991) *Drug Deliv. Syst.*, **6**, 119.
26. Kuzuya, M., Ito, H., Noda, N., Yamakawa, I., and Watanabe, S. (1991) *Drug Deliv. Syst.*, **6**, 437.
27. Yamakawa, I., Watanabe, S., Matsuno, Y., and Kuzuya, M. (1993) *Biol. Pharm. Bull.*, **16**, 182.
28. Ishikawa, M., Matsuno, Y., Noguchi, A., and Kuzuya, M. (1993) *Chem. Pharm. Bull.*, **41**, 1626.
29. Ishikawa, M., Noguchi, T., Niwa, J., and Kuzuya, M. (1995) *Chem. Pharm. Bull.*, **43**, 2215.
30. Kuzuya, M., Ishikawa, M., Noguchi, T., Niwa, J., and Kondo, S. (1996) *Chem. Pharm. Bull.*, **44**, 192.
31. Ishikawa, M., Hattori, K., Kondo, S., and Kuzuya, M. (1996) *Chem. Pharm. Bull.*, **44**, 1232.
32. Kuzuya, M., Ito, K., Kondo, S., and Makita, Y. (2001) *Chem. Pharm. Bull.*, **49**, 1586.
33. Ito, K., Kondo, S., and Kuzuya, M. (2001) *Chem. Pharm. Bull.*, **49**, 1615.
34. Kondo, S., Ito, K., Sasai, Y., and Kuzuya, M. (2002) *Drug Deliv. Syst.*, **17**, 127.
35. Sasai, Y., Kondo, S., Nagato, M., and Kuzuya, M. (2005) *J. Photopolym. Sci. Technol.*, **18**, 281.
36. Kuzuya, M., Nakagawa, T., Kondo, S., Sasai, Y., and Makita, Y. (2002) *J. Photopolym. Sci. Technol.*, **15**, 331.
37. Kondo, S., Nakagawa, T., Sasai, Y., and Kuzuya, M. (2004) *J. Photopolym. Sci. Technol.*, **17**, 149.
38. Nakagawa, T., Kondo, S., Sasai, Y., and Kuzuya, M. (2006) *Chem. Pharm. Bull.*, **54**, 514.
39. Singh, B.N. and Kim, K.H. (2000) *J. Controlled Release*, **63**, 235.
40. Streubel, A., Siepmann, J., and Bodmeier, R. (2006) *Curr. Opin. Pharmacol.*, **6**, 501.
41. Sasai, Y., Sakai, Y., Nakagawa, T., Kondo, S., and Kuzuya, M. (2004) *J. Photopolym. Sci. Technol.*, **17**, 185.
42. Kuzuya, M., Sasai, Y., Mouri, M., and Kondo, S. (2002) *Thin Solid Films*, **407**, 144.

43. Kuzuya, M., Yamashiro, T., Kondo, S., and Tsuiki, M. (1997) *Plasmas Polym.*, **2**, 133.
44. Kuzuya, M., Sawa, T., Yamashiro, T., Kondo, S., and Takai, O. (2001) *J. Photopolym. Sci. Technol.*, **14**, 87.
45. Kuzuya, M., Sawa, T., Mouri, M., Kondo, S., and Takai, O. (2003) *Surf. Coat. Technol.*, **169**, 587.
46. Sasai, Y., Matsuzaki, N., Kondo, S., and Kuzuya, M. (2008) *Surf. Coat. Technol.*, **202**, 5724.
47. Ramsay, G. (1998) *Nat. Biotechnol.*, **16**, 40.
48. Halliwell, C.M. and Cass, A.E.G. (2001) *Anal. Chem.*, **73**, 2476.
49. Kondo, S., Sawa, T., and Kuzuya, M. (2003) *J. Photopolym. Sci. Technol.*, **16**, 71.
50. Kondo, S., Sasai, Y., and Kuzuya, M. (2007) *Thin Solid Films*, **515**, 4136.
51. Sasai, Y., Kondo, S., Yamauchi, Y., and Kuzuya, M. (2006) *J. Photopolym. Sci. Technol.*, **19**, 265.
52. Sasai, Y., Oikawa, M., Kondo, S., and Kuzuya, M. (2007) *J. Photopolym. Sci. Technol.*, **20**, 197.
53. Patten, T.E., Xia, J., Abernathy, T., and Matyjaszewski, K. (1996) *Science*, **272**, 866.
54. Yao, J., Hori, S., Minamitani, K., Hashimoto, T., Yoshimura, H., Nomura, N., Ishida, T., Fukuda, H., Tomoda, K., and Nakamura, H. (1996) *Nippon Igaku Hoshasen Gakkai Zasshi*, **56**, 19.
55. Osuga, K., Hori, S., Kitayoshi, H., Khankan, A.A., Okada, A., Sugiura, T., Murakami, T., Hosokawa, K., and Nakamura, H. (2002) *J. Vasc. Interv. Radiol.*, **13**, 1125.
56. Osuga, K., Khankan, A.A., Hori, S., Okada, A., Sugiura, T., Maeda, M., Nagano, H., Yamada, A., Murakami, T., and Nakamura, H. (2002) *J. Vasc. Interv. Radiol.*, **13**, 929.
57. Nakazawa, T., Osuga, K., Hori, S., Mikami, K., Higashihara, H., Maeda, N., Tomoda, K., and Nakanura, H. (2006) *Jpn. J. Interv. Radiol.*, **21**, 393.
58. Kuzuya, M., Izumi, T., Sasai, Y., and Kondo, S. (2004) *Thin Solid Films*, **457**, 12.

11
Targeting Dendritic Cells with Carbon Magnetic Nanoparticles Made by Dense-Medium Plasma Technology*

Heidi A. Schreiber, Jozsef Prechl, Hongquan Jiang, Alla Zozulya, Zsuzsanna Fabry, Ferencz S. Denes, and Matyas Sandor

Vaccine strategies often require efficient delivery of antigen to dendritic cells (DCs). DCs, a rare population of white blood cells, play a central role in the initiation of immune responses as the only antigen-presenting cell (APC) capable of both activating naïve T cells and efficiently initiating a recall T-cell response [1]. Despite their low frequency during steady-state conditions, DCs can be found in most tissues as tissue-resident DCs, as well as in lymphoid organs [2]. The primary role of tissue-resident DCs is to sample antigen, mature, and traffic to lymph nodes to initiate a T-cell response. It has been suggested that the size of the immunizing antigen affects which APC prefers to sample and process it [3]. Bacterial-sized ($>1\ \mu m$) uptake is favored by macrophages, whereas viral-sized (<100 nm) particles are preferentially phagocytosed by DCs [4]. Additionally, unlike the larger particles, nano-sized particles can efficiently migrate through the lymphatics to reach DCs residing in the lymph nodes [5]. Nanoparticle (NP) traffic through the lymphatics further supports the role of DCs as the preferred sampler of nano-sized particles, as DCs utilize the lymphatics as their favored route to the lymph nodes. Targeting DCs as a method of vaccination is a strategy that has been gaining increasing interest [6–8]. It has been previously demonstrated that antigen conjugated to NPs is effective at targeting DCs and generating strong cellular and humoral immune responses *in vivo* [9–14]. Iron-based NPs have allowed for the *in vivo* imaging of DCs via magnetic resonance imaging and for their targeted delivery to specific areas by using a magnetic gradient field [15–17]. An ideal DC-targeting NP would harbor all of the following functionalities together: DC-specific uptake preference, antigen conjugation capabilities, *in vivo* imaging, targeted delivery, and biocompatibility. Here, we describe the generation of a NP that possesses all of these functionalities, which makes it an ideal DC-targeting vector and provides a tool for influencing DC functions *in vivo*.

* Work supported by National Institutes of Health funding R21-A1072638 (M. Sandor) and OTKA K68617 (J. Prechl).

Industrial Plasma Technology. Edited by Yoshinobu Kawai, Hideo Ikegami, Noriyoshi Sato, Akihisa Matsuda, Kiichiro Uchino, Masayuki Kuzuya, and Akira Mizuno

ISBN: 978-3-527-32544-3

Color Fig.: **11.1**

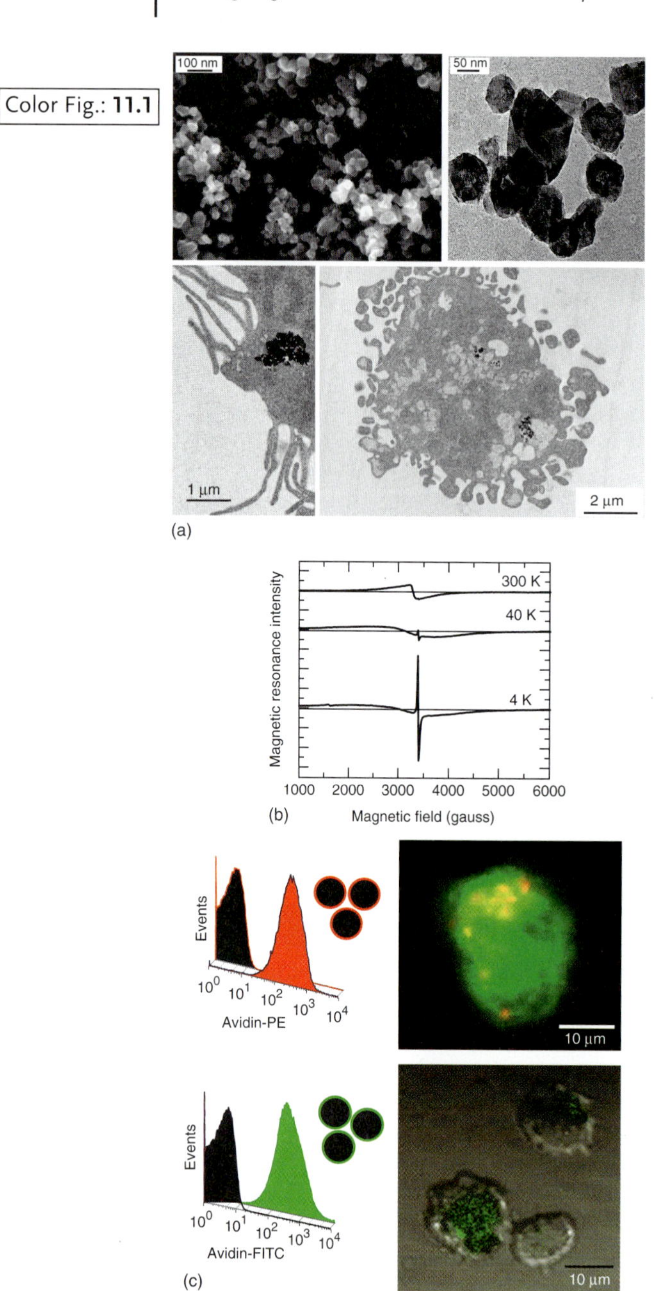

Figure 11.1 Characterization of CMNPs. (a) Top, TEM image of CMNPs. Bottom, EM images of DCs containing CMNPs. (b) FMR. (c), Left, flow cytometry histograms of fluorescent FITC- and PE-CMNPs. Right top, GFP DC containing PE-CMNPs and bottom, DC containing FITC-CMNPs.

Using dense-medium plasma technology, we have synthesized carbon-based magnetic nanoparticles (CMNPs) by sustaining benzene between two iron electrodes, as previously described in detail [18]. These CMNPs are carbon based, have an irregular, crystalline-like surface, and are 40–80 nm in size (Figure 11.1a top panels). To test whether CMNPs are preferentially endocytosed by DCs, we fed bone-marrow-derived DC cultures with CMNPs. By TEM image analysis, we observed that DCs efficiently endocytose CMNPs *in vitro* (Figure 11.1a lower panels). Importantly, TEM images also demonstrate that CMNPs within the DCs are contained within endocytic vesicles, cellular compartments that are associated with loading antigen-presenting molecules. Ferromagnetic resonance spectroscopy confirmed the magnetism of the CMNPs (Figure 11.1b). It also suggested that the magnetism was due to metallic iron; however, some magnetite or maghemite may have also been present. Binding of avidin-FITC or -PE allowed for both fluorescent monitoring of the particles and the indirect binding of biotinylated proteins to the CMNPs (Figure 11.1c histograms). The binding of fluorochromes to the CMNPs enabled imaging of both PE- and FITC-CMNPs inside DCs (Figure 11.1c right-hand side images).

The large surface area of the CMNPs, along with the plasma-functionalized carboxyl or amino surface, allows for high-capacity binding of proteins. Single or multiple proteins can be conjugated by ester bonds, potentially displaying a complex array of proteins. Figure 11.2a demonstrates the efficient binding of both antigen (hen egg lysozyme (HEL)) and monoclonal IgG proteins. The selectivity and efficiency of DC uptake of CMNPs can be further promoted by conjugation of DC-specific antibodies to the particles. DEC205/Ag-conjugated CMNPs were endocytosed even more readily by DCs compared to CMNPs conjugated with antigen alone (Figure 11.2b). As previously described, the composition of the irregular, crystalline surface of the CMNPs allows for efficient binding of protein (Figure 11.2d). We next tested whether the CMNPs can deliver specific antigen to DCs and subsequently activate their cognate-antigen-specific T cells *in vivo*. Biotinylated HEL was bound to avidin-labeled CMNPs. HEL-conjugated and control CMNPs were subcutaneously injected into mice at the base of the tail and 3×10^5 CFSE-labeled HEL-specific T-cell receptor transgenic CD4+ T cells were adoptively transferred via retroorbital injection. One week later, activation and CFSE dilution of the transferred cells was assessed (Figure 11.2c). High CFSE intensity indicated that control CMNPs did not induce activation of transferred cells. On the contrary, HEL-conjugated CMNPs resulted in the proliferation of the TCR-specific CD4+ T-cell population, demonstrated by the decreased fluorescence of CFSE from each round of cell division. However, for many vaccines to become biologically effective, certain cues must be delivered to the DC in addition to the antigen to allow it to shape the immune response. The CMNPs presented here also allow for binding of biotinylated CpG (an instructional danger signal for the DC) along with HEL protein (Figure 11.2d). The binding and delivery of this signal resulted in the DC favoring a proinflammatory T-cell response, indicated by the production of the cytokine interferon-gamma (IFNγ). This is a favorable response to fight off many intracellular infections. These data suggest that conjugation of antigen to CMNPs is an efficient method of vaccine delivery.

Color Fig.: **11.2**

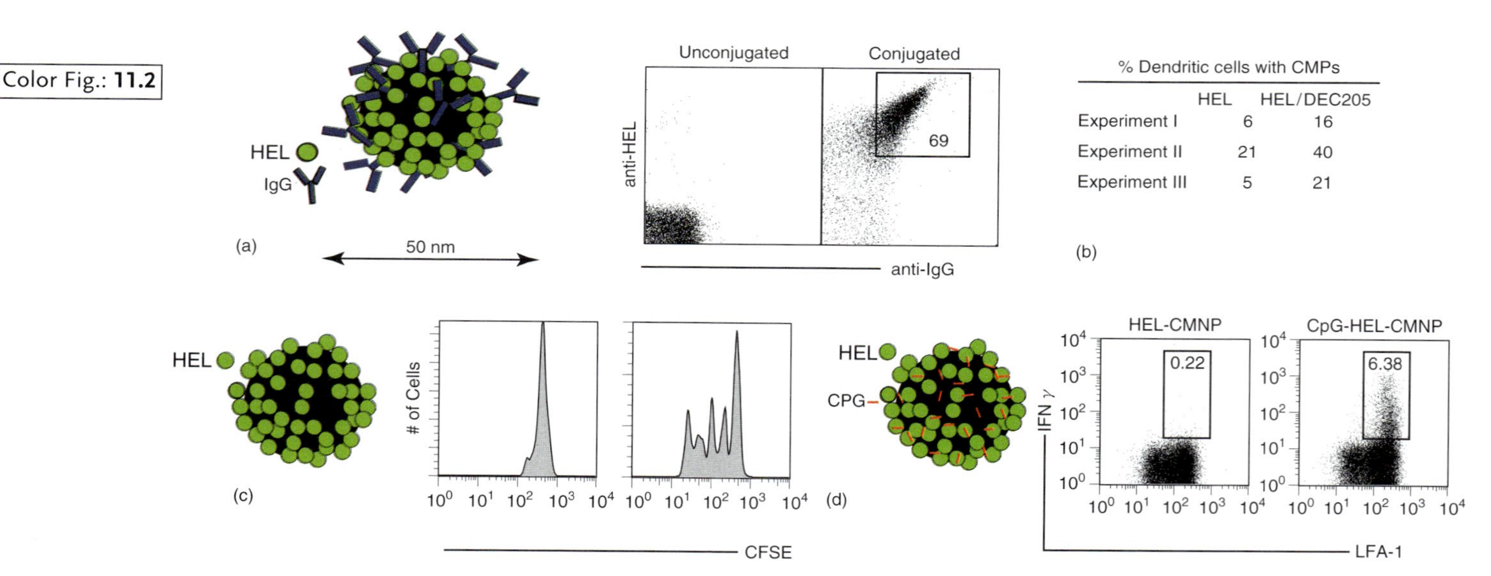

Figure 11.2 Delivery of proteins and cues to DCs using CMNPs. (a) Direct conjugation of protein; detected using fluorochrome-labeled antibodies against protein with flow cytometry. (b) Enhanced targeting of CMNPs to DCs by conjugation of DC-specific (DEC205) antibodies. (c) *In vivo* HEL-specific CD4+ T-cell proliferation by antigen-conjugated CMNPs. (d) T-cell commitment induced by binding TLR ligand (CpG) to CMNPs in addition to HEL protein.

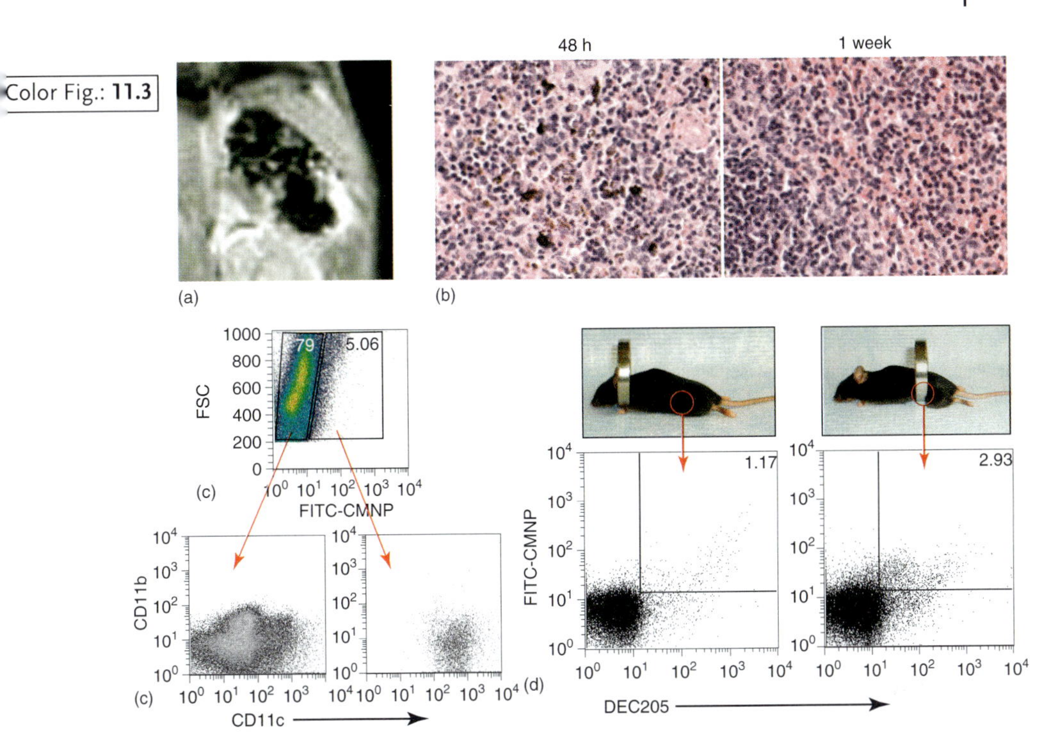

Figure 11.3 CMNPs provide a tool for monitoring and targeting DCs *in vivo*. (a) MRI scan (4.7 T) of mouse spleen 15 minutes post-i.v. injection of CMNPs. (b) Hematoxylin and Eosin staining of formalin-fixed spleen two days and one week post-i.v. injection of CMNPs. (c) Flow cytometry analysis of splenocytes five days post-i.v. injection of FITC-CMNPs. Samples stained with CD11b (macrophage marker) and CD11c (DC marker), demonstrating CMNPs preferentially target DCs *in vivo*. (d) Analysis of nonmagnet-exposed inguinal LN (left) and magnet-exposed inguinal LN (right) 30 minutes after i.v. injection of CMNPs. DC staining (DEC205) shows enhanced particle uptake and migration to magnet-exposed LN.

In addition to delivering protein and various cues to the DCs, CMNPs can also be traced *in vivo*. Owing to the iron core, we are able to trace CMNPs *in vivo* using a 4.7T MRI (Figure 11.3a). It is important to note that the CMNPs do not induce inflammation or toxicity following intravenous injection, and nearly all extracellular particles are cleared after one week (Figure 11.3b). As found *in vitro*, CMNPs were also preferentially endocytosed by DCs *in vivo*. Flow cytometric analysis of the spleen from mice that were i.v. injected with FITC-CMNPs demonstrated that FITC+ cells are entirely CD11c + CD11b−, suggesting their ability to selectively target DCs over other APCs *in vivo* (Figure 11.3c). *In vitro* and *in vivo* uptake of FITC-CMNPs by DCs permitted the DCs to become magnetic and fluorescent. Using these newly acquired features of the DCs, we tested whether *in vivo* trafficking of FITC-CMNP-loaded DCs could be altered by an external magnetic field. To test this, FITC-CMNPs were i.v. injected and a ring-magnet was positioned to encompass either the cervical or inguinal lymph nodes for 30 minutes (Figure 11.3d); then, inguinal lymph nodes were stained with DC-specific antibody, DEC205. This assay demonstrated that an exterior magnet is often able to enrich CMNP-loaded DCs in the neighboring, magnet-exposed lymph node, suggesting the potential ability to influence DC trafficking *in vivo*.

Collectively, the CMNPs presented here may provide a better vaccine strategy. Their preferential uptake by DCs based on their nanometer size distribution and their large surface area, which allows for high concentrations of protein conjugation, make them an ideal antigen delivery vector. Additionally, the magnetic properties of the CMNPs allow for visualization, isolation, and probable influential trafficking of the DCs *in vivo*. These latter properties distinguish them from other DC-targeting NPs investigated to date. Having a vaccine vehicle like these, CMNPs could potentially enable us to influence the immune response with a higher level of specification.

References

1. Banchereau, J. and Steinman, R.M. (1998) Dendritic cells and the control of immunity. *Nature*, **392**, 245–252.
2. Guermonprez, P., Valladeau, J., Zitvogel, L., Thery, C., and Amigorena, S. (2002) Antigen presentation and T cell stimulation by dendritic cells. *Annu. Rev. Immunol.*, **20**, 621–667.
3. Xiang, S.D. *et al.* (2006) Pathogen recognition and development of particulate vaccines: does size matter? *Methods*, **40**, 1–9.
4. Fifis, T. *et al.* (2004) Size-dependent immunogenicity: therapeutic and protective properties of nano-vaccines against tumors. *J. Immunol.*, **173**, 3148–3154.
5. Reddy, S.T. *et al.* (2007) Exploiting lymphatic transport and complement activation in nanoparticle vaccines. *Nat. Biotechnol.*, **25**, 1159–1164.
6. Reddy, S.T., Swartz, M.A., and Hubbell, J.A. (2006) Targeting dendritic cells with biomaterials: developing the next generation of vaccines. *Trends Immunol.*, **27**, 573–579.
7. Steinman, R.M. and Banchereau, J. (2007) Taking dendritic cells into medicine. *Nature*, **449**, 419–426.
8. Tacken, P.J., de Vries, I.J., Torensma, R., and Figdor, C.G. (2007) Dendritic-cell immunotherapy: from ex vivo loading to in vivo targeting. *Nat. Rev. Immunol.*, **7**, 790–802.
9. Fifis, T., Mottram, P., Bogdanoska, V., Hanley, J., and Plebanski, M. (2004) Short peptide sequences containing MHC class I and/or class II epitopes linked to nano-beads induce strong immunity and inhibition of growth of antigen-specific tumour challenge in mice. *Vaccine*, **23**, 258–266.
10. Elamanchili, P., Lutsiak, C.M., Hamdy, S., Diwan, M., and Samuel, J. (2007) "Pathogen-mimicking" nanoparticles for vaccine delivery to dendritic cells. *J. Immunother. (1997)*, **30**, 378–395.
11. Reddy, S.T., Rehor, A., Schmoekel, H.G., Hubbell, J.A., and Swartz, M.A. (2006) In vivo targeting of dendritic cells in lymph nodes with poly(propylene sulfide) nanoparticles. *J. Controlled Release*, **112**, 26–34.
12. Uto, T. *et al.* (2007) Targeting of antigen to dendritic cells with poly(gamma-glutamic acid) nanoparticles induces antigen-specific humoral and cellular immunity. *J. Immunol.*, **178**, 2979–2986.
13. Wang, X., Uto, T., Akagi, T., Akashi, M., and Baba, M. (2007) Induction of potent CD8+ T-cell responses by novel biodegradable nanoparticles carrying human immunodeficiency virus type 1 gp120. *J. Virol.*, **81**, 10009–10016.
14. Ochoa, J. *et al.* (2007) Protective immunity of biodegradable nanoparticle-based

vaccine against an experimental challenge with Salmonella Enteritidis in mice. *Vaccine*, **25**, 4410–4419.

15. Baumjohann, D. *et al.* (2006) In vivo magnetic resonance imaging of dendritic cell migration into the draining lymph nodes of mice. *Eur. J. Immunol.*, **36**, 2544–2555.
16. Lee, J.H. *et al.* (2007) Artificially engineered magnetic nanoparticles for ultra-sensitive molecular imaging. *Nat. Med.*, **13**, 95–99.
17. Dames, P. *et al.* (2007) Targeted delivery of magnetic aerosol droplets to the lung. *Nat. Nanotechnol.*, **2**, 495–499.
18. Denes, F.S. *et al.* (2003) Dense medium plasma synthesis of carbon/iron-based magnetic nanoparticles. *J. Appl. Phys.*, **94**, 3498–3508.

12
Applications of Pulsed Power and Plasmas to Biosystems and Living Organisms

Hidenori Akiyama, Sunao Katsuki, and Masahiro Akiyama

12.1
Introduction

Research on discharge plasmas produced by pulsed power and their action on biosystems [1, 2] has been advanced, and, with it, a new discipline called *bioelectrics* has been formed. Bioelectronics is a similar field. In this discipline, the characteristic of the living body is considered at the molecular level from the point of view of its electronics, and the research results have been applied to electronic devices.

Bioelectrics [3, 4] is a new multidisciplinary field which aims to make the action of pulsed power to biosystems clear and to apply those results to environment, food, and a medical care. Here, the pulsed power includes a short pulse electric field, burst sinusoidal electric field, nonequilibrium pulse plasma, and others, which are used for operating a biological cell, a biological tissue, and living things. The bioelectrics field has the possibility that can develop many applications such as taking out the undesired cell, treating bacteria and virus, physical treatment of food, fields of environment, a medical care, and the organogenesis control by using the action to a tissue stem cell and an embryonic stem cell.

The technology to produce the pulsed power is described here. The repetitive pulsed-power source using a magnetic pulse compression (MPC) system [5] is one of the powerful candidates that will be used in industrial applications. Then, characteristics of discharge plasmas in atmospheric gases and water are described. Finally, the action of the pulse power on biosystems and their applications are described.

12.2
Pulsed-Power Source Using Magnetic Pulse Compression System

Figure 12.1 shows the pulsed-power source driven by a solar cell. The electric energy from a solar cell is stored in the battery and is converted to 200 V of the

Industrial Plasma Technology. Edited by Yoshinobu Kawai, Hideo Ikegami, Noriyoshi Sato, Akihisa Matsuda, Kiichiro Uchino, Masayuki Kuzuya, and Akira Mizuno

ISBN: 978-3-527-32544-3

commercial frequency with inverter. After the high-voltage power source charges C_0 at 3.5 kV, a thyristor is operated. The current flows through SI_0 and a transformer. The SI_0 acts as a magnetic switch and protects the thyristor. The transformer PT_1 amplifies the voltage to 20 kV, and C_1 is charged to its voltage. The magnetic saturation at SI_1 occurs when the C_1 is charged completely. The SI_1 changes from an off state to an on state as an electric switch, since the SI_1 changes from high impedance to low impedance by the magnetic saturation. The pulse compression system using magnetic switch is called the *magnetic pulse compression system*. The output from the MPC is directly connected to a load. In the case of necessity of a rectangular pulse, a pulse forming network using the Blumlein line (BPFN) is added between the MPC and the load. The capacitors of BPFN are charged by the energy transfer from C_1 through SI_1. A rectangular pulse of voltage 20 kV is applied on the primary side of the transformer PT_2, just after the SI_2 changes from an off state to an on state by the charging of capacitors of BPFN. The rectangular pulse is furthermore amplified by the transformer PT_2. An important technique with this device is to use saturable inductors as electric switches. An inductor is an off state at the nonsaturation and becomes an on state at the saturation.

In the case of linear solenoidal coil, the inductance Lu of nonsaturation is

$$L_u \propto \frac{\mu_0 \mu_u A N^2}{l} \tag{12.1}$$

where A, N, l, μ_u are a cross section of the solenoid, number of turns, length and relative permeability, respectively. From

$$\frac{\partial (N\Phi)}{\partial t} = V \tag{12.2}$$

$$\frac{1}{N}\int_0^T V(t)dt = \int_A B \cdot dS = 2B_s A \tag{12.3}$$

where B_s is the saturation magnetic flux density, and T is time till the magnetic saturation. If it is assumed that V is a step function

$$T = \frac{2NB_s A}{V} \tag{12.4}$$

When N is selected small to make a small inductance after the saturation, a necessary cross section A of the magnetic substance is determined from the values of V and T. The impedance after the saturation of magnetic substance is

$$L_s \propto \frac{\mu_0 \mu_s A N^2}{l} \tag{12.5}$$

where μ_s is a relative permeability at the time of saturation. Because the inductance decreases from Equations 12.1–12.5, the magnetic switch changes from an off state to an on state.

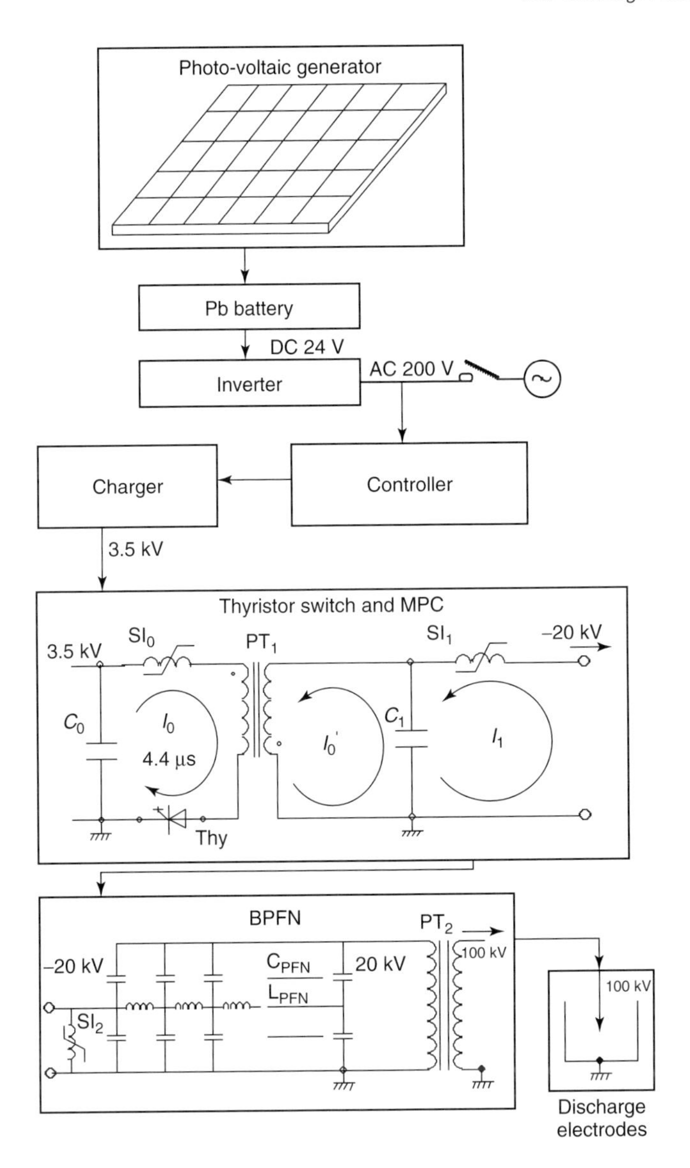

Figure 12.1 Pulsed-power generator driven by solar battery.

12.3 Discharge Plasmas by Pulsed Power

12.3.1 Large Volume Discharge Plasmas in Atmospheric Gases

Figure 12.2 shows the photograph of discharge plasmas when the pulsed power is applied between the wire and plane electrodes put in an atmospheric pressure air.

Figure 12.2 Photograph of discharge plasmas in atmospheric pressure air.

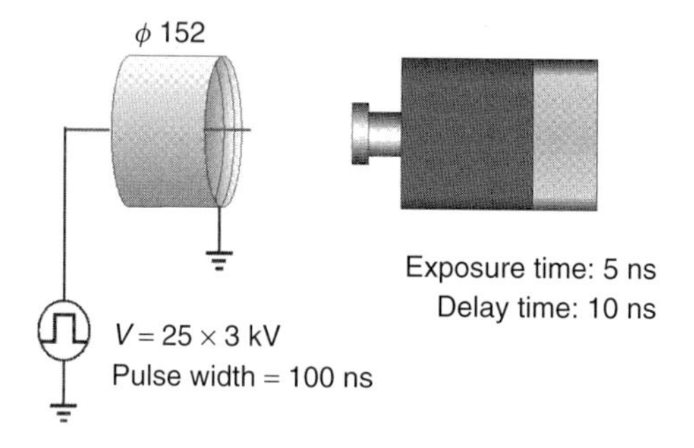

Figure 12.3 Experimental diagram to observe the development of discharge plasmas.

A diameter of wire electrode, the electrode separation, and the wire electrode length are 0.5 mm, 38 mm, and 4.5 m, respectively. The wire electrode length is limited by a laboratory size. Though it is imagined from Figure 12.2 that a uniform discharge plasma is produced, 10 discharge plasmas per 1 cm are observed in the vicinity of electrode by increasing the resolution of the photograph. When a risetime of voltage is shorter than a development time of discharge plasmas from initiation of discharges to arriving at the plane electrode, many discharge plasmas progress. On the other hand, when the rise time is longer than the development time, an arc discharge that seems to be lightning occurs.

The apparatus shown in Figure 12.3 is used to observe the development of the discharge plasmas [6]. A pulsed power of voltage, 75 kV, and pulse width, 100 ns, is applied between the wire and toric electrodes. The diameter and width of the toric electrode are 152 and 10 mm, respectively, and a diameter of the wire electrode is 0.5 mm. The exposure time of the high-speed camera is 5 ns, and a time between frames is 10 ns. Figure 12.4 shows the framing photographs of streamer discharges in atmospheric pressure air. Streamer discharges progress from the wire electrode put in the center. Propagation velocity of the streamer tip is 0.8–2.5 mm ns^{-1}. A tip of the streamer discharge strongly shines. Electrons are accelerated in that strong

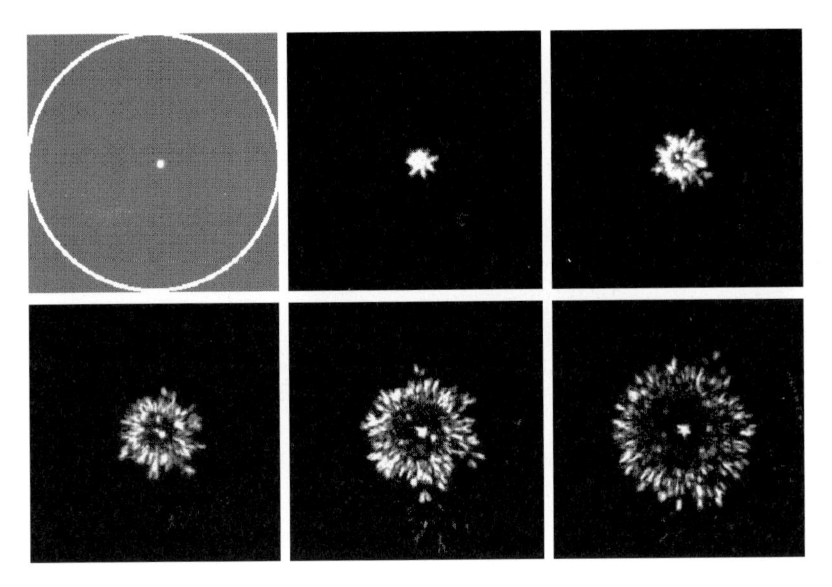

Figure 12.4 Framing photographs of streamer discharges in atmospheric pressure air.

electric field, and produce many radicals. According to the computer simulation, the electric field of the streamer tip is beyond 200 kV cm^{-1}. When a negative pulse voltage is applied on the wire electrode, the discharge behaves in a complicated way. Depending on a case, a streamer discharge begins at the toric electrode.

In industrial applications such as effluent gas treatment [7] and the ozone generation [8], it causes the increase of treatment efficiency and generation efficiency to reduce the pulse width of pulse power. In the case of 7 ns pulse width, the progress distance of streamer discharge, 5.6–17.5 mm, is calculated from 0.8–2.5 mm $ns^{-1} \times 7$ ns. This might suggest that the electrode separation should be decreased to fill the discharge plasma between electrodes. The pulsed power of 7 ns is applied between electrodes with 37 mm separation to check whether this assumption is true. Within the 7 ns pulse duration, the tip of streamer discharge reaches the toric electrode regardless of the polarity of the applied voltage. The propagation velocity of streamer discharge is 6–8 mm ns^{-1} which is several times larger than that in the case of 100 ns pulsed power.

A high-energy density phenomena such as high-electric field, high-speed electron generation, and productions of O, OH, O_3, and ultraviolet rays occurs near the tip of a streamer discharge, and has been used in effluent gas treatment, ozone generation, dioxin treatment, VOC treatment, sterilization, deodorization, light source, and active atom/molecule source.

12.3.2 Large Volume Discharge Plasmas in Water

A study began from interest whether it was possible to produce large volume discharge plasmas in water like the discharge plasmas in atmospheric pressure

Figure 12.5 Streamerlike discharge plasmas in water with 30 cm diameter.

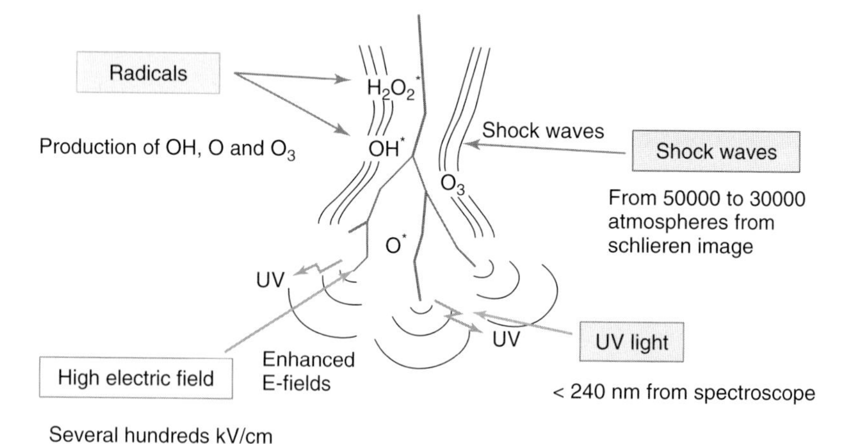

Figure 12.6 High-energy density phenomena caused by discharge plasmas in water.

gases [9]. The innumerable streamer kind discharges progress when pulsed power of about 100 kV is applied on a bar electrode put underwater as shown in Figure 12.5. The visible light from discharge plasmas in water is much stronger than that from the streamer discharges in atmospheric pressure air, and electric discharges progress vaporizing water.

Figure 12.6 shows the high-energy density phenomena to be generated while streamer discharge plasmas progress. The shock wave occurs by the process that varies from water to plasma, and its pressure reaches 30 000 atm. The electric field near the tip of streamer discharge plasmas reaches several hundred kilovolts per centimeter. The radicals and ultraviolet rays are generated, too. Such a high-energy density phenomena acts on bacteria in the water and sterilizes it. Also, a chemical compound is decomposed. The development speed of the streamer discharge plasmas in water is about 0.03 mm ns^{-1}, and is slower about two orders than the development speed in the atmospheric pressure air. If the range where the high-electric field of the tip acts on 10 times of the discharge plasma's diameter, 0.1 mm, is assumed, the high-electrical field stressing time at a point is only 30 ns.

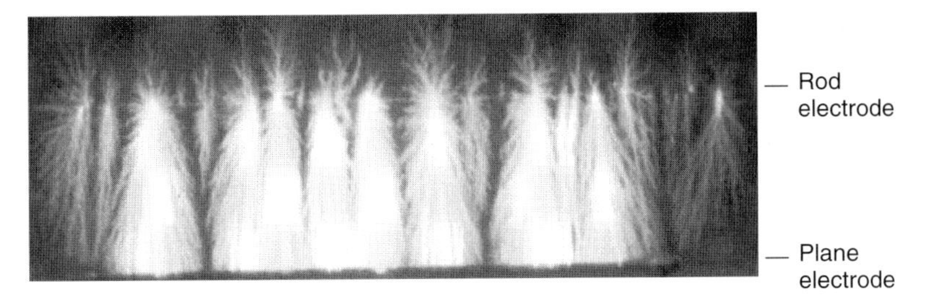

Figure 12.7 Streamer discharge plasmas in water between rod and plane electrodes.

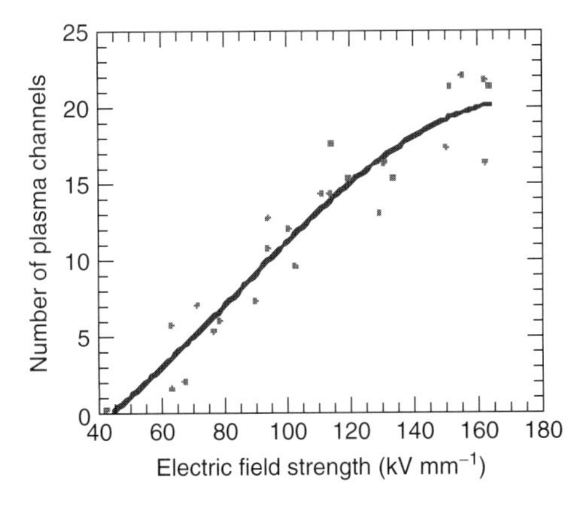

Figure 12.8 Dependence of the number of plasma channel on electric field intensity at the surface of rod electrode.

Figure 12.7 is a photograph of the discharge plasmas in water in the case of the wire to plane electrode. The length of plane electrode is 15 cm, electrode separation is about 5 cm, and a diameter of the wire electrode is 1 mm. Many plasma channels are generated in parallel. Figure 12.8 shows the dependence of the number of plasma channels near the wire electrode on the electric field strength of the wire electrode surface, which is calculated from the electrode separation and the diameter of wire electrode. The diameter and electrode separation are changed to change the electric field strength. The number of plasma channels is almost in proportion to the electrode surface field strength of wire electrode. The framing photographs of discharge plasmas in water, which is measured by an image converter camera, shows that the propagation speed is 0.03–0.04 mm ns^{-1}.

Figure 12.9 shows a shock wave produced by plasma of about 100μm. A schlieren method is used for measurement. A microplasma is produced at the needle electrode tip, and a spherical shock wave is generated. The dependence of the propagation velocity of the shock wave on the distance from the needle electrode is obtained from taking the streak photograph by setting a slit axially. From a relationship of Hugoniot

$$u_s = A + Bu_p \tag{12.6}$$

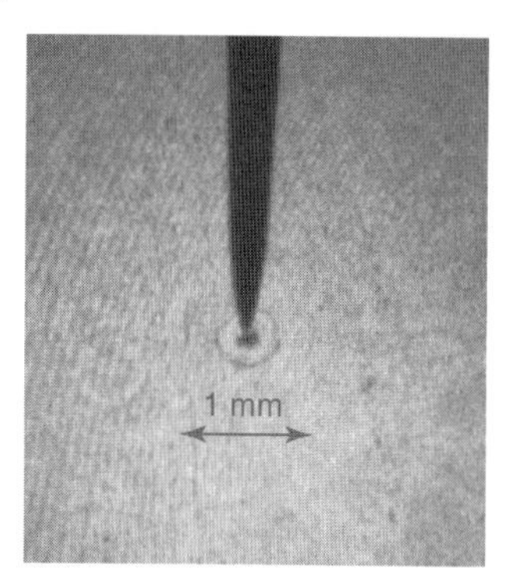

Figure 12.9 Shock wave produced by plasma of about 100 μm.

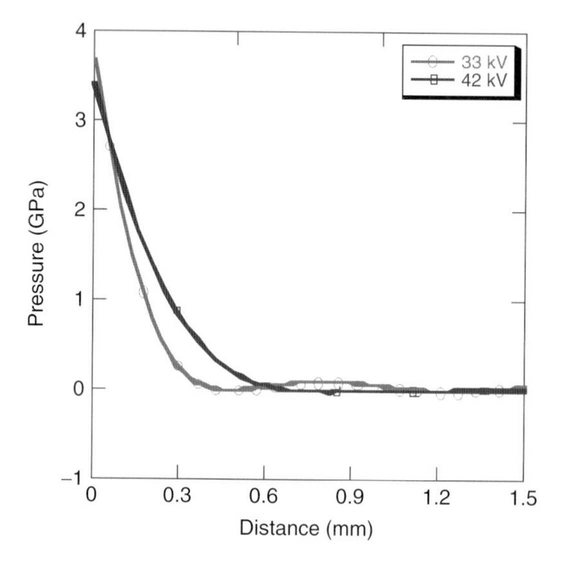

Figure 12.10 Dependence of shock wave pressure on distance from the needle tip.

$$P = \rho_0 u_s u_p \tag{12.7}$$

Here, P, ρ_0, u_s,u_p, A, and B are shock wave pressure, density of the water, shock wave speed, from acoustic velocity to particle speed in the water, and fixed numbers, respectively. If shock wave speed is measured experimentally, shock wave pressure is obtained from

$$P = \rho_0 u_s \frac{u_s - A}{B} \tag{12.8}$$

Figure 12.10 is the dependence of shock wave pressure on the distance from the needle electrode. The shock wave pressure is calculated from Equation 12.8 using a shock wave speed obtained experimentally. The parameter is applied voltage. The shock wave pressure is over 3 GPa. Figure 12.11 shows the framing photographs of a discharge light and the schlieren image. The discharge light disappears and

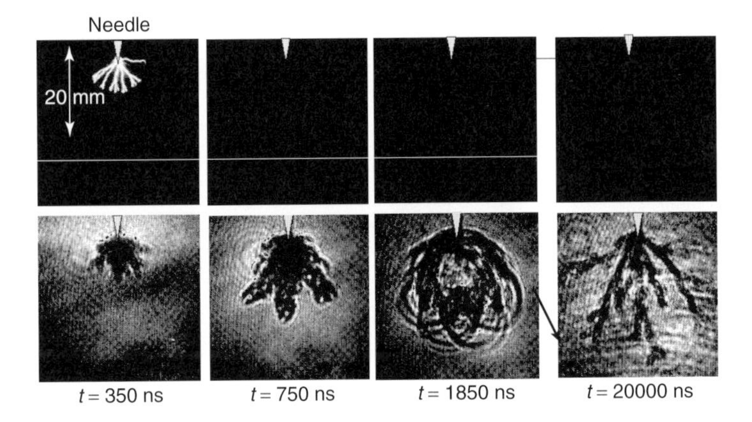

Figure 12.11 Upper figures: radiation from discharge plasmas in water, lower figures: Schlieren images of shock waves.

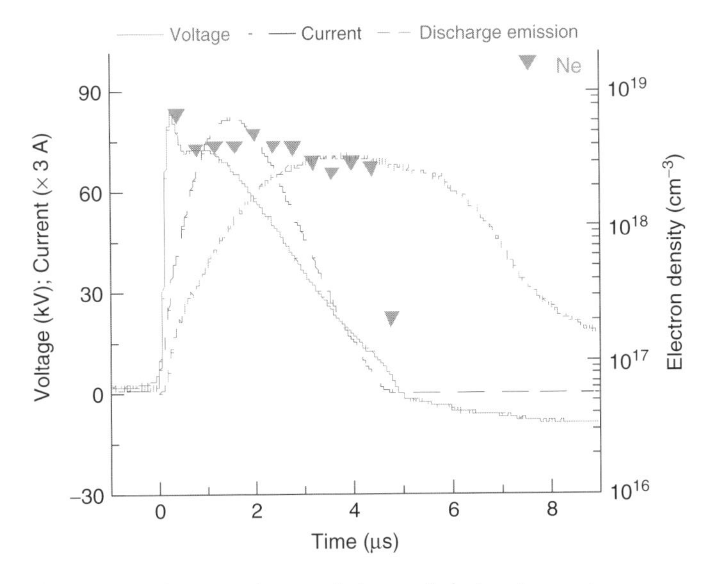

Figure 12.12 Current, voltage, radiation, and electron temperature.

the schlieren image continues after the pulsed power fades away. The bottom and right photograph is expanded in space, and its schlieren image continues for a long time.

Plasma temperature and density are measured near the bar electrode of Figure 12.5 [10]. The plasma temperature is measured from the emission line strength ratio measurement of copper in conjunction with the electrode material, and is about 1.3 eV. The electron density is measured from Stark spreading of Hα line, and is $3–7 \times 10^{18}\text{cm}^{-3}$. Figure 12.12 shows waveforms of current, voltage, emission light, and electron density.

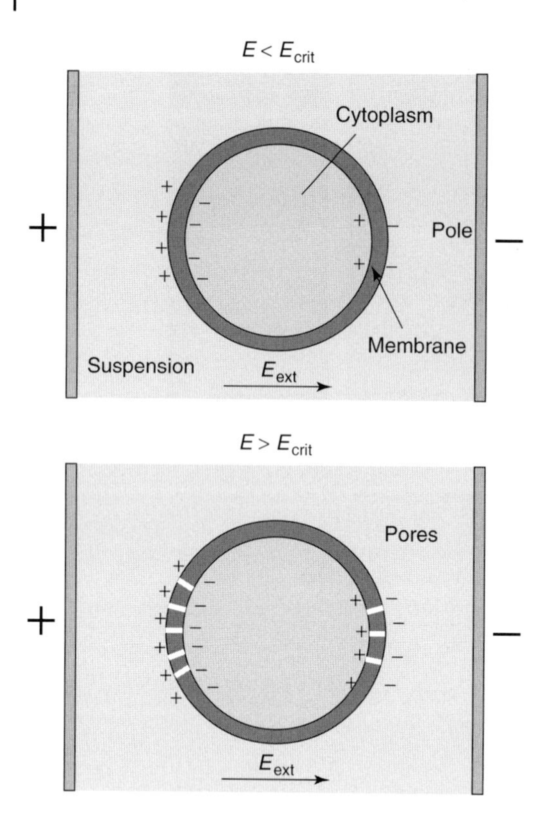

Figure 12.13 Electroporation of cell by pulsed electric field.

12.4
Action of Pulsed Power and Discharge Plasma to Biosystems

12.4.1
Action of Pulsed Power to Biosystems

Electroporation to leave the aperture in a cell membrane has been used widely. Figure 12.13 shows a state of the electroporation. Charge of the antipolarity is stored on both sides of the cell membrane, and apertures are opened by voltage of about 1 V. When an aperture is small, the cell is restored, but the cell cannot be restored and dies if the aperture is large. An organization in the cell goes out when an irreversible puncture occurs, and restoration becomes difficult. Figure 12.14 shows the equivalent circuit of a cell [11]. The C_s and R_s are the capacitance and resistance of the medium which a cell is in, C_m is the capacitance of the cell membrane, R_{c1} and R_{c2} are resistances of the cytoplasm, C_n is the capacitance of the nuclear membrane, and R_n is resistance in the cell nucleus. The electroporation occurs in the case of the pulsed voltage with low frequency because this pulsed voltage is applied to the cell membrane with a large impedance. In the case of the

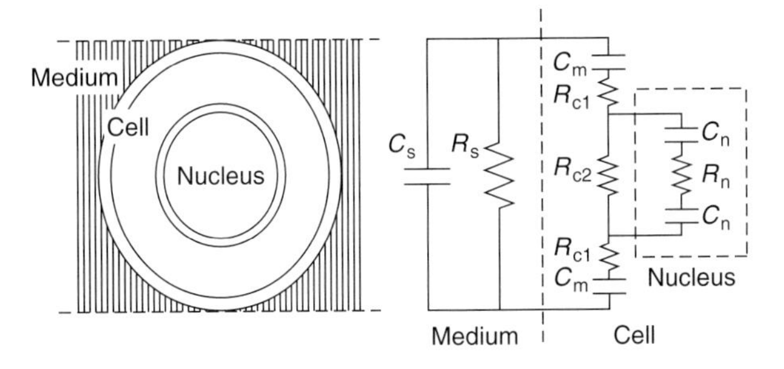

Figure 12.14 Equivalent circuit of a cell.

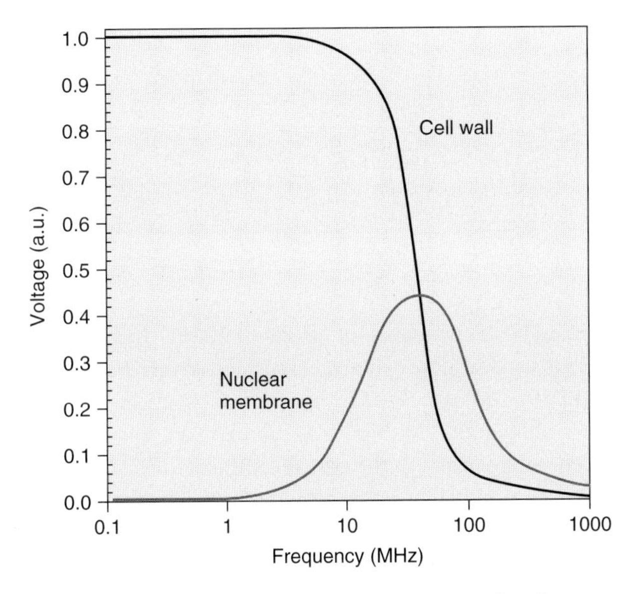

Figure 12.15 Dependence of applied voltages on cell wall and nuclear membrane on frequency.

pulsed voltage with high frequency over 10 MHz, voltage is applied not only to the cell membrane but also to the organism inside cell. Depending on the kind of the cell, an aperture of the nuclear membrane can be made without breaking a cell membrane from Figure 12.15.

There are many applications using action of pulsed power to biosystems, such as gene delivery, drug delivery, sterilization, and the organism control inside a cell without heating of cell. A pulsed power with extremely short pulse gives an influence to an organism in cells. This is a new study domain, and there is a possibility to treat cancer by causing apoptosis. A fundamental experiment has been done using an intense burst sinusoidal electric field (IBSEF) in substitution for the pulsed electric field. The IBSEF is applied in various cells placed between plane electrodes with a separation of about 100 μm, and the action of IBSEF on various cells has been examined [4]. The experiment goes under a microscope.

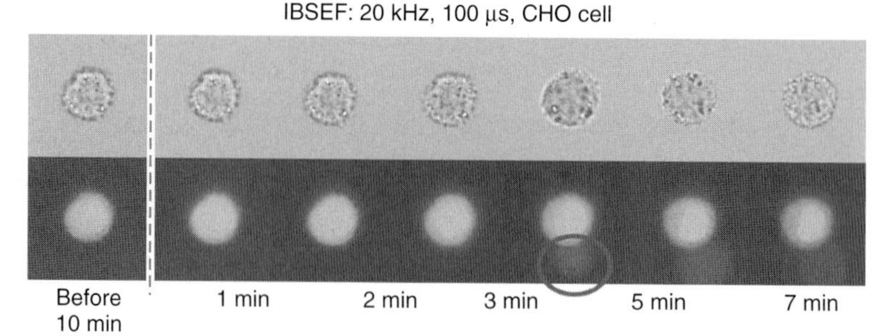

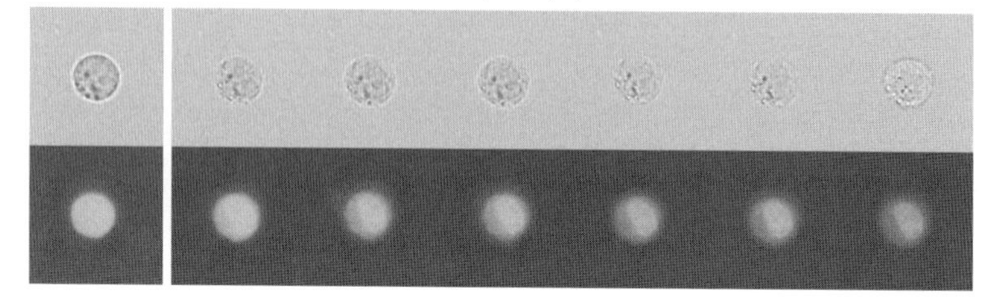

Figure 12.16 Frequency dependence of effects of burst sinusoidal electric field on cells.

Figure 12.16 shows the frequency dependence of effects of IBSEF on cells. The IBSEF of 20 kHz or 50 MHz is used, and the Chinese hamster ovary (CHO) is used as a cell. The duration of IBSEF is 100 μs. The Acridine orange (AO) is used to see DNA. The upper and lower photographs in Figure 12.16 are taken by an optical microscope and a fluorescent microscope. In the case of IBSEF of 20 kHz, there is an aperture in the cell membrane, and DNA leaks forth as shown by a circle. In the case of IBSEF with 50 MHz, there is no leak of the DNA to the outside of the cell, and the aperture is not open in a cell membrane. The fluorescence intensity in the cell decreases in the whole, and fragmentation of the DNA is caused. These results show that the IBSEF can control the organism inside cell.

12.4.2 Cleaning of Lakes and Dams

As one of the applications of the large volume discharge plasma in water, cleaning of lakes and dams is described [12]. Polluted lakes and dams are increasing by agriculture wastewater or domestic wastewater. A salt water red tide seems to have a reddish tinge, and Microcystis is a kind of algae consisting of a green powder, which is called *water bloom*. These are caused by heterology of the plankton, which occurs by the eutrophication of water caused by higher levels of nitrogen or phosphorus. This produces a pungent smell. Depending on the kind of plankton, this could lead to the death of many fishes and pose a serious problem.

Figure 12.17 shows the photograph of algae outbreak in a dam. The surface of the water appears like green paint. The increase in Microcystis is the cause of green

Figure 12.17 Algae in dam.

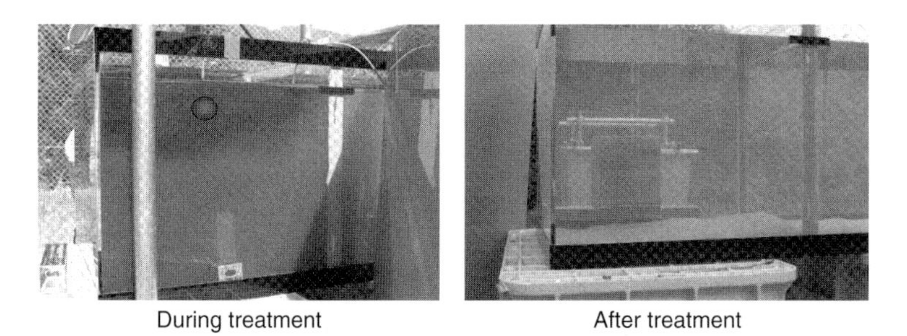

Figure 12.18 Effect of streamer discharge plasmas in water on algae.

powder that is observed on microscopic examination. Microcystis has a gas vesicle in a cell and floats within about 10 cm from the surface of water forming colony. The water of the dam including Microcystis was poured into a container of the 50 cm cube, and the discharge plasma in water is produced as shown in Figure 12.5. The bar electrode is moved to draw a ring. After treatment, Microcystis sinks at the bottom of the container, and water becomes transparent as shown in Figure 12.18. Figure 12.19 shows the states before and after the treatment. Microcystis floats by gas vesicle before treatment, but sinks after treatment. Microcystis sinks by extinction of gas vesicle which corresponds to the black spot in discharge plasma.

12.5 Summary

Bioelectrics is a new multidisciplinary field which aims to study the action of pulsed power on biosystems and to apply those results to various fields such as environment, food, and medical care. The pulsed power includes a short pulse

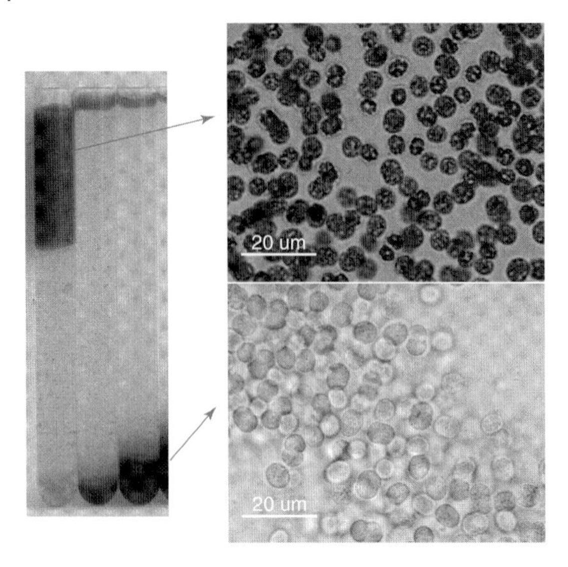

Figure 12.19 Microscopic photographs of algae before and after applying pulsed power.

electric field, IBSEF, nonequilibrium pulse plasma, shock waves, and others, which are used for operating biological cells, biological tissues, and living things.

In the near future, many fields of application such as water treatment, cancer treatment, sterilization of food, and the differentiation control of tissue stem cells and embryonic stem cells will be considered.

References

1. Akiyama, H., Sakai, S., Sakugawa, T., and Namihira, T. (2007) Environmental applications of repetitive pulsed power. *IEEE Trans. Dielectr. Electr. Insul.*, **14** (4), 825–833.
2. Akiyama, H., Sakugawa, T., Namihira, T., Takaki, K., Namihira, Y., and Shimomura, N. (2007) Industrial applications of pulsed power technology *IEEE Trans. Dielectr. Electr. Insul*, **14** (5), 1051–1054.
3. Schoenbach, K., Katsuki, S., Stark, R., Buescher, E.S., and Beebe, S.J. (2002) Bioelectrics – new applications for pulsed power technology. *IEEE Trans. Plasma Sci.*, **30** (1), 293–300.
4. Katsuki, S., Nomura, N., Koga, H., Akiyama, H., Uchida, I., and Abe, S.-I. (2007) Biological effects of narrow band pulsed electric fields. *IEEE Trans. Dielectr. Electr. Insul.*, **14** (3), 663–668.
5. Choi, J., Namihira, T., Sakugawa, T., Katsuki, S., and Akiyama, H. (2007) Simulation of 3-staged MPC using custom characteristics of magnetic cores. *IEEE Trans. Dielectr. Electr. Insul.*, **14** (4), 1025–1032.
6. Namihira, T., Wang, D., Katsuki, S., Hackam, R., and Akiyama, H. (2003) Propagation velocity of pulsed streamer discharges in atmospheric air. *IEEE Trans. Plasma Sci.*, **31** (5), 1091–1094.
7. Hackam, R. and Akiyama, H. (2000) Air pollution control by electrical discharges. *IEEE Trans. Dielectr. Electr. Insul.*, **7** (5), 654–683.
8. Samaranayake, W.J.M., Miyahara, Y., Namihira, T., Katsuki, S., Sakugawa, T., Hackam, R., and Akiyama, H. (2001) Pulsed power production of ozone using nonthermal gas discharges. *IEEE Electr. Insul. Mag.*, **17** (4), 17–25.

9. Akiyama, H. (2000) Streamer discharges in liquids and their applications. *IEEE Trans. Dielectr. Electr. Insul.*, **7** (5), 646–653.
10. Namihira, T., Sakai, S., Yamaguchi, T., Yamamoto, K., Yamada, C., Kiyan, T., Sakugawa, T., Katsuki, S., and Akiyama, H. (2007) Electron temperature and electron density of underwater pulsed discharge plasma produced by solid-state pulsed-power generator. *IEEE Trans. Plasma Sci.*, **35** (3), 614–618.
11. Deng, J., Schoenbach, K.H., Buescher, E.S., Hair, P.S., Fox, P.M., and Beebe, S.J. (2003) The effects of intense sub-microsecond electrical pulses on cells. *Biophys. J.*, **84**, 2709–2714.
12. Li, Z., Sakai, S., Yamada, C., Wang, D., Chung, S., Lin, X., Namihira, T., Katsuki, S., and Akiyama, H. (2006) The effects of pulsed streamerlike discharge on cyanobacteria cells. *IEEE Trans. Plasma Sci.*, **34** (5), 1719–1724.

13
Applications of Plasma Polymerization in Biomaterials

David A. Steele and Robert D. Short

13.1
Introduction

In the last several decades, the health sector has seen a constant increase in the use of medical devices. These devices, from a "simple" disposable contact lens to multicomponent heart assist devices, are used to support or replace failing organs to the benefit of millions of patients. Today biomaterials and their use in the US healthcare market has an estimated worth in excess of $1.4 trillion dollars – a figure typically multiplied by a factor of between 2 and 3 when considering the global market [1]. The single factor common to all these devices is the biomaterial-tissue interface and, while initial material selection is made upon one or more bulk properties, it is all too often failure at this interface that is key to device performance.

One such example can be seen in cardiovascular surgery and the use of mechanical prosthetic valves. In developed nations, poor diet and lifestyle choices have resulted in a large increase in the incidence of cardiovascular disease (CVD) with 80 000 replacement valves implanted per annum in the United States. Whereas this operation can and does improve a patient's life expectancy – a 50% chance of survival after 3 years without replacement improving to a 70% chance of survival after 10 years with – issues with efficacy still exist with as many as 60% of these patients undergoing surgery suffering serious complications within 10 years [1]. Here, the dominant cause of device failure, despite anticoagulation treatment with warfarin derivatives, is valve occlusion resultant from a buildup of thromogenic deposits on the valve surface. The outcome serves to illustrate an aspect of the immunological response triggered by implantation of a biomaterial, medical device, or prosthesis. This healing response incorporates a cascade of reactions including initial protein adsorption, matrix formation, inflammation, and fibrous encapsulation.

In the late 1960s and early 1970s, a popular opinion developed that surface modification would serve to improve a materials "biocompatibility." This term was originally defined by Williams as:

> Biocompatibility is the ability of a material to perform with an appropriate host response in a specific application

Industrial Plasma Technology. Edited by Yoshinobu Kawai, Hideo Ikegami, Noriyoshi Sato, Akihisa Matsuda, Kiichiro Uchino, Masayuki Kuzuya, and Akira Mizuno

ISBN: 978-3-527-32544-3

and recognized the fact that the response of a material and the appropriateness of such a response was dependent upon the situation in which it was used [2]. However, as the field matured and the variety of applications expanded, the limitations of a single definition applied to such a large number of situations were revealed such that, at a workshop held in Seattle in 2003, a redefinment of the term *biocompatibility* based on the application rather than on the device was discussed. As a result three definitions were proposed, each specific to the intended application of the biomaterial [3]. With respect to the biocompatibility of long term medical devices:

> The biocompatibility of a long-term implantable medical device refers to the ability of the device to perform its intended function, with the desired degree of incorporation in the host, without eliciting any undesirable local or systemic effects in that host.

for short term devices:

> The biocompatibility of a medical device that is intentionally placed within the cardiovascular system for transient diagnostic or therapeutic purposes refers to the ability of the device to carry out its intended function within flowing blood, with minimal interaction between device and blood that adversely affects device performance, and without inducing uncontrolled activation of cellular or plasma protein cascades.

and, for tissue engineering products:

> The biocompatibility of a scaffold or matrix for a tissue-engineering product refers to the ability to perform as a substrate that will support the appropriate cellular activity, including the facilitation of molecular and mechanical signalling systems, in order to optimise tissue regeneration, without eliciting any undesirable effects in those cells, or inducing any undesirable local or systemic responses in the eventual host.

In 1971 Robert Baier presented a study on the role of surface energy in thrombogenesis at a symposium to address the issues of blood contacting biomaterials [4]. In reviewing the research into cardiovascular medical devices he sought to correlate a material property, the surface energy, with the degree of thrombogenicity detected after 30–50 days implantation. These observations led Baier to propose that the single most important criteria for good cardiovascular prosthesis was surface energy. He further tentatively proposed that this "magic parameter" of surface tension could be used as a guide with a hypothetical zone of biocompatibility being in the region of 22 dyn cm^{-1} (Figure 13.1). So, it was proposed that biomaterials, with surface energies in this region, were to be considered biocompatible and further, that those materials with surface energies greater than this could be considered as candidates for situations requiring good bioadhesion. However, it must be noted that Baier further qualified his observations stating that while a low-critical surface

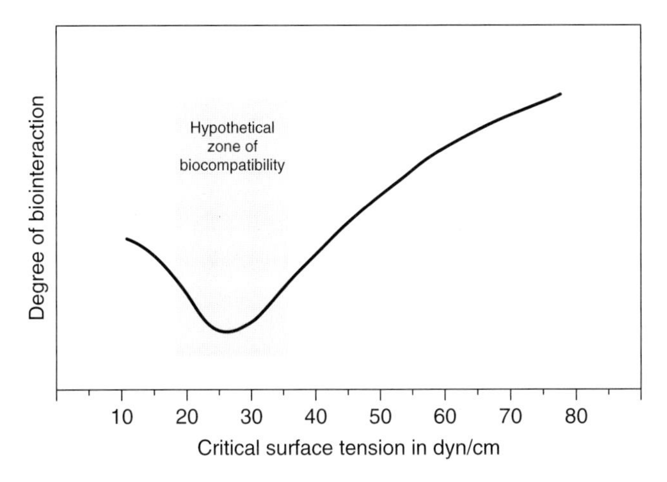

Figure 13.1 A measure of material surface energy in relation to biointeraction identifying a hypothetical critical surface energy of around 22 dyn cm^{-1} for "good" biocompatibility. (Reproduced with permission from page 767 Ratner, B. D., Hoffman, A. S. *et al.* (2004) *Biomaterials Science: an Introduction to Materials in Medicine*. Elsevier, Amsterdam. Copyright Elsevier 2004.)

energy was a useful material characteristic this was not the only criterion that could be addressed when designing improved biomaterials. He further recognized, particularly in relation to blood contacting devices, that proteins deposited onto the surface were the potential controlling factors and questioned how the material surface energy might be influencing this protein deposition.

The article by Baier illustrates the school of thought that developed during this time when the belief arose that many of these observed limitations of implantable biomaterials could be overcome by the surface modification of a commodity engineering material; one such method was the direct modification of the surface by plasma treatment with an inert gas or alternatively, the plasma polymer deposition of new surface coating. This concept was fixated upon to the exclusion of Baier's qualifying observations such that many groups studied this correlation in isolation of other factors. In doing so a hypothesis developed that a particular biological response could be correlated with a specific physical or chemical surface characteristic. An example of this was the study of cellular attachment to a plastic and the correlation with surface energy as determined by measurement of the surface-water contact angle. This *in vitro* observation, that the degree of cellular attachment increases with increasing water contact angle, has been observed by many [5–9]. For example, Neumann *et al.* used whole blood and a static platelet adhesion test to study polymer surfaces. The surfaces demonstrated a range of surface energies and platelet adhesion was observed to increase with increasing surface energy. The same polymers were also precoated with plasma proteins and the observations made were that more platelets adhered to fibrinogen coated surfaces than those coated with albumin; and that the high-energy polymer surfaces, once coated with protein, showed more platelet adhesion than the low-energy polymer surfaces coated with protein. It is worth reinforcing that Neumann and others noted that the simple relationship between surface energy and cellular attachment was *unsustained* in the presence of protein. Given this it is disappointing to note the many *in vitro* studies, where groups tailored the surface wettability of a polymer employing a wide variety of methods and, having determined the cellular response, declared the surface "biocompatible" [10–12].

Apart from the simplicity of this argument, in adopting this approach what has often been overlooked is that in changing one physico-chemical feature of a surface, a host of other surface characteristics may also have been changed, and perhaps in a nonsystematic manner. For example, plasma modification of surfaces with air, oxygen, or water plasma has often been used to changed surface hydrophilicity [13–18]. Yet, plasma modification also has a marked effect on surface functional group chemistry, surface molecular weight, and topography which, in combination, give rise to the change in, for example, water contact angle [19–21]. Whereas to a first approximation wettability may correlate with cell attachment, the changes in surface chemistry/topography may produce secondary, and undesirable, effects.

It is beyond the scope of this chapter to consider all the examples where plasma surface modification/plasma deposition have been used to improve a surfaces biocompatibility, and therefore we have chosen just three examples as illustrative of what has happened over the preceding three decades. The first two examples seek to reason why apparently good results in the laboratory have failed to translate into a successful clinical outcome while the third example illustrates how such success can be achieved.

13.2
Example 1: Improving Surfaces for Blood Biocompatibility

As previously mentioned, CVD is a growing health and socioeconomic burden. Of the 17.5 million people who died of CVDs in 2005, over 80% of these deaths occurred in low and middle income countries [22]. Incidences of peripheral vascular diseases (PVDs) that are typically present in low-flow, small-diameter (below 6 mm) vasculature present a particular challenge [23, 24]. The biomaterials of choice, selected primarily for their mechanical properties, are currently Dacron (polyethylene terephtalate fiber) for the larger vessels and expanded polytetrafluoroethylene (ePTFE) for the smaller, although polyurethane and polyurethane copolymers continue to attract significant interest. However, as with all synthetic substrates, thrombogenicity remains an issue and a great deal of research continues to improve or correlate blood response with surface properties. Of the strategies employed, researchers have sought to develop new synthetic graft materials that are less reactive to plasma proteins and blood cells [25].

In recent years, researchers have sought to elicit the reasons why such encouraging results *in vitro* and even in small animal studies should fail clinically and, in a more sophisticated study in 2001, Sefton *et al.* asked the question: does surface chemistry affect thromogenicity of surface modified polymers? [26] In this study, the group sought to measure the performance of several polymers typically used as cardiovascular biomaterials, both untreated and surface modified, with respect to their hemocompatability using a number of physical and clinical tests. The test regime employed sought to re-evaluate the accepted tests for hemocompatability in light of newly developed ELISA assays. With such a comprehensive examination, it was felt that a number of correlations between material and performance might be drawn.

In this study, commercially available tubing of polyethylene (PE), Pellethane® (a thermoplastic polyurethane, PU), latex, nylon and Silastic® (a silicone elastomer) were used. Additionally, the PE and PEU samples were subjected to a range of surface modifications: RF plasma modification/coating with water, ammonia, carbon tetraflouride, and fluorocarbon enrichment in addition to treatment with PE imine and heparinization to give a total of 12 samples. The inner surfaces were characterized by XPS and contact angle and, following human serum and blood exposure, subjected to a range of assays designed to assess material thrombogenicity. The assays used were blood coagulation assessed by measurement of the partial thromboplastin time (PTT); complement activation determined by adsorbed C3a and sC5b-9 activation fragments; platelet and leukocyte activation analyzed by flow cytometry; platelet/leukocyte aggregation measured by CD61 expression and SEM analysis to quantitatively determine activation of platelets, leukocytes, and thrombi.

XPS results confirmed a high concentration of fluorine in the samples having plasma fluorination and small concentrations of nitrogen and trace levels of silicon in the as-received PEU. The PU plasma modified samples, including the fluorinated sample, showed a range of oxygen and nitrogen concentrations present as several carbon–oxygen and carbon–nitrogen functionalities. Contact angle results, as expected, demonstrated that the materials ranged from strongly hydrophobic (in the case of PE, PU, and fluorinated PU) through hydrophilic following surface modification to strongly hydrophilic in the case of nylon. However, the results of the assays for thrombogenicity (summarized for selected materials in Table 13.1) were less straightforward. In the extremes PE, without a positive response, demonstrated good thrombogenicity, while PE–CF_4, with a strongly

Table 13.1 A summary of material thrombogenicity assays with (+) indicating an increase in thrombogenicity.

Materials	Platelet deposits	Coagulation (PTT)	Complement activation	Microparticle formation	Thrombi	Platelet activation
PE	–	–	–	–	–	–
Nylon	±	–	–	–	–	–
PU–F	+	–	–	–	–	–
PU–H_2O	–	–	±	–	–	–
PE–CF_4	+	+	±	+	+	+

From Sefton *et al.* [26].

positive response in most assays, demonstrated poor thrombogenicity. For the remaining samples, particularly those that were plasma modified, there was little if any correlation between surface chemistry, surface energy, and the assays such that nylon, PU–F, and PU–H_2O, despite their obviously differing surface properties demonstrated no difference in terms of thrombogenicity.

The conclusions drawn by Sefton and colleagues demonstrated that while some assays, such as microparticle formation, could distinguish material differences the flow cytometry results were less sensitive and so a degree of variance regarding material performance was noted. Certainly, significant differences in surface chemistry elucidated differing levels of biological surface interaction but the more subtle surface differences remained concealed. With the exception of the heparinized surface however, surface modification/coating, regardless of both chemistry and surface energy, failed to improve thrombogenicity *in vitro*. The authors concluded that further study *in vivo* was required to identify which, if any, of the assays employed *in vitro* would be indicative of a materials clinical performance.

13.3
Example 2: Foreign Body Response

Any implanted material will invoke a reaction from the host, termed the *foreign body response* – a cascade of specific events. Upon implantation, the foreign body response immediately begins with protein adsorption followed by formation of a provisional matrix, macrophange, and leukocyte adhesion. The end result of this response is usually seen as a fibrous encapsulation of the implant. The reduction and/or control of this response is a challenge faced in the development of any biomaterial and in this respect one long-held hypothesis has been that to prevent a foreign body response the implant could be coated with a surface that resists cell adhesion proteins – in fact all proteins. The criteria are that truly non-fouling surfaces that completely inhibit protein adsorption are required to control (prevent) biological interactions with biomaterials [27]. Indeed, *in vitro* studies of polystyrene surfaces preadsorbed with normal, afibrinogenemic, and fibrinogen replenished afibrinogenemic blood plasmas have shown that surfaces adsorbing as little as $<10\,ng\,cm^{-2}$ of fibrinogen still promote platelet adhesion as determined by the lactate dehydrogenase method [28]. The Holy Grail has therefore been the production by plasma processes of surfaces that adsorb $<10\,ng\,cm^{-2}$ of protein (typically fibrinogen), or more preferably none at all. *In vitro* studies with monocytes have shown that when this criteria is met monocyte adhesion is significantly reduced [29].

To this end a number of studies have shown, *in vitro*, that polyethylene oxide (PEO) surfaces significantly reduce protein adsorption [27, 30, 31]. Using suitable glycol monomers, PEO-like plasma polymer coatings can be prepared using low-power RF plasma. Coatings have been prepared from triglyme [32, 33], tetraglyme [29, 34, 35], and crown ethers [36], and with crown ethers and a pulsed power source [37].

In 2002, a significant paper of M Shen *et al.* further explored this relationship between protein adsorption and fibrous encapsulation after four weeks subcutaneous implantation in mice [34]. PEO-like plasma polymerized surfaces were produced from tetraglyme onto fluorinated ethylene propylene (FEP) copolymer surfaces. *In vitro,* these plasma polymer surfaces resisted protein adhesion after 2 hours contact with human whole blood; however, after one day or four weeks subcutaneous implantation in mice, XPS analysis (Figure 13.2) showed significant changes in their C1s core lines with the presence of 6.3% nitrogen on the plasma polymerized surfaces indicating the adsorption of a substantial amount of proteinaceous material.

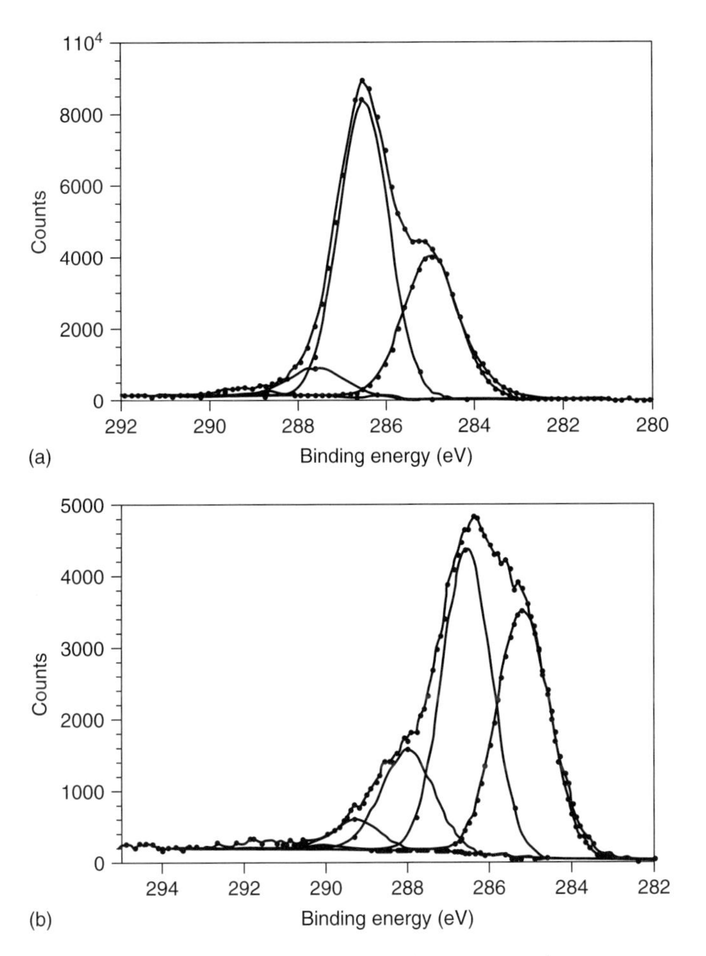

Figure 13.2 High-resolution C1s spectra of plasma polymerized tetraglyme (a) after 2 hours contact with human whole blood and (b) after one day subcutaneous implantation. (Reproduced with permission from pages 380–381, Shen, M., Martinson, L., Wagner, M.S., Castner, D.G., Ratner, B.D., and Horbett, T.A. (2002) *J. Biomater. Sci. Polymer Edn*, **13** (4), 367. Permission kindly granted by Koninklijke Brill NV.)

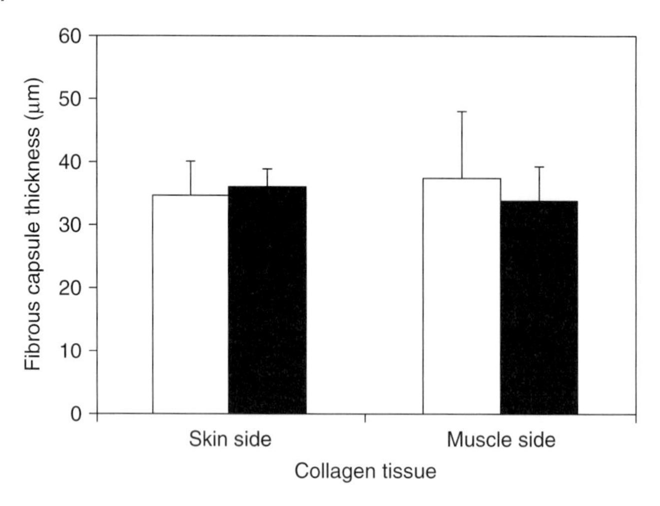

Figure 13.3 The thickness of fibrous tissue capsules around four-week implanted FEP (white bars) and polymerized tetraglyme (black bars) was measured with light microscopy. The capsule thickness from the skin and muscle side was measured. The data represents mean ± SD; $n = 4$ FEP or polymerized tetraglyme samples that were implanted in four mice. No differences in capsule thickness were found. (Reproduced with permission from page 374, Shen, M., Martinson, L., Wagner, M.S., Castner, D.G., Ratner, B.D., and Horbett, T.A. (2002) *J. Biomater. Sci. Polymer Edn*, **13** (4), 367. Permission kindly granted by Koninklijke Brill NV.)

Following implantation of the plasma polymer tetraglyme surfaces after four weeks, a comparable fibrous encapsulation to naked FEP was seen (Figure 13.3). Thus, while a significant reduction of fibrinogen adsorption (<10 ng/cm^2) resulted in a significantly reduced level of platelet adhesion *in vitro*, it did not reduce polymorphonuclear leukocytes (PMNs) or macrophage adhesion *in vivo*. This again identifies a limitation of the *in vitro* material characterization and testing methodologies which fail, *in vivo*, to respond in the predicted manner. In this instance this was attributed to loosely bound proteins, present at almost (initially) negligible amounts. This study serves to highlight that plasma polymer tetraglyme surfaces that have "excellent" non-fouling performance *in vitro* may be fabricated but such surfaces fail to perform *in vivo*. Whereas, one day implantation studies show that these surfaces are not truly non-fouling, the authors also conclude that;

> such surfaces may not be appropriate in vivo unless further understanding leads to their improvement

However, they quite rightly recognize that their *in vitro* stability, ease of application, and non-adhesive nature make them viable candidates for use in non-implantable applications such as diagnostics, biosensors, MEMs, and array technologies; and indeed such applications are being realized [38–41].

The above two examples may paint a very bleak prospect for plasma modification and plasma deposition in the field of biomaterial science. However, this certainly is not the case. Plasma technology has already been successfully integrated into products that are well established on the market.

13.4
Example 3: Extended Wear Contact Lenses

For example, in the ophthalmic area there now exists a new generation of high permeability silicone hydrogel contact lenses that can transmit previously unknown levels of oxygen to the extent that the lenses can be worn for extended periods of time (up to 30 days) without removal. The requirements for an ocular biomaterial are that the material is comfortable to wear, maintains a tear film, and is permeable to oxygen and ions. With conventional hydrogel materials oxygen, required by the cornea to maintain a normal level of metabolism, is transported via the water. While the oxygen permeability of these is sufficient for use during the day, at night, when the wearer is asleep, the eye receives insufficient oxygen to maintain the required level of metabolic activity. Silicone elastomer lenses have a very much higher oxygen permeability; however, there are issues with their use. Fluid is unable to flow through the material and the surfaces are extremely hydrophobic, which result in considerable lipid fouling. In silicone hydrogel materials, silicon-based monomers are combined with conventional hydrogel monomers [42]. The silicone component facilitates high oxygen permeability while the hydrogel component provides fluid transport. The remaining issue of wettability is addressed by plasma treatment and/or coating which, in addition to patient comfort, reduces the lipid-based biofilm formation.

Of the few brands of silicone hydrogel currently available only two have FDA approval; Night & Day from Ciba Vision which is a proprietary plasma polymer coating and PureVision from Bausch & Lomb which is oxygen plasma modified [43]. There are obviously a number of advantages that biomaterials as contact lenses have over other, implantable devices. Here, the material performance with respect to biocompatibility is short-term over a period of a few days unlike implantables where, in many instances, the criteria needs to be met for many years, the conditions regarding Williams' definition of biocompatibility for short-term devices are met.

Additionally the tear film, the biological matrix the lens contacts, is less complex than that of say blood and unlike many implantables the performance is readily monitored; when this falls below the level deemed acceptable the lens can be removed for cleaning or be replaced. In fact, the simpler the interfacial performance demanded the more appropriate plasma treatment and/or plasma deposition become for modifying the non-ideal surface properties of an engineering thermoplastic. Another example being in the product Myskin™ where a plasma polymer layer is used as support for the culture *in vitro* of a layer of keratinocyte cells prior to clinical delivery of these cells to a wound bed [44].

But this is not the limitation of plasma modification/deposition in the context of the field. Plasma surfaces may be used as the platform on which to engineer a more biologically "acceptable" surface, by the immobilization of biomolecules. This is elegantly demonstrated in the following example [45].

13.5
Example 4: Platform for Immobilizing a Biomolecule

Griesser and colleagues, with an interest in protein adsorption and subsequent cell colonization, sought to explore the preparation of non-biofouling coatings. They noted that contradictions existed in the literature regarding the performance of such coatings, prepared from such materials as PEO and polysaccharides [46, 47]. In an attempt to elucidate the importance of molecular structure and substrate attachment of such coatings the group chose to use carboxymethyl dextrans (CMDs) – a derivative of a naturally occurring biopolymer.

Two plasma polymer-based strategies were employed to immobilize the CMDs onto FEP tape (Figure 13.4): the first, direct covalent attachment to an RF generated plasma polymer layer of heptylamine (HApp) and the second, a plasma polymer layer of acetylaldehyde (AApp) onto which a polyamide (PLys) spacer was covalently attached. Dextrans with a range of molecular weights (70 kDa, 500 kDa, and 5–40 MDa) were carboxymethyl functionalized to varying degrees of substitution.

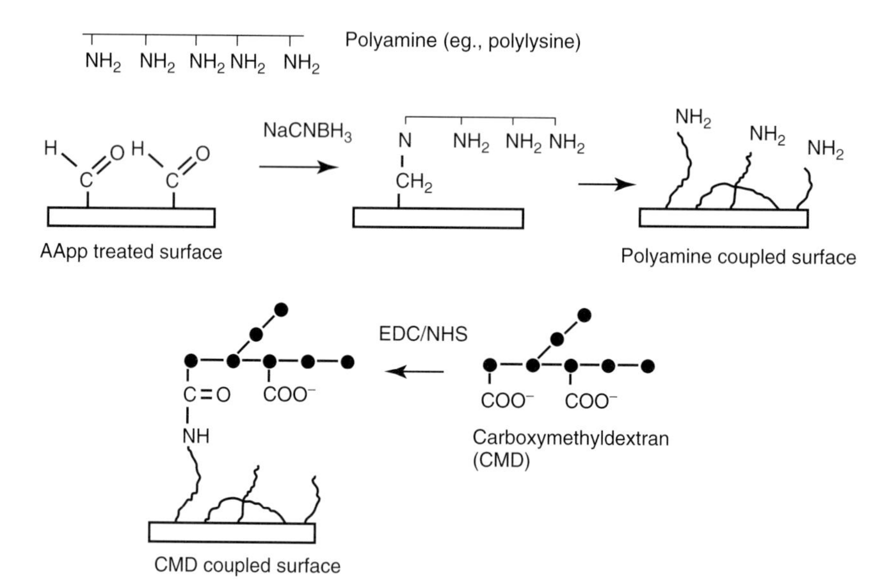

Figure 13.4 Schematic diagrams of the methods used for the covalent interfacial immobilization of carboxymethyl substituted dextrans via an acetalaldehyde plasma polymer coating and a hydrated polyamine spacer layer. (Reproduced with permission from page 223 of McLean, K.M., Johnson, G., Chatelier, R.C., Beumer, G.J., Steele, J.G., Griesser, H.J. (2000) *Colloids and Surfaces B: Biointerfaces*, **18**, 221. Permission kindly granted by Elsevier. Copyright Elsevier 2000.)

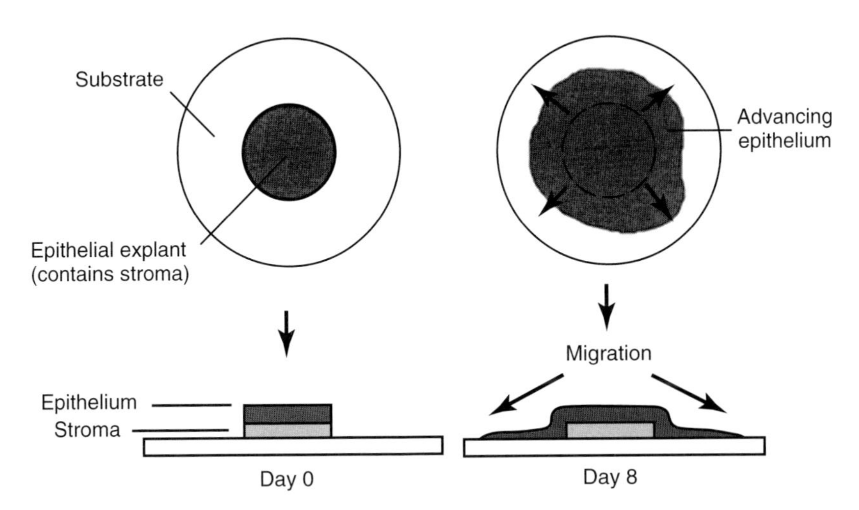

Figure 13.5 Schematic diagram of the corneal tissue outgrowth assay. (Reproduced with permission from page 226 of McLean, K.M., Johnson, G., Chatelier, R.C., Beumer, G.J., Steele, J.G., and Griesser, H.J. (2000) *Colloids and Surfaces B: Biointerfaces*, **18**, 221. Permission kindly granted by Elsevier. Copyright Elsevier 2000.)

For cell attachment and spreading assays, bovine corneal epithelial cells (BCEps) were employed. These cells were selected for their ability, unlike many cell lines, to attach and proliferate on untreated polystyrene. The cells were seeded and cultured in complete medium for a period of seven days at which time the ability of the coatings to support the attachment and migration of the BCEp cells was evaluated (Figure 13.5).

The results of the cell migration assays were revealing. Surfaces prepared where the CMDs were immobilized directly to the heptylamine plasma polymer surface demonstrated reduced but not complete inhibition to cell outgrowth, although their performance was much better than the heptylamine plasma polymer alone. The more telling results were observed with the CMDs immobilized via the polyamide spacer layer (Figure 13.6). Here, regardless of molecular weight and the degree of carboxylmethyl substitution, the tissue outgrowth was completely arrested. XPS surface analysis indicated that the total amount of CMD immobilized onto surfaces was greater when the polyamide spacer was employed than without it. The authors proposed that the microstructure of protein and cell resistant polysaccharide coatings was of importance and that the use of a hydrophilic spacer afforded a thicker, more effective bio-resistant coating.

Yet, if one wants to predict where plasma technology in the field of biomaterials will make the greatest impact, it is where technology is being applied to mitigate the types of problems encountered *in vivo*, but over much shorter periods of times such as assays, sensing devices, and microfluidics [38–40]. In the field of diagnostics, technological advances are being driven by the desire for improvements

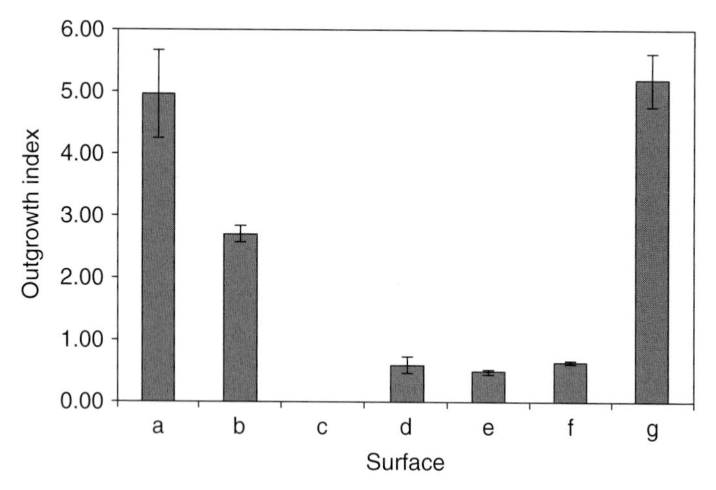

Figure 13.6 Corneal epithelial tissue outgrowth on (a) FEP + AApp; (b) FEP + AApp + PLys; (c) FEP + AApp + PLys + CMD 1: MW 70 kDa; (d) FEP + AApp + PLys + CMD 1 : 14 MW 70 kDa; (e) FEP + AApp + PLys + CMD 1 : 1 MW 500 kDa; (f) FEP + AApp + PLys + CMD 1 : 1 MW 5–40 MDa; and (g) TCPS. (Reproduced with permission from page 229 of K. M. McLean, G. Johnson, R. C. Chatelier, G. J. Beumer, J. G. Steele, H. J. Griesser, *Colloids and Surfaces B: Biointerfaces*, 18, 221, 2000. Permission kindly granted by Elsevier. Copyright Elsevier 2000.)

to the efficiency, accuracy, and speed of detection. Miniturization of devices offer several potential advantages including a reduced consumption of costly reagents and the potential for highly automated, *in situ* diagnostics. A comparable scheme exists in the field of drug development where high-throughput screening is employed to screen potential candidates. The development of such "lab on a chip" devices is exemplified by the Affymetrix Genechip® Technology in the late 1980s where a microarray can be employed to determine which genes exist in a sample – a single chip being capable of performing thousands of assays in parallel.

What is clear in all the advances being explored is that the surface interactions are becoming increasingly important. Once again materials selection is driven by the bulk properties such as optical clarity and transmissibility, rigidity, and the potential for low-cost, high-volume mass production. The surfaces of such devices are not necessarily compatible with miniaturization where events such as surface degradation caused by manufacturing or application, unwanted chemical reactions between the device and reagents, and the unspecific surface adsorption of reagents can all result in poor performance, reduced device functionality, and increased cost. Thus, the developments of plasma surface modification/deposition in the preceding decades finds application in the need to prepare improved surfaces for these most recent technological advances.

13.6
Example 5: An Improved Surface Plasmon Resonance Biosensor

Briefly, surface plasmon resonance (SPR) is a technique for detecting changes at the surface of a sensor. The sensor comprises a glass substrate and a thin metal (gold) coating such that polarized light can pass through the substrate and is reflected off the gold coating. At certain angles of incidence, a portion of the light energy couples through the gold coating and creates a surface plasmon wave at the sample and gold surface interface. The angle of incident light required to sustain the surface plasmon wave is very sensitive to changes in the refractive index at the surface (due to mass change) and these changes can be used to monitor the adsorption of biomolecules.

In a study by Mar and colleagues, the surface of an SPR sensor was modified with an RF plasma polymer coating [48]. Following cleaning of the sensor surface with an argon gas plasma, a PEO-like plasma polymer film of triglyme was deposited. XPS and FTIR analysis confirmed the presence of high concentrations of ether groups – a characteristic of PEO-like biofouling-resistant film. Subsequently, a 1 mg ml^{-1} solution of bovine serum albumin (BSA) in 0.15 M PBS was used and the performance of the sensor compared with that of an uncoated sensor (Figure 13.7). It can be clearly seen that adsorption of BSA on both the uncoated and plasma polymer coated was readily detected. However, upon rinsing the sensor surface with buffer solution, the adsorbed protein layer on the uncoated sensor remained whereas that on the plasma polymer coated surface was very quickly removed with a negligible amount of BSA remaining. The authors concluded that triglyme

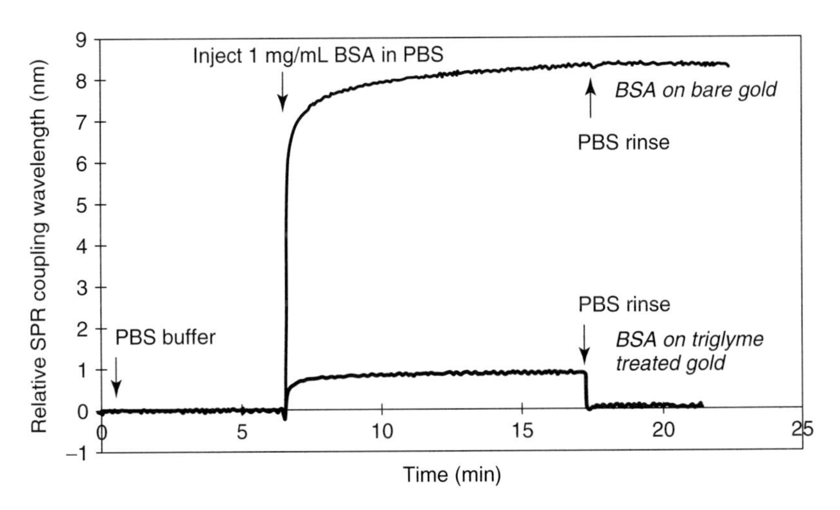

Figure 13.7 Comparison of BSA adsorption for triglyme and unmodified gold SPR sensor surfaces. (Reproduced with permission from page 130 of Mar, M.N., Ratner, B.D., and Yee, S.S. (1999) *Sensors and Actuators B*, **54**, 125. Permission kindly granted by Elsevier. Copyright Elsevier 1999.)

plasma polymer coated sensors could be employed as reference in multichannel SPR systems.

13.7 Conclusions

In summary, we can see that the application of biomaterials in the manufacture of medical devices is well established but, in respect of efficacy the long-term performance is still lacking. In addressing these shortcomings, there is substantial and continued interest in the preparation of novel surfaces to mediate the biomaterial-tissue interface. *In vitro* assays have and continue to be developed. However, there remains a gap in the translation of these *in vitro* observations to *in vivo* performance. Where plasma treatments and plasma polymer coatings have seen successful application is at the interface of nano and bio technologies; particularly in areas of diagnostics and biosensors. Indeed, an almost cyclic process of research and development is beginning to emerge. The techniques and skills for the preparation of plasma polymer coatings were initially gained with a view to application in the development of biomaterials and medical devices. When their performance in respect of these applications failed to meet expectations, the focus turned to their use in allied areas with less demanding requirements. Through their use the information gathered here is serving to develop a more complete and sophisticated assay toolbox. This in turn is now leading to the development of second and third generation coatings in the belief that one day the biomaterial-tissue interface may be fully understood.

References

1. Ratner, B.D., Hoffman, A.S. *et al.* (2004) *Biomaterials Science: An Introduction to Materials in Medicine*, Elsevier, Amsterdam.
2. Williams, D.F. (1989) A model for biocompatibility and its evaluation. *J. Biomed. Eng.*, **11** (3), 185–191.
3. Williams, D.F. (2008) On the mechanisms of biocompatibility, *Ratner Symposium 2006*, Elsevier Science Ltd, Maui.
4. Baier, R.E. (1972) The role of surface energy in thrombogenesis. *Bull. N. Y. Acad. Med.*, **48** (2), 257–272.
5. Mohandas, N., Hochmuth, R.M. *et al.* (1974) Adhesion of red cells to foreign surfaces in the presence of flow. *J. Biomed. Mater. Res.*, **8** (2), 119–136.
6. Chang, S.K., Hum, O.S. *et al.* (1977) Platelet adhesion to solid surfaces. The effect of plasma proteins and substrate wettability. *Med. Progr. Technol.*, **5**, 57–66.
7. Yasuda, H., Yamanashi, B.S. *et al.* (1978) The rate of adhesion of melanoma cells onto nonionic polymer surfaces. *J. Biomed. Mater. Res.*, **12** (5), 701–706.
8. Neumann, A.W., Moscarello, M.A. *et al.* (1979) Platelet adhesion from human blood to bare and protein-coated polymer surfaces. *J. Polym. Sci. Polym. Symp.*, **66** (1), 391–398.
9. Bruil, A., Brenneisen, L.M. *et al.* (1994) In-vitro leukocyte adhesion to modified polyurethane surfaces 2. Effect of wettability. *J. Colloid Interface Sci.*, **165** (1), 72–81.
10. Yasuda, H. and Gazicki, M. (1982) Biomedical applications of plasma polymerisation and plasma treatment of

polymer surfaces. *Biomaterials*, **3** (2), 68–77.

11. Lee, S.D., Hsiue, G.H. *et al.* (1996) Plasma-induced grafted polymerization of acrylic acid and subsequent grafting of collagen onto polymer film as biomaterials. *Biomaterials*, **17** (16), 1599–1608.
12. Webb, K., Hlady, V. *et al.* (1998) Relative importance of surface wettability and charged functional groups on NIH 3T3 fibroblast attachment, spreading, and cytoskeletal organization. *J. Biomed. Mater. Res.*, **41** (3), 422–430.
13. Hatada, K., Kobayashi, H. *et al.* (1982) The glow-discharge treatment of poly(vinyl chloride) tube. *Org. Coat. Appl. Polym. Sci. Proc.*, **47**, 391–396.
14. Kirkpatrick, C.J., Mueller-Schulte, D. *et al.* (1991) Surface modification of polymers to permit endothelial-cell growth. *Cells Mater.*, **1**, 93.
15. Wittenbeck, P. and Wokaun, A. (1993) Plasma treatment of polypropylene surfaces – characterisation by contact-angle measurements. *J. Appl. Polym. Sci.*, **50** (2), 187–200.
16. Ikada, Y. (1994) Surface modification of polymers for medical applications. *Biomaterials*, **15** (10), 725–736.
17. Elbert, D.L. and Hubbell, J.A. (1996) Surface treatments of polymers for biocompatibility. *Annu. Rev. Mater. Sci.*, **26**, 365–394.
18. Lin, J.C. and Cooper, S.L. (1996) In vitro fibrinogen adsorption from various dilutions of human blood plasma on glow discharge modified polyethylene. *J. Colloid Interface Sci.*, **182**, 315–325.
19. Gerenser, L.J. (1993) XPS studies of in-situ plasma-modified polymer surfaces. *J. Adhes. Sci. Technol.*, **7** (10), 1019–1040.
20. Sheu, G.S. and Shyu, S.S. (1994) Surface modification of Kevlar 149 fibers by gas plasma treatment. 1. Morphology and surface characterization. *J. Adhes. Sci. Technol.*, **8** (5), 531–542.
21. France, R.M. and Short, R.D. (1997) Plasma treatment of polymers – effects of energy transfer from an argon plasma on the surface chemistry of poly(styrene), low density poly(ethylene), poly(propylene) and poly(ethylene terephthalate). *J. Chem. Soc., Faraday Trans.*, **93** (17), 3173–3178.
22. World Health Organization (2007) Fact Sheet 317, Cardiovascular Diseases, February 2007.
23. Zilla, P., Vonoppell, U. *et al.* (1993) The endothelium – a key to the future. *J. Card. Surg.*, **8** (1), 32–60.
24. Teebken, O.E. and Haverich, A. (2002) Tissue engineering of small diameter vascular grafts. *Eur. J. Vasc. Endovasc. Surg.*, **23** (6), 475–485.
25. Zhang, Z., King, M.W. *et al.* (1994) In vivo performance of the polyesterurethane Vascugraft® prosthesis implanted as a thoraco-abdominal bypass in dogs: an exploratory study. *Biomaterials*, **15**, 1099–1112.
26. Sefton, M.V., Sawyer, A. *et al.* (2001) Does surface chemistry affect thrombogenicity of surface modified polymers? *J. Biomed. Mater. Res.*, **55** (4), 447–459.
27. Cima, L.G. (1994) Polymer substrates for controlled biological interactions. *J. Cell. Biochem.*, **56** (2), 155–161.
28. Tsai, W.B., Grunkemeier, J.M. *et al.* (1999) Human plasma fibrinogen adsorption and platelet adhesion to polystyrene. *J. Biomed. Mater. Res.*, **44** (2), 130–139.
29. Shen, M.C., Pan, Y.V. *et al.* (2001) Inhibition of monocyte adhesion and fibrinogen adsorption on glow discharge plasma deposited tetraethylene glycol dimethyl ether. *J. Biomater. Sci., Polym. Ed.*, **12** (9), 961–978.
30. Andrade, J.D., Nagaoka, S. *et al.* (1987) Surfaces and blood compatibility current hypotheses. *Trans. Am. Soc. Artif. Intern. Organs*, **33**, 75–84.
31. Lee, J.H., Kopecek, J. *et al.* (1989) Protein-resistant surfaces prepared by PEO-containing block copolymer surfactants. *J. Biomed. Mater. Res.*, **23** (3), 351–368.
32. Beyer, D., Knoll, W. *et al.* (1997) Reduced protein adsorption on plastics via direct plasma deposition of triethylene glycol monoallyl ether, *23rd Annual Meeting of the Society-for-Biomaterials, New Orleans*, John Wiley & Sons, Inc.
33. Hendricks, S.K., Kwok, C. *et al.* (2000) Plasma-deposited membranes for controlled release of antibiotic to prevent

bacterial adhesion and biofilm formation. *J. Biomed. Mater. Res.*, **50** (2), 160–170.
34. Shen, M.C., Martinson, L. *et al.* (2002) PEO-like plasma polymerized tetraglyme surface interactions with leukocytes and proteins: in vitro and in vivo studies. *J. Biomater. Sci., Polym. Ed.*, **13** (4), 367–390.
35. Cao, L., Ratner, B.D. *et al.* (2007) Plasma deposition of tetraglyme inside small diameter tubing: optimization and characterization. *J. Biomed. Mater. Res.*, **81A** (1), 12–23.
36. Johnston, E.E., Bryers, J.D. *et al.* (2005) Plasma deposition and surface characterization of oligoglyme, dioxane, and crown ether nonfouling films. *Langmuir*, **21** (3), 870–881.
37. Wu, Y.L.J., Timmons, R.B. *et al.* (2000) Non-fouling surfaces produced by gas phase pulsed plasma polymerization of an ultra low molecular weight ethylene oxide containing monomer. *Colloids Surf., B: Biointerfaces*, **18** (3-4), 235–248.
38. Bouaidat, S., Berendsen, C. *et al.* (2004) Micro patterning of cell and protein non-adhesive plasma polymerized coatings for biochip applications. *Lab Chip*, **4** (6), 632–637.
39. Bretagnol, F., Valsesia, A. *et al.* (2006) Surface functionalization and pPatterning techniques to design interfaces for biomedical and biosensor applications. *Plasma Processes Polym.*, **3** (6-7), 443–455.
40. Favia, P., Sardella, E. *et al.* (2007) Plasma assisted surface modification processes for biomedical materials and devices, *Conference of the NATO-Advanced-Study-Institute on Plasma Assisted Decontamination of Biological and Chemical Agents, Cesme-Izmir, Turkey*, Springer.
41. Salim, M., Mishra, G. *et al.* (2007) Non-fouling microfluidic chip produced by radio frequency tetraglyme plasma deposition. *Lab Chip*, **7** (4), 523–525.
42. Nicolson, P.C. and Vogt, J. (2001) Soft contact lens polymers: an evolution. *Biomaterials*, **22** (24), 3273–3283.
43. Lopez-Alemany, A., Compan, V. *et al.* (2002) Porous structure of Purevision (TM) versus Focus (R) night & day (TM) and conventional hydrogel contact lenses. *J. Biomed. Mater. Res.*, **63** (3), 319–325.
44. Haddow, D.B., MacNeil, S. *et al.* (2006) A cell therapy for chronic wounds based upon a plasma polymer delivery surface. *Plasma Processes Polym.*, **3** (6-7), 419–430.
45. McLean, K.M., Johnson, G. *et al.* (2000) Method of immobilization of carboxymethyl-dextran affects resistance to tissue and cell colonization. *Colloids Surf. B: Biointerfaces*, **18** (3-4), 221–234.
46. Frank, B.P. and Belfort, G. (1997) Intermolecular forces between extracellular polysaccharides measured using the atomic force microscope. *Langmuir*, **13** (23), 6234–6240.
47. Harder, P., Grunze, M. *et al.* (1998) Molecular conformation in oligo(ethylene glycol)-terminated self-assembled monolayers on gold and silver surfaces determines their ability to resist protein adsorption. *J. Phys. Chem. B*, **102** (2), 426–436.
48. Mar, M.N., Ratner, B.D. *et al.* (1999) An intrinsically protein-resistant surface plasmon resonance biosensor based upon a RF-plasma-deposited thin film. *Sens. Actuators B: Chem.*, **54** (1-2), 125–131.

14
Plasma Sterilization at Normal Atmospheric Pressure

Tetsuya Akitsu, Siti Khadijah Za aba, Hiroshi Ohkawa, Keiko Katayama-Hirayama, Masao Tsuji, Naohiro Shimizu, and Yuichirou Imanishi

14.1
Introduction

The recent development of high-power semiconductors has expanded the use of solid-state devices in various fields in the pulsed-power technology. One of the effective applications is the replacement of on control switching in the capacitive energy storage, and more unique features of some semiconductor switching devices enable novel applications in the control of inductive energy storage systems. The atmospheric pressure glow (APG) was realized by Okazaki, University of Sophia, in 1987 [1–3]. Original APGs used dielectric barrier electrodes to prevent arcing in the atmospheric pressure discharge. Recently, we reported a comparative study on the plasma sterilization and showed the antibacterial effect for gram-positive bacteria, gram-negative bacteria, fungus, and yeast. Our effort was dedicated to establish the use of *Geobacillus stearothermophilus* and *Bacillus atrophaeus* [4–8]. Sterilization of medical instruments is one of the research fields increasingly attracting attention, as a part of integrated prevention systems against infection. The limitation of conventional sterilization methods, such as high-temperature pressurized steam sterilization, does not suit modern integrated medical instruments. Sterilization and inactivation of end-toxins are studied in comprehensive researches by Kong *et al.* [9–19] and Shintani *et al.* [20], but no work confirms the differences in self-destructive activities in conventional pulsed-high-frequency atmospheric pressure glow (HF–APG) and pulse discharges.

Major modern medical devices are thermo- and hydrosensitive. Limitations and sometimes drawbacks of other sterilization schemes, such as autoclaving, have been accelerating the development of low-temperature plasma sterilization. A type of gas plasma sterilization using low-temperature hydrogen peroxide gas plasma, is commercially sold under the brand name Sterrad® by Johnson & Johnson, Irvine CA, USA. Currently, Sterrad® is one of the few realized cases of gaseous plasma sterilization systems that FDA has approved, and is increasingly attracting attention since the discovery of environmental pollution and the delayed carcinogenic effect of ethylene oxide (EtO) used in the low-temperature chemical

Industrial Plasma Technology. Edited by Yoshinobu Kawai, Hideo Ikegami, Noriyoshi Sato, Akihisa Matsuda, Kiichiro Uchino, Masayuki Kuzuya, and Akira Mizuno

ISBN: 978-3-527-32544-3

sterilization of health-care products. Uses of other types of gas plasma have been studied. Majority of researches are based on oxidation using helium-diluted oxygen or related compounds, such as O^*, O_3, or HO_2. The oxidation affects the carrier of microbes, the surface of health-care products and sterile packages, causing pinholes and deteriorates the strength that might allow postexposure pollution of health-care materials. Another problem is the process of antibacterial treatment based on the oxidation. The quenching of oxygen radicals by hydrogen atoms covering polymer surface shows dependence of survival curve on the types of carriers: glass plate and polypropylene. A sterilization process in inert gas, helium, and nitrogen shows antibacterial effect comparable to oxidizing process and smaller dependence on the carrier material as a result of the nonthermal excitation by a novel inductive energy storage pulse-power supply controlled by power semiconductor switching device. A sterilization process using HF–APG plasma, APG at 27.12 MHz, has been demonstrated. This work was carried out using plain dielectric barrier electrodes, 3–5 mm gap filled with He/O_2 mixture. In the HF–APG excitation process, there are two problems. A serious problem is the control of temperature. Because the current quenching by dielectric plate no longer acts as a control of the neutral gas temperature, the pulse modulation technique was the correct choice. Characteristics of the neutral gas temperature as functions of the discharge gap, power, and modulation pulse width were studied.

14.2 Experimental Schemes

14.2.1 Inductive Energy Storage Pulse-Power Source

Pulsed discharges in air or various gases under atmospheric pressure are of great interest to industrial applications. Because of the high-pressure nature and the uniformity requirement, these discharges require pulsed-high voltage with very short pulse duration. Recently, the high-power semiconductor switching devices changed the high-voltage generation technique from large single-shot to compact and repetitive pulse generation scheme, benefiting the replacement of a part of industrial requirement for the atmospheric pressure plasma applications. Homogeneous streamer discharge was generated along a high-voltage electrode powered by inductive energy storage device. The voltage increase, 10^{11} V s^{-1} maximum, was generated by a semiconductor device: static induction thyristor. In this work, experiment was carried out using asymmetric electrode structure that consists of Re/W fine wire high-voltage electrode, 0.6 mm diameter surrounded symmetrically by a dielectric barrier electrode, quartz tube attached with perforated aluminium ground electrode, 20 mm in diameter, and horizontally aligned high-voltage electrodes facing a plain dielectric barrier electrode for nitrogen streamer discharge. This asymmetric structure of the dielectric barrier discharge

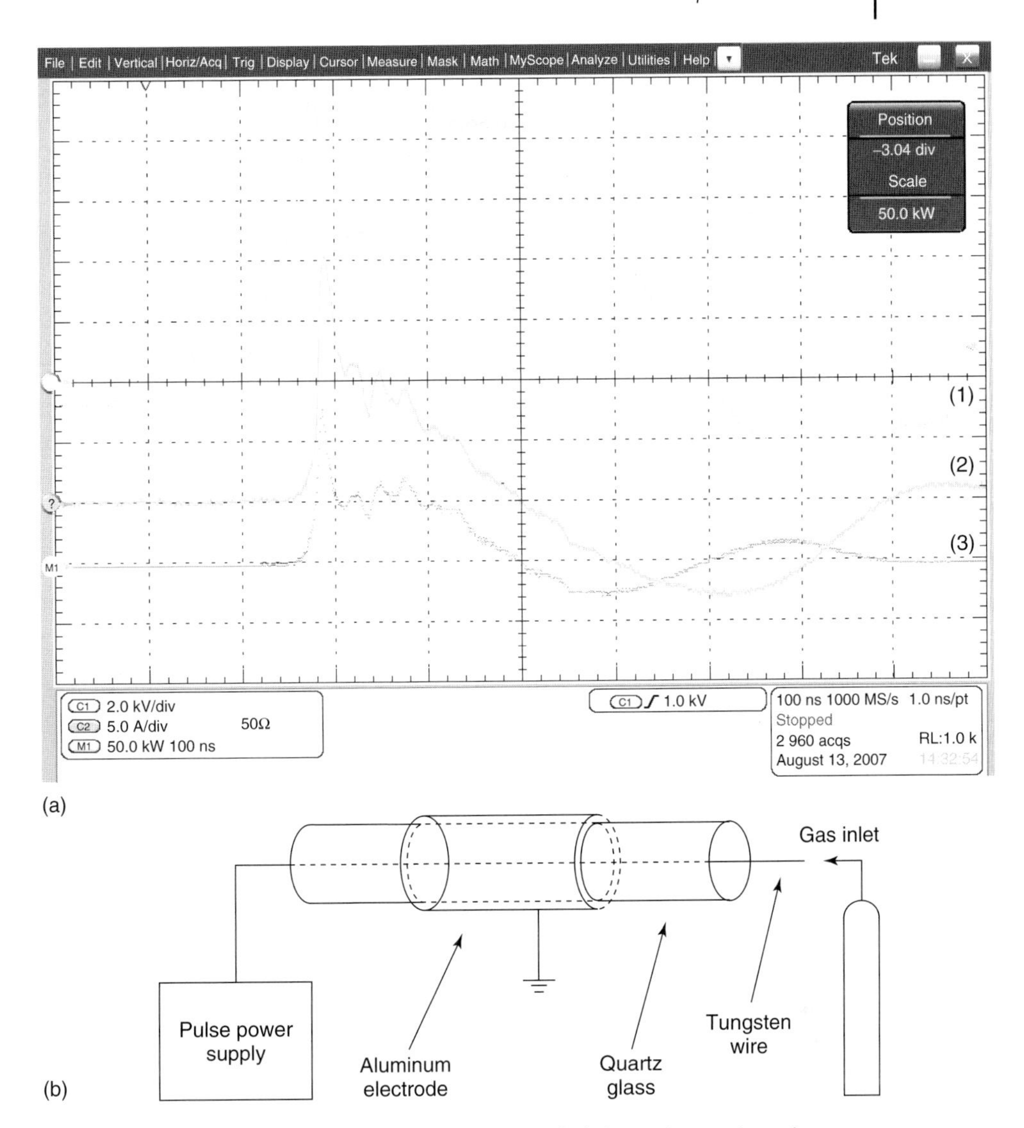

Figure 14.1 Experimental setup for pulse-discharge. (a) Temporal evolution of the fast rising pulse. Breakdown, sustain and reversed voltage phase. (b) Cylindrical configuration, wire electrode surrounded by a coaxial, dielectric barrier electrode, diameter 20 mm, length 100 mm, at ground potential. Curve 1, voltage; curve 2, current; curve 3, instantaneous power generated by multiplication.

was necessary to keep the sustain phase in both configurations. Figure 14.1 shows a typical example for the pulse voltage (5 kV div^{-1}, yellow), current (5 A div^{-1}, red), and power (50 kW div^{-1}, blue), 2 k pulse s^{-1}, gap distance 10 mm, electrode: cylindrical dielectric barrier surrounding high-voltage electrode, used for helium-based oxygen and nitrogen mixtures. The apparatus for nitrogen streamer discharge has three important factors to attain sufficient sterilization efficiency: production of molecular nitrogen radicals, UV–C irradiation, and high rising-up voltage (high dV/dt). The apparatus was operated at raised temperature at around 60 °C under atmospheric pressure nitrogen, with a 4.5 mm gap between the high-voltage electrode and dielectric barrier electrode at the ground potential. In both cases, positively polarized pulse was applied to wire electrodes and the discharge was sustained for 300–400 ns, then reversed polarity was applied to the electrode.

14.2.2
Antibacterial Test

The antibacterial effect of atmospheric plasma was tested on the basis of mortality of microbes exposed to the discharge. Spore-forming bacteria, *Geobacillus stearothermophilus* ATCC 7953, were selected for this work. In the incubation and sterility judgments, we followed the sterility test in the Japanese Pharmacopoeia, the JP 14th edition.

Sterility judgments in the previous work were made based on the turbidity examination and pH indicator. In this work, we studied the decreasing constant of the survival curve by cascade dilution and colony counting method. The biological indicator on special carrier was prepared in the following way. Stock of *G. stearothermophilus* was incubated in liquid soybean casein digest. A quantity of 0.1 ml was sampled and transplanted to plain SCD agar and incubated. One scoop with platinum wire was diluted in sterilized water. The density of this dilution was adjusted to 10^6 CFU 0.1 ml^{-1}, and then the suspension was applied to carriers, such as a glass-slide and a small tablet of polypropylene sheet. For the estimation of the initial population and the number of survivors, cascade dilution was carried out in the following way. A quantity of 10 ml and 9 ml sequence of sterilized water was measured with a pipette. After the exposure, the spore on each carrier was expanded in 10 ml of sterile water by a supersonic mixer, and then diluted in sterile water to 10, 10^2, 10^3, 10^4 times dilutions. A quantity of 1 ml was sampled from appropriate dilution and incubated. Same maneuver were repeated for untreated control indicators to determine the initial population of spores.

In the experiment, dried biological indicator was extracted from vial and set in a sterilized petri dish. At the same time, control indicator was extracted from vial and expanded in 10 ml of sterilized water in supersonic bath. About 1 ml of the suspension was sampled with measuring pipette and added to 9 ml of newly sterilized water and mixed. Similarly, 1 ml was sampled and cascade dilution was formed. About 1 ml of dilution was sampled from appropriate stage and mixed with autoclaved nonselective meat-broth agar. This recovery medium was heated again in a boiling water bath and cooled down to 40–50 °C, prior to the mixing. About 1 ml of suspension was mixed with this medium inside sterilized petri dish of 95 mm in diameter, and kept still until the medium formed solid gel. Solidified dish was shielded with paraffin tape (NOVIX), and then incubated at

57 °C for 48 hours. The antibacterial effect was evaluated using spore-forming bacteria, *G. stearothermophilus*. This species is frequently used in the evaluation of the plasma sterilization because *G. stearothermophilus* survive high temperatures, 115 °C, and form easily identifiable colony on nonselective agar. The number of survivors was estimated by multiplying the count of colony with the dilution ratio.

14.3
Experimental Result

Experimental result compares typical survival curves for *G. stearothermophilus* applied to two types of carriers, in He/O_2 mixture excited by pulsed-high frequency and repetitive pulse discharge. Figure 14.2 shows a comparison with RF–APG and pulse-discharge. CW–APG is sustained by 13.56 MHz, 250 W and the average power of the pulse-discharge treatment is 30 mJ per pulse at 10^3 Hz repetition, approximately 30 W can excite almost same discharge volume and the process, sterilization time indicates almost comparable results. This pulse discharge is sustained by significantly lower power but the resulting sterilization time was of the same order and demand on forced cooling system was smaller. Also shown in Figure 14.2 is a comparison of survival curve for He/N_2 1% mixture, showing smaller dependence on carrier materials. Filled symbols indicate the population for glass carrier and empty symbols that for polypropylene carrier. The survival curve shows two stages of the sterilization process. In the initial stage, the population of microorganism decreases faster than the second stage where the rate saturates. This result shows the organic loading effect in the cases of polymer materials. In the vicinity of polymer materials, the sterilization biological indicator is retarded. A sufficiently wide gap, 10 mm in He and 4.5 mm in N_2, was excited by pulse discharges, below the required level of <60 °C, by natural convection and gas flow without forced circulation of coolant. Sterilization using nitrogen mixture shows minor dependence on the carrier materials: polypropylene and glass plate. Pulsed excitation of nitrogen discharge shows a strong antibacterial effect. Figure 14.3 shows optimized case of N_2 and He/O_2 streamer discharges. Figure 14.4 shows the decrease in the population during the initial phase of direct exposure to streamer discharge. Initial spore density of 4.7×10^4 CFU was sterilized within 30 seconds, as shown in Figures 14.3 and 14.4. The experiment was repeated changing the gas temperature and disinfection time from 15 to 300 seconds. The decade decreasing time constant D was as small as 5–10 seconds. This sterilization is based on the strong radiation of UV lines by N_2 and NO. A few cases of unsatisfactory sterilization were caused by the presence of cold spot, probably a shadowed region in the biological indicators.

The validation was carried out in an isolated area in the laboratory and the Japanese Pharmacopoeia was followed [21].

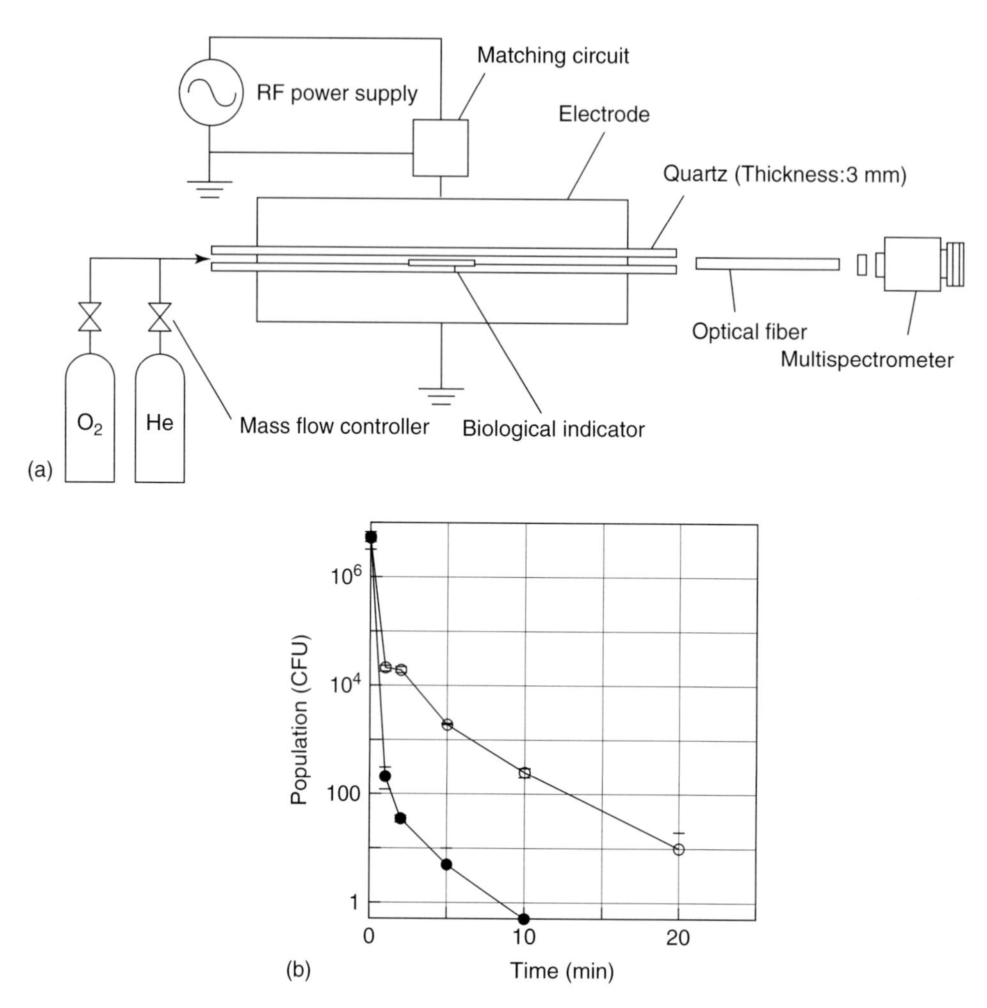

Figure 14.2 Experimental setup of HF–APG in He/O_2 experiment and a typical example for survival curve. (a) HF–APG plasma experiment, (b) survival curve for HF–APG, incident power: 250 W, He 1.5 l min^{-1}, O_2 ratio 0.6%, 5 ml min^{-1}, electrodes: 50 mm × 150 mm in width and length, gap 2 mm. Empty symbols: polypropylene carrier BI, filled symbols: glass-slide cover, control: $1.0 \pm 0.1 \times 10^6$ CFU, *Geobacillus stearothermophilus*.

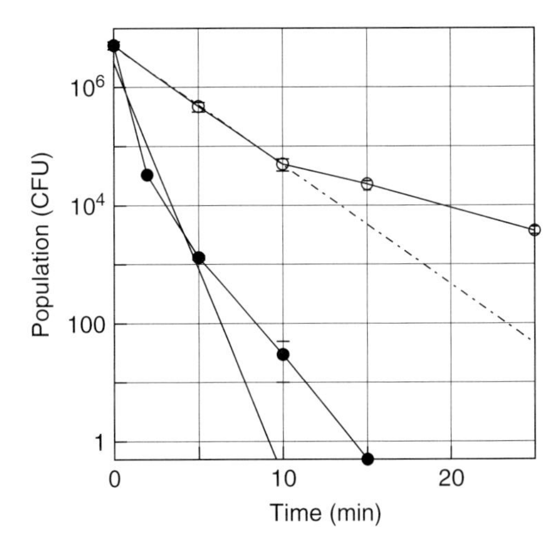

Figure 14.3 Survival curves for streamer discharge. Comparison of pulse-excited discharge in He 1 l min^{-1}. O_2 ratio: 3%. Average power: 30 W, cylindrical configuration, empty symbols: polypropylene carrier BI, filled symbols: glass-slide cover BI, 1.0 ± 0.1 × 10^6 CFU, *Geobacillus stearothermophilus*.

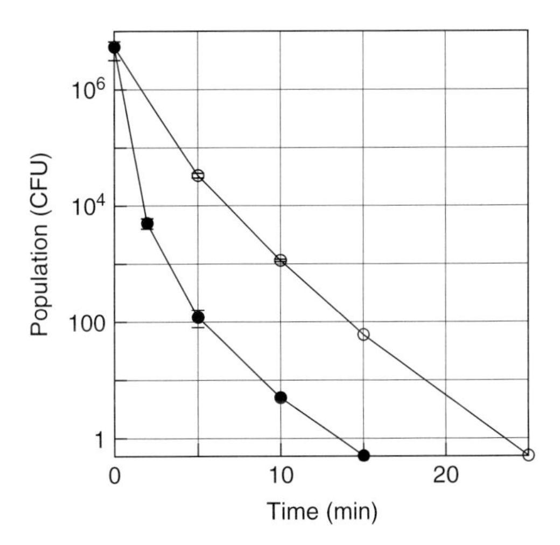

Figure 14.4 Survival curves for streamer discharge. Comparison of pulse-excited discharge in He 1 l min^{-1}. N_2 ratio: 1%. Average power: 30 W. Empty symbols: polypropylene carrier BI, filled symbols: glass-slide cover, control: 1.0 ± 0.1 × 10^6 CFU, *Geobacillus stearothermophilus*.

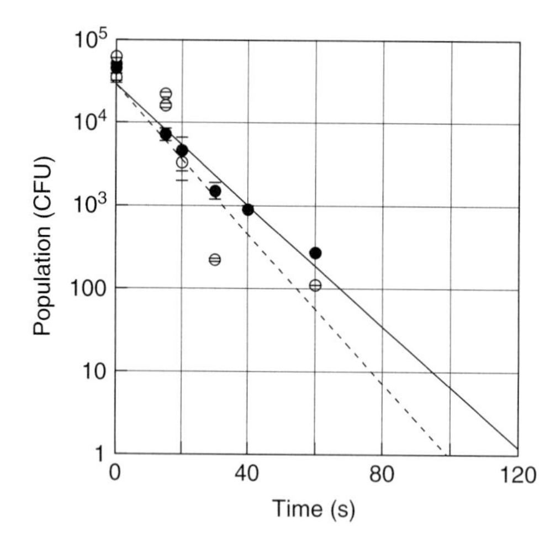

Figure 14.5 Survival curves for streamer discharge. N_2 60 °C, pulse width 150 ns, peak voltage 16 kV, parallel wire configuration, three wire electrodes, 600 μm in diameter, gap 4.5 mm separated from a dielectric barrier electrode at ground level, control: $4.5 \pm 1 \times 10^4$ CFU, and disinfection time 120 seconds. Filled symbols and empty symbols indicate the result of independent trials after decontaminations. The solid and broken lines indicate extrapolations of the least square fit for each result.

Figures 14.5, 14.7, and 14.8 show the survival curve in atmospheric pressure nitrogen streamer discharge measured with filter paper base-biological indicator of *Geobacillus stearothermophilus*. The regression curves in Figure 14.5 indicate that the SAL6 level sterilization is completed in 120 seconds. This result is obtained at 60 °C gas temperature. The sterilization is completed in shorter time than the helium/oxygen mixture shown in Figure 14.6, in that the SAL6 level sterilization needs 300 seconds. The experiment in Figure 14.7 was carried out using one year older shelf stock of the biological indicator. In this case the SAL6 level sterilization is completed in 30 seconds. The dependence on the gas temperature is a key point for the reproducibility of this experiment. Figure 14.8 shows that SAL6 level sterilization is completed in 500 seconds when the gas temperature is slightly lower. Figure 14.9 shows the most important part of this work: the making process of the biological indicators of various microorganisms. We carried out this experiment carefully, but the result is case sensitive and the experimental result may vary depending on the condition of the preparation of the microorganism. The microorganism is separated by centrifugal separation or filtration process. The result also depends on the choice of the recovery media and incubation temperature.

14.4
Conclusion

Plasma sterilization at the normal atmospheric pressure was examined using pulse discharge excited by an inductive energy storage source controlled with a

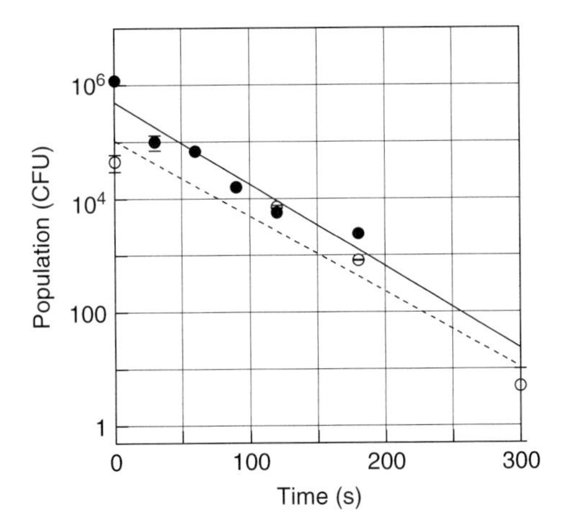

Figure 14.6 Survival curve for streamer discharge, in He/O_2, 60 °C, He 1 l min^{-1}, O_2 10 ml min^{-1}, cylindrical electrode, gap 10 mm, length 100 mm, control: 4.5 ± 1 × 10^4 CFU, and disinfection time 300 seconds. Filled symbols and empty symbols indicate the result of independent trials after decontaminations. The solid and broken lines indicate extrapolations of the least square fit for each result.

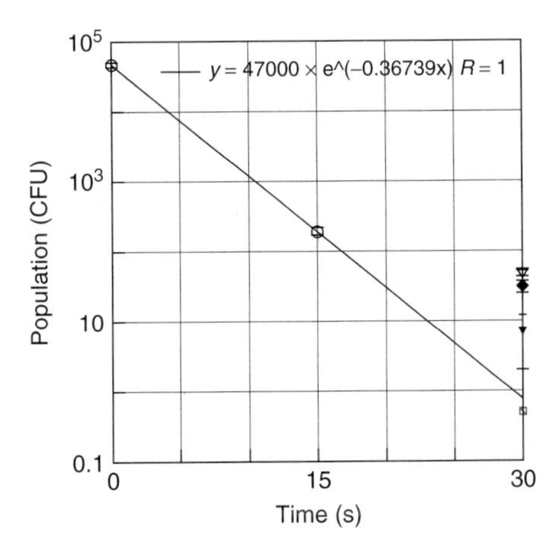

Figure 14.7 Decrease of population during the initial phase in atmospheric pressure nitrogen. Gap 4.5 mm, 60 mJ per pulse, 2 kHz. The solid line indicates extrapolations of the least square fit for the result and corresponding formula.

semiconductor switching device: static induction thyristor. A measurement using cascade dilution and colony counting scheme showed that the disinfection time varies depending on the species of carrier of biological indicators. The mechanism of the sterilization in oxygen mixtures depends on the oxygen radicals. On the other hand, the emission of UV radiation seems to be the primary process of sterilization in nitrogen mixtures. The compatibility with other medical-care materials and research on wider antibacterial spectrum will be necessary to establish the state

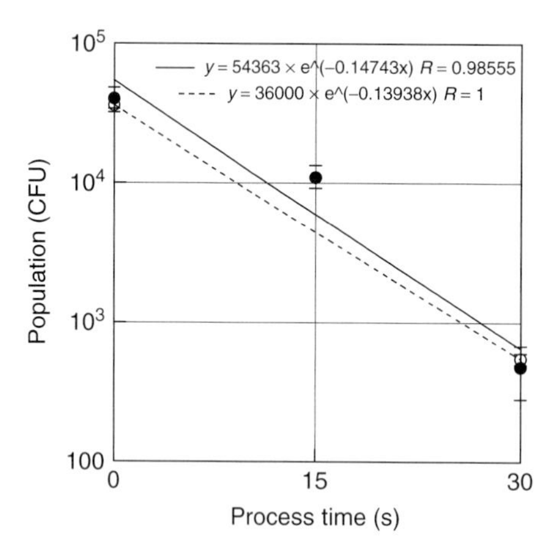

Figure 14.8 Survival curves for streamer discharge N_2 49 °C, pulse width 150 ns, peak voltage 16 kV, parallel wire configuration, gap 4.5 mm, control: $3.6 \pm 0.2 \times 10^4$ CFU, D > 96 seconds. Filled symbols and empty symbols indicate the result of independent trials after decontaminations. The solid and broken lines indicate extrapolations of the least square fit for each result and corresponding formula.

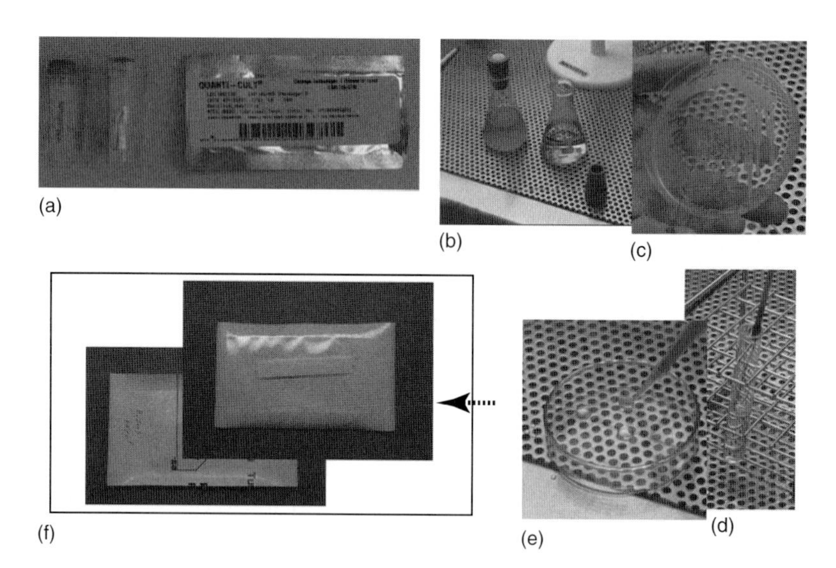

Figure 14.9 Preparation of biological indicator: (a) culture sample, (b) cultivation in SCD liquid medium, (c) dish cultivation on SCD agar, (d) sampling and water dissolution using platinum wire disseminator, (e) application to carriers, and (f) sterile packaging in Tyvek. Commercially available biological indicators of spore-forming bacteria: *Attest*™ 1262 and *Attest*™ 1264, from 3M Co., were also used for screening test. *Attest*™ type 1262 vial contained *Geobacillus stearothermophilus* ATCC 7953 at 8.6×10^5 to 1.4×10^6 CFU, type 1264 *Bacillus atrophaeus* ATCC 9372 at 5.0×10^5 to 4.8×10^6 CFU.

of this antibacterial treatment in the production and daily care of precautionary sterilization.

Acknowledgments

We would like to cordially express thanks to all organizers of the summer school in Valenna 2008. The work was supported financially in part by a Grant-in-Aid for Scientific Research on Priority Area (Grant No. 15075204) from the Ministry of Education, Culture, Sports, Science and Technology.

References

1. Kanazawa, S., Kogoma, M., Moriwaki, T., and Okazaki, S. (1987) Carbon film formation by cold plasma at atmospheric pressure. Proceedings of the 8th International Symposium on Plasma Chemistry (ISPC-8), Vol. 3, Tokyo, p. 1839.
2. Okazaki, S., (1993) Development of Atmospheric Pressure Glow Plasma and Its Application on a Surface with Curvature, J. Photopolymer Sci & Technol, **6**, 393.
3. Yokoyama, T., Kogoma, M., Moriwaki, T., and Okazaki, S., (1990) The Mechanism of the stabilization of Glow Plasma at Atmospheric Pressure, *J. Phys. D: Appl. Phys.*, **23**, 1125–1128.
4. Laroussi, M., (1996), Sterilization of Contaminated Matter with an Atmospheric Pressure Plasma, *IEEE Trans. Plasma Sci.*, **24**, 1188–1191.
5. Laroussi, M., (2002) Non-Thermal Decontamination of Biological Media by Atmospheric Pressure Plasmas: Review, Analysis, and Prospects, *IEEE Trans. Plasma Sci.*, **30** (4), 1409–1415.
6. Akitsu, T., Ohkawa, H., Tsuji, M., Kimura, H., and Kogoma, M. (2005) Plasma sterilization using glow discharge at atmospheric pressure. *Surf. Coat. Technol.*, **193** (1-3) 29–34.
7. Ohkawa, H., Akitsu, T., Tsuji, M., Kimura, H., and Fukushima, K. (2005) Initiation and microbial-disinfection characteristics of wide-gap atmospheric-pressure glow discharge using soft X-ray ionization. *Plasma Processes Polym.*, **2**, 120–126.
8. Ohkawa, H., Akitsu, T., Tsuji, M., Kimura, H., Kogoma, M., and Fukushima, K. (2006) Pulse-modulated, high-frequency plasma sterilization at atmospheric pressure. *Surf. Coat. Technol.*, **200**, 6829–5835.
9. Vleugels, M., Shama, G., Deng, X.T., Greenacre, E., Brocklehurst, T., and Kong, M.G. (2005) Atmospheric plasma inactivation of biofilm-forming bacteria for food safety control. *IEEE Trans. Plasma Sci.*, **33** (2), 824–828.
10. Deng, X., Shi, J., and Kong, M.G. (2006) Physical mechanisms of inactivation of Bacillus subtilis spores using cold atmospheric plasmas. *IEEE Trans. Plasma Sci.*, **34** (4), 1310–1316.
11. Deng, X.T., Shi, J.J., Shama, G., and Kong, M.G. (2005) Effect of microbial loading and sporulation temperature on atmospheric plasma inactivation of Bacillus subtilis spores. *Appl. Phys. Lett.*, **87** (1–3), 153901.
12. Yu, H., Perni, S., Shi, J.J., Wang, D.Z., Kong, M.G., and Shama, G. (2006) Effect of cell loading and phase of Growth in cold atmospheric gas plasma inactivation of Escherichia coli K12. *J. Appl. Microbiol.*, **101**, 1232–1330.
13. Walsh, J.L., Shi, J.J., and Kong, M.G. (2006) Sub-microsecond pulsed atmospheric Glow discharges sustained without dielectric barriers at kilohertz frequencies. *Appl. Phys. Lett.*, **89** (1–3), 161505.
14. Perni, Stefano., Shama, Gilbert., Hobman, J.L., Lund, P.A., Kershaw, C.J., Hidalgo-Arroyo, G.A.,

Penn, C.W., Deng, X.T., Walsh, J.L., and Kong, M.G. (2007) Probing bactericidal mechanisms induced by cold atmospheric plasmas with Escherichia-coli mutants. *Appl. Phys. Lett.*, **90** (1–3), 073902.

15. Deng, X.T., Shi, J.J., Chen, H.L., and Kong, M.G. (2007) Protein destruction by atmospheric pressure glow discharges. *Appl. Phys. Lett.*, **90** (1–3), 013903.
16. Deng, X.T., Shi, J.J., and Kong, M.G. (2007) Protein destruction by a helium atmospheric pressure glow discharge: capability and mechanisms. *J. Appl. Phys.*, **101** (1–9), 074701.
17. Walsh, J.L. and Kong, M.G. (2007) 10 ns pulsed atmospheric air plasma for uniform treatment of polymeric surfaces. *Appl. Phys. Lett.*, **91** (1–3), 251504.
18. Perni, S., Liu, D.W., Shama, G., and Kong, M.G. (2008) Cold atmospheric plasma decontamination of the pericarps of fruit. *J. Food Prot.*, **71** (2), 302–308.
19. Perni, S., Shama, G., and Kong, M.G. (2008) Cold atmospheric plasma disinfection of fruit surfaces contaminated with migrating microorganisms. *J. Food Prot.*, **71** (8), 1619–1625.
20. Shintani, H., Shimizu, N., Imanishi, Y., Sekiya, T., Tamazawa, K., Taniguchi, A., and Kido, N. (2007) Inactivation of microorganisms and endotoxins by low temperature nitrogen gas plasma exposure. *Biocontrol Sci.*, **12** (4), 131–143.
21. The Japanese Pharmacopoeia JP14th, Tokyo, Japan, B-628 (2001), *1*, pp. 87–89 (in Japanese).

15
Elimination of Pathogenic Biological Residuals by Means of Low-Pressure Inductively Coupled Plasma Discharge

Ondřej Kylián, Hubert Rauscher, Ana Ruiz, Benjamin Denis, Katharina Stapelmann, and François Rossi

15.1
Introduction

The possibility to use low-pressure, nonequilibrium plasma discharges for decontamination and sterilization of surfaces of medical equipment is nowadays gaining increased attention. The primary interest of this technique is related mainly to its capability to inactivate effectively infectious microorganisms including highly resistant bacterial spores [1–3] without using toxic substances and at low-temperature process conditions required for the treatment of heat degradable materials. Moreover, recent results revealed the ability of plasma discharges to inactivate or also eliminate other kinds of biological pathogens such as bacterial endotoxins [3–5] or proteins [3, 6–8]. Especially, the latter makes the plasma treatment a real alternative to the commonly used decontamination techniques that are in many cases insufficient to assure complete elimination of residual biological contamination.[1)]

Naturally, distinct characteristics of biological pathogens imply different strategies for their destruction: living microorganisms can be readily inactivated by intense UV radiation emitted by plasma discharges [1], in contrast to the elimination of pathogenic biomolecules from surfaces, where physicochemical removal is needed [10]. However, it is clear that any process that is intended to be effective and universal for complete elimination of biological contamination has to combine both pathways. Recently, it has been suggested that this can be fulfilled by using an inductively coupled plasma (ICP) discharge sustained in a ternary $Ar/O_2/N_2$ discharge mixture combining high emission of UV radiation in the spectral range suitable for the sterilization of bacterial spores with fast erosion and removal of biomolecules [11, 12]. Nevertheless, the latter has been, up to

1) For instance, according to recent studies in UK, more than 60% of medical instruments exhibit traces of protein soiling after routine cleaning and sterilization in hospital services [9], which represent a potential risk of transmission of diseases such as Creutzfeld-Jakob disease.

Industrial Plasma Technology. Edited by Yoshinobu Kawai, Hideo Ikegami, Noriyoshi Sato, Akihisa Matsuda, Kiichiro Uchino, Masayuki Kuzuya, and Akira Mizuno

ISBN: 978-3-527-32544-3

now, demonstrated solely on bovine serum albumin and no effort has been devoted to test whether it is possible to eliminate a wider range of biological agents, that is, an aspect important for the demonstration of the universality of this sterilization/decontamination approach. Therefore, in this study to fill this experimental gap, the feasibility of low-pressure $Ar/O_2/N_2$ ICP discharge to remove other biological systems differing significantly in their properties, namely, bacterial spores and biomolecules in general, is investigated. Moreover, to highlight the advantageous properties of $Ar/O_2/N_2$ mixture, the results are compared directly with the results obtained by using Ar and Ar/O_2 plasma treatment.

15.2 Experimental

15.2.1 Plasma Treatment

The processing of biological samples was performed using a low-pressure, double-coil, ICP reactor depicted in Figure 15.1, which was used and described in detail in previous studies [8, 11, 12]. It consists of a cylindrical stainless-steel vacuum chamber (chamber diameter = 200 mm and height = 100 mm) connected to a pumping stage and a gas inlet system. The discharge is fed by two spiral RF coils connected to the RF generator working at 13.56 MHz through the tunable matching network.

In this study, the plasma treatment was performed in discharges sustained in pure Ar, Ar/O_2 20 : 2, and $Ar/O_2/N_2$ 20 : 1 : 1 discharge mixtures at a pressure of 10 Pa, at an applied RF power of 200 W, and at a total gas flow of 22 sccm. The intensity of UV radiation was determined using an Avantes AVS-PC2000 monochromator equipped with a 2048-element linear CCD array.

15.2.2 Samples Preparation and Evaluation of the Effect of a Plasma Treatment

As a model of bacterial contamination of the surface, spores of *Geobacillus Stearothermophilus* deposited on stainless-steel disks (provided by Raven Biological Laboratories, Inc., declared spore's population 2.5×10^6) were used in this study. After the plasma treatment, a thin gold layer (~10 nm) was deposited on the samples, which were subsequently examined by a scanning electron microscope (SEM) (LEO 435VP) in order to evaluate the degree of spores' destruction caused by the plasma treatment.

For the demonstration of the elimination of various biomolecules, three different homopolymers of amino acids, that is, building blocks of proteins, were used in this study: poly-L-histidine, poly-L-arginine, and poly-L-lysine. The samples were

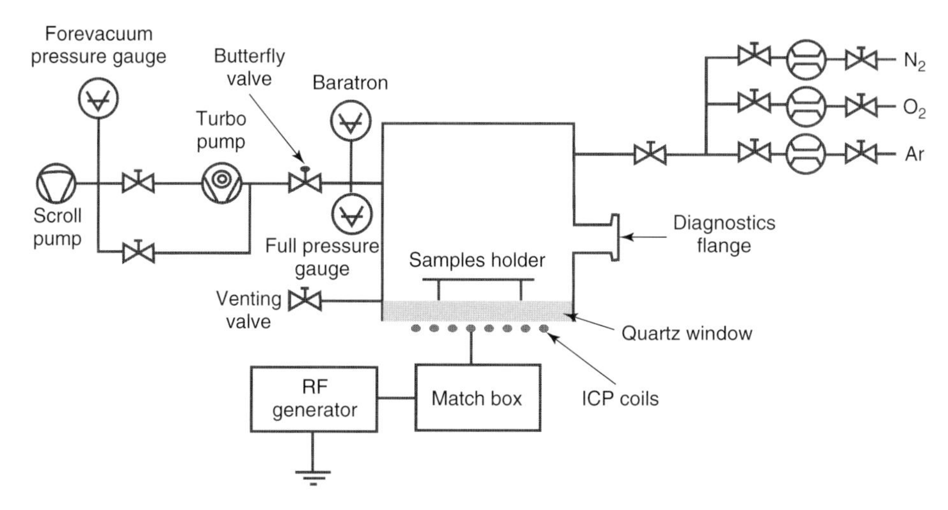

Figure 15.1 Experimental setup.

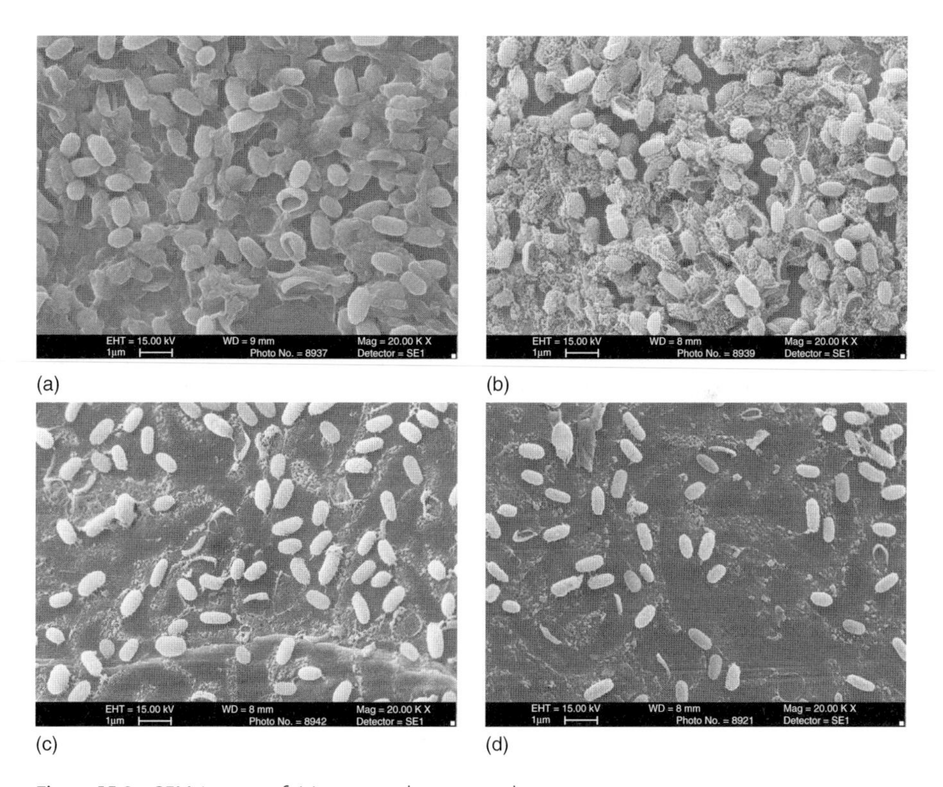

Figure 15.2 SEM images of (a) untreated spores and spores treated 60 seconds in (b) Ar plasma, (c) $Ar/O_2/N_2$ plasma, and (d) Ar/O_2 plasma.

prepared following the protocol introduced in the previous study [13], which consists in spotting small droplets of aqueous solution (0.1% wt) on Si wafers using a programmable automatic piezoelectric spotter S3 sciFLEXARRAYER (Scienion AG), followed by drying the deposited droplets in a common flow hood. For the visualization of the samples before and after the plasma treatment, a variable angle multiwavelength imaging ellipsometer (EP3, Nanofilm Surface Analysis GmbH) was employed, which offers reliable and noncontact determination of the thickness of the deposit after different treatment steps and, in turn, allows evaluating the treatment efficiency. All measurements reported in this chapter were performed in air at room temperature at an angle of incidence of 42° and using a monochromatized Xe-arc lamp tuned to 554.3 nm as a light source. The 2D $2000\,\mu m \times 2000\,\mu m$ maps of the ellipsometric angles Δ and Ψ were acquired using a conventional PCSA (polarizer–compensator–sample–analyzer) null-ellipsometric procedure, and the thickness of the deposits was calculated by means of optical modeling.

15.3 Results

15.3.1 Bacterial Spores

Figure 15.2 shows SEM images of untreated spores and spores exposed to different plasma discharges for 60 seconds. As can be seen, the morphology of the spores is significantly altered after the plasma treatment, showing both erosion of spores and reduction in their sizes. This finding shows, on the contrary to the statements made in a previous study [14], that ICP plasma discharges are capable to etch bacterial spores in relatively short time, which can in turn explain fast inactivation of bacterial spores also observed in discharge mixtures offering only limited UV radiation intensity (e.g., Ar/O_2 plasma).

However, it is evident that the effect of plasma on spores is more pronounced in both oxygen-containing mixtures. According to the analysis of a statistically relevant number of spores done on the SEM pictures (60 spores at three different samples regions), it has been found that the mean width of spores after 60 seconds of the treatment is reduced only by 3% in the case of an argon discharge, whereas Ar/O_2 and $Ar/O_2/N_2$ plasmas caused width reductions of 40 and 35%, respectively, showing importance of oxygen for the etching process. Nevertheless, these results also clearly demonstrate that partial substitution of oxygen in an Ar/O_2 mixture by nitrogen does not lead to a significant decrease in the ability of such a discharge to etch bacterial spores. Furthermore, as can be seen in Figure 15.3, partial replacement of O_2 by N_2 leads to a substantial increase in the emission of UV radiation below 300 nm, that is, radiation in the spectral range favorable for the inactivation of bacterial spores.

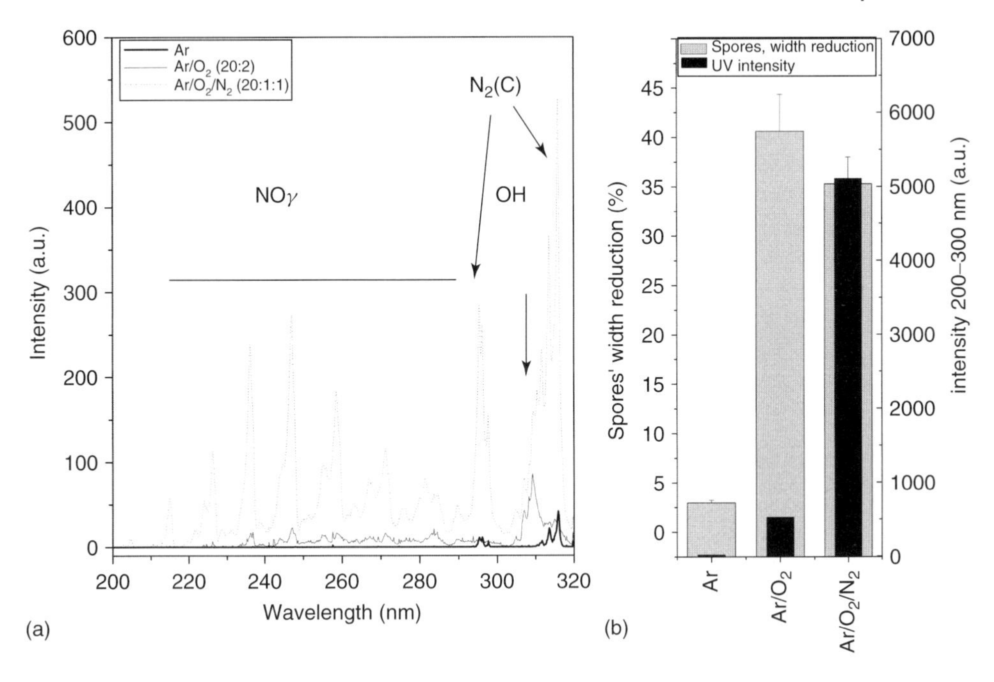

Figure 15.3 UV part of emission spectra of employed discharges (a) and comparison of spores' width reduction with UV radiation intensity in the 200–300 nm spectral range (b).

15.3.2
Homopolymers of Amino Acids

As depicted in Figure 15.4, plasma treatment also leads to a significant decrease in the thickness of all three studied homopolymers. However, in analogy to the previous case of bacterial spores, the rates of this process differ significantly depending on the discharge used, again showing markedly faster elimination of the test substances in both oxygen-containing mixtures as compared to the treatment in Ar discharge. Moreover, also in this case, there is only a slight difference between results obtained by using a ternary $Ar/O_2/N_2$ mixture and an Ar/O_2 mixture (for instance, poly-L-histidine is etched at rate of 460 nm min^{-1} in the ternary mixture, which is only by 12% lower than the removal rate observed in the argon–oxygen mixture that reached a value of 520 nm min^{-1}).

15.4
Conclusions

As it was demonstrated in this chapter, bacterial spores as well as homopolymers of amino acids can be readily etched using an $Ar/O_2/N_2$ ternary mixture. This finding, together with the results of previous studies using protein samples

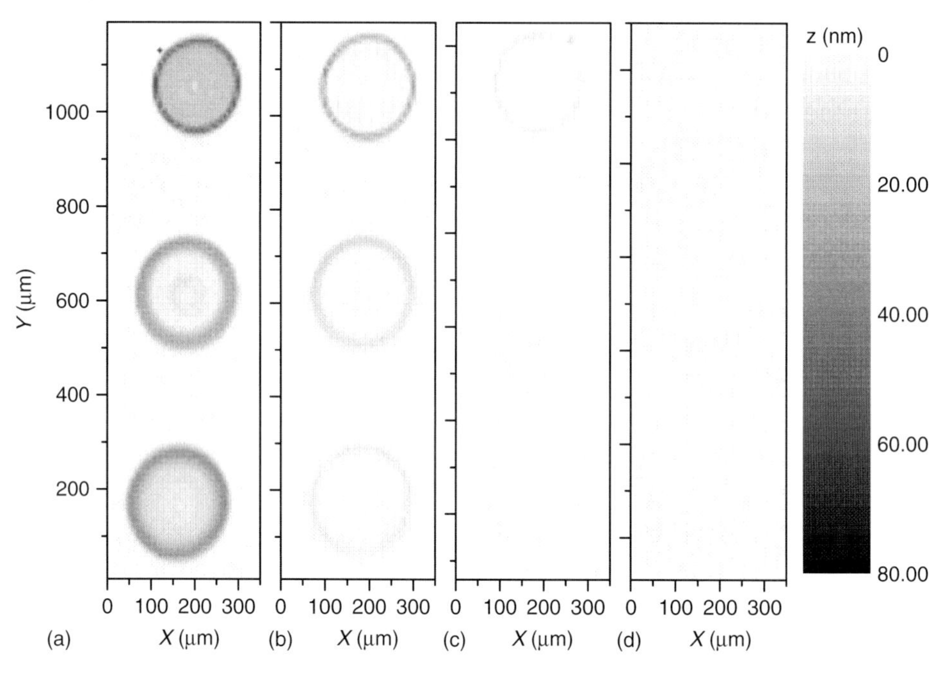

Figure 15.4 2D thickness maps of poly-L-histidine (top row), poly-L-arginine (middle row), and poly-L-lysine (bottom row): (a) untreated sample and samples treated for 5 seconds in (b) argon, (c) Ar/O_2 (20 : 2), and (d) $Ar/O_2/N_2$ (20 : 1 : 1) plasma discharges.

[11, 12], has important consequences for the optimization of plasma-based sterilization and decontamination of surfaces. First, it has been confirmed that this type of plasma discharge has a similar effect as an Ar/O_2 plasma not only on proteins but also for biological systems based in similar building blocks. This also allows presuming similar results for other biomolecules or microorganisms. Second, it has been shown that the application of $Ar/O_2/N_2$ ternary mixture leads to a much higher intensity of emitted UV radiation as compared to the Ar/O_2 plasma, which is rather advantageous, since it enhances the process of inactivation of bacterial spores via alteration of their DNA strands induced by energetic photons. This, in turn, should result in a faster sterilization process and, therefore, lowering treatment time necessary to achieve complete sterility of the processed surface, which conforms to the available experimental results in the literature [14, 15].

Acknowledgments

This work has been supported by the FP6 2005 NEST project "Biodecon" and the JRC Action 15008: NanoBiotechnology for Health.

References

1. Moisan, M., Barbeau, J., Moreau, S., Pelletier, J., Tabrizian, M., and Yahia, L.H. (2001) *Int. J. Pharm.*, **226**, 1.
2. Lerouge, S., Fozza, A.C., Wertheimer, M.R., Marchand, R., and Yahia, L.H. (2001) *Plasma Polym.*, **5**, 31.
3. Rossi, F., Kylián, O., and Hasiwa, M. (2006) *Plasma Processes Polym.*, **3**, 431.
4. Kylián, O., Hasiwa, M., and Rossi, F. (2006) *IEEE Trans. Plasma Sci.*, **34**, 2606.
5. Hasiwa, M., Kylián, O., Hartung, T., and Rossi, F. (2008) *Innate Immun.*, **14**, 89.
6. Baxter, H.C., Campbell, G.A., Whittaker, A.G., Jones, A.C., Aitken, A., Simpson, A.H., Casey, M., Bountiff, L., Gibbard, L., and Baxter, R.L. (2005) *J. Gen. Virol.*, **86**, 2393.
7. Bernard, C., Leduc, A., Bardeau, J., Saoudi, B., Yahia, L.Y., and De Crescenzo, G. (2006) *J. Phys. D: Appl. Phys.*, **39**, 3470.
8. Kylián, O., Rauscher, H., Gilliland, D., Brétagnol, F., and Rossi, F. (2008) *J. Phys. D: Appl. Phys.*, **41**, 5201.
9. Lipscomb, I.P., Sihota, A.K., and Keevil, C.W. (2006) *J. Clin. Microbiol.*, **44**, 3728.
10. Rossi, F., Kylián, O., Rauscher, H., Gilliland, D., and Sirghi, L. (2008) *Pure Appl. Chem.*, **80**, 1939.
11. Stapelmann, K., Kylián, O., Denis, B., and Rossi, F. (2008) *J. Phys. D: Appl. Phys.*, **41**, 2005.
12. Kylián, O. and Rossi, F. *J. Phys. D: Appl. Phys.*, in press.
13. Kylián, O., Rauscher, H., Denis, B., Ceriotti, L., and Rossi, F., *Plasma Processes Polym.*, submitted.
14. Halfmann, H., Bibinov, N., Wunderlich, J., and Awakowicz, P. (2007) *J. Phys. D: Appl. Phys.*, **40**, 4145.
15. Halfmann, H., Denis, B., Bibinov, N., Wunderlich, J., and Awakowicz, P. (2007) *J. Phys. D: Appl. Phys.*, **40**, 5907.

16
Sterilization and Protein Treatment Using Oxygen Radicals Produced by RF Discharge

Nobuya Hayashi and Akira Yonesu

16.1
Introduction

In the field of medicine, sterilization has been one of the important procedures for the disinfections of medical equipments. In hospitals, autoclaves using high temperature and high-pressure steam and EOG (ethylene oxide gas) sterilizers have been utilized for sterilization of medical equipments. Recently, a new sterilization method, plasma sterilization, has been developed as a secure inactivation method for medical equipment and biomaterials [1–7]. A plasma sterilization device using hydrogen peroxide (H_2O_2) as a sterilization agent has been put to practical use in hospitals as an alternative to conventional equipment such as autoclaves and EOG sterilizers. To enhance the popularity of plasma sterilization at small hospitals and clinics, it is necessary to simplify device handling and reduce running costs. We investigate the use of oxygen gas or water vapor RF plasma for the sterilization of medical equipment with tiny gaps. The proposed method is able to inactivate bacillus spores and it only uses a small amount of oxygen gas or liquid water without the need for special chemical agents.

16.2
Experimental Procedure

Figure 16.1 shows the schematic diagram of experimental apparatus. The plasma chamber of the device is made of stainless steel and has dimensions of 450 mm in length, 200 mm in diameter, and 20 l of capacity. Oxygen gas or liquid water is introduced into the chamber through a needle bulb. Liquid water is evaporated rapidly in the chamber. An inductively coupled plasma (ICP) antenna set inside the chamber is constructed in a kind of wavy shape to ensure that the generation of oxygen radicals is spatially uniform. When RF power (13.56 MHz) is applied to the antenna, a glow discharge plasma with high uniformity is produced below the antenna. The RF input power is 60 W throughout the experiment. The generation of oxygen radicals such as excited oxygen atom or molecule, and hydroxyl radicals

Industrial Plasma Technology. Edited by Yoshinobu Kawai, Hideo Ikegami, Noriyoshi Sato, Akihisa Matsuda, Kiichiro Uchino, Masayuki Kuzuya, and Akira Mizuno

ISBN: 978-3-527-32544-3

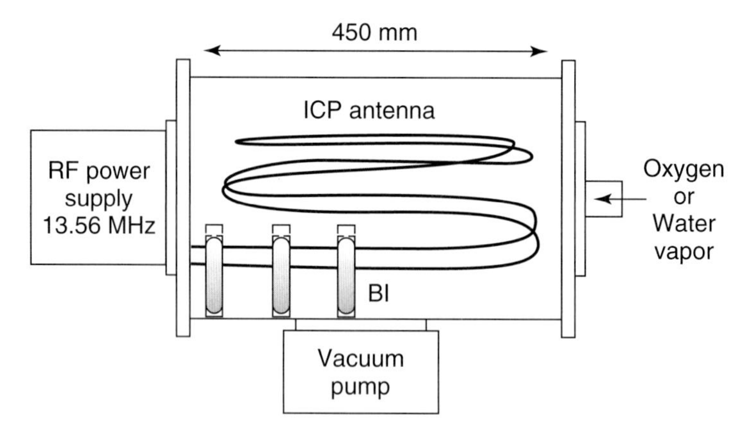

Figure 16.1 Schematic diagram of experimental apparatus.

(OH) during plasma operation was confirmed based on UV–vis light emission spectra and a strip-type chemical indicator.

The device performance in sterilizing medical equipment was confirmed using vial-type biological indicator (BI) containing heat-resistive spore of *Geobacillus stearothermophilus* (ATCC 7953). This BI imitates tiny structures of medical equipments, and a successful result indicates a proof of the complete sterilization of medical equipments. The decimal reduction value (D value) of bacilli was determined based on the colony count method. For colony count, the D values were estimated from survive curve where the number of bacilli changes from 100 to 10.

The characteristics of removing protein from the medical equipments were investigated using casein, albumin, and heat-resistive avidin proteins. The treatment effect of proteins is determined by the peak height of chemical bonds of C–H, N–H, C–O, C = O, and O–H appeared on FTIR spectra.

16.3 Generation of Oxygen Radicals

Because medical equipments may have small gaps or scratches on the surface, it is important to ensure penetration of radicals into the above tiny structures. To evacuate tiny gaps from the medical equipments such as catheters and syringes, the pressure in the vacuum chamber has been reduced to 3 Pa. Also, oxygen and OH radicals are produced effectively at the lower pressure regime. And then the pressure was increased to 10^3 Pa to penetrate oxygen radicals into the tiny gaps of the medical equipments. In this phase, due to the increasing pressure, the bacilli come in contact with oxygen and OH radicals, and are inactivated by the oxygen radicals.

The UV–vis light emission spectrum of the oxygen and water vapor plasma indicate that the intense peak at 777 and 309 nm proves the generation of excited triplet atomic oxygen $O(^3P)$ or OH radicals, respectively, that are able to destroy atomic bonds such as C–H, C–N, and N–H bonds within the bacillus spores. The peak intensity of the oxygen radicals increases with RF power and decreases rapidly as gas pressure increases from 10 to 300 Pa [7]. In the case of water vapor plasma, the red color originates from dissociated hydrogen ($H\alpha$) at 656 nm, which is the circumstantial evidence of OH radical generation. The intensity of the emission is almost uniform in the radial direction although there is slight localization around the RF antenna. This result implies that oxygen radicals are produced uniformly in the plasma chamber. The uniformity of the radical is important for a sterilizer.

To confirm the penetration of oxygen radicals in the tiny structure of medical equipments, a chemical indicator was adopted, which responds to sufficient amount of oxygen radicals for sterilization. The chemical indicator embedded in the BI vessel was enclosed in the surgical cloth of several layers, and then wrapped in the sterilization bag. When the chemical indicator was placed in the plasma for four hours, the indicator turned into yellowish color from its original red color. Therefore, the oxygen radicals were confirmed to pass through the several layers of cloths and then contact the bacilli. This result indicates that the proposed sterilization method could succeed in the AAMI (The Association for the Advancement of Medical Instrumentation) Challenge Test ST-41 for sterilization of tiny gaps.

16.4 Sterilization of Medical Equipments

The sterilization performance of the oxygen radicals was investigated using vial-type BI. The BIs were located at 50 mm intervals in the chamber. The BI contains spores of *Geobacillus stearothermophilus*, which is a heat-resistant spore. All BIs indicated successful sterilization within 3 hours, including cold points. When the pressure in the sterilization chamber was changed from 3 to 300 Pa introducing the oxygen gas for 160 ms, the radicals were accelerated by the pressure variation, and the velocity of injected gas including oxygen neutral gas and oxygen radicals was estimated to be approximately 32 ms^{-1}. Then, the flying range of the oxygen radical was estimated to be 30 cm in this experiment, assuming the lifetime of the oxygen radical as 0.01 s. This range enables the oxygen and OH radicals to reach the tiny gaps of medical equipments from the RF antenna where the oxygen radicals were produced. When the pressure inside the BI was changed to 3 Pa and the collision of oxygen radicals is significantly low, the oxygen and OH radicals can reach the inside of BI sufficiently in the low-pressure regime. Temperature of the gas and materials in the chamber during the inactivation was 60 °C at maximum without any cooling system, which was measured using both a thermo-level on the material and an IR emission temperature meter. Spore of *Geobacillus stearothermophilus* used in this experiment is alive under 120 °C. Also, sterilization cannot be achieved by Argon gas that does not create radicals. Therefore, the sterilization was achieved by oxygen radicals produced by low-temperature plasma.

Bacilli on the medical scissors subjected to the plasma treatment were collected and cultivated for 48 hours using the medium sheet for microbial detection, and

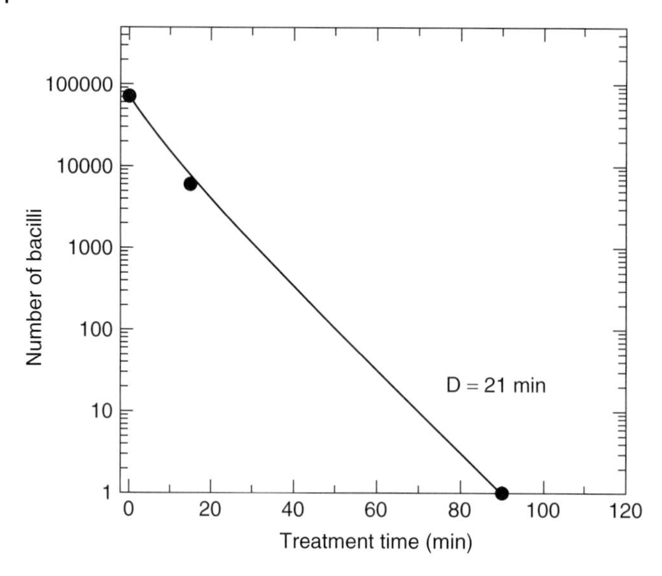

Figure 16.2 Typical survivorship curve of *Geobacillus stearothermophilus* using oxygen plasma.

the grown colonies were counted to evaluate the inactivation ability. Figure 16.2 illustrates a typical survivorship curve using oxygen plasma. Time evolution of the colony population obtained at different times during the 3 hours indicated a D value of 21 and 26 minutes using oxygen and water vapor, respectively. It can be concluded that OH and O radicals generated from the water vapor plasma with pressure-varying driving are effective for the sterilization of medical equipments. The performance in removing proteins from the medical equipment was investigated using water vapor plasma.

16.5
Decomposition of Protein's Structure

A kind of protein that can cause infections in the living body has been a serious problem in the field of medicine. Since this infectious protein called *prion* has a robust β-sheet in the second order structure, it possesses significantly high receptivity to the EOG, autoclave, and formalin as well as heat and radiation. To inactivate the robust protein, all bonds of atoms in the protein particle must be decomposed completely. Even though the sufficient heating and radiation is able to destruct the protein completely, medical equipments tend to be damaged by them. In this experiment, the oxygen radicals produced by the RF discharge are adopted to remove protein on medical equipments. The removal effect is estimated changing the parameters of RF power, gas pressure RF, and treatment period. The optimum parameters to remove protein from medical equipments are found out avoiding the damage to the equipment.

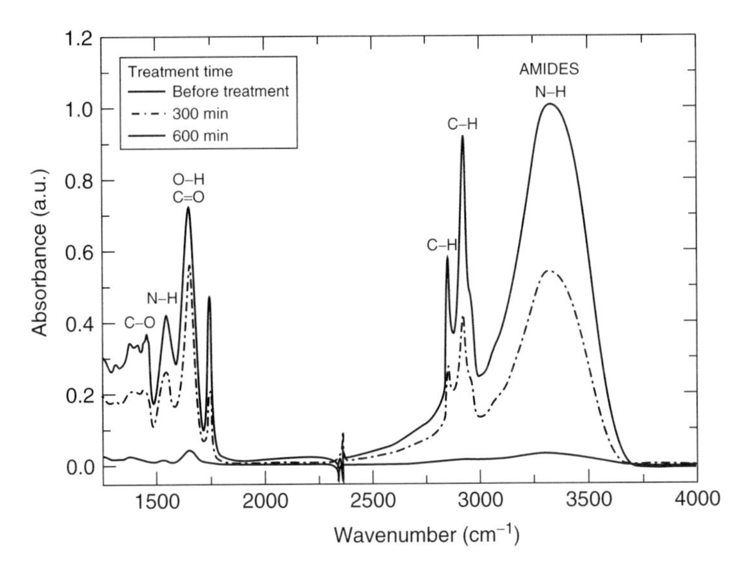

Figure 16.3 FTIR spectra of casein varying the treatment time.

Figure 16.3 shows FTIR spectra of casein protein on a CaF_2 substrate varying during treatment period. The density of the casein on the plate is 5.3 mg cm^{-2}, which is approximately same as the remnant protein on medical equipments after the first stage wash and before the sterilization sequence in hospitals. In the spectra, C–O peak at 1460 cm^{-1}, amide peaks ranged from 1600 to 1700 cm^{-1}, C–H peak at 2800 cm^{-1}, and N–H peak at 3300 cm^{-1} are observed. The peaks from 1600 to 1700 cm^{-1} include the second order structure of protein such as α-helix and β-sheet. The β-sheet structure of protein that appears at 1635 cm^{-1} is significantly robust and resistive to the heat and chemical agents due to hydrogen bonds. When the protein is rich in the β-sheet structure, the protein is difficult to destroy using conventional treatment methods of autoclave and formalin avoiding the damage to the medical equipment.

FTIR spectra of the casein protein applied at an initial coverage of 5.3 mg cm^{-2} show that the amount of protein reduces over time and is removed completely after 10 hours. Both peaks of C–H and N–H bindings of amides structure in amino acid at 2925 and 3332 cm^{-1} are reduced with the treatment time. Spectral peaks around 1635 cm^{-1} attributed to the second order structure also decrease with time. The rate of the temporal decrease of the amides structure is found to be faster than that of the second order structure, as shown in Figure 16.4. This fact implies that the amides structure of amino acid is decomposed rapidly, while the second order structure of protein has resistance against the oxygen radicals, which is same as another sterilization method. Structures of amino acid almost diminished after treatment for eight hours, while the second order structure remains approximately 15% of original amount.

The protein that is rich in β-sheet, such as fibrin and prion, is difficult to destroy by chemical agents and heat due to the strong bond of a sheet-type second order

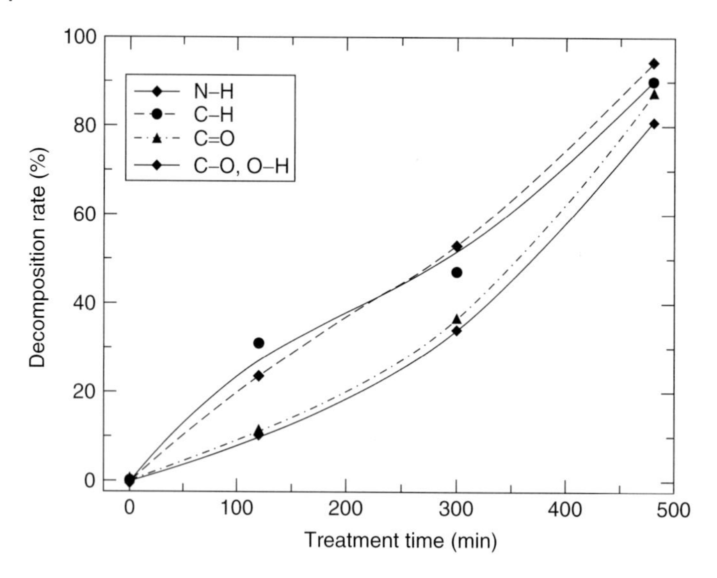

Figure 16.4 Temporal evolution of decomposition rate of each bonds of casein protein.

structure. The second order structure of protein is confirmed by FTIR spectra. α-helix and β-sheet can be identified clearly as the peaks of 1655 and 1635 cm^{-1}, respectively. When the protein of the egg albumin and avidin that is rich in the β-sheet structure is immersed in the oxygen plasma that is produced by the low-pressure RF discharge, both the α-helix and β-sheet in the FTIR second derivative spectra of both proteins are reduced after the several hours. Therefore, the oxygen radicals such as atomic oxygen and hydroxyl radical enable to destruct the β -sheet structure.

References

1. Lerouge, S., Werttheimer, M.R., Marchand, R., Tabrizian, M., and Yahia, L.H. (2000) *J. Biomed. Mater. Res.*, **51**, 128.
2. Hury, S., Vidal, D.R., Desor, F., Pelletier, J., and Lagarde, T. (1988) *Lett. Appl. Microbiol.*, **26**, 417.
3. Moreau, S., Moisan, M., Tbrizian, M., Barbeau, J., Pelletier, J., Ricard, A., and Yahia, L.H. (2000) *J. Appl. Phys.*, **88**, 1166.
4. Purevdorj, D., Igura, N., Ariyada, O., and Hayakawa, I. (2003) *Lett. Appl. Microbiol.*, **37**, 31.
5. Park, B.J., Lee, D.H., Park, J.-C., Lee, I.-S., Lee, K.-Y., Hyun, S.O., Chun, M.-S., and Chung, K.-H. (2003) *Phys. Plasmas*, **10**, 4539.
6. Lei, X., Rui, Z., Peng, L., Li-Li, D., and Ru-Juan, Z. (2004) *Chinese Phys.*, **13**, 913.
7. Hayashi, N., Guan, W., Tsutsui, S., Tomari, T., and Hanada, Y. (2006) *Jpn. J. Appl. Phys.*, **45**, 8358.

17
Hydrophilicity and Bioactivity of a Polyethylene Terephthalate Surface Modified by Plasma-Initiated Graft Polymerization

Nagahiro Saito, Takahiro Ishizaki, Junko Hieda, Syohei Fujita, and Osamu Takai

17.1
Introduction

Polyethylene terephthalate (PET) film is widely used in a variety of applications because it has some excellent material properties, such as a high melting point and high tensile strength; other characteristics include very good barrier properties, crease resistance, solvent resistance, and resistance to fatigue. However, bonding and finishing of the PET film presents a problem because of its low surface hydrophilicity. For example, this affects wettability, biocompatibility, adhesion, and various other surface treatments. Among these, biocompatibility is very important for biomedical applications of PET film, since it greatly affects protein adsorption and cell cultures. Thus, controlling the hydrophilicity of the PET surface is crucial for biomedical applications. To impart hydrophilicity to the PET surface, many methods have been devised and used commercially. Among them, plasma treatment is a promising means of enhancing the hydrophilicity of the polymer surface [1–4]. For example, surface oxidation by plasma treatment (e.g., O_2 and H_2O) improves the wettability of polymers such as polyethylene (PE) and poly(methyl methacrylate) [5]. However, these treatments lack permanence because of surface rearrangement. Indeed, previous attempts to make polysulfone membranes hydrophilic by plasma treatment, primarily with O_2 plasma, resulted in only transient hydrophilicity, as demonstrated by contact angle changes within 24 hours of plasma treatment.

Plasma is a complex mixture of charged (electrons and ions) and neutral (atoms, molecules, radicals) species. The excited molecular and atomic species in the plasma, in turn, can emit photons over a very broad span of the electromagnetic spectrum, ranging from X-rays to the infrared. The radiation in the vacuum ultraviolet or ultraviolet region is sufficiently energetic to cause modification of surface layers by breaking bonds or initiating photochemical reactions in the polymer [6]. Many researchers have explained that improvement in surface hydrophilicity was generated by introducing new oxygen-containing functional groups, such as –OH and –OOH, onto the surface. These functional groups are hydrophilic [3, 4, 7–9], so the plasma surface modification should be able to increase the hydrophilicity of

Industrial Plasma Technology. Edited by Yoshinobu Kawai, Hideo Ikegami, Noriyoshi Sato, Akihisa Matsuda, Kiichiro Uchino, Masayuki Kuzuya, and Akira Mizuno

ISBN: 978-3-527-32544-3

the PET surface. However, surface modification by the plasma process introduces two complications. First, polymers are damaged by plasma irradiation [10]; this induces polymer chain scission, thereby contaminating the polymer surfaces. Second, surface hydrophilicity decreases over time because of surface reorientation [11]. To overcome these problems when using plasma treatments, it is necessary to introduce organic molecules with hydrophilic functional groups without causing surface damage. Surface-wave plasma (SWP) is a promising method for this purpose, because it allows treatment of polymer surfaces on a large scale at a low electron temperature [12]. In addition, plasma-initiated graft polymerization can attach functional groups or long alkyl chains to polymer surfaces.

In this chapter, we demonstrate surface modification of the PET film to provide hydrophilicity through plasma-initiated graft polymerization using SWP. The wettability of the modified PET surface was estimated using contact angle measurements. The surface composition was analyzed using X-ray photoelectron spectroscopy (XPS). The roughness of the polymer surface was observed with atomic force microscopy (AFM). Bioactivity was investigated by culturing mouse fibroblast cells on the modified PET film.

17.2 Experimental

17.2.1 Materials

PET film was purchased from Tsutsunaka Plastic Industry Co., Ltd., Japan. The surface of the film was cleaned with ethanol and subsequently rinsed with ultrapure water for 5 minutes before use. Acrylic acid (AA), (2-hydroxyethyl) methacrylate (HEMA), and styrene (St) monomers were purchased from WAKO Pure Chemical Industries, Ltd., Japan.

17.2.2 Surface Modification

The PET surface was modified by SWP excited by microwave radiation. The SWP system consisted of a waveguide attached to a microwave generator, a coaxial waveguide, and a stainless steel chamber. A curved reflective plate was attached to the waveguide to efficiently introduce microwaves to the coaxial waveguide. The top of the chamber was sealed with a quartz plate 12 mm in thickness and 14 cm in diameter. The coaxial waveguide was connected vertically to the quartz plate as the dielectric. The generated microwave passed through the waveguide and reached the quartz plate, leading to the formation of the plasma. The cleaned PET substrates were placed at the center of the substrate stage in the chamber. The distance between the quartz plate and substrate stage was kept constant at 5 cm. The chamber was evacuated to 0.5 Pa with a rotary pump prior to surface modification and Ar gas was subsequently introduced. The total pressure in the chamber varied from 20 to 100 Pa. A microwave power of 200 W was applied to generate plasma for 5 minutes using a 2.45-GHz generator. The PET surface was irradiated with Ar plasma that generated radicals on the polymer surface [13]. The polymer films were then exposed to AA, HEMA, or St monomer in the vapor phase. The gas pressures for AA, HEMA, and St monomer were 2000, 80, and 400 Pa, respectively. Each gas pressure was kept constant for 60 minutes, leading to the graft polymerization of the monomer. After exposure to the gas, the polymer films were ultrasonically cleaned in ultrapure water or ethanol for 10 minutes.

17.2.3
Surface Characterization

Topographic images of PET surfaces before and after plasma irradiation were acquired with an AFM in the tapping mode; a Micro Cantilever type SI–DF3 with a spring constant of 1.1 N m^{-1} and a resonance frequency of 24 kHz was used for all the measurements. The scanning area was a square of side 10 μm. The hydrophilicity of the polymer surface was characterized using a static water contact angle measurement. A 2-μl water droplet was placed on the sample surfaces. The chemical bonding states and chemical composition of the polymer surfaces were analyzed using XPS. The MgKα X-ray source was operated at 10 mA and 12 kV.

17.2.4
Cell Culture

To characterize the bioactivity of the modified PET surface, mouse fibroblast cells (NIH-3T3) were cultured onto both untreated and modified PET surfaces by immersing them into the culture medium (DMEM, pH 7.0) in a humidified atmosphere containing 5.0% CO_2 at 37.0 °C for three days. The initial cell-seeding number was 5000 cells cm^{-2}. The cultured cells were observed with optical and phase-contrast microscopes at intervals of 12 hours and were counted using a blood-cell counting chamber.

17.3
Results and Discussion

To provide hydrophilicity to the PET surface without incurring damage, surface modification of the PET film was conducted using varied gas pressures and treatment times. Figure 17.1a shows the water contact angles of the PET surface modified at different Ar gas pressures. The water contact angle of the untreated PET surface was estimated to be 76.9°, while those modified by Ar gas pressures from 20 to 100 Pa ranged from 34° to 37°. The Ar plasma treatments greatly improved the hydrophilicity of the PET surface. However, the hydrophilicity of the modified PET surface decreased over a period of 70 hours after the plasma treatment, which

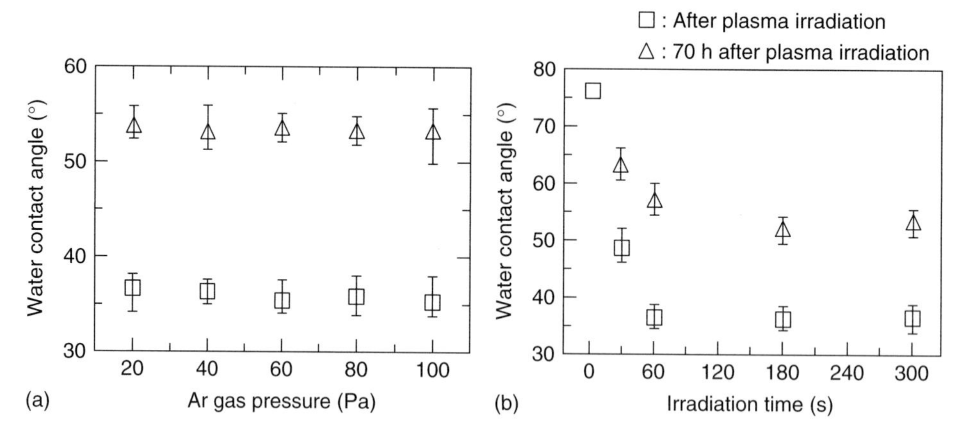

Figure 17.1 (a) Water contact angles of the PET surface modified at different Ar gas pressures. (b) Effect of plasma irradiation time on water contact angles of the PET surface modified by Ar plasma at 20 Pa.

could have been caused by surface reorientation of the PET film. It should be noted that Ar gas pressures have no effect on the chemical properties of the modified PET film surface. Figure 17.1b shows the effect of the plasma irradiation time on the water contact angles of the PET surface modified by Ar plasma at 20 Pa. The hydrophilicity of the PET surface was greatly improved by the Ar plasma treatment for 60 seconds. The hydrophilicity of the modified PET film remained almost constant with treatment time longer than 60 seconds. The decrease in the degree of hydrophilicity with time after the treatment of the modified PET surfaces remained almost the same, independent of the plasma treatment time.

Figure 17.2 shows topographic images of (a) the untreated PET surface, (b) the PET surface modified by Ar plasma at 20 Pa, and (c) the PET surface modified by Ar plasma at 100 Pa. The surface of the untreated PET film appeared to be comparatively smooth with a root-mean-square roughness (R_{rms}) of 1.2 nm (Figure 17.2a). No change in the topography was observed on the PET surface treated by Ar plasma at 20 and 100 Pa, and the R_{rms} were estimated to be 1.0 and 0.7 nm, respectively. This indicated that Ar plasma treatments using SWP did not etch the PET surface. Thus, we were successful in altering the hydrophilicity of the PET surface without any physical damage. However, there was still a lack of permanence of the hydrophilicity of the PET surface modified by Ar plasma.

Plasma-initiated graft polymerization was carried out to enhance the permanence of the hydrophilicity on the PET surface. Figure 17.3 shows a time course of the water contact angles of the PET surface modified by (i) Ar plasma treatment and (ii) plasma-initiated graft polymerization using hydrophilic AA monomer. The water contact angles of the PET surface after Ar plasma treatment and Ar plasma followed by hydrophilic AA modification were 35° and 23°, respectively, indicating that plasma-initiated graft polymerization slightly improved the hydrophilicity. In addition, there was a noticeable difference between the treatments in the

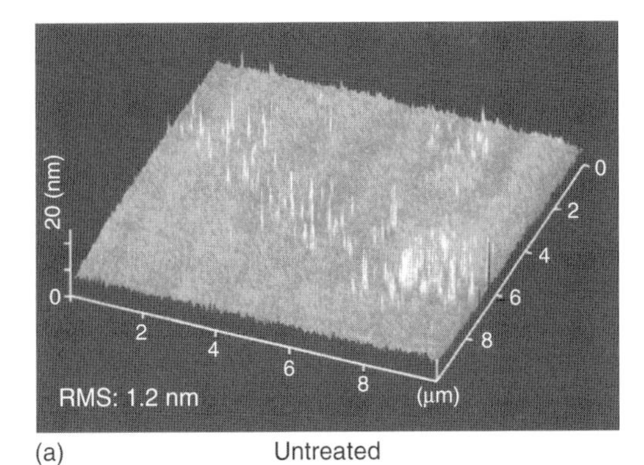

(a) Untreated

RMS: 1.0 nm

(b) Plasma irradiation (Ar pressure: 20 Pa)

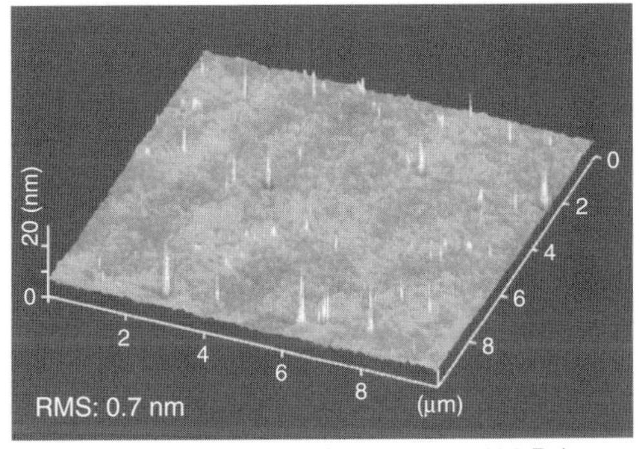

(c) Plasma irradiation (Ar pressure: 100 Pa)

Figure 17.2 Topographic images of (a) the untreated PET surface, (b) the PET surface modified by Ar plasma at 20 Pa, and (c) the PET surface modified by Ar plasma at 100 Pa.

permanence of the hydrophilicity. The water contact angles of the Ar plasma-treated PET surface increased over time as a result of the instability of hydrophilic groups, such as –OH and –COOH groups, on the polymer surface. In contrast, those on the PET surface after Ar plasma followed by hydrophilic AA modification remained constant at approximately 20°, thereby showing stable hydrophilicity.

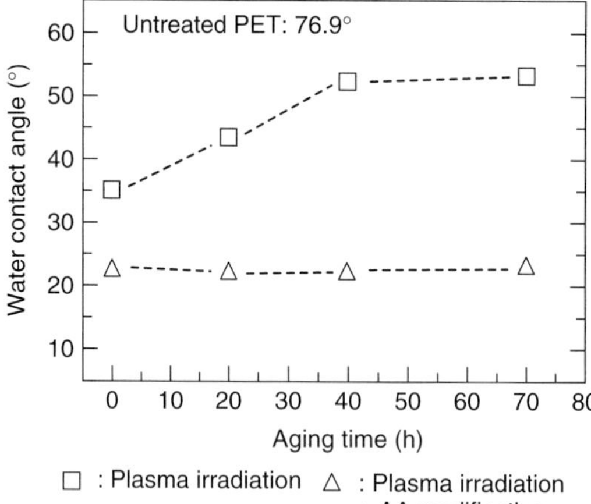

Figure 17.3 Time course of water contact angles of the PET surface modified by Ar plasma treatment and plasma-initiated graft polymerization using hydrophilic AA monomer.

To test the chemical bonding state of the PET surface, XPS measurements were performed. Figure 17.4 shows XPS C 1s spectra obtained from (a) the untreated PET surface, (b) the PET surface after Ar plasma treatment, and (c) the PET surface after Ar plasma followed by hydrophilic AA modification. The C 1s spectrum of the untreated PET surface (Figure 17.4a) is deconvoluted into three peaks corresponding to carbon atoms of the benzene rings unbonded to the ester group (peak C1 at 284.7 eV), carbon atoms singly bonded to oxygen (peak C2 at 286.5 eV), and ester carbon atoms (peak C3 at 289.1 eV) [13–23]. The relative component concentrations determined from these peak areas were 58.9% for C1, 21.9% for C2, and 19.2% for C3. However, the C 1s spectrum (Figure 17.4b) reveals that the PET surface was chemically altered by the Ar plasma treatment. The relative component concentrations determined from these peak areas were 54.8% for C1, 34.1% for C2, and 11.1% for C3. The C1–C3 peaks of the untreated PET were broadened by the Ar plasma treatment, indicating that each peak includes more than one unique species. These species are ascribed to be formed through a chemical reaction of the polymer chains with activated ion species and radicals. The broadening of C1 and C2 peaks has been associated with the destruction of aromatic rings in PET [19, 20, 24]. These peaks are assumed to include signals from polar groups such as –C–C=O, –C–COO, or –C–C–O [17, 19]. The C 1s spectrum (Figure 17.4c) reveals that the intensity of the C2 and C3 peaks originating from polyacrylic acid has become stronger than that of the plasma-treated PET surface. This indicates that many AA monomers were adsorbed onto the PET surface and then polymerized by plasma-initiated radicals. Thus, polyacrylic acid was successfully grafted onto the PET surface, resulting in increased hydrophilicity. The modified PET surfaces maintained stable hydrophilicity for 70 hours because two functional groups in polyacrylic acid (the –COOH group and the hydrocarbon chain) strongly affect the surface properties.

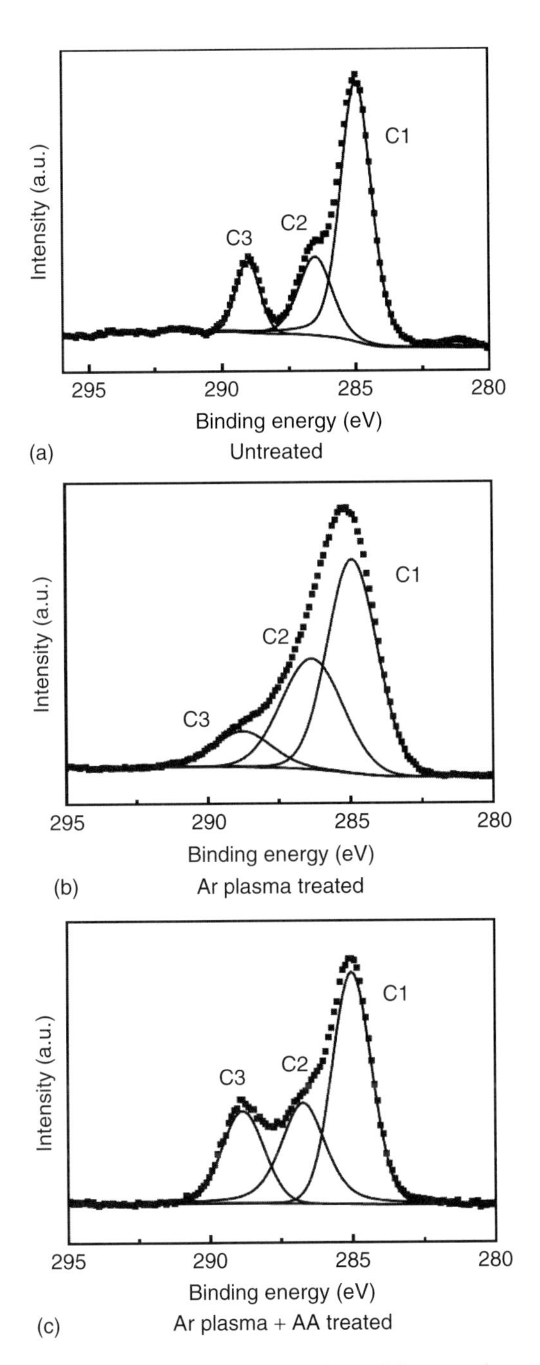

Figure 17.4 XPS C 1s spectra obtained from (a) the untreated PET surface, (b) the PET surface after Ar plasma, and (c) the PET surface after Ar plasma followed by hydrophilic AA modification.

Thus, we successfully provided a PET surface with stable hydrophilicity using plasma-initiated graft polymerization. However, to control the bioactivity of the PET surface, it is necessary to produce PET surfaces with various other physicochemical properties. Hence, we performed additional surface modifications of the PET film through the plasma-initiated graft polymerization using two different monomers, that is, HEMA and St.

Figure 17.5a shows a time course of the water contact angles of the PET surface modified by Ar plasma treatment and plasma-initiated graft polymerization using hydrophilic HEMA monomer. The water contact angles of the PET surface after Ar plasma treatment and Ar plasma followed by hydrophilic HEMA modification were 35° and 29°, respectively. The plasma-initiated graft polymerization of HEMA also slightly improved the hydrophilicity of the PET surface. We observed a noticeable difference between the treatments in the permanence of the hydrophilicity. The water contact angles of the Ar plasma-treated PET surface increased over time, while those on the PET surface after Ar plasma followed by hydrophilic HEMA modification remained almost constant, ranging from 30° to 35°. The XPS C 1s spectrum of the PET surface after Ar plasma followed by hydrophilic HEMA modification (Figure 17.5d) is deconvoluted into three peaks corresponding to

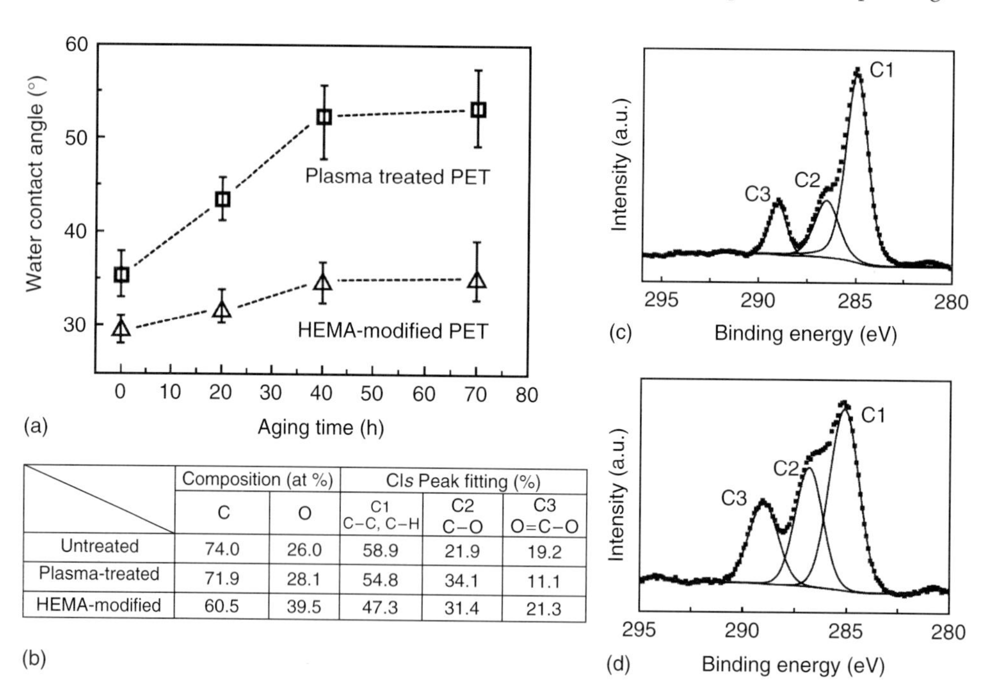

	Composition (at %)		Cls Peak fitting (%)		
	C	O	C1 C–C, C–H	C2 C–O	C3 O=C–O
Untreated	74.0	26.0	58.9	21.9	19.2
Plasma-treated	71.9	28.1	54.8	34.1	11.1
HEMA-modified	60.5	39.5	47.3	31.4	21.3

Figure 17.5 (a) Time course of water contact angles of the PET surface modified by Ar plasma treatment and plasma-initiated graft polymerization using hydrophilic HEMA monomer. (b) Chemical composition and area rates of C1, C2, and C3 peaks obtained from C 1s peak fitting. (c) XPS C 1s spectrum of the untreated PET surface. (d) XPS C 1s spectrum of the PET surface after Ar plasma followed by hydrophilic HEMA modification.

the carbon atoms of the benzene rings unbonded to the ester group (peak C1 at 284.7 eV), carbon atoms singly bonded to oxygen (peak C2 at 286.5 eV), and ester carbon atoms (peak C3 at 289.1 eV), while Figure 17.5c shows the XPS C 1s spectrum of the untreated PET surface. The relative component concentrations determined from these peak areas were 47.3% for C1, 31.4% for C2, and 21.3% for C3. The intensity of the C2 peak attributed to HEMA increased compared to that of the untreated PET surface. This indicates that many HEMA monomers were grafted and polymerized on the PET surface, leading to the formation of a stable hydrophilic surface. Figure 17.6a shows a time course of the water contact angles of the PET surface modified by Ar plasma treatment and plasma-initiated graft polymerization using the hydrophobic St monomer. The water contact angle of the PET surface after Ar plasma followed by hydrophobic St modification was 53° and increased up to more than 60° over time. The St-modified PET surface was hydrophilic even though the St monomer is hydrophobic. This indicates that the St monomers might not have been sufficiently adsorbed at the reaction sites on the PET surface because of the steric hindrance between the St monomers. On the basis of the Cassie–Baxter equation, the ideal water contact angles of a surface covered by St polymers and the Ar plasma-treated PET surface should be 90° and 37°, respectively; the St surface

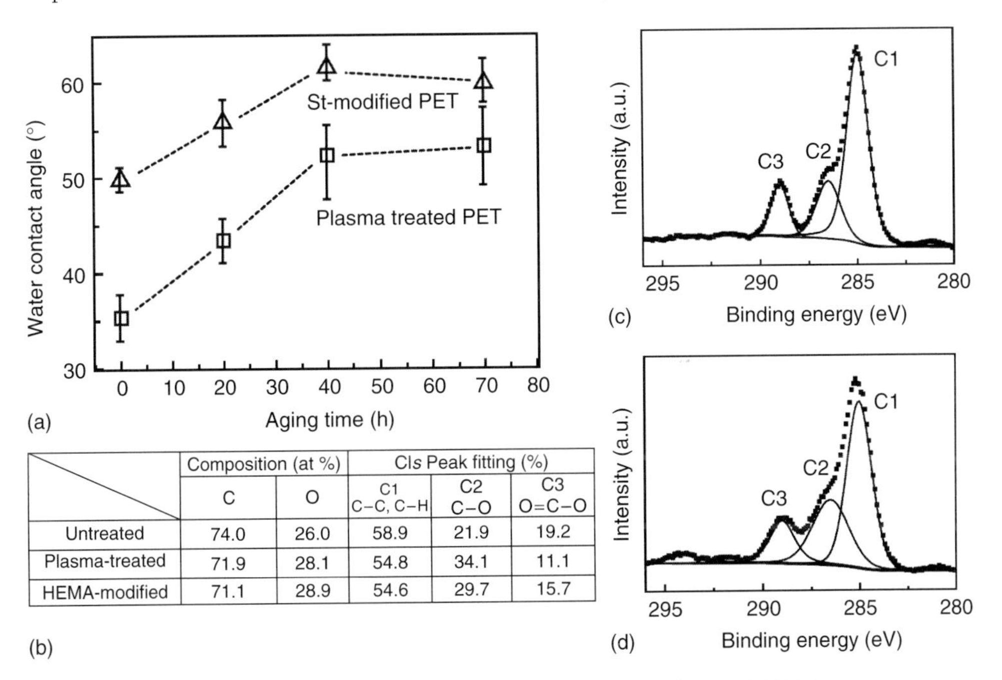

	Composition (at %)		Cls Peak fitting (%)		
	C	O	C1 C–C, C–H	C2 C–O	C3 O=C–O
Untreated	74.0	26.0	58.9	21.9	19.2
Plasma-treated	71.9	28.1	54.8	34.1	11.1
HEMA-modified	71.1	28.9	54.6	29.7	15.7

Figure 17.6 (a) Time course of water contact angles of the PET surface modified by Ar plasma treatment and plasma-initiated graft polymerization using hydrophobic St monomer. (b) Chemical composition and area rates of C1, C2, and C3 peaks obtained from C 1s peak fitting. (c) XPS C 1s spectrum of the untreated PET surface. (d) XPS C 1s spectrum of the PET surface after Ar plasma followed by hydrophobic St modification.

coverage was estimated to be about 20%. This surface coverage might be insufficient to provide the chemical properties of the St polymer. Thus, the water contact angle of the St-modified PET surface was lower than the expected value and increased greatly over time. Figure 17.6c,d shows the XPS C 1s spectra of the untreated PET surface and the PET surface after Ar plasma followed by hydrophilic St modification, respectively. The XPS C 1s spectrum after St modification is deconvoluted into three peaks corresponding to the carbon atoms of the benzene rings unbonded to the ester group (peak C1 at 284.7 eV), carbon atoms singly bonded to oxygen (peak C2 at 286.5 eV), and ester carbon atoms (peak C3 at 289.1 eV), as well as a weak peak at around 294.0 eV resulting from a $\pi \rightarrow \pi^*$ shake-up process. The peak resulting from the $\pi \rightarrow \pi^*$ transition, attributable to the phenyl groups, appeared as a result of the reaction of St monomers on the PET surface. Thus, three types of monomers with hydrophilic or hydrophobic functional groups were polymerized on the PET surface. The water contact angle of the PET surface modified with hydrophilic AA and HEMA monomers decreased from approximately 80° before treatment to less than 35°. The hydrophilicity of the PET surface modified with hydrophilic AA and HEMA monomers was maintained for 70 hours.

Finally, 3T3 fibroblast cells were cultured on the untreated and modified PET surface to investigate bioactivity. Figure 17.7 shows the number of cells cultured

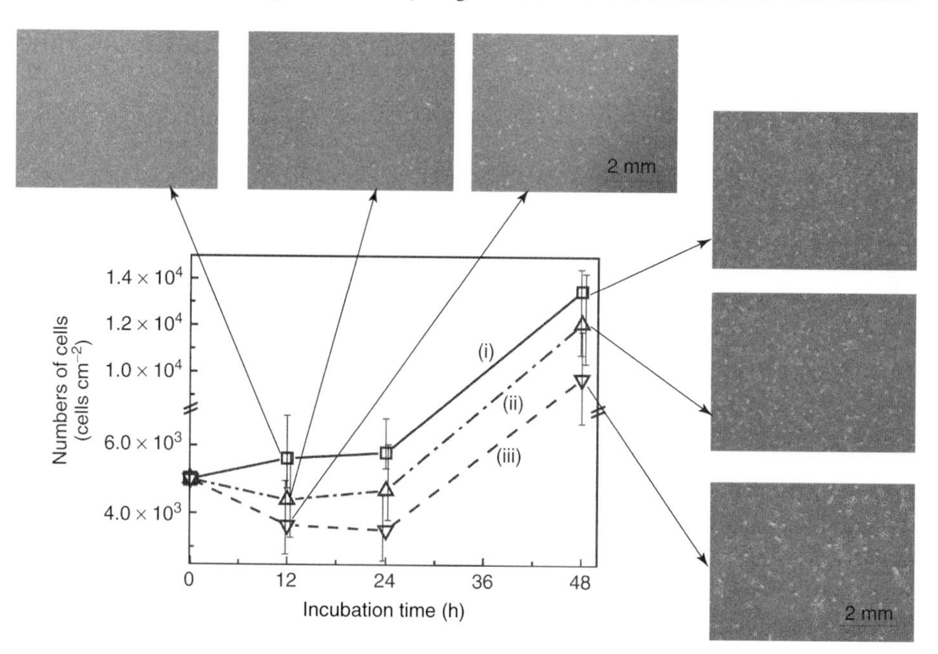

Figure 17.7 Numbers of cells cultured on the Ar plasma or AA-modified PET surface and the phase-contrast microscopic images for the behaviors of the cell growth at 12 and 48 hours after the cell culture. (i) Untreated PET surface. (ii) Ar plasma-treated PET surface. (iii) AA-modified PET surface.

on the Ar plasma or AA-modified PET surface and the phase-contrast microscopic images for the cell growth behavior at 12 and 48 hours after the cell culture. Initial cell growth was depressed on the AA-modified PET surface compared with the untreated and Ar plasma-treated PET surfaces. Phase-contrast microscopic images showed that the 3T3-fibroblast cells did not adhere well to the AA-modified surface. The number of cells cultured for 12 hours correlated to some extent with the hydrophilicity of the PET surface. The cells increased as the water contact angles of the PET surface increased. After the cells were cultured for 48 hours, the cell numbers increased on all PET surfaces, that is, untreated, Ar plasma-treated, and AA-treated PET surfaces. The growth rate remained almost constant on all PET surfaces. This indicates that the number of cultured cells might depend on the cell adhesion behavior after cell seeding. Figure 17.8 shows the number of 3T3 fibroblast cells cultured for 48 hours on the untreated, Ar plasma-treated, AA-modified, HEMA-modified, and St-modified PET surfaces. On the untreated and St-modified PET surfaces, the number of cultured cells was higher than the initial number of seeded cells. Despite the low coverage of the St monomer, the number of cultured cells increased on the St-modified PET surface. This indicates that hydrophobic functional groups in the polystyrene could affect adsorption of proteins that are necessary for cell adhesion. These proteins have hydrophobic functional groups in their molecular structure, so hydrophobic interaction between grafted polystyrene and proteins must have occurred, resulting in an increase in the number of cultured cells. In contrast, the number of cells cultured on the AA-modified and

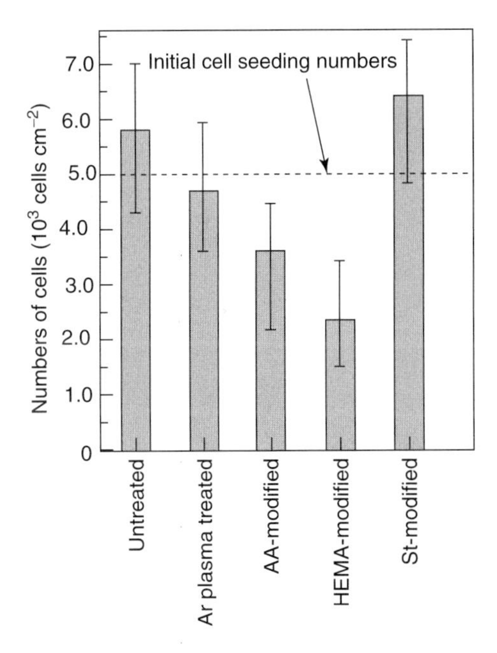

Figure 17.8 Numbers of the 3T3 fibroblast cells cultured for 48 hours on the untreated, Ar plasma-treated, AA-modified, HEMA-modified, and St-modified PET surfaces.

HEMA-modified surfaces was much lower than the initial number of seeded cells. It should be noted that the AA-modified and HEMA-modified surfaces had polar groups (−C−C=O, −C−COO, or −C−C−O) and maintained a stable hydrophilicity. These results suggest that these negative polar groups might suppress adsorption of proteins resulting from repulsive electrostatic interaction.

17.4
Conclusions

We successfully produced stable hydrophilicity on the PET surface through plasma-initiated graft polymerization using a SWP process. The water contact angles of the PET surface modified with hydrophilic AA and HEMA monomers decreased from approximately 80° before treatment to less than 35°. The hydrophilicity of the PET surface modified with hydrophilic AA and HEMA monomers was maintained for 70 hours. In contrast, the water contact angle of the PET surface after Ar plasma followed by hydrophobic St modification was 53° and increased up to more than 60° over time. 3T3 fibroblast cells were cultured on the modified PET surface. After the cells were cultured for 48 hours on the untreated and St-modified PET surfaces, the number of cultured cells was higher than the initial number of seeded cells. In contrast, the number of cells cultured on AA-modified and HEMA-modified surfaces was much lower than the initial number of seeded cells. We believe that our surface modification method would provide a way to control hydrophilicity and bioactivity of various polymer surfaces.

References

1. Noeske, M., Degenhardt, J., Strudthoff, S. *et al.* (2004) *Int. J. Adhes. Adhes.*, **24**, 171–177.
2. Han, S., Lee, Y., Kim, H. *et al.* (1997) *Surf. Coat. Technol.*, **93**, 261–264.
3. Cui, N.-Y. and Brown, N.M.D. (2002) *Appl. Surf. Sci.*, **189**, 31.
4. Johnston, E.E. and Ratner, B.D. (1996) *J. Electron Spectrosc.*, **81**, 303.
5. Egitto, F.D. and Matienzo, L.J. (1994) *IBM J. Res. Dev.*, **38**, 423.
6. Li, J. and McConkey, J.W. (1996) *J. Vac. Sci. Technol. A*, **14**, 2102.
7. Keil, M., Rastomjee, C.S., Rajagopal, A., and Sotobayashi, H. (1998) *Appl. Surf. Sci.*, **105**, 273.
8. Sakudo, N., Mizutani, D., Ohmura, Y. *et al.* (2003) *Nucl. Instrum. Methods B*, **206**, 687–690.
9. Toufik, M., Mas, A., Shkinev, V., Nechaev, A., Elharfi, A., and Schue, F. (2002) *Eur. Polym. J.*, **38**, 203–209.
10. Sprang, N., Theirich, D., and Engemann, J. (1995) *Surf. Coat. Technol.*, **74-75**, 689.
11. Ballauf, M. and Borisov, O. (2006) *Curr. Opin. Colloid Interface Sci.*, **11**, 316.
12. Kudela, J., Terebessy, T., and Kando, M. (2000) *Appl. Phys. Lett.*, **76**, 1249.
13. Llanos, G.R. and Seastom, M.V. (1993) *J. Biomed. Mater. Res.*, **27**, 1383.
14. Lippitz, A., Friedrich, J.F., Unger, W.E.S., Schertel, A., and Wöll, Ch. (1996) *Polymer*, **37**, 3151.
15. Chataib, M., Roberfroid, E.M., Novis, Y., Pireaux, J.J., Caudano, R., Lutgen, P., and Feyder, G. (1989) *J. Vac. Sci. Technol. A*, **7**, 3233.
16. Gerenser, L.J. (1990) *J. Vac. Sci. Technol. A*, **8**, 3682.

17. Wong, P.C., Li, Y.S., and Mitchell, K.A.R. (1995) *Appl. Surf. Sci.*, **84**, 245.
18. Cueff, R., Band, G., Benmalek, M., Besse, J.P., Butruille, J.R., and Jacquet, M. (1997) *Appl. Surf. Sci.*, **115**, 292.
19. Paynter, R.W. (1998) *Surf. Interface Anal.*, **26**, 674.
20. Koprinarov, I., Lippitz, A., Friedrich, J.F., Unger, W.E.S., and Wöll, Ch. (1998) *Polymer*, **39**, 3001.
21. Sandrin, L. and Sacher, E. (1998) *Appl. Surf. Sci.*, **135**, 339.
22. Ektessabi, A.M. and Yamaguchi, K. (2000) *Thin Solid Films*, **377**, 793.
23. Vasquez-Borucki, S., Achete, C.A., and Jacob, W. (2001) *Surf. Coat. Technol.*, **138**, 256.
24. Médard, N., Soutif, J.C., and Poncin-Epaillard, F. (2002) *Langmuir*, **18**, 2246.

18
Strategies and Issues on the Plasma Processing of Thin-Film Silicon Solar Cells

Akihisa Matsuda

18.1
Introduction

Amorphous silicon (hydrogenated amorphous silicon: a-Si : H) and microcrystalline silicon (hydrogenated microcrystalline silicon: μc-Si : H) are promising thin-film materials for low-cost and high-performance solar cells (1–3). Band diagrams of single-junction a-Si : H-based solar cell and tandem-type a-Si : H/μc-Si : H-stacked cell are shown in Figure 18.1. For the reduction of marketing cost, growth of high-quality materials by plasma-enhanced chemical vapor deposition (PECVD) method is an essential issue to be realized.

High-quality thin-film silicon for solar-cell application simply means the materials including low dangling-bond-defect density even after photoinduced degradation in amorphous silicon (a-Si : H) and low dangling-bond-defect density in microcrystalline silicon (μc-Si : H) deposited even at an extremely high growth rate ($\sim$10 nm s^{-1}) [2]. This is because dangling-bond defects in those materials act as recombination centers for photoexcited electrons and holes. As schematically shown in the upper part of Figure 18.1, photoexcited carriers in the solar-cell structure are preferentially recombined through the dangling-bond-defect sites, causing a severe reduction in carrier-collection efficiency toward both electrodes in solar cells.

18.2
Growth Process of a-Si : H and μc-Si : H by PECVD

The main difference in process conditions in a-Si : H growth and μc-Si : H growth is hydrogen (H_2) dilution ratio ($[H_2]/[SiH_4]$: R) of source-gas materials, that is, low R (0–5) for a-Si : H and high R (>10) for μc-Si : H [2]. Initial event in the PECVD process for the growth of a-Si : H and μc-Si : H is electron-impact dissociation of source-gas molecules in the plasma, that is, electronic excitation of ground-state electrons in the source-gas molecule to its electronic excited state (dissociating state) [2]. A variety of radicals, reactive species, emissive species, and ions (positive

Industrial Plasma Technology. Edited by Yoshinobu Kawai, Hideo Ikegami, Noriyoshi Sato, Akihisa Matsuda, Kiichiro Uchino, Masayuki Kuzuya, and Akira Mizuno

ISBN: 978-3-527-32544-3

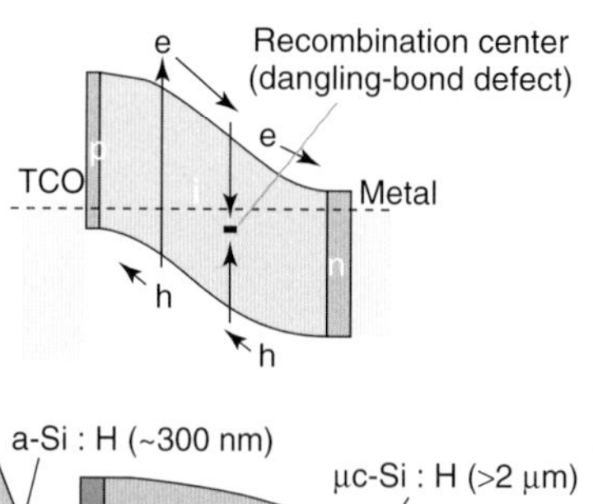

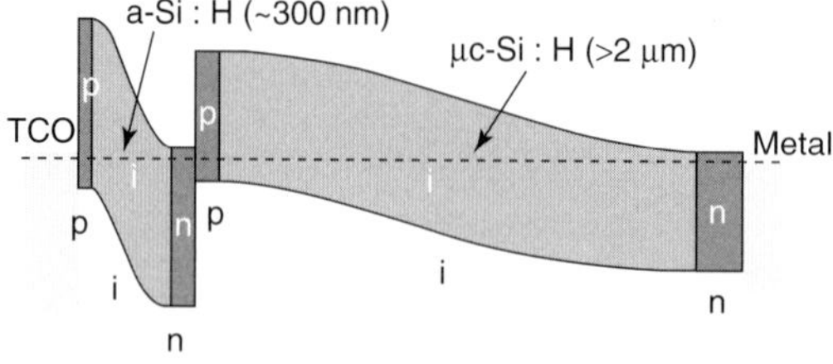

Figure 18.1 Band diagrams of single-junction p–i–n and tandem-type p–i–n–p–i–n thin film silicon solar cells. Transparent Conducting Oxide (TCO), e, and h represent transparent conductive oxide, photoexcited electron, and hole, respectively.

and negative) are produced almost simultaneously in the plasma. These species collide and react mostly with parent molecules reaching a steady state. Steady-state densities of reactive species determined using a variety of diagnostic techniques in realistic plasmas both for a-Si : H and μc-Si : H depositions are summarized in Figure 18.2.

In the deposition process of a-Si : H, silyl radical (SiH_3) shows the highest steady-state density among a variety of reactive species produced in the plasma, since short-lifetime reactive species (SLS) such as silylene (SiH_2), silylidine (SiH), and silicon (Si) react well with parent SiH_4 and change their form, whereas SiH_3 has no reactivity with SiH_4 (2, 4). Therefore, low-defect density a-Si : H is rather easily prepared when the temperature of film-growing surface (substrate temperature) is optimized under sufficient SiH_4-flowing conditions (non-SiH_4-depletion conditions). Substrate-temperature dependence of dangling-bond density in the resulting films is shown in Figure 18.3 both for a-Si : H and μc-Si : H.

However, as is well known, a-Si : H shows photoinduced degradation (Staebler–Wronski effect: S–W), in which dangling-bond-defect density is very much increased by prolonged light soaking [3]. Therefore, high-quality a-Si : H here means a-Si : H having low initial defect density and showing less S–W effect.

On the other hand, in the deposition process of μc-Si : H, depletion condition of SiH_4 (high R) is popularly used for the survival of atomic hydrogen (H) due to its important reaction on the film-growing surface for the formation process of μc-Si : H [2]. Under SiH_4-depletion condition, SLS such as SiH_2, SiH, and Si tend to survive (as is seen in the steady-state density of reactive species shown in Figure 18.2) and contribute to film growth, leading to high defect density in the

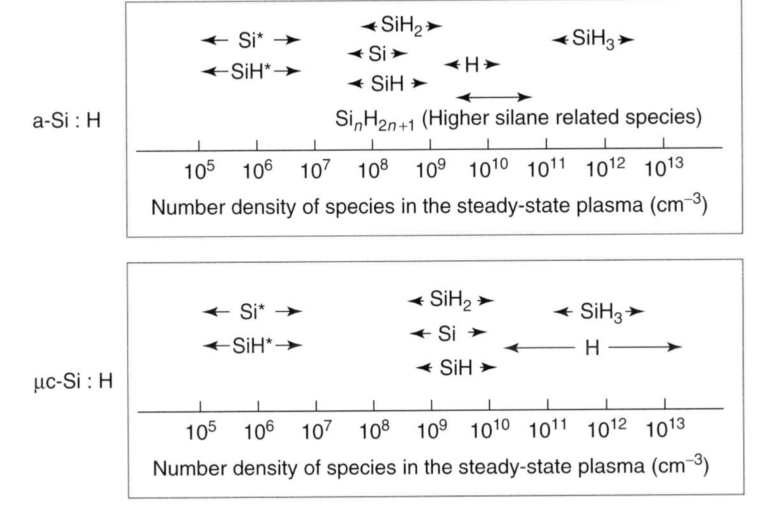

Figure 18.2 Number density of reactive species and emissive species in the steady-state plasma both for a-Si : H and μc-Si : H depositions.

resulting μc-Si : H especially when the generation rate of film precursor (SiH_3) is set at high rate in the plasma (high-growth-rate condition).

18.3 Growth of High-Quality a-Si : H

As is mentioned above, SLS react well with SiH_4 molecules in the a-Si : H-deposition process, leading to easier preparation of low-defect density a-Si : H. However, the reaction of SLS with SiH_4 produces higher order silane molecules such as disilane (Si_2H_6), trisilane (Si_3H_8), tetrasilane (Si_4H_{10}), and so on [4]. Electron-impact-dissociation process on these higher order silane molecules in the plasma produces higher silane-reactive species (HSRS) such as Si_4H_9 and particulates at an extreme case [2]. Figure 18.4 shows the relationship between Si-H2 (di-hydride) density in the resulting film and mass-signal–intensity ratio of $(m/e = 90)/(m/e = 30)$ as a proportional quantity to the contribution ratio of representative HSRS or particulates to the film growth taken under a variety of growth conditions [3]. Result of mass spectroscopy has shown that HSRS or particulates are the cause of excess Si-H2 (di-hydride) bonding configuration in the resulting a-Si : H, although it is still controversial whether HSRS or particulates are more plausible chemical species for the origin of excess Si-H2. Since Si-H2 bonding has been indicated as the responsible network structure for the photoinduced degradation of a-Si : H as shown in Figure 18.5 [3], one of the most important issues in the preparation process of high-quality a-Si : H showing low-defect density even after

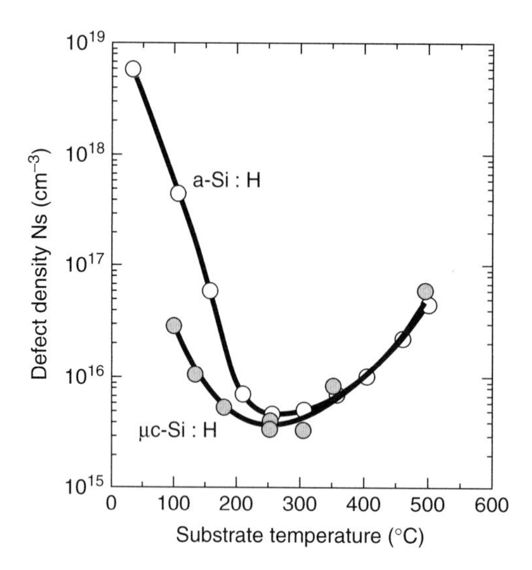

Figure 18.3 Initial dangling-bond-defect density measured by electron-spin resonance in high-quality a-Si : H and μc-Si : H plotted against substrate temperature during film growth.

long-time light soaking (S–W less high-quality a-Si : H) is to control the formation and contribution processes of HSRS or particulates during film growth.

Contribution ratio of HSRS (here, Si_4H_9) to film growth is expressed using a couple of reaction-rate equations as

$$\frac{[Si_4H_9]}{[SiH_3]} \propto \frac{N_{e2}^3 \tau_x^3 k_{2a}^3}{N_{e3}\{(K_{2a} + K_{2b}) + k_1 R\}^3} \tag{18.1}$$

where N_{e2} and N_{e3} show electron densities in the plasma to produce SiH_2 and SiH_3, respectively. R is H_2-dilution ratio of source-gas molecules ($[H_2]/[SiH_4]$). τ_x is representative time constant related to gas-flow rate and/or residence time. Reaction-rate constants k_{2a}, k_{2b}, and k_1 are the temperature-dependent constants for $SiH_2 + SiH_4 = Si_2H_6$ (a), $SiH_2 + SiH_4 = 2SiH_3$ (b), and $SiH_2 + H_2 = SiH_4$ (c) reactions, respectively. Therefore, the contribution ratio of HSRS is sensitive to many process parameters, that is, electron density (N_e), electron temperature (N_{e2}/N_{e3}) as the plasma parameters, gas-flow rate and residence time (τ_x) including gas-flow direction, gas temperature (k_{2a} and k_{2b} show strong ambient-temperature dependences), gas pressure (through electron temperature), and H_2-dilution ratio (R).

Figure 18.6 shows the degree of photoinduced degradation (same definition as in Figure 18.5) in the film as a function of electron temperature (emission intensity ratio of I_{Si*}/I_{SiH*} is a good measure of electron temperature in SiH_4/H_2 plasma [3]) in the plasma during film growth. As is clearly seen in the figure, the degree of photoinduced degradation in a-Si : H is well controlled when changing the electron temperature during film growth, as far as the temperature on the film-growing surface (substrate temperature) is kept the same.

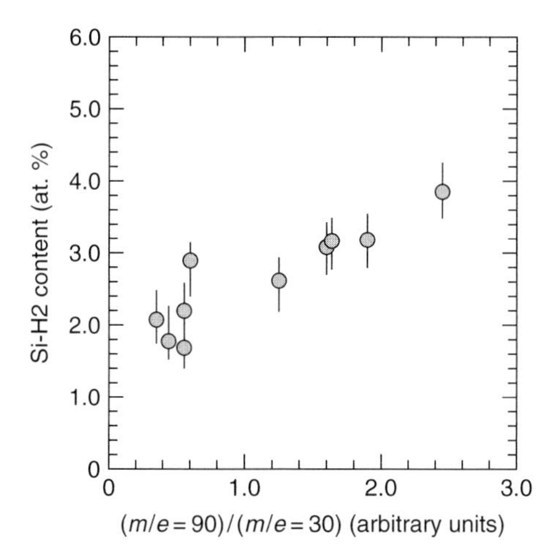

Figure 18.4 Si-H2 content measured by infrared absorption spectroscopy in the resulting a-Si : H plotted against signal–intensity ratio of $(m/e = 90)/(m/e = 30)$ measured by mass spectrometry during film growth.

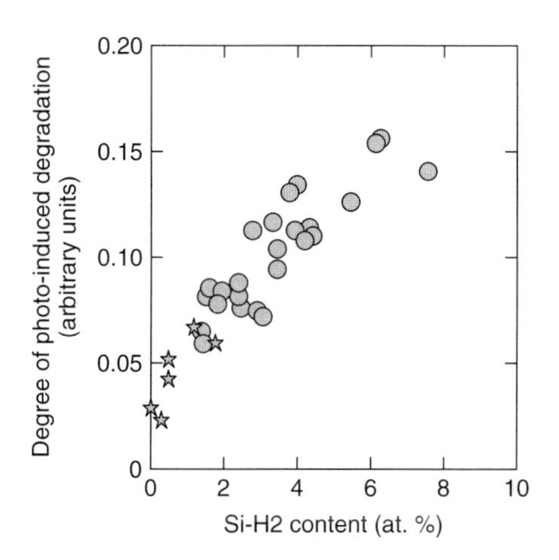

Figure 18.5 Degree of photoinduced degradation in a-Si : H prepared under a variety of deposition conditions plotted against Si-H2 content. Star symbols represent films prepared using triode plasma apparatus.

When we use a triode configuration to expect an additional removal reaction of HSRS with SiH_4, resulting a-Si : H film involves 0 at% of Si-H2 bonding configuration and shows very small degree of photoinduced degradation [2] as indicated by star symbols in Figure 18.5.

As shown in Equation 18.1, contribution ratio of HSRS to film growth is dependent on a variety of process conditions, which is indeed sensitive to the

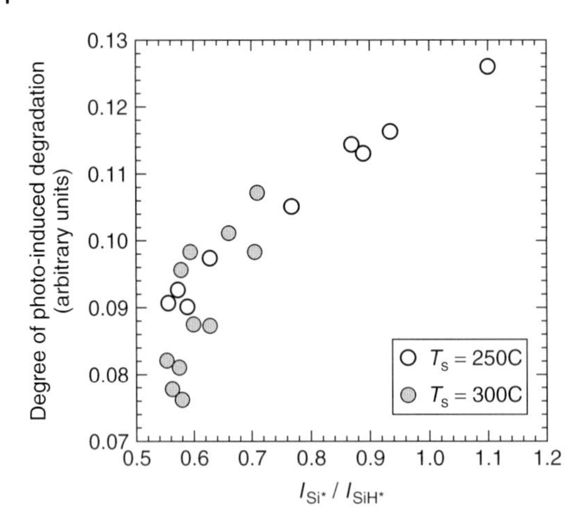

Figure 18.6 Degree of photoinduced degradation (corresponding to the increment of dangling-bond-defect density after light soaking) in a-Si : H films plotted against signal–intensity ratio of I_{Si*}/I_{SiH*} (corresponding to the electron temperature in the plasma) during film growth.

structure of plasma apparatus such as cathode–anode distance, gas-flow direction (toward cathode or toward side of the plasma), cathode temperature (as well as substrate temperature), and so on, and as a matter of course, sensitive to power density applied to the plasma, gas pressure, H_2–dilution ratio, plasma-excitation frequency, and so on. It is concluded from above-mentioned discussions that optimal design is required in the PECVD system for industrial production of large-area and high-efficiency a-Si : H-based solar cells using high-quality a-Si : H. In other words, once the PECVD system including reaction apparatus is constructed, stable efficiency (conversion efficiency after photoinduced degradation) of a-Si : H solar cells fabricated using this is fatally limited even changing the controllable parameters (pressure, power, H_2–dilution ratio, etc).

18.4
Concept of Protocrystal a-Si : H (Stable a-Si : H)

When the H_2–dilution ratio R is increased, keeping other conditions (substrate temperature, power applied to the plasma, total working pressure, etc.) the same, the structure of resulting film is changed from amorphous (a-Si : H) to microcrystalline (μc-Si : H) [5]. Figure 18.7 shows structural (defect density measured by electron-spin resonance: Ns) and opto-electronic properties (photo- and dark conductivities: $\Delta\sigma_{ph}$ and σ_d) of deposited materials as a function of H_2-dilution ratio during film growth. Ns, $\Delta\sigma_{ph}$, and σ_d after light soaking are also shown by closed circles in the figure. Gray zone in the figure indicates the transition region from a-Si : H to μc-Si : H, where a very small amount of crystallites (<30

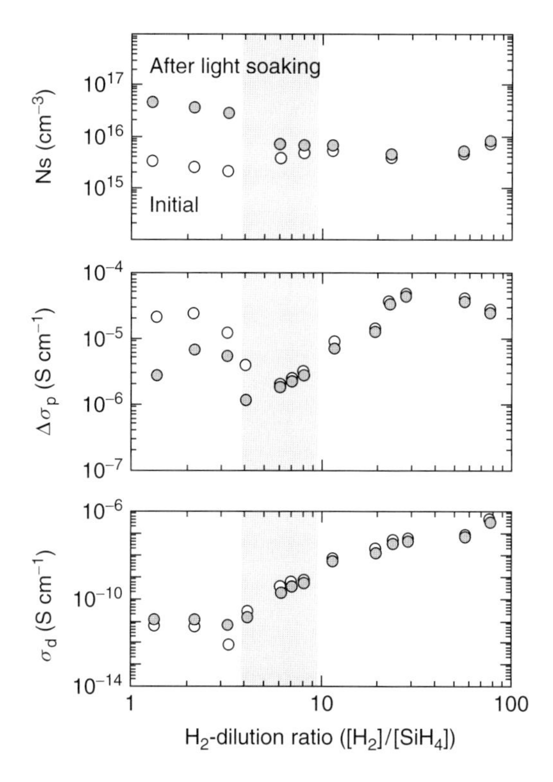

Figure 18.7 Dangling-bond-defect density Ns, photoconductivity $\Delta\sigma_{ph}$, and dark conductivity σ_d of amorphous (left side of the gray zone), protocrystal (gray zone), and microcrystalline (right side of the gray zone) silicon films plotted against hydrogen-dilution ratio (*R*) during film growth. Open circles show initial values and closed gray circles show values after light soaking (using sunlight for 1 hour).

volume%) and/or nanometer-size crystallites (ordered structure is subtly observed under the high-resolution transmission electron microscope HRTEM images [6]) are involved in the resulting films; therefore, materials deposited under these conditions are named *protocrystal silicon* or *nanocrystal-embedded amorphous silicon* (hereafter terminology of "protocrystal silicon" is used [6]). It is interesting to note that photoinduced degradation is not observed in the films prepared under protocrystal silicon-formation conditions as shown in Figure 18.7 [7]. In other words, dangling-bond-defect density (Ns) of $\sim 10^{15}$ cm^{-3} and photoconductivity ($\Delta\sigma_{ph}$) are not deteriorated even after prolonged light soaking in the protocrystal silicon even when the crystalline-volume fraction is less than 10%. This stable behavior is simply explained by the presence of crystallites in the amorphous silicon network structure. In other words, since the bandgap of crystal silicon ($\sim$1.1 eV) is narrower than that of amorphous silicon ($\sim$1.8 eV), photoexcited electrons and holes are preferentially trapped and recombined at crystallite regions where a rigid/stable structure is presented against structural change such as S–W effect, leading to no change in Ns and $\Delta\sigma_{ph}$ even after prolonged light soaking [7].

However, it should be emphasized here that crystallite sites do act as recombination centers for photoexcited electrons and holes (with huge recombination-cross sections) in the protocrystal materials; therefore, initial value (before light soaking) of $\Delta\sigma_{ph}$ in this material shows much lower value as compared to pure a-Si : H as is clearly seen in Figure 18.7, causing inferior initial solar-cell performance (even before light soaking) when using protocrystal silicon as the active layer (i-layer) material for conventional p–i–n solar-cell structure.

18.5 Growth of High-Quality μc-Si : H

When the crystallite-volume fraction exceeds 30%, photoexcited electrons and holes flow through the crystallite-path way by percolation manner. This μc-Si : H is popularly used as the bottom-cell i-layer material in the tandem-type p–i–n–p–i–n stacked cell structure by utilizing narrower bandgap properties of μc-Si : H as compared to a-Si : H top cell material as shown in the bottom part of Figure 18.1. Since μc-Si : H shows indirect optical transition properties same as that of single crystal Si, thick layer as thick as 2–3 μm is required to absorb sufficient longer wavelength light transmitted through a-Si : H top layer; then, high rate growth of high-quality μc-Si : H is crucial issue to reduce the production cost of tandem-type thin film silicon solar cells.

Under μc-Si : H growth conditions, steady-state density of SLS shows higher value as compared to a-Si : H growth conditions (as shown in Figure 18.2 [2]) due to the use of SiH_4-depletion regime for accumulation of large amount of hydrogen atoms to the film-growing surface. Therefore, contribution of SLS to the film growth is thought as an origin of higher defect density in μc-Si : H especially under high-growth-rate condition [2]. For the growth of high-quality μc-Si : H at high rate, contribution ratio of SLS to the film growth, which is given by following equation, should be minimized:

$$\frac{[SiH_x]}{[SiH_3]} \propto \frac{\left(\frac{N_{ex}}{N_{e3}\tau_3}\right)}{\{(k_{2a+}K_{2b})[SiH_4]+k_1[H_2]\}} \tag{18.2}$$

where N_{ex} and N_{e3} show electron densities in the plasma to produce SLS denoted here as SiH_x and SiH_3, respectively. τ_3 is specific lifetime of SiH_3, $[SiH_4]$ and $[H_2]$ are number density of SiH_4 and H_2 molecules, and reaction-rate constants (k_{2a}, k_{2b}, and k_1) are the same as used in Equations 18.1. To minimize the contribution ratio of SLS to film growth for obtaining high-quality μc-Si : H, electron temperature in the plasma, gas-flow rate, gas-residence time, and number density of SiH_4 and H_2 molecules should be controlled. To reduce the electron temperature, to increase the specific lifetime of SiH_3, and to increase the number density of H_2 molecules (being SiH_x scavenger instead of SiH_4 molecule under SiH_4-depletion condition), high-pressure condition as high as 1–3 kPa is popularly used [2].

However, in the conventional plasma production, high-pressure condition needs narrow gap between radio-frequency (RF) or very high frequency (VHF) electrode (cathode) and substrate electrode (anode). For the industrial fabrication of μc-Si : H-based solar cell, very large-area PECVD system more than 2 m^2 is required. Therefore, uniform plasma production in large area with narrow gap (as narrow as several mm) space is the most important demand for obtaining high-quality μc-Si : H at high growth rate of $\sim$10 nm s^{-1}. Heating effect of film-growing surface by the heat transfer from high density and high pressure (close to thermal plasma condition) should be controlled to prepare high-quality μc-Si : H, because overheating of film-growing surface gives rise to the thermal removal event of surface-covering hydrogen. Similar to the case of a-Si : H-based solar cells, optimal design is required in the PECVD system for industrial production of large-area and high-efficiency μc − Si : H-based solar cells at high rate.

18.6 Summary

Strategies and issues on the plasma processing are discussed to obtain high-quality a-Si : H and μc-Si : H for realizing high-efficiency thin-film silicon-based solar cells. It is concluded that the optimal design of the PECVD system is highly expected both for *high-quality* a-Si : H showing less photoinduced degradation and for high-quality μc-Si : H grown at high rate.

References

1. Spear, W.E. and LeComber, P.G. (1975) *Solid State Commun.*, **17**, 1193.
2. Matsuda, A. (2004) *J. Non-Cryst. Solids*, **338**, 1.
3. Matsuda, A. (2003) *Sol. Energy Mater. Sol. Cells*, **78**, 3.
4. Perrin, J., Leroy, O., and Bordage, M.C. (1996) *Contrib. Plasma Phys.*, **36**, 3.
5. Matsuda, A. (1983) *J. Non-Cryst. Solids*, **59-60**, 767.
6. Tsu, D.V., Chao, B.S., Ovshinsky, S.R., Guha, S., and Yang, J. (1997) *Appl. Phys. Lett.*, **71**, 1317.
7. Kamei, T., Stradins, P., and Matsuda, A. (1999) *Appl. Phys. Lett.*, **744**, 1707.

19
Characteristics of VHF Plasma with Large Area

Yoshinobu Kawai, Yoshiaki Takeuchi, Yasuhiro Yamauchi, and Hiromu Takatsuka

19.1
Introduction

Microcrystalline silicon (μc-Si : H) has been intensively investigated because of the many advantages it has for application in solar cells [1–3]. To reduce production costs of solar cells, high deposition rates of μc-Si : H with large area ($>1\ m^2$) have to be achieved [4]. Usually, VHF plasma sources have been adopted to increase the deposition rate because it provides high electron density plasma [5, 6]. Furthermore, higher deposition rates of μc-Si : H were achieved by a short gap discharge at high pressure [1–3]. On the other hand, Paschen law suggests that the higher the pressure becomes, the shorter should be the spacing gap between the electrodes. As a result, discharge voltages become greater because of larger losses of plasma. Furthermore, since high quality μc-Si : H can be obtained at high pressure, the sheath potential is expected to be very low.

VHF plasma is characterized by electron trapping, that is, discharge frequency should satisfy the following condition:

$$f_{pi} \ll f < f_{pe}$$

Here, f is the discharge frequency, and f_{pi} and f_{pe} are the plasma frequencies of ions and electrons, respectively. In addition, the electron displacement x_m should be shorter than a spacing gap between discharge electrodes for $\omega \ll \upsilon_m$ [7]:

$$x_m = qE_0/(m_e\omega\upsilon_m) \ll L/2$$

Here, L is a spacing gap between discharge electrodes, ω and υ_m are the angular frequencies of VHF power source and electron collision frequency, respectively, and q, m_e, and E_0 are electron charge, electron mass, and the amplitude of the electron oscillation in the VHF electric field. Electron trapping effect provides better confinement of electrons and, as a result, the electron density becomes high. Thus, the ion saturation current that is proportional to the electron density is considered to peak at a certain condition where electron trapping is most effective. When VHF powers are increased, the amplitude of the electron oscillation in the VHF electric field E_0 increases and, as a result, the condition for electron trapping, $x_m \ll L/2$,

Industrial Plasma Technology. Edited by Yoshinobu Kawai, Hideo Ikegami, Noriyoshi Sato, Akihisa Matsuda, Kiichiro Uchino, Masayuki Kuzuya, and Akira Mizuno

ISBN: 978-3-527-32544-3

is not valid. Therefore, it is required to increase υ_m by increasing the pressure for VHF plasma discharge to occur at high powers. Thus, examining the characteristics of the VHF plasma produced with a short gap discharge is an important aspect in the production of microcrystalline silicon solar cells. However, there is no report that discusses the VHF plasma at high pressure from the point of view of electron trapping effect.

In the VHF range, it is hard to deposit uniform microcrystalline silicon films over a large area ($>1\ m^2$) since the electrical voltage distribution caused by the standing wave effects is not uniform on the surface of the conventional parallel plate electrodes [8–12]. In most of the experiments, a coaxial cable is used as feeder lines to feed VHF power to the power electrode of the parallel plate electrodes. However, it is difficult to avoid the leakage of electric fields of VHF in the vacuum chamber in the case where a coaxial cable is used as feeder lines and, as a result, VHF discharges occur outside the parallel plate electrodes. According to antenna theory, it is well known that when a coaxial cable is used as feeder lines in the VHF and UHF range, the screen of the coaxial cable emits electromagnetic fields. To avoid such leaking of the electric fields, a balanced power feeding method should be adopted. Many kinds of balanced power feeding methods are proposed in antenna fields, such as the so-called baluns. We developed a balanced power feeding method to use a power divider consisting of two outputs with 180° phase difference and succeeded in producing a stable VHF plasma [13]. As described later, a balanced power feeding method was found to be indispensable for short gap discharges.

19.2 Development of Balanced Power Feeding Method

Figure 19.1 is a schematic of the experimental apparatus, which consists of a stainless steel vacuum vessel (height: 420 mm, width: 1350 mm, and depth: 470 mm), a multirod electrode [13] of 1200 mm $\times$ 114 mm, and a VHF power supply. The multirod electrode consisted of five stainless steel rods. The distance between the multirod electrode and the glass substrate was 34 mm, where a punched electrode (stainless-steal-disk plate) of 1200 mm $\times$ 114 mm was used as a substrate holder to look at plasma uniformity. VHF powers of discharge frequencies up to 100 MHz were supplied to the feeding point on the multirod electrode and a punched electrode through an impedance matching box. Figure 19.2 shows a schematic of how to feed VHF powers to the multirod electrode and the punched electrode. Figure 19.2a shows a conventional method, that is, VHF power is directly fed to the multirod electrode using a coaxial cable and the punched electrode is grounded. On the other hand, Figure 19.2b shows a schematic of a balanced power feeding method developed here. As shown in Figure 19.2b, the VHF power is divided with a power divider consisting of two outputs with 180° phase difference and the output powers are fed to the multirod electrode and the punched electrode by using semirigid coaxial cables, respectively. In this case, both outer conductors of the semirigid cables are soldered to keep the same potential, providing two parallel

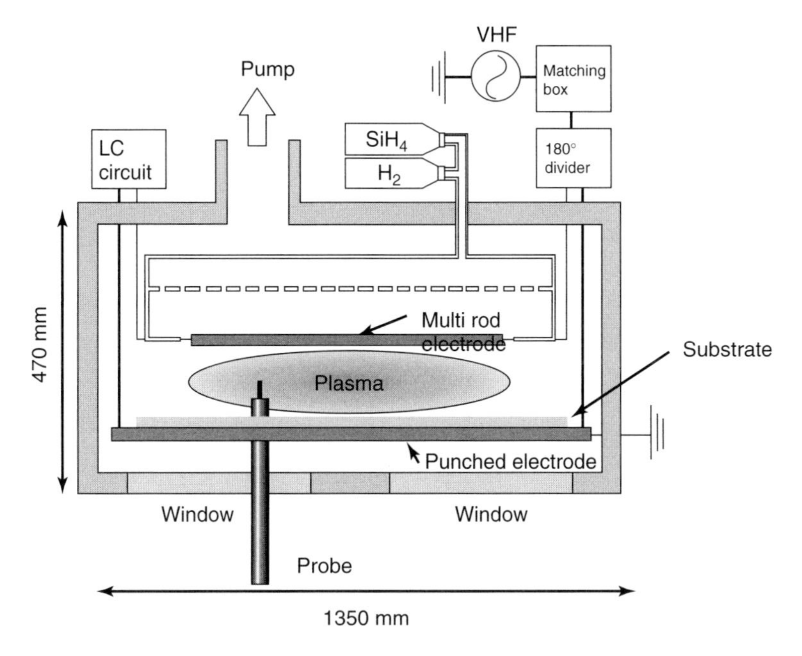

Figure 19.1 Schematic of the experimental apparatus.

transmission lines (Lecher wires) that are commercially used for TV antenna. The matching box was adjusted in such a way that reflected powers become as near zero as possible, but power of a few watts was reflected. Thus, we succeeded in realizing a stable VHF discharge between the multirod electrode and punched electrode. The forward and reflected VHF power in front of the vacuum chamber was measured with power meters to examine the characteristics of the balanced power feeding method developed here. The gas used was pure hydrogen gas and the experiments were carried out in the pressure range from 30 mTorr to 1.2 Torr, keeping the gas flow rate 100 sccm. The plasma parameters and the profile of the ion saturation current were measured with a movable cylindrical Langmuir probe with a 0.2 mm-diameter-tungsten wire. The Langmuir probe was positioned at 11 mm from the multirod electrode. In addition, the electron density was estimated from the ion saturation current, which shows a smaller effect of the RF fields on $I-V$ curves.

To examine the effect of the new power divider developed here on the VHF plasma production, we measured the forward and reflected VHF power as a function of pressure. Figure 19.3 indicates that the sum of the forward power and reflected power on the multirod electrode is nearly equal to that of the VHF power source (150 W) at low pressure. This is one of the advantages to using the balanced power feeding method. Furthermore, the forward power on the multirod electrode is considered to be dissipated for VHF discharges at low pressure. However, as the pressure was increased, less power dissipation was observed. Figure 19.3 also suggests that higher powers are needed for VHF plasma production at high

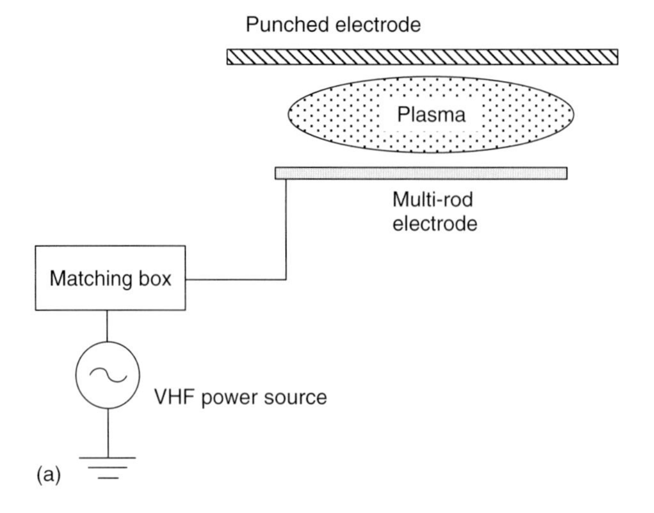

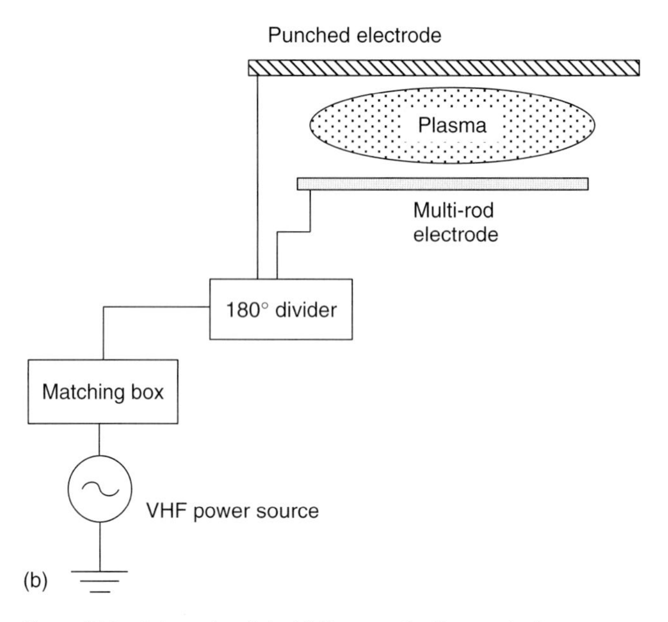

Figure 19.2 Schematic of the VHF power feeding method: (a) conventional method and (b) new power feeding method.

pressure. In fact, VHF power more than 300 W was necessary to produce a uniform plasma over 1 m at pressures as high as a few Torr. In this case, the discharge gap distance between the multirod electrode and the punched electrode had to be shorter, which is understood from Paschen law. On the other hand, as seen in Figure 19.3, the forward power fed to the punched electrode is very low compared with that on the multirod electrode and almost all the forward power is reflected.

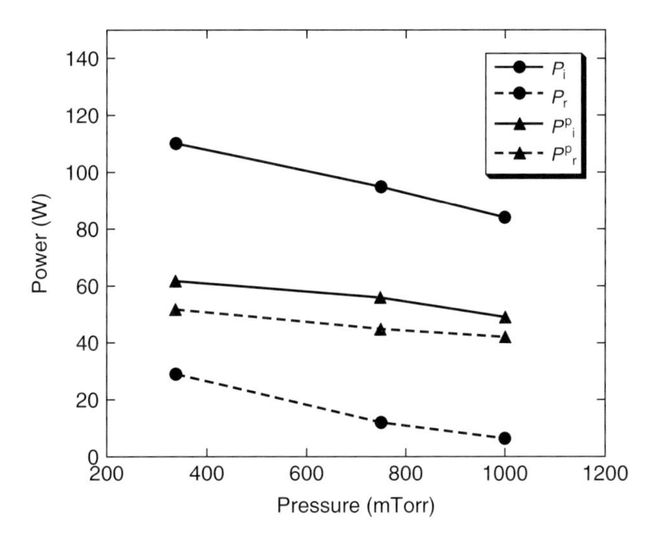

Figure 19.3 The dependence of the forward and reflected power on the pressure at 60 MHz, where the power of the VHF source is 150 W. In the figure, P_i and P_r is the forward power and reflected power to the multirod electrode, and $P^p{}_i$ and $P^p{}_r$ is the forwarded power and reflected power to the punched electrode, respectively.

This means that the punched electrode does not act as the power electrode to produce the VHF plasma. Generally, a glass substrate is placed on the power electrode in plasma CVD so that it is hard to realize perfect balanced power feeding.

Then, we examined the characteristics of the plasma parameters as a function of pressure for the different power feeding methods in Figure 19.2a,b. In the case of Figure 19.2a, VHF discharge occurred outside the multirod electrode while in the case of Figure 19.2b, VHF discharge occurred only in the region between the multirod electrode and the punched electrode, providing the experimental results with good reproducibility. Figure 19.4a shows the dependence of the ion saturation current density I_{is} on the pressure at $z = 0$, which corresponds to the end of the multirod electrode. Here, the frequency and power of the VHF power source is 60 MHz and 150 W, respectively. As seen in Figure 19.4a, the balanced power feeding method provides an ion saturation current density higher than that of the unbalanced power feeding method at high pressure, that is, higher deposition rate will be expected at the high pressure. There is the tendency that microcrystalline silicone films are fabricated at pressures higher than 1 Torr to increase the deposition rate [1, 3, 14]. Therefore, the balanced power feeding method is suitable for the fabrication of microcrystalline silicone films. Figure 19.4a also shows that the ion saturation current density peaks around 150 mTorr, which is due to the electron trapping effect in the VHF electric fields. The electron temperature T_e, which is one of the important parameters in the plasma CVD, was measured as a function of pressure. As shown in Figure 19.4b, the electron temperature does not depend on the power feeding method.

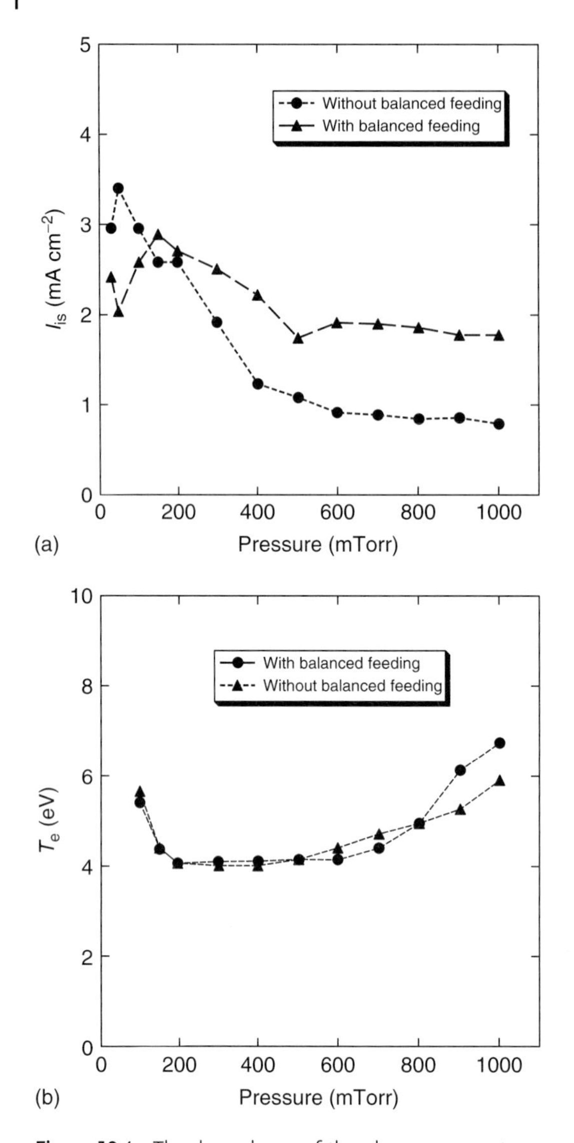

Figure 19.4 The dependence of the plasma parameters on the pressure: (a) the ion saturation current density I_{is} and (b) the electron temperature T_e. Here, the frequency and power is 60 MHz and 150 W, respectively.

VHF discharges of 60 MHz have been mostly used to fabricate microcrystalline silicone films [4]. However, this deposition rate is not enough for the reduction of cost of solar cells. To increase the frequency of VHF power source is one of the candidates that respond to such demand. Thus, we increased the discharge frequency from 60 to 120 MHz, keeping constant power of 150 W. The results at 100 MHz are shown in Figure 19.5. The dependence of both the ion saturation

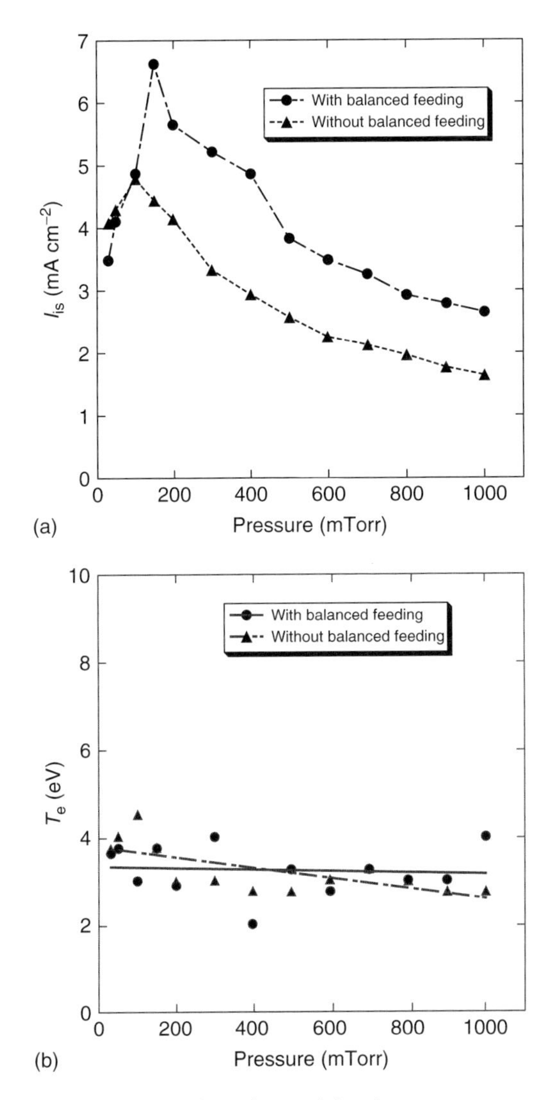

Figure 19.5 The dependence of the plasma parameters on the pressure: (a) the ion saturation current density I_{is} and (b) the electron temperature T_e. Here, the frequency and power is 100 MHz and 150 W, respectively.

current density and the electron temperature on the pressure is similar to that of 60 MHz. As seen in Figure 19.5a, the ion saturation current density also peaks around 150 mTorr, amounting to $1.5 \times 10^{10}\,cm^{-3}$. Note that when the power divider developed here was not used, the VHF plasma was also produced outside the multirod electrode, leading to lack of reproducibility of the experimental results. On the other hand, as seen in Figure 19.5b, the electron temperature decreases to around 3 eV, which is favorable for fabrication of high-quality microcrystalline silicone films. However, the plasma became nonuniform compared with the case of 60 MHz. Thus, we focused on the VHF plasma production at 60 MHz, where the observed VHF plasma is almost uniform over 1 m.

19.3 Characteristics of VHF Plasma

VHF plasmas were produced with the multirod electrode at high pressure, and the plasma parameters were measured as a function of pressure for different VHF powers at 60 MHz.

19.3.1 H_2 Plasma Characteristics

A schematic diagram of the experimental apparatus [15] is shown in Figure 19.1. The pressure was ranged from 0.3 to 3.75 Torr. The gas flow rate was 50 sccm. The plasma parameters were measured with a tiny cylindrical Langmuir probe (diameter 0.2 mm, length 1.2 mm) which was placed at a distance of 3 mm from the multirod electrode.

The plasma parameters were examined as a function of pressure for different VHF powers. Figure 19.6a shows that when the pressure is increased, the ion saturation current density I_{is} peaks at a certain pressure and finally decreases. This figure also shows that two peaks are observed at 300 and 450 W. It was found for the first time that the pressure at which I_{is} peaks depends on the VHF power. Figure 19.6a also shows that when the VHF power is increased, I_{is} increases at high pressure. On the other hand, as seen in Figure 19.6b, the electron temperature T_e is around 10 eV. The fact that T_e became as high as 10 eV is not surprising because the distance between the electrodes for VHF discharge is very short. Generally, T_e is proportional to the electric field between the electrodes for discharge [16]. In fact, when the discharge gap of 34 mm was used, T_e decreased to about 4 eV, as shown in Figure 19.5.

In this experiment, the electron mean free path λ_m is appreciably smaller than the discharge gap because $\lambda_m \sim 0.4$ mm at $T_e = 9$ eV for 1 Torr. Thus, we can estimate the mean power density P of the discharge supplied by the electric fields E_0 using the diffusion-controlled model [7, 17]. The mean power density P absorbed by the electrons is written as follows [7, 17]:

$$P = \frac{nq^2 E_0^2}{2m_e} \frac{\nu_m}{\nu_m^2 + \omega^2} \tag{19.1}$$

In this experiment, $\omega/2\pi = 60$ MHz, $\upsilon_m/2\pi \sim 1$ GHz at 1 Torr, that is, $\omega \ll \upsilon_m$, so that P should decrease in proportion to υ_m^{-1}. Looking at Figure 19.6a carefully, the ion saturation current exponentially decreases at high pressure rather than the inverse of the pressure (υ_m^{-1}). This suggests that there is another loss such as electron attachment. Figure 19.6a also suggests that higher VHF power will be necessary for increasing the ion saturation current at high pressure.

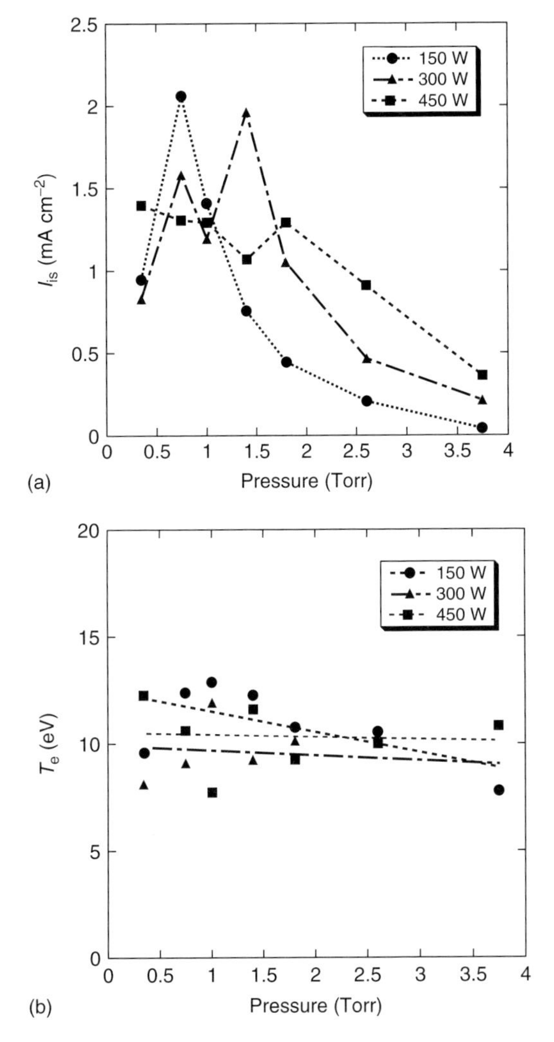

Figure 19.6 The dependence of the plasma parameters on the pressure for different VHF powers: (a) the ion saturation current density I_{is} and (b) the electron temperature T_e.

The sheath potential, which is estimated from the difference between the plasma potential and the floating potential, is a key parameter in plasma CVD because when the sheath potential is high, the ion bombardment energy increases and, as a result, the film quality obtained becomes bad. According to the probe theory [18] for the plasma consisting of electrons and ions with Maxwellian distribution, the sheath potential V_w is given by

$$V_w \approx \frac{\kappa T_e}{2q} \ln\left(\frac{2m_i}{\pi m_e}\right) \tag{19.2}$$

Here, κ is the Boltzmann constant. As seen from Eq. 19.2, since the electron temperature is almost constant at high pressure in this experiment, the sheath potential should be constant at high pressure. The sheath potential was measured as a function of pressure for different powers and the result is shown in Figure 19.7, where the theoretical values calculated from Eq. 19.2 are plotted as open circles (150 W), open triangles (300 W), and open squares (450 W). As seen in Figure 19.7, observed sheath potentials are anomalously lower than the theoretical values for high VHF powers. Especially, the sheath potential at high pressure becomes lower for higher VHF powers: the sheath potential of 2.6 V is achieved for 3.75 Torr and 450 W.

It is well known [19–21] that when negative ions exist in plasma, the electron saturation current of the V–I curve of the Langmuir probe becomes anomalously small. Thus, the ratio of I_{es}/I_{is} provides a measure of the existence of negative ions, where I_{es} is the electron saturation current density proportional to the electron density. The dependence of I_{es}/I_{is} on the pressure for different powers is shown in Figure 19.8, indicating that observed I_{es}/I_{i} is anomalously low compared with the theoretical value, $I_{es}/I_{is} = (\exp)^{1/2}(m_i/2\pi m_e)^{1/2}$ [18], where electrons and ions are assumed to have Maxwellian distribution. In fact, assuming that dominant ion species are H^+ for H_2 plasma, the ratio $I_{es}/I_{is} \sim 28$ is obtained. In most of the experiments, the observed electron saturation current is lower than the theoretical

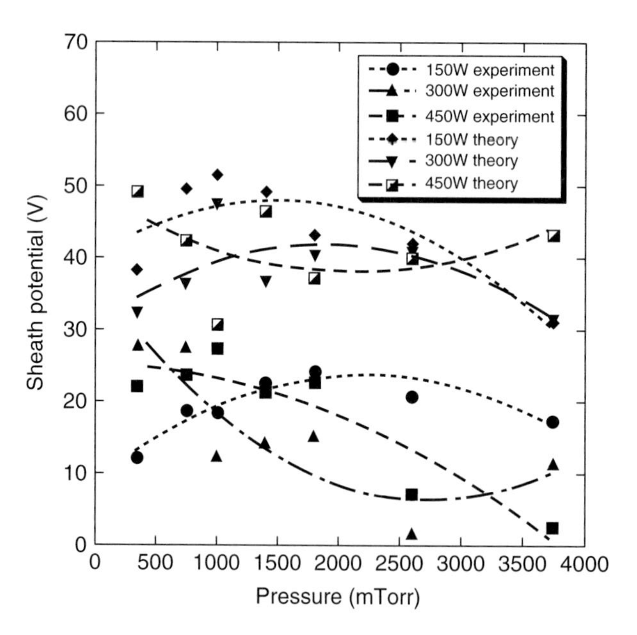

Figure 19.7 The dependence of the sheath potential on the pressure for different VHF powers. Here, open circles, open triangles, and open squares are the theoretical values calculated from Equation 19.2 for 150, 300, and 450 W, respectively.

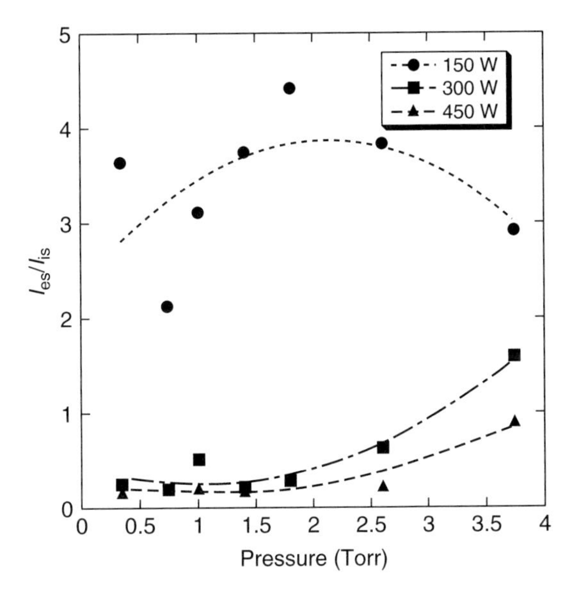

Figure 19.8 The dependence of I_{es}/I_{is} on the pressure for different VHF powers.

value because some amount of electrons around the Langmuir probe are absorbed by the probe and, as a result, the electron density becomes low, leading to the reduction of the electron saturation current. However, the observed reduction of the electron saturation current is much larger than the above-mentioned result. Now, the reason for the negative ion density to be very high has not been understood. Direct measurements of both the negative ion density [22] and negative ion species will be necessary in the future.

As seen in Figure 19.6a, the ion saturation current density peaks at a certain pressure and the peak pressure shifts to higher pressures as the VHF power is increased. The shift of the peak pressure is qualitatively understood as follows. When VHF powers are increased, the amplitude of the electron oscillation in the VHF electric field E_0 increases and, as a result, the condition for electron trapping, $x_m \ll L/2$, is not valid, as discussed in Section 19.1. Thus, it is required to increase υ_m by increasing the pressure for VHF plasma discharge to occur at high powers. In fact, it was found [1, 2] that the deposition rate with VHF plasma CVD peaks at a certain VHF power or a certain pressure. Although this may not be explained directly by the plasma parameters, there should be a correlation between the deposition rate and the plasma parameters. This will be the most important aspect to be solved to achieve VHF plasma production with high pressure.

19.3.2
SiH_4/H_2 Plasma Characteristics

Microcrystalline silicon is deposited by introducing a small amount of SiH_4 gas into H_2 plasmas. Here, we investigated the characteristics of SiH_4/H_2 VHF plasma by measuring the plasma parameters with a heated Langmuir probe to avoid the deposition of contaminations to the probe [23]. The same experimental apparatus as in the case of H_2 plasma described in Section 19.3.1 was used. The pressure ranged from 1 to 3 Torr and the gas flow rate of SiH_4/H_2 was 50 sccm.

The plasma parameters were measured with a heated Langmuir probe which was placed at a distance of 3 mm from the multirod electrode. To reduce the disturbance of the Langmuir probe to the plasma, a tungsten wire of 0.3 mm in diameter was used as the heated Langmuir probe. Before measuring SiH_4/H_2 VHF plasma parameters, the current range that does not disturb plasma generation due to heating was confirmed, and calibration for the surface area of the heated probe was performed by measuring argon and hydrogen plasma using an ordinary Langmuir probe. The plasma parameters estimated from the Langmuir probe $I-V$ characteristics contain some errors which mainly arise from the evaluation of plasma sheath. Especially, when negative ions exist in plasma, the sheath behaves differently from the Bohm sheath [19]. In this experiment, the plasma parameters were obtained assuming the Bohm sheath behavior.

At first, the plasma parameters were examined as a function of pressure for different gas mixture rates of SiH_4/H_2. Figure 19.9a shows that when the pressure is increased, the ion saturation current density I_{is} decreases independent of gas mixture rate SiH_4/H_2. Here, VHF power and gas flow rate were 450 W and 50 sccm, respectively. This figure does not show clearly the trapping effect where I_{is} takes a peak value because of the low resolution of the heated probe. Furthermore, Figure 19.9a indicates that when the gas mixture rate SiH_4/H_2 is increased, I_{is} decreases. On the other hand, as shown in Figure 19.9b, when the pressure is increased, the electron temperature T_e begins to increase at high pressures. As described before, the distance between the discharge electrodes was very short in this experiment, so that T_e is considered to become relatively high [16]. In addition, the production of negative ions may be one of the reasons. In fact, as described later, the sheath potential was lower than that of theoretical value at 3 Torr.

The sheath potential is a key value in plasma CVD because when the sheath potential is high, the energy of the ion bombardment becomes high and, as a result, the quality of the film obtained drops. Figure 19.10 is a typical result of the sheath potential for different concentrations. As seen in Figure 19.10a, when the VHF powers are increased, the sheath potential for SiH_4/H_2 = 0% tends to increase at high pressures. However, as seen in Figure 19.10b, the sheath potential for SiH_4/H_2 = 10% decreases with increasing VHF powers. Furthermore, the sheath potential observed at high pressures for both concentrations was lower than the theoretical value estimated from Eq. 19.2 In the case of SiH_4/H_2 = 10%, SiH_3^+ should contribute to dominant ions, leading to higher sheath potentials. It is well known that when there are negative ions in plasma, the sheath potential as well as the ion saturation current decreases [24]. Thus, the measurement of negative ions is necessary to clarify the reduction mechanism of both the ion saturation current and the sheath potential observed at high pressure.

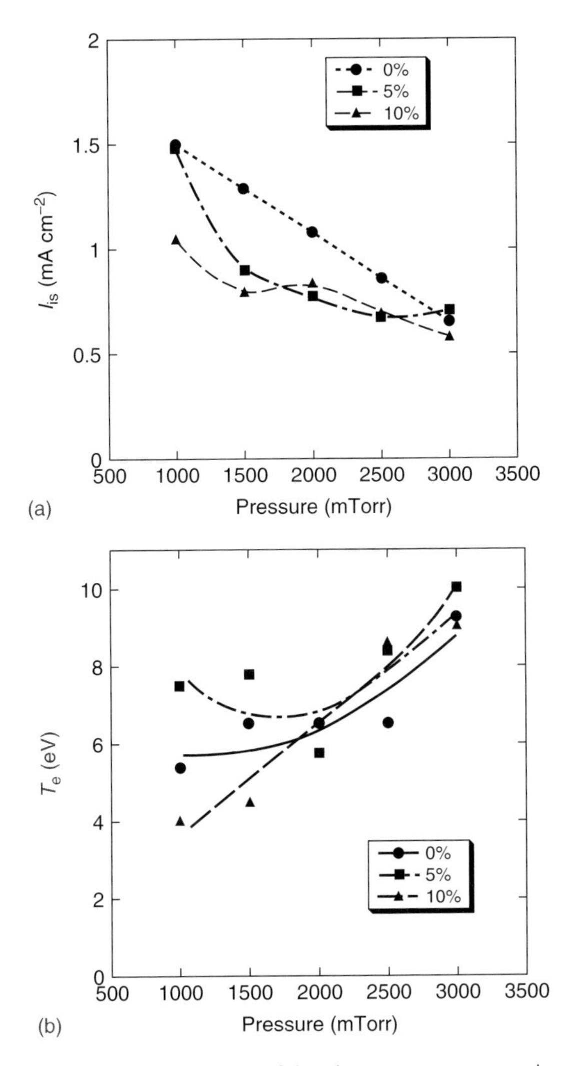

Figure 19.9 Dependence of the plasma parameters on the pressure for different gas mixture rates SiH_4/H_2: (a) the ion saturation current density I_{is} and (b) the electron temperature T_e. Here, VHF power is 450 W.

Here, we estimated the plasma density, n, in pure hydrogen plasma from Figure 19.9; $n \sim 3.8 \times 10^9\,\text{cm}^{-3}$ was achieved at 3 Torr, where dominant ion species were assumed to be H_3^+ [25, 26]. The plasma density obtained here is higher than what is expected in the case of a short discharge gap, which is considered to be due to electron trapping effect [7]. Thus, the above-mentioned results suggest that a short gap VHF discharge provides a high-density plasma with low-sheath potential at high pressures, that is, high deposition rate and high-quality films.

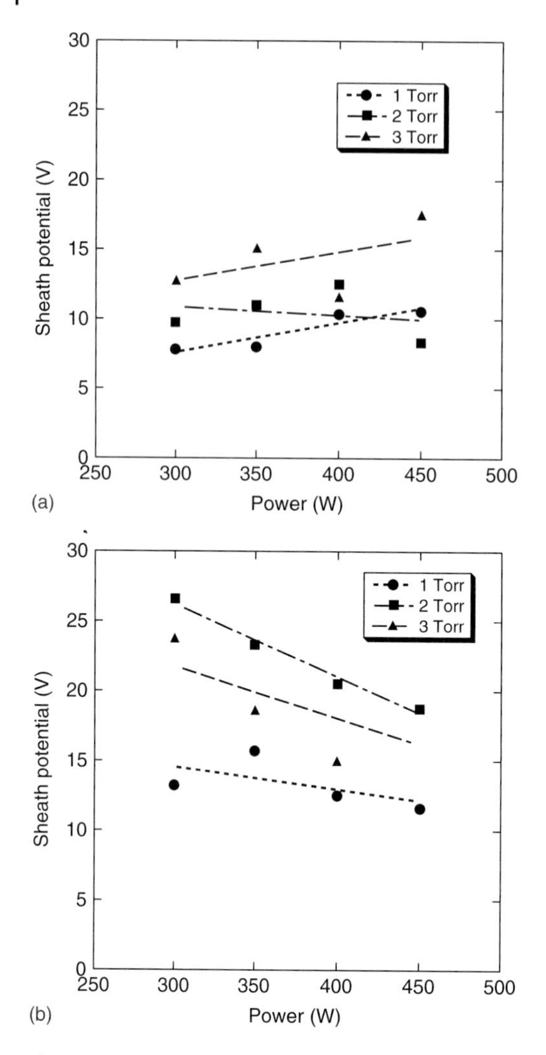

Figure 19.10 Dependence of the sheath potential on VHF power for different pressures, where (a) SiH_4/H_2 = 0% and (b) SiH_4/H_2 = 10%.

19.4 Summary

To realize a stable VHF discharge over a large area (>1 m^2), we developed a balanced power feeding method and produced VHF H_2 plasma and SiH_4/H_2 plasma using this balanced power feeding method. The main results are listed as follows:

1) The ion saturation current density of the H_2 plasma peaked at a certain pressure and the peak pressure depended on VHF powers. These results are understood qualitatively by electron trapping effect in collisional plasma. The

sheath potential at high pressures was much lower than the theoretical value. The reduction of the electron saturation current density was also observed, suggesting that there were a lot of negative ions.

2) When the pressure was increased, the ion saturation current density of the SiH_4/H_2 plasma decreased independent of the concentration ratio of SiH_4/H_2 while the electron temperature tended to increase at high pressures. When the VHF power was increased, the sheath potential decreased, leading to low ion bombardment. These results suggest that VHF SiH_4/H_2 plasma at high pressure provides high-quality microcrystalline silicon films.

References

1. Kondo, M., Fukawa, M., Guo, L., and Matsuda, A. (2000) *J. Non-Cryst. Solids*, **266-269**, 84.
2. Kondo, M. (2003) *Sol. Energy Mater. Sol. Cells*, **78**, 543.
3. Graf, U., Meier, J., Kroll, U., Bailat, J., Droz, C., Vallat-Sauvain, E., and Shah, A. (2003) *Thin Solid Films*, **427**, 37.
4. Takatsuka, H., Noda, M., Yonekura, Y., Takeuchi, Y., and Yamauchi, Y. (2004) *Sol. Energy*, **77**, 951.
5. Howling, A.A., Dorier, J.-L., Hollenstein, Ch., Kroll, U., and Finger, F. (1992) *J. Vac. Sci. Technol. A*, **10**, 1080.
6. Takeuchi, Y., Mashima, H., Murata, M., Uchino, S., and Kawai, Y. (2001) *Jpn. J. Appl. Phys.*, **40**, 3405.
7. Brown, S.C. (1966) *Introduction to Electrical Discharges in Gases*, Chapter 10, John Wiley & Sons, Inc.
8. Schwarzenbach, W., Howling, A.A., Fivaz, M., Brunner, S., and Hollenstein, Ch. (1996) *J. Vac. Sci. Technol. A*, **14**, 132.
9. Sansonnens, L., Pletzer, A., Magni, D., Howling, A.A., Hollenstein, Ch., and Schmitt, J.P.M. (1997) *Plasma Sources Sci. Technol.*, **6**, 170.
10. Kuske, J., Stephan, U., Steinke, O., and Rohlecke, S. (1995) *Mat. Res. Soc. Symp. Proc.*, **27**, 377.
11. Lieberman, M.A., Booth, J.P., Chabert, P., Rax, J.M., and Turner, M.M. (2002) *Plasma Sorces Technol.*, **11**, 283.
12. Schmidt, H., Sansonnens, L., Howling, A.A., Hollenstein, Ch., Elyaakoubi, M., and Schmitt, J.P.M. (2004) *J. Appl. Phys.*, **95**, 4559.
13. Nishimiya, T., Takeuchi, Y., Yamauchi, Y., Takatsuka, H., Shioya, T., Muta, H., and Kawai, Y. (2008) *Thin Solid Films*, **516**, 4430.
14. Isomura, M., Kondo, M., and Matsuda, A. (2002) *Jpn. J. Appl. Phys.*, **41**, 1947.
15. Yamauchi, Y., Takeuchi, Y., Takatsuka, H., Yamashita, H., Muta, H., and Kawai, Y. (2008) *Contrib. Plasma Phys.*, **48**, 326.
16. von Engel, A. (1965) *Ionized Gases*, Chapter 4, The Clarendon Press, Oxford.
17. Lieberman, M.A. and Lichtenberg, A.J. (1994) *Principles of Plasma Discharges and Materials Processing*, Chapter 4, John Wiley & Sons, Inc.
18. Lochte-Holtgreven, W. (1968) *Plasma Diagnostics*, Chapter 11, North-Holland.
19. Amemiya, H. (1990) *J. Phys. D*, **23**, 999.
20. Bacal, M. (2000) *Rev. Sci. Instrum.*, **71**, 3981.
21. St. Brithwaite, N. and Allen, J.E. (1988) *J. Phys. D*, **21**, 1733.
22. Noguchi, M., Hirao, T., Shindo, M., Sakurauchi, K., Yamagata, Y., Uchino, K., Kawai, Y., and Muraoka, K. (2003) *Plasma Sources Sci. Technol.*, **12**, 403.
23. Yamauchi, Y., Takeuchi, Y., Takatsuka, H., Kai, Y., Muta, H., and Kawai, Y. (2008) *Surf. Coat. Technol.*, **202**, 5668.
24. Shindo, H. and Horiike, Y. (1991) *Jpn. J. Appl. Phys.*, **30**, 161.
25. Salabas, A., Marques, L., Jolly, J., Goussel, G., and Alves, L.L. (2004) *J. Appl. Phys.*, **95**, 4605.
26. Nunomura, S. and Kondo, M. (2007) *J. Appl. Phys.*, **102**, 093306.

20
Deposition of a-Si : H Films with High Stability against Light Exposure by Reducing Deposition of Nanoparticles Formed in SiH_4 Discharges

Kazunori Koga, Masaharu Shiratani, and Yukio Watanabe

20.1
Introduction

Although thin-film Si solar cells such as hydrogenated amorphous silicon (a-Si : H) single-junction solar cells and a-Si : H/μc-Si : H tandem solar cells have been commercially mass-produced, they still have an important issue to be solved that large area cells of a stable efficiency more than 10% should be fabricated at a low cost of less than 1.2 $/W in order to play a significant role in the long-term photovoltaic market [1]. This issue has been found to be closely related to light-induced defects created in a-Si : H films [2, 3], and hence their formation mechanism during deposition has been discussed in many literatures [4–9]. Matsuda *et al.* have reported that the stability against light exposure is improved by reducing Si$-H_2$ bonds in the films [10]. This result reminds us that it is important to identify the kinds of particle species responsible for the formation of Si$-H_2$ bonds and to develop a method for depositing a-Si : H films of high stability by reducing their density. A candidate of the species for the Si$-H_2$ bonds formation is nanoparticles (small particles below about 10 nm in size) generated in SiH_4 discharges. Their formation mechanism has been experimentally revealed from their nucleation phase [11–18]. On the basis of the results, methods for suppressing deposition of the nanoparticles on the films have been developed [19–21], and such "almost nanoparticle-free" a-Si : H films have been shown to be highly stable against light exposure [20].

In this chapter, after summarizing formation mechanism of nanoparticles in silane discharges in Section 20.2, particle species responsible for the light-induced degradation are identified in Section 20.3, and effects of nanoparticles on a-Si : H film qualities are described in Section 20.4. Finally, high rate deposition of highly stable a-Si : H films using a multihollow discharge plasma chemical vapor deposition (CVD) method is demonstrated in Section 20.5.

Industrial Plasma Technology. Edited by Yoshinobu Kawai, Hideo Ikegami, Noriyoshi Sato, Akihisa Matsuda, Kiichiro Uchino, Masayuki Kuzuya, and Akira Mizuno

ISBN: 978-3-527-32544-3

20.2
Formation Mechanism of Nanoparticles in Silane Discharges

In this section, we briefly describe formation mechanism of amorphous silicon nanoparticles in a size range below about 10 nm, since a-Si : H films with a low volume fraction of the nanoparticles in the size range incorporated into the films have been found to show better film qualities.

Formation mechanism of nanoparticles from their nucleation phase has been studied observing dependence of their size, density, and volume fraction V_f on discharge duration T_{on} by using a double-pulse discharge method [14]. The fraction V_f is defined as a total volume of nanoparticles per unit volume in the discharge space where they exist. One of the results is shown in Figure 20.1. The nanoparticles of about 0.5 nm in size (hereafter, referred to as *nanoparticles of the small size group*) exist 3 ms after the discharge initiation ($T_{on} = 3$ ms). For $T_{on} = 10$ ms, in addition to this small size group, nanoparticles for which their size increases with T_{on} are observed (hereafter, referred to as *nanoparticles of the large size group*). This indicates that nucleation of nanoparticles of the large size group takes place 3–10 ms after the discharge initiation. The nanoparticles of the small size group have been identified to be almost composed of higher order silane (HOS) molecules of Si_mH_n

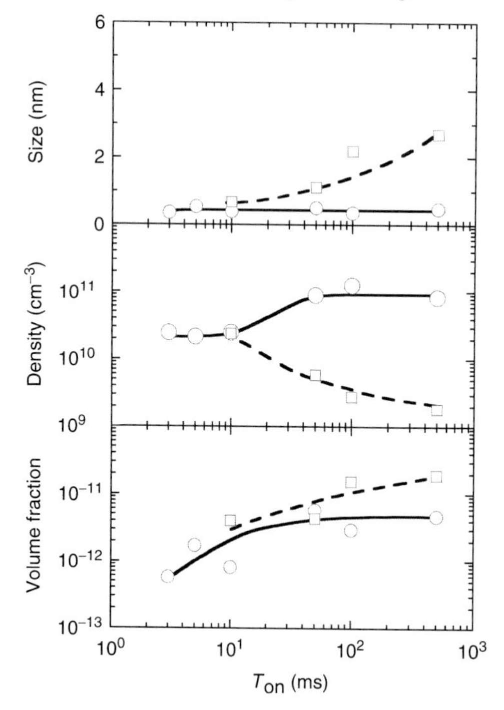

Figure 20.1 Time evolution of size, density, and volume fraction V_f of nanoparticles formed in silane discharges. Discharge conditions: SiH_4 5 sccm, 13.3 Pa, 13.56 MHz, 10 W, T_s = RT.

($m < 4, n \leq 2m + 2$). Once the nanoparticles of the large size group nucleate, they consume particle species to be led to their nucleation. Such consumption results in suppressing further nucleation of nanoparticles. The HOS molecules have a nearly constant average size of 0.5 nm for $T_{on} = 3$–500 ms, while their density increases from 2×10^{10} cm^{-3} at $T_{on} = 3$ ms to a quasisteady-state value of 1×10^{11} cm^{-3} for $T_{on} > 50$ ms. According to the time evolution of size and density of HOS molecules, their volume fraction V_f increases from 5×10^{-13} at $T_{on} = 3$ ms to a quasisteady-state value of 5×10^{-12} for $T_{on} > 50$ ms. The size of the nanoparticles of the large size group increases from 0.7 nm at $T_{on} = 10$ ms to 2.7 nm at $T_{on} = 500$ ms at an average growth rate of 4 nm s^{-1} in a monodisperse way, because their further nucleation is suppressed. Their density decreases from 2×10^{10} cm^{-3} at $T_{on} = 10$ ms to 2×10^{9} cm^{-3} at $T_{on} = 500$ ms. The volume fraction of nanoparticles of the large size group monotonically increases with T_{on}. This increase indicates that the nanoparticles of the large size group (hereafter, referred to as *nanoparticles*) mainly grow due to deposition of monoradicals or HOS radicals, that is, due to CVD, while the decrease in their density due to coagulation among them is appreciable for $T_{on} > 10$ ms. Since the densities of HOS molecules and nanoparticles are higher than that of the positive ions 1×10^{9} cm^{-3}, most of the HOS molecules and the nanoparticles are neutral, and hence they can be incorporated into films. Such incorporation can lead to degradation of qualities of a-Si : H films.

Figure 20.2 shows a nanoparticle formation model of interest, which is based on a series of experimental results in silane high-frequency discharges [12, 14, 15]. Monoradicals such as SiH_x ($0 \leq x \leq 3$) are mainly generated by electron impact dissociation of SiH_4 in the plasma/sheath boundary region near

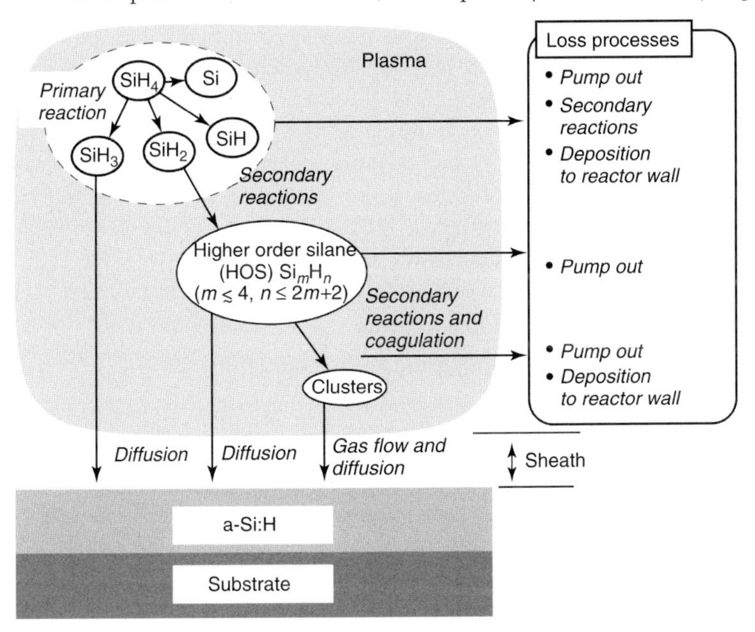

Figure 20.2 Reaction model in silane discharges.

the electrodes (for asymmetric discharges, their generation rate is higher in the region near the powered electrode side). Long lifetime SiH_3 radicals, which are the precursors for deposition of high-quality a-Si : H films, diffuse to the substrate almost without gas-phase reactions and deposit on it. Short lifetime SiH_x ($x \leq 2$) radicals react rapidly with SiH_4, triggering formation of HOS molecules [22]. SiH_x ($x \leq 2$) radicals or HOS molecules then lead to nucleation of nanoparticles.

20.3 Contribution of Higher Order Silane Molecules and Nanoparticles to SiH_2 Bond Formation in Films

The HOS molecules and nanoparticles are candidate species responsible for light-induced degradation. Matsuda *et al.* have proposed the contribution of HOS molecules to the degradation [22]. As shown in Figure 20.1, the HOS molecules coexist with the nanoparticles. This means that the latter also have a possibility to contribute toward the degradation. We have studied which of the HOS molecules and nanoparticles contribute more effectively to the degradation. On the basis of the results obtained, we have proposed that the nanoparticles are mainly responsible for the degradation.

Experiments were carried out using a capacitively coupled high-frequency discharge reactor [21]. The reactor is of the nanoparticle-suppressed plasma CVD type; the growth of nanoparticles is suppressed by utilizing gas viscous and thermophoretic forces exerted on nanoparticles together with reducing the regions of gas flow stagnation. To measure the density of SiH_4, $[SiH_4]$, as well as that of Si_2H_6, $[Si_2H_6]$, and Si_3H_8, $[Si_3H_8]$, in the discharges, the gas passed through the powered electrodes was sampled through an orifice of 2 μm in diameter for analysis with a quadrupole mass spectrometer (QMS).

To obtain information on the contribution of the HOS molecules to films growth, signal intensities of $[m/e] = 30$, 60, and 84 were measured with a QMS. These intensities give information of the densities of $[SiH_4]$, $[Si_2H_6]$, and $[Si_3H_8]$ by using a calibration procedure of them described in [23]. The contribution of HOS molecules was evaluated under the following assumption. The rate equation of the steady-state density of HOS molecules $[Si_mH_n]$ ($n \leq 2m + 1$) may be given by

$$\frac{d[Si_mH_n]}{dt} = k_{dm} n_e [Si_mH_{2m+2}] - \frac{[Si_mH_n]}{\tau_m}$$

where k_{dm} is the rate coefficient of Si_mH_n generation due to electron impact dissociation of Si_mH_{2m+2} and τ_m the characteristic lifetime of Si_mH_n. The steady-state density of Si_mH_n can be written in the form

$$[Si_mH_n] = \tau_m k_{dm} n_e [Si_mH_{2m+2}]$$

Since the SiH_3 molecules are the main precursors of film growth in SiH_4 discharges, the contribution ratio of HOS molecules to film growth is given by

$$\frac{[Si_mH_n]}{[SiH_3]} \propto \frac{[Si_mH_{2m+2}]}{[SiH_4]}$$

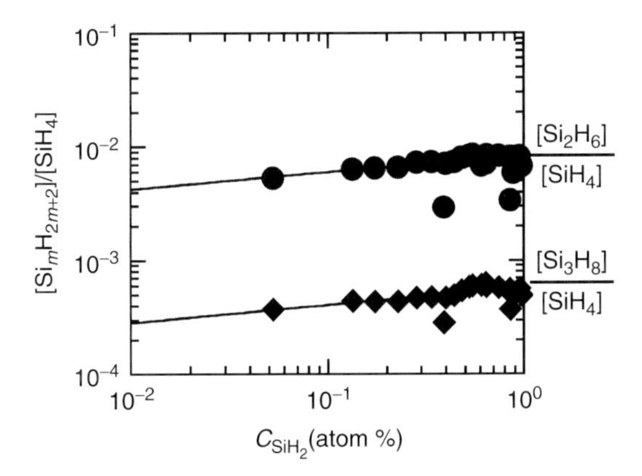

Figure 20.3 C_{SiH_2} dependence of $[Si_mH_{2m+2}]/[SiH_4]$ ($m = 2$ and 3).

Figure 20.3 shows dependence of density ratios of $[Si_2H_6]/[SiH_4]$ and $[Si_3H_8]/[SiH_4]$ on the hydrogen content associated with Si–H_2 bonds, C_{SiH_2}, in the a-Si : H films. Even with increasing the C_{SiH_2} value by more than one and a half orders of magnitude, these ratios increase just a little. This weak C_{SiH_2} dependence strongly suggests that the HOS molecules of Si_2H_x and Si_3H_x *do not* contribute seriously to the Si–H_2 formation.

This suggestion is also supported by the following discussion. Taking into account that most of Si atoms in the films originated from SiH_3 molecules impinging on the surface and also that the C_{SiH_2} of interest is in a range of 0.1–1 atom % as shown in Figure 20.3, the HOS density necessary for the Si–H_2 bond formation is of the order of magnitude of $C_{SiH_2}[SiH_3]$. Therefore, the following relation should be satisfied if HOS molecules were to mainly contribute to the formation of Si–H_2 bonds.

$$\frac{[Si_mH_n]}{[SiH_3]} \approx \frac{[Si_mH_{2m+2}]}{[SiH_4]} \geqslant C_{SiH_2} \approx 10^{-3} - 10^{-2}$$

Taking into account the results shown in Figure 20.3 together with the fact that $[Si_mH_{2m+2}]$ sharply decreases with the increase of m in a range of $m = 1$–5 [24], any radicals except Si_2H_x molecules ($x = 0$–5) cannot satisfy the condition. However, a-Si : H films of $C_{SiH_2} = 1$ atom % have been deposited using Si_2H_6 discharges [25] while Si_2H_x molecules are precursors of film growth in the discharge. It suggests that maximum Si–H_2 bond density formed by incorporation of Si_2H_x molecules into films should be 10^{-3}–10^{-2} atom %, because the density ratio of Si_2H_6 to SiH_4 is 10^{-3}–10^{-2} as shown in Figure 20.3. Therefore, the Si_2H_x molecules do not mainly contribute to the Si–H_2 formation.

Figure 20.4 shows dependence of the C_{SiH_2} on the volume fraction of nanoparticles in the films V_f. While the data points in the figure are rather scattered, all of them exist around the dashed line $V_f \propto C_{SiH_2}$. On the basis of our present and previous results regarding the surface hydride composition of a-Si : H films, the Si–H_2 bonds in the films are considered to be generated through two processes:

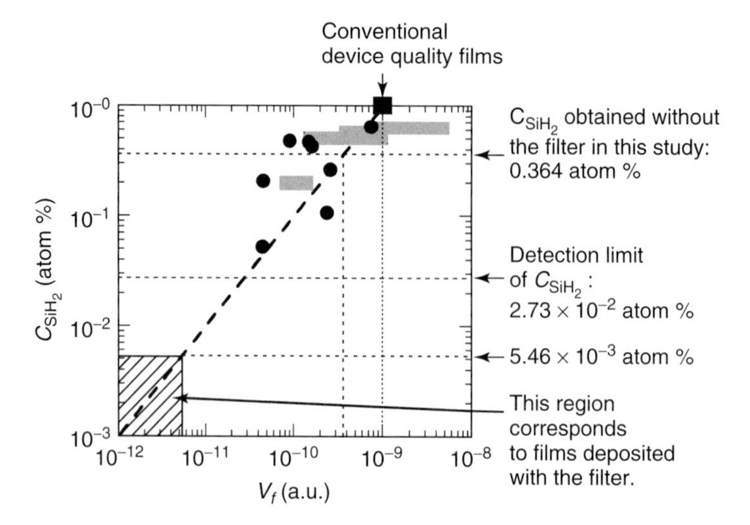

Figure 20.4 Dependence of C_{SiH_2} on volume fraction V_f of nanoparticles incorporated into *a*-Si : H films. Solid circles and gray rectangles are results of films deposited by nanoparticle-suppressed plasma CVD without nanoparticle-eliminating filter reported in [18]. Solid squares correspond to conventional device quality films.

(i) incorporation of nanoparticles into the films and (ii) hydrogen elimination reactions on the surfaces of films and in their subsurfaces [26]. Fourier-transform infrared (FTIR) measurements of nanoparticles show that many $Si-H_2$ bonds exist in them [21]. It is difficult for the hydrogen elimination reactions to proceed in a nanoparticle as compared to those on the film surface, because it is not easy for the radicals to enter a nanoparticle. For the same reason, it is also difficult for the reaction to take place at the interface between the a-Si : H film and a nanoparticle deposited on the film. Therefore, the $Si-H_2$ bonds in nanoparticles deposited in the films tend to remain in the films.

20.4
Effects of Nanoparticles on a-Si : H Qualities

The light-induced degradation of a-Si : H films is a major drawback in realizing thin-film Si solar cells of high efficiency. Matsuda *et al.* have reported that a-Si : H films of a lower C_{SiH_2} show better stability against light exposure [10]. To study contribution of nanoparticles in the formation of $Si-H_2$ bonds in films, the volume fraction V_f is plotted as a function of C_{SiH_2} [17, 27]. As shown in Figure 20.4, the V_f is nearly proportional to the C_{SiH_2}. This tendency indicates that the incorporation of nanoparticles into a-Si : H films is an important origin of the $Si-H_2$ bond formation of interest. This result motivates us to deposit the nanoparticle-free a-Si :

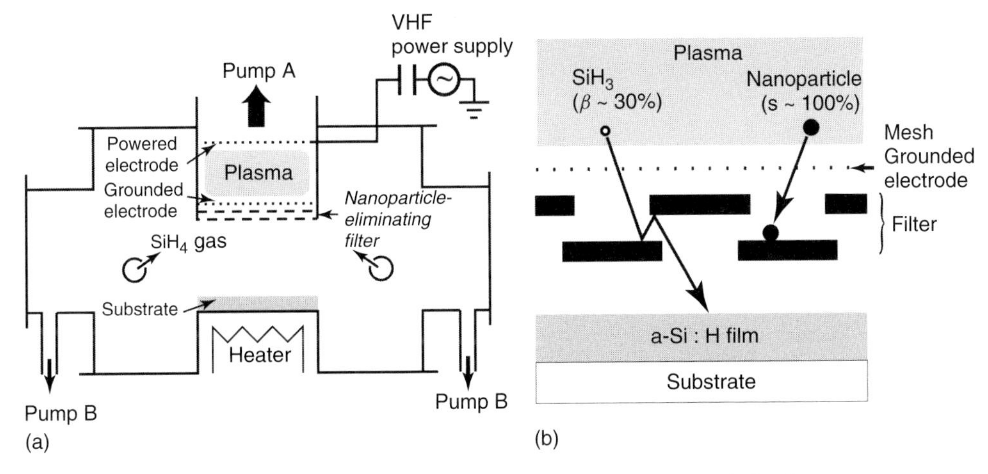

Figure 20.5 (a) The nanoparticle-suppressed triode plasma CVD reactor equipped with the nanoparticle-eliminating filter. (b) Concept of the nanoparticle-eliminating filter. The open and solid circle shows an SiH_3 radical and a nanoparticle, respectively. Symbols of β and s indicate the surface reaction probability of SiH_3 and the sticking probability of a nanoparticle, respectively.

H films using a nanoparticle-eliminating filter to reveal effects of nanoparticles on film quality.

Experiments were carried out using a nanoparticle-suppressed triode plasma CVD reactor as shown in Figure 20.5a. The suppression methods have been described in detail elsewhere [21]. A nanoparticle-eliminating filter was placed between the grounded electrode and the substrate holder. Figure 20.5b shows a cross-sectional view of the filter [19], which has two stainless silts placed at a distance less than mean free paths of nanoparticles and SiH_3 radicals. The surface reaction probability of SiH_3 is above 30% [28], whereas the sticking probability of nanoparticles is close to 100% [29]. Therefore, most nanoparticles are trapped by the filter, while an appreciable proportion of SiH_3 radicals can pass through the filter. The V_f value can be reduced by a factor of 1/100 by using the filter.

The C_{SiH_2} value of the almost nanoparticle-free a-Si : H films obtained employing this filter was obtained by FTIR spectroscopy. As a C_{SiH_2} value of interest is estimated to be less than the lower detection limit of the FTIR measurement, its value was found using the transmittance of the filter and the V_f dependence of C_{SiH_2} as shown in Figure 20.4. The C_{SiH_2} tends to decrease linearly with V_f. Using this dependence, a C_{SiH_2} with the filter is estimated to be less than 3.6×10^{-3} atom % taking into account that a C_{SiH_2} without the filter is 3.6×10^{-1} atom % and the transmittance of the filter is 1/100. Thus, values of V_f and C_{SiH_2} can be reduced by a factor of 1/300 compared to those for the conventional films by using the filter.

Dependence of the defect density on light exposure time has been studied to evaluate stability of the films using light of intensity 240 mW cm^{-2} (2.4 SUN). The result is shown in Figure 20.6. During the light exposure, the temperature of films was kept at 50 °C. The defect density of the films was obtained using electron spin

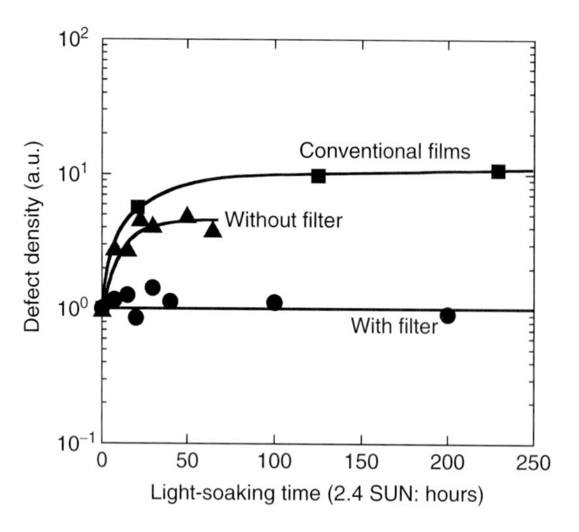

Figure 20.6 Light exposure time dependence of defect density of a-Si : H films. Condition of light exposure is 240 mW cm^{-2} (2.4 SUN) of AM1.5 spectrum, 50 °C.

resonance (ESR) spectroscopy. For the conventional films of device quality, a value of defect density after light soaking increases by one order of magnitude compared to its initial value. a-Si : H films deposited using the nanoparticle-suppressed triode plasma CVD without the filter have the better stability than that of the conventional films. In contrast, the defect density for the films deposited by using the filter is nearly constant. In other words, our "almost nanoparticle-free a-Si : H films" show little light-induced degradation.

20.5
High Rate Deposition of a-Si : H Films of High Stability against Light Exposure Using Multihollow Discharge Plasma CVD

While nanoparticle-free and stable a-Si : H films have been realized by using the nanoparticle-eliminating filter, the deposition rate of 0.006 nm s^{-1} is much lower than the conventional deposition rate of about 0.5 nm s^{-1}. To overcome this problem, we have developed a multihollow discharge plasma CVD method.

The multihollow discharge plasma CVD reactor is shown in Figure 20.7. Two electrodes of 75-mm diameter having 24 holes of 5-mm diameter were placed 2 mm apart in a stainless steel tube of 79-mm inner diameter. Gas of SiH_4 diluted with H_2 was supplied from a tube ring and was pumped out through the electrodes by a vacuum system. Flow rates of SiH_4 and H_2 were 10 and 40 sccm, respectively, and the total pressure was 66.5 Pa. Discharges were sustained in the holes by applying a 60-MHz high-frequency voltage to the powered electrode. SiH_3 radicals, HOS molecules, and nanoparticles are generated in the discharges. The nanoparticles

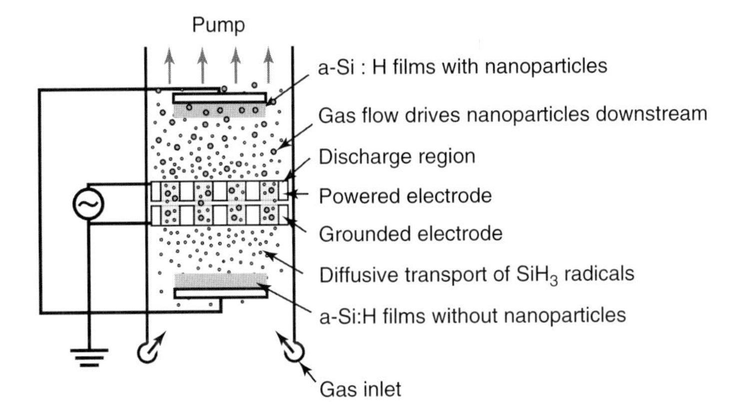

Figure 20.7 Schematic of multihollow discharge plasma CVD method.

are transported toward the downstream region from the discharge region because their diffusion velocity is less than the gas velocity. On the other hand, SiH_3 radicals and HOS molecules are transported toward both the upstream region and the downstream region due to their fast diffusion. Therefore, incorporation of nanoparticles into the films deposited in the upstream region can be significantly suppressed using the reactor. Substrates were set 31 mm upstream and 72 mm downstream from the powered electrode.

In order to evaluate stability of the films against light exposure, we have measured dependence of their defect density on light exposure time *t*. The results are shown

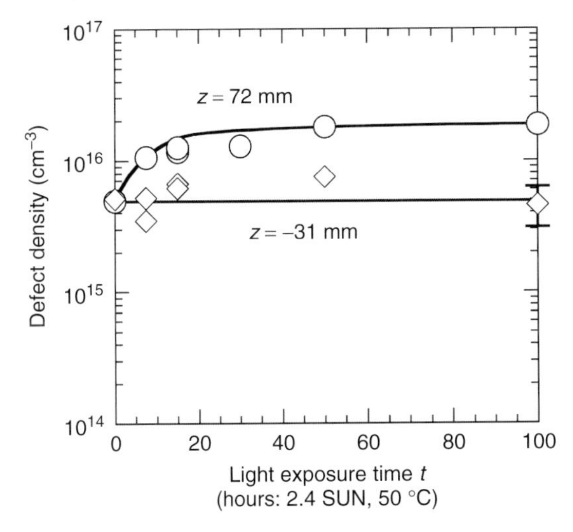

Figure 20.8 Dependence of defect density on light exposure time *t*. The conditions for light exposure are 240 mW cm^{-2} (2.4 SUN) of AM1.5 spectrum, 50 °C. Open diamonds indicate results for film at $z = -31$ mm. Open circles denote results for film at $z = 72$ mm. Deposition conditions: SiH_4 10 sccm, H_2 40 sccm, 66.5 Pa, 60 MHz, 30 W, $T_s = 250$ °C.

in Figure 20.8. The defect density of the film of 2-μm thickness, prepared 31 mm upstream from the powered electrode ($z = -31$ mm) at a deposition rate of 0.12 nm s^{-1}, keeps a low value of 5×10^{15} cm^{-3} for $t = 0$–100 hours. The defect density of the film of 0.5-μm thickness, prepared 72 mm downstream from the electrode ($z = 72$ mm) at a deposition rate of 0.04 nm s^{-1}, increases from an initial value of 5×10^{15} to 2×10^{16} cm^{-3} at $t = 100$ hours. Such increase in the defect density due to light exposure is common for the conventional a-Si : H films. While our present deposition experiments are preliminary, we have realized a considerably high deposition rate of 0.12 nm s^{-1}. We expect that a higher rate can be achieved by optimizing discharge conditions such as gas pressure, gas flow velocity, and hydrogen dilution rate.

20.6 Conclusions

In this chapter, we have discussed formation mechanism of nanoparticles in silane discharges and their effects on qualities of a-Si : H films. Deposition of nanoparticles of size below 10 nm into the a-Si : H films is responsible for the light-induced degradation while the HOS molecules such as Si_2H_x and Si_3H_x have little effect on the degradation. Deposition of stable a-Si : H films is realized at a rate of 0.12 nm s^{-1} using the multihollow discharge plasma CVD method for which incorporation of nanoparticles into the films is reduced by employing gas viscous force. We are now carrying out experiments for realizing a higher deposition rate by optimizing discharge conditions in the holes of the developed CVD reactor.

Acknowledgments

We are grateful to Prof. A. Matsuda for discussions. We also would like to acknowledge the assistance of Messrs T. Kinoshita and H. Matsuzaki who contributed greatly to the preparation of the experimental setup. This work was partially supported by a grant from the Japan Society of the Promotion of Science and from New Energy and Industrial Technology Development Organization.

References

1. Marti, A. and Luque, A. (eds) (2004) *Next Generation Photovoltaics*, IOP Publishing, Ltd., p. 133.
2. Staebler, D.L. and Wronski, C.R. (1977) *Appl. Phys. Lett.*, **31**, 292.
3. Schropp, R.E.I. and Zeman, M. (1998) *Amorphous and Microcrystalline Silicon Solar Cells*, Kluwer Academic Publishers, Boston, p. 99.
4. Matsuda, A. (1987) *J. Non-Cryst. Solids*, **59/60**, 767.
5. Matsuda, A. and Tanaka, K. (1987) *J. Non-Cryst. Solids*, **97/98**, 1367.
6. Yan, H. (1992) *Phys. Rev. Lett.*, **68**, 3048.
7. Li, Y.M., An, I., Nguyn, H.V., Wronski, C.R., and Colins, R.W. (1992) *Phys. Rev. Lett.*, **68**, 2814.

8. Gangly, G. and Matsuda, A. (1993) *Phys. Rev. B*, **47**, 3661.
9. Matsuda, A. (1998) *J. Vac. Sci. Technol. A*, **16**, 365.
10. Nishimoto, T., Takai, M., Miyahara, H., Kondo, M., and Matsuda, A. (2002) *J. Non-Cryst. Solids*, **299**, 1116.
11. Shiratani, M., Fukuzawa, T., Eto, K., and Watanabe, Y. (1992) *Jpn. J. Appl. Phys.*, **31**, L1791.
12. Shiratani, M. and Watanabe, Y. (1998) *Rev. Laser Eng.*, **26**, 449.
13. Shiratani, M., Fukuzawa, T., and Watanabe, Y. (1999) *Jpn. J. Appl. Phys.*, **38**, 4525.
14. Koga, K., Matsuoka, Y., Tanaka, K., Shiratani, M., and Watanabe, Y. (2000) *Appl. Phys. Lett.*, **77**, 196.
15. Shiratani, M., Maeda, S., Koga, K., and Watanabe, Y. (2000) *Jpn. J. Appl. Phys.*, **39**, 287.
16. Watanabe, Y., Shiratani, M., Fukuzawa, T., and Koga, K. (2000) *J. Tech. Phys.*, **41**, 505.
17. Watanabe, Y., Shiratani, M., and Koga, K. (2002) *Plasma Sources Sci. Technol.*, **11**, A229.
18. Watanabe, Y. (2006) *J. Phys. D Appl. Phys.*, **39**, R329.
19. Koga, K., Kaguchi, N., Bando, K., Shiratani, M., and Watanabe, Y. (2005) *Rev. Sci. Instrum.*, **76**, 113501.
20. Koga, K., Inoue, T., Bando, K., Iwashita, S., Shiratani, M., and Watanabe, Y. (2005) *Jpn. J. Appl. Phys.*, **44**, L1430.
21. Shiratani, M., Koga, K., Kaguchi, N., Bando, K., and Watanabe, Y. (2006) *Thin Solid Films*, **506/507**, 17.
22. Takai, M., Nishimoto, T., Takagi, T., Kondo, M., and Matsuda, A. (2000) *J. Non-Cryst. Solids*, **266**, 90.
23. Turban, G., Drevillon, B., Mataras, D.S., Rapakoulias, D.E. (1995) in *Plasma Deposition of Amorphous Silicon-Based Materials* (eds Bruno, G., Capezzuto, P., and Madan, A.), Academic Press, p. 83.
24. Takai, M., Nishimoto, T., Kondo, M., and Matsuda, A. (2001) *Thin Solid Films*, **390**, 83.
25. Azuma, K., Tanaka, M., Nakatani, M., and Shimada, T. (1993) *Sol. Energy Mater. Sol. Cells*, **29**, 233.
26. Toyoshima, Y., Arai, K., Matsuda, A., and Tanaka, K. (1991) *J. Non-Cryst. Solids*, **137-138**, 765.
27. Koga, K., Kaguchi, N., Shiratani, M., and Watanabe, Y. (2004) *J. Vac. Sci. Technol. A*, **22**, 1536.
28. Perrin, J., Shiratani, M., Kae-Nune, P., Videlot, H., Jolly, J., and Guillon, J. (1998) *J. Vac. Sci. Technol. A*, **16**, 278.
29. Koga, K., Iwashita, S., and Shiratani, M. (2007) *J. Phys. D Appl. Phys.*, **40**, 2267.

21
Diagnostics and Modeling of SiH_4/H_2 Plasmas for the Deposition of Microcrystalline Silicon: the Case of Dual-Frequency Sources

Eleftherios Amanatides and Dimitrios Mataras

21.1
Introduction

Interest in the application of hydrogenated microcrystalline silicon (μc-Si : H) thin films deposited from highly diluted silane plasmas in thin-film solar cells has been increasing in recent years [1–3]. Nevertheless, the increase in the deposition rate of the intrinsic μc-Si : H layer remains a key issue in reducing the cost and in the widespread use of tandem devices containing μc-Si : H [4]. Many alternatives to the conventional 13.56-MHz capacitively coupled SiH_4/H_2 discharges have been proposed for solving this problem, which include radical, completely different processes as well as improvements in the established industrial process. In the latter category, increase in plasma excitation frequency toward very high frequency (VHF) and high-pressure – high-power conditions are recognized to have a predominantly beneficial effect on the film deposition rate [5–8]. A common indirect conclusion of the existing studies on the subject is that for a certain SiH_4/H_2 gas composition, a proper combination of frequency, pressure, and power may optimize the ion bombardment on the growing film surface, resulting in device quality material. This is mainly based on theoretical and experimental results related to the contribution of ion flux and energy to the deposition rate [9, 10] and film properties [11, 12].

In reality, the only straightforward method that can alter the flux and energy of ions striking the surface of the growing film is the application of an RF or DC bias on the substrate holder. Recently, a number of studies have appeared in the literature concerning this effect on the deposition rate as well as on the structure and optoelectronic properties of both a-Si : H and μc-Si : H. However, these studies include many controversies regarding the effect of ion bombardment on the film growth, indicating the complexity of the processes involved. In some cases, negative values of DC bias were found to be beneficial for the initial growth stage and the following crystallization of the growing material. In other cases, a positive bias and a reduction of the plasma potential was found to improve the film growth rate, the morphology, and the electronic properties, while there are also studies that report almost no effect of the bias voltage on the deposition process [13–19].

Industrial Plasma Technology. Edited by Yoshinobu Kawai, Hideo Ikegami, Noriyoshi Sato, Akihisa Matsuda, Kiichiro Uchino, Masayuki Kuzuya, and Akira Mizuno

ISBN: 978-3-527-32544-3

Therefore, the open questions that this work attempts to answer are whether (i) there is a possibility to improve the deposition rate of μc-Si : H thin films through the application of an external bias voltage and (ii) substrate biasing is beneficial for the film crystallinity. The study focused on high-pressure – high-power capacitively coupled SiH_4/H_2 discharges using different substrate holder configurations (positive, negative AC and DC biasing – grounded and floating) with subsequent monitoring of film deposition rate and crystallinity, in conditions known to be close to the transition from a-Si : H to μc-Si : H growth. The interpretation of these results is supported by the application of *in situ* plasma diagnostics as electrical measurements, 2D) plasma emission spectroscopy as well as by the results of self-consistent modeling of SiH_4/H_2 discharges.

21.2
Experimental

Silicon thin-film deposition studies were performed in a capacitively coupled ultrahigh vacuum (UHV) parallel plate, having a base vacuum of 10^{-9} mbar. The reactor is equipped with a load lock system for the introduction of substrates and with four quartz windows suitable for spectroscopic observations.

The setup used for the substrate holder biasing is schematically shown in Figure 21.1a,b and has been presented in detail in [20]. The substrate holder (12 cm in diameter) is mounted on an UHV linear motion feedthrough, permitting continuous variation of the interelectrode space. In the present study, the distance between the two electrodes was fixed at 1.5 cm. The holder is electrically isolated from the ground chamber using a ceramic break flange allowing for the application of a low frequency and/or a DC voltage on it, while it can be either grounded or floating. We have investigated the following configurations: (i) $100 + 200\ \sin\omega_l t$ (volts), (ii) grounded, (iii) floating, and (iv) $-100 + 200\ \sin\ \omega_l t$ (volts). The frequency ω_l was 20 kHz in all cases, while the main plasma frequency, applied on the RF electrode was the conventional frequency of 13.56 MHz. The glass (8×6 cm) substrate is held at the top and bottom by the electrode, while there is a gap of approximately 1 mm between the glass and the conductive part of the electrode on the left and right sides (Figure 21.1b). This configuration permits a distribution of the applied bias voltage on the substrate either via electric field conduction (top and bottom) or via plasma-induced conduction (left and right sides). Points 1, 2, and 3 in Figure 21.1b denote the locations where postdeposition Raman measurements were performed for determining the film crystallinity. The highlighted area in Figure 21.1a denotes the area from which 2D emission images were taken and also the geometry used for the 2D self-consistent modeling. Moreover, the high-frequency power (13.56 MHz) actually consumed in the discharge as well as the discharge impedance were determined using Fourier transform voltage and current analysis. The voltage and current waveforms were measured on the powered electrode, using a high impedance 1 : 1000 attenuation voltage probe (Lecroy PPE 1 : 1000), and a $0.1 - \Omega$ transfer impedance RF current

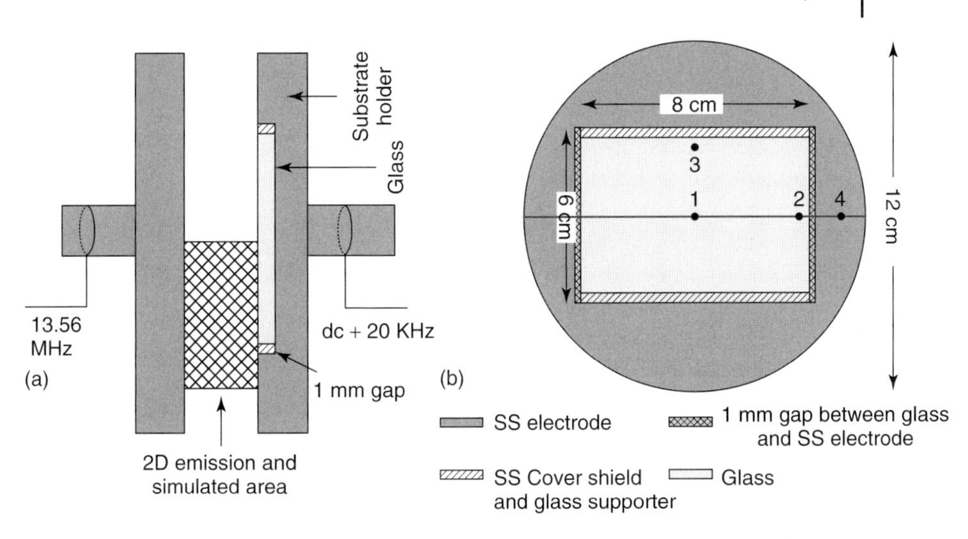

Figure 21.1 (a) Cross section of the setup used for the deposition experiments. The region where 2D emission measurements and plasma simulation were performed is also highlighted. (b) The configuration of the substrate holder used for the biasing of the glass substrate. Points 1, 2, 3, and 4, and the horizontal line denote where either Raman measurements were performed or where emission and simulation results are presented.

probe (FCC model F-35-1). The detailed method used for the measurement of the power consumed in 13.56 MHz discharges can be found in [21]. The setup used to record the emission spectra and 2D emission images consists of a cylindrical or a focusing achromatic lens, an imaging spectrograph, and an intensified Charged Coupled Device (iCCD) detector (Andor, iStar734) [22]. Emission 1D and 2D images were recorded for SiH^*, H_α, and H_β using suitable interference filters.

The films were deposited on common glass substrates, heated to a temperature of 250 °C. The deposition rate was measured *in situ* using laser reflectance interferometry. The structural properties of films were measured with the T64000 Raman system (J. Y.), which was excited by the 514.5-nm line of an air-cooled Ar^+ laser (Spectra Physics) [23]. Finally, atomic force microscope (AFM) topography images were taken using an NT-MDT Scanning Probe Microscope (Ntegra Spectra) operating in the noncontact mode.

21.3 Model Description

A 2D, time-dependent fluid model was used to study the effect of substrate bias on the deposition of μc-Si : H thin films and to understand the experimental observations related to the variation of the deposition rate and crystallinity with the application of bias. Briefly, the model describes the discharge using a combination of particle, momentum, and electron energy conservation equations derived from the Boltzmann equation, coupled with Poisson's equation for a self-consistent calculation of the electric field. A detailed description of the model can be found in [24]. A set of 25 species together with a total number of 112 reactions (85 in the gas phase and 27 on the surface) are included in this version of the model, comprising electron-impact reactions with molecular species, neutral–neutral, and ion–neutral reactions. Cross-sectional data are used for the electron-impact reactions, whereas the rate constants for the rest of the reactions involving neutral species are either calculated or taken from experimental measurements. Plasma surface interaction of radicals leading to the growth of μc-Si : H are handled using sticking coefficients.

The required input data for this model besides the general process conditions are the geometry of the reactor and the applied voltage, while the results include all the important properties of SiH_4/H_2 discharges, such as the distribution of the electric field and voltage in the plasma, electron, ion and species densities, electron temperature, radical fluxes toward the substrate surface, and the deposition rate.

In the present case, the model results of particular interest were mostly the variation of the electric field distribution, ion flux, and energy and species fluxes related to the changes of the substrate holder bias configuration.

21.4
Results and Discussion

As stated above, the main targets of the present study are to investigate if it is possible to enhance the μc-Si : H thin-film deposition rate and/or the percentage of film crystallinity through the application of an external bias voltage to the surface. Therefore, the starting point was to determine the plasma conditions that lead to the transition from μc-Si : H to a-Si : H growth, when the substrate holder is grounded. The total gas pressure was set to 2.5 torr as this is the pressure that optimizes the deposition rate in the specific reactor [25]. A set of experiments was then performed with variable SiH_4 percentage fraction and discharge power in order to reach the parameters that lead to the transition at a reasonably high deposition rate. The determined macroscopic parameters were 4% SiH_4 in H_2 mole fraction, discharge power 35 W, frequency 13.56 MHz, electrode gap 1.5 cm, and total gas pressure 2.5 torr. At these conditions, the deposition rate was 7.3 Å s^{-1} and a very a small shoulder appeared in the Raman spectra around 518 cm^{-1}. Then, using the same parameters we applied a positive bias ($+100 + 200 \sin \omega_l t$) and a negative bias ($-100 + 200 \sin \omega_l t$) voltage, while in the final experiment, the substrate holder was left floating. Special care was taken to maintain the real power that was dissipated in the discharge constant as we applied the different substrate bias voltages. Figure 21.2a–d summarizes the results of the electrical measurements that lead to the same power dissipation. The error bars in the figures were obtained from about 10 measurements performed on different days. This exhaustive check of reproducibility was very important because the application of the external bias voltage induces some distortion in the voltage and current waveforms measured at the RF electrode.

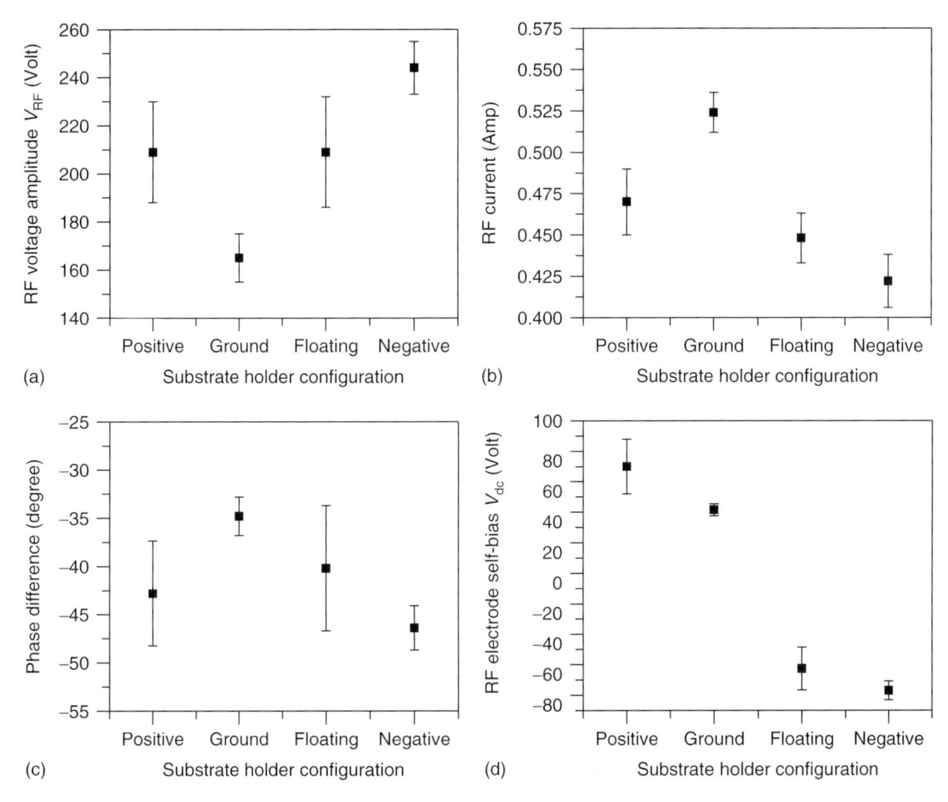

Figure 21.2 (a) The applied RF voltage amplitude; (b) the total discharge current; (c) the DC self-bias in the RF electrode; and (d) the phase difference between RF voltage and current that correspond to conditions of constant RF power dissipation at positive and negative biasing, grounded, and floating conditions of the of the conductive part of the substrate holder.

In fact, the change from the grounded conditions imposes an increase in the applied 13.56-MHz voltage amplitude (V_{RF}) to maintain the same power, and this increase is higher for the negative bias conditions (Figure 21.2a). On the other hand, the discharge current presented in Figure 21.2b shows the opposite behavior compared to that of voltage. It presents a drop in any configuration different than the grounded one. Again, the drop is stronger for the negative biasing conditions. In addition, the discharge phase impedance, as plotted in Figure 21.2c, indicates that for the grounded conditions, the discharge exhibits the most resistive character, since in absolute values the phase is reduced. Finally, the self-bias voltage V_{dc} that is developed on the RF electrode is strongly affected by the application of an external bias voltage on the substrate holder. Actually, the application of a positive bias induces an increase of V_{dc} to more positive values, while the application of a negative bias leads to more negative values of V_{dc}. As expected, the floating conditions lead to similar values of V_{dc} as in the case of negative bias, indicating that the holder is negatively charged. It is also worth noting that, in the grounded conditions, the

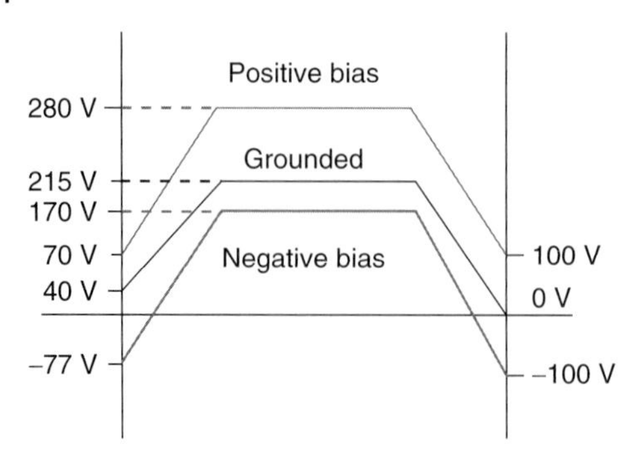

Figure 21.3 Simplified time-averaged voltage distribution between the two electrodes according to the measured values of V_{RF}, V_{dc}, and V_b at different substrate holder bias conditions.

value of V_{dc} is positive, meaning that in these conditions of relatively high gas pressure and power, the discharge has a pronounced electronegative character. The variation of RF voltage amplitude, self-bias voltage of the RF electrode and substrate holder biasing conditions indicates that voltage distribution between the two electrodes is significantly altered. Figure 21.3 presents a simplified scheme of the time-averaged voltage distribution in the discharge according to the measured values of V_{RF}, V_{dc}, and V_b assuming a constant voltage in the bulk of the plasma. We can observe that the voltage drop in front of the substrate holder is reduced as we go from negative to grounded and finally to positive bias. So, in the case of positive bias, a reduction of the energy of the ions striking the surface is expected. In addition, the method allows control of the voltage drop and, consequently, of the ion energy by varying the amplitude of the bias voltage. Another important observation is that, in the case of positive bias, the discharge is closer to electrical symmetry (voltage drop in front of the RF electrode is about the same to the drop in front of the substrate holder) and this is also discussed later when the maps of emission (Figure 21.5) measurements are presented.

To summarize the results of the electrical measurements, we conclude that in the case of grounded substrate holder, a lower voltage, a higher current flow, and a more resistive discharge is required to maintain the same discharge power as in the other configurations. Typically, these are conditions of higher plasma density and lower electron temperature, which, in turn, have been identified in previous studies [25] as beneficial for the deposition rate, the species production, and the film crystallinity. In addition, the variation of the V_{dc} clearly indicates that with the application of the substrate bias, the voltage distribution in the discharge is redistributed as to maintain the time-averaged plasma potential (i.e., the plasma bulk potential) as the most positively charged part of the plasma. Finally, the results of the electrical measurements have clearly pointed out that the application of substrate biasing strongly affects all the plasma properties and this influence cannot be limited to the sheath of the substrate holder and the expected alteration of ion bombardment.

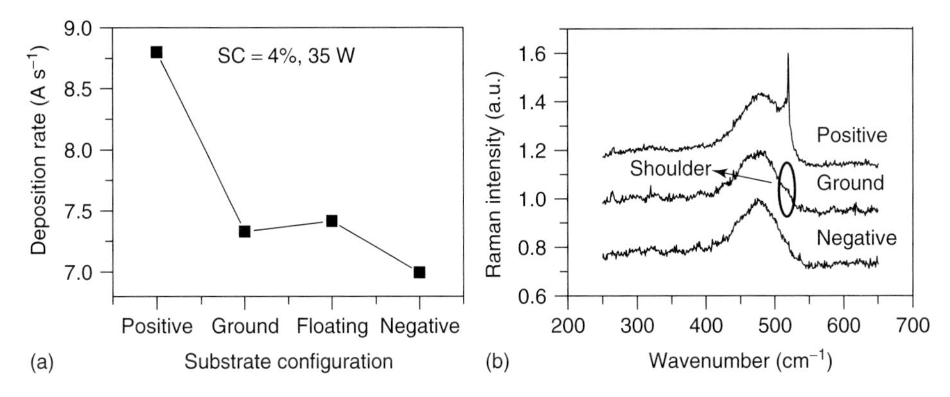

Figure 21.4 (a) Deposition rate and (b) Raman spectra of the films deposited at different substrate holder conditions.

Figure 21.4a presents the variation of the silicon thin films' deposition rate for the different substrate holder configurations. In fact, the change from positive biasing to grounded, floating, and negative biasing results in a significant decrease of the deposition rate. It is remarkable that deposition rates of $\sim$9 Å s^{-1} can be obtained for the films prepared with positive biasing. The deposition rate is determined by the balance between the deposition of radicals and ions and the etching via ion bombardment and hydrogen atoms. In turn, these parameters depend on the number of active species generated in the plasma and the electric field distribution in the case of the charged species. As we discuss later, both these parameters are affected by the different bias configurations resulting in the observed variations of the film growth rate.

With regard to the film structure, the Raman scattering spectra show characteristic changes in the crystallinity of the films deposited under different substrate bias conditions (Figure 21.4b): the films deposited with positive bias have a sharp peak at $\sim$519.6 cm^{-1}, which is due to the TO phonon mode in the crystalline phase. In contrast, the films deposited with negative bias are clearly amorphous as indicated by the broad Raman feature peaked around $\sim$480 cm^{-1}. The same results were obtained for the films deposited on a floating substrate. Finally, the samples prepared on the grounded substrate present only a small "shoulder" at a long wave number around 518 cm^{-1}, indicating that the material is in the microcrystalline to amorphous silicon growth transition regime. Thus, the results so far suggest that an appropriate positive substrate bias during the film growth can improve both the deposition rate and the crystallinity of μc − Si : H thin films. We have to note here that the Raman spectra presented in Figure 21.4b were obtained from the middle of the glass (point 1, Figure 21.1). As already discussed in the experimental section, the application of an external bias in an insulator like glass and the achievement of the desired voltage drop in the sheath of the substrate holder are difficult tasks that depend on the setup around the substrate. In order to check the effectiveness of the configuration that we applied in order to induce the desired voltage drop across the glass, we used Raman spectroscopy as a probing tool. Thus besides the spectra

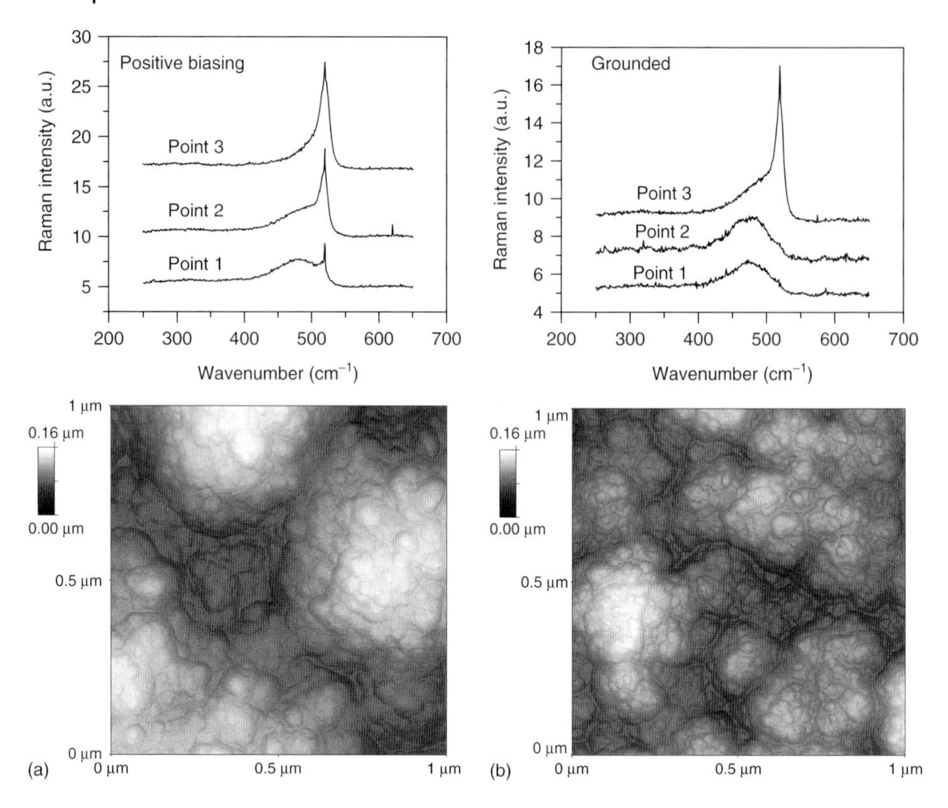

Figure 21.5 Raman spectra and AFM images at three different points of the films deposited (a) under positive biasing and (b) under ground conditions of the conductive part of the substrate holder.

obtained from the center of the samples, two more measurements were made, one close to right edge of the substrate (point 2, Figure 21.1) and one close to the cover shield of the substrate (point 3, Figure 21.1). The results of these measurements are presented in Figure 21.5a,b for the films deposited under positive and grounded conditions, respectively. Measurements for the films deposited under negative and floating modes were also performed but these films were purely amorphous in all positions. The Raman measurements of the film deposited with positive bias have shown that the crystallinity varies with the position, and for point 2, corresponding to the area of the substrate biasing through plasma-induced conduction [19], the amorphous phase is reduced. On the other hand, for point 3, corresponding to the area of substrate biasing through electric field conduction [26], the deposited film has almost no amorphous phase. The same trend is also maintained for the film deposited with grounded substrate although the crystallinity of this film at each of the measured points is much lower compared to the one in Figure 21.5a. Furthermore, the topographies of the films deposited under positive biasing as well as with grounded substrate are also included in Figure 21.5. In both cases, we

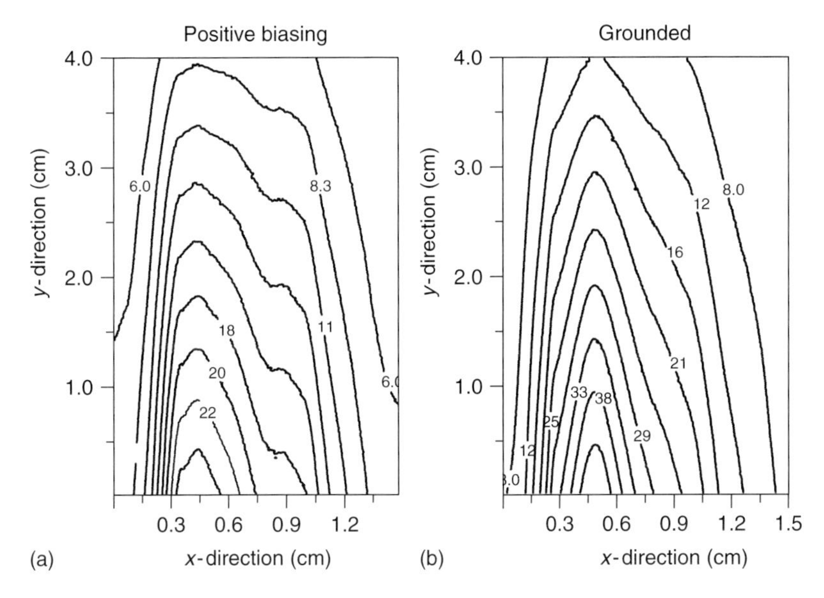

Figure 21.6 H_β emission intensity contours recorded for (a) positive and (b) grounded, substrate holder. The data are from the area highlighted in Figure 21.1b.

observe the formation of nanocrystallites that are connected to form larger grains with an amorphous phase between them, the grain size being much larger in the case of positive biasing.

In order to get a better understanding of the effect of the substrate holder configuration on both the deposition rate and the film structure, we have applied 2D emission measurements of excited species SiH^*, H_α, and H_β. The density of these species in the discharge is much lower compared to ground-state radicals and ions so they cannot play an important role in the film growth. However, the emission measurements can clearly give an indication of the discharge efficiency in producing radicals, the population and distribution of high-energy electrons, and also information concerning the changes in the discharge structure and uniformity. In many studies, emission spectroscopy has been used for the prediction of amorphous to microcrystalline silicon growth transition [5–8], the reduction of the incubation layer [27, 28], and also for monitoring the substrate heating during the film growth [29–32].

Thus, Figure 21.6a,b show the contours of H_β emission intensity for different substrate holder configurations, as recorded in the discharge area that is marked in Figure 21.1a. The lifetime of H_β species is very short (5 nanoseconds) and therefore the de-excitation of these species can be directly related to their production rate. In Figure 21.6, the x-direction is along the axis of the discharge, from the 13.56-MHz electrode to the substrate holder, while y is the radial direction from the center of the glass substrate to the holder's edge. The $x = 1.5$ cm and $y = 2$ cm coordinates coincide with the border of the glass with the conductive part of the electrode as shown in Figure 21.1b.

We have observed that the use of different substrate holder configurations induces drastic changes in both the emission intensity and its distribution in space. Actually, in the grounded conditions (Figure 21.6b) emission intensity is much higher compared to all other cases. For grounded- and positive-biasing conditions, the maximum intensity is located at a distance of 0.45 cm from the RF electrode, while for floating and negative biasing, the maximum shifts toward the RF electrode and is at 0.35 cm. The production of excited species near the substrate holder is, in all cases, rather low but is enhanced by the application of either a positive or a negative bias voltage. The arms in the contours that can be observed for $x = 0.9$ cm in Figure 21.6a can be attributed to species production in the bulk/sheath interface of the substrate holder. Thus, we can say that under biasing conditions, the electron-heating mechanism in the sheath of the substrate holder will play a more important role in the species production compared to the grounded configuration.

Moreover, the emission intensity drops in the radial direction, that is, as we go from the center of the electrodes to the edges, and this is more important for the grounded conditions. It is also clear that in the area close to the substrate holder (for $x = 1.2-1.5$ cm), the production of excited species is higher in front of the glass ($y < 2$) compared to the production in front of the conductive part of the electrode ($y > 2$). Therefore, the sheath length is not constant but increases as we go from the center of the electrode to the edges, indicating that the voltage drop and the electric field are much higher in front of the conductive part of the electrode.

Taking into account that positive biasing had a significant effect on the discharge properties and structure, the production of species closer to the surface and a beneficial effect on both film growth rate and crystallinity, we focused our further investigation on the positive substrate biasing case. A set of experiments were then performed by varying the positive substrate bias from 50 to 125 V while the other parameters were 4.5% SiH_4 in H_2 mole fraction, discharge power 35 W, high frequency 13.56 MHz, low-frequency bias and voltage 20 kHz, and 200 V, peak to peak respectively, electrode gap 1.5 cm, and total gas pressure 2.5 torr.

Figure 21.7 presents the experimental measurements of the deposition rate as a function of the positive bias voltage. The percentage crystalline volume fraction, as determined from Raman measurements, are also denoted for each condition. The increase of the DC bias voltage from 0 to 125 V results in a significant enhancement of the film growth rate from 7 to 12 Å s^{-1} as the DC bias voltage is varied. The film deposited at grounded conditions was amorphous and the same was true for the film deposited for the highest applied DC voltage (125 V). All other films had percentage crystallinity around 50%.

In order to get a better understanding of the significant enhancement of the deposition rate with the positive bias, the self-consistent fluid simulation was applied in these conditions and the simulation outputs related to the time-averaged electrical properties are presented below.

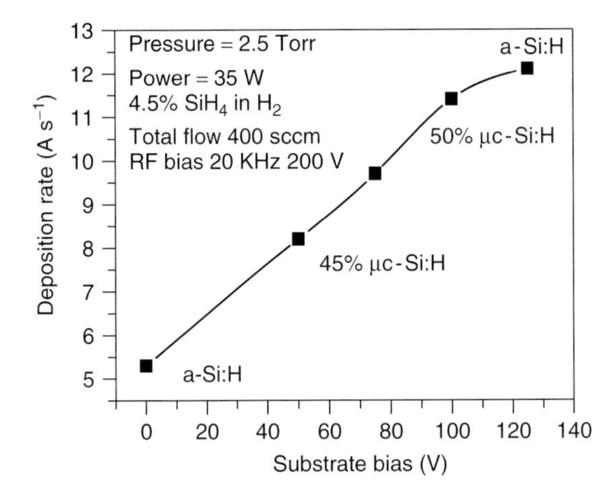

Figure 21.7 Deposition rate as a function of positive substrate bias voltage for 4.5% SiH_4 in H_2 discharges, pressure 2.5 torr, and RF power 35 W.

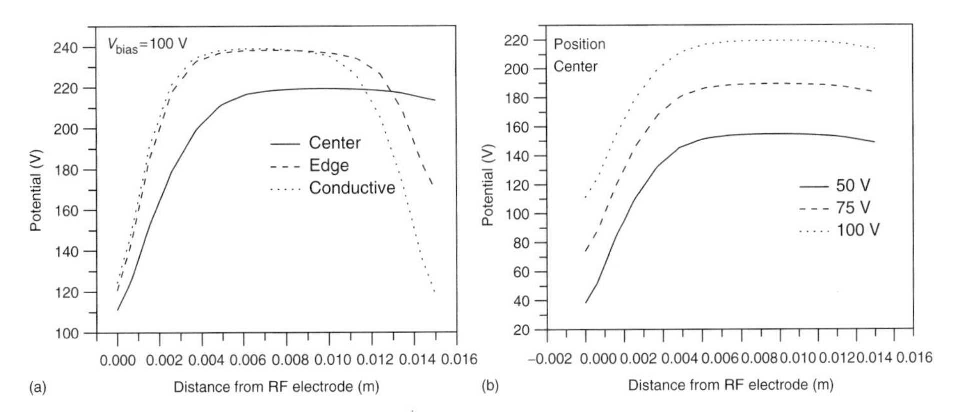

Figure 21.8 (a) Voltage distribution in the discharge in the center and the edge between glass and the conductive part of the electrode. (b) Voltage distribution in the center of the plasma for three positive bias conditions.

More precisely, Figure 21.8a presents the time-averaged voltage distribution for $V_b = 100$ V at three different points in the discharge, in the center (point 1, Figure 21.1b), in the glass – SS electrode (point 2, Figure 21.1b), and in the SS part (point 4, Figure 21.1b). The voltage distribution in the discharge at each of these points, and especially close to the substrate holder, is significantly different and this partly explains the crystallinity nonuniformities that were recorded in Raman measurements. The rather small voltage drop in front of the substrate is slightly affected by the value of the bias voltage. This is demonstrated in Figure 21.8a, where the voltage distribution in the center of the discharge is presented for three different bias values. The application of a DC bias on the substrate holder leads to

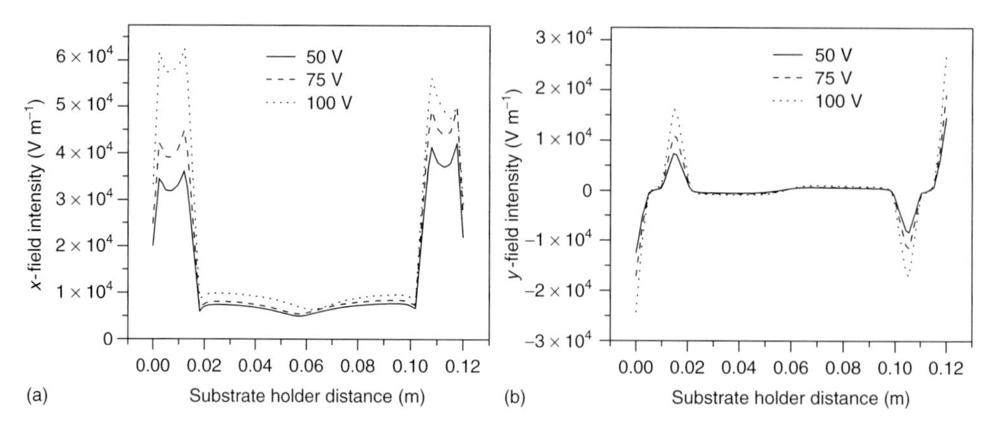

Figure 21.9 (a) The *x*-electric field distribution across the substrate holder for three positive bias conditions. (b) The *y*-electric field distribution across the substrate holder for three positive bias conditions.

an increase of the potential at each point of the discharge. However, the voltage profile is not significantly altered with the increase of bias voltage, and thus strong changes in the electric field intensity are not expected at least in the center of the plasma.

This is better illustrated in Figures 21.9a,b where the *x*- and *y*-electric field intensity distribution at the substrate holder (line, Figure 21.1b) is presented for different values of the substrate bias. It has to be mentioned that the *x*-direction field intensity is related to the motion of charged species from and toward the substrate holder, while the *y*-direction field is related to the radial or the tangential motion of charged species close to the substrate. As can be observed, the increase of the DC bias voltage results in an enhancement of both *x*- and *y*-electric field intensity almost at any point of the substrate holder. This increase is much smaller in the center of the holder, which is the glass covered area, and this is in agreement with the previous discussion concerning the voltage drop (Figure 21.2b). The sign of the electric field in Figure 21.9a indicates that positive ions are attracted at each point of the substrate holder. On the other hand, the sign for the *y*-direction electric field (Figure 21.9b) shows that positive ions move from the glass toward the conductive part of the substrate holder. Therefore, we can say that the application of a positive substrate bias will result in a change of the angle of incidence of ions and consequently will alter the amount of kinetic energy transferred to the surface. Moreover, the increase of the DC bias will tend to attract ions toward the conductive parts of the electrode, thus reducing the energy with which the ions strike the surface. As has been experimentally observed, this reduction will have a beneficial effect on both the growth rate and the crystallinity, probably through a reduction of sputtering effects and the enhancement in mobility of the surface species. However, for higher applied values of the DC bias voltage, although the film growth rate increases, amorphization takes place again, since probably the amount of energy transferred in the growth zone is not enough to induce crystallization.

21.5 Conclusions

A study of the effect of various substrate bias configurations on the deposition of μc-Si : H thin films was performed with the purpose of investigating the possibility to enhance deposition rate and/or control film crystallinity.

Electrical and optical measurements show that microscopic plasma parameters as well as species' production are significantly affected by the change of the substrate bias. In fact, grounded substrate always required the lowest applied voltage, the highest discharge current, and the higher resistive character for constant power dissipation. In addition, the results of the spatially resolved emission profiles of excited species in the discharge have shown that the rate of species production in the discharge is higher when the substrate is grounded. On the other hand, the application of a positive bias voltage on the substrate holder resulted in more symmetric emission profiles, indicating an important enhancement of the plasma potential in these conditions.

Moreover, the deposition rate and the film crystallinity were favored by the positive bias. A deposition rate increase was observed as we changed from the grounded to positive bias, while the film crystallinity also improved. The negative bias and the floating configuration, both gave the worst results. Raman measurements at different points of the deposited films show structure nonuniformities and higher crystalline volume fractions in the areas close to the conductive part of the substrate holder for both positive and grounded substrate configurations.

Further investigation of the effect of the positive DC bias voltage on the deposition process has shown that the increase of substrate bias amplitude significantly enhances the μc-Si : H deposition rate but for high DC voltages (>100 V) amorphization takes place again. According to the numerical simulation results, the increase of the substrate bias enhances the potential everywhere in the discharge without affecting the voltage distribution. The spatial and radial electric field intensities are both favored by the increase of substrate bias voltage and the same is true for the ion flux. The direction of the radial field developed in front of the substrate due to the holder bias tends to attract ions toward the conductive part of the electrode resulting in a change of the angle of incidence of the ions, consequently leading to a reduction of the kinetic energy transferred to the growing film surface. This reduction appears to have a beneficial effect on both the growth rate and crystallinity probably through a reduction of etching effects and the enhancement of surface species mobility. Finally, the increase of the deposition rate cannot be simply explained by the variation of the ion flux and therefore the effect of bias on the production of species in the discharge needs to be further clarified.

With regard to the application of this technique in industrial conditions, it is expected that it will lead to an enhancement of the deposition rate comparable to or higher than that in the use of VHF, by widening the deposition conditions window. At the same time, the method can give more control of the film structure around the transition from amorphous to microcrystalline silicon growth. The only parameter that needs to be optimized in this case is film homogeneity through some process modifications, which may be considered minor compared to the redesigning needed for the use of other plasma sources.

Acknowledgments

Dr A. Soto and Dr G. Voyatzis are gratefully acknowledged for the Raman spectra of the samples and discussion of the results.

References

1. Vetterl, O., lamberts, A., Dasgupta, A., Finger, F., Rech, B., Kluth, O., and Wagner, H. (2001) *Sol. Energy Mater. Sol. Cells*, **66**, 345.
2. Feitknecht, L., Kluth, O., Ziegler, Y., Niquille, X., Torres, P., Meier, J., Wyrsh, N., and Shah, A. (2001) *Sol. Energy Mater. Sol. Cells*, **66**, 397.
3. Finger, F., Hapke, P., Luysberg, M., Carius, R., Wagner, H., and Scheib, M. (1994) *Appl. Phys. Lett.*, **65**, 2588.
4. Kroll, U., Meier, J., Torres, P., Pohl, J., and Shah, A. (1998) *J. Non-Cryst. Solids*, **227-230**, 68.
5. Roschek, T., Repmann, T., Muller, J., Rech, B., and Wagner, H. (2002) *J. Vac. Sci. Technol. A*, **20**, 492.
6. Yamamoto, K., Yoshimi, M., Suzuki, T., Tawada, Y., Okamoto, Y., and Nakajima, A. (1998) *Mater. Res. Soc. Symp. Proc.*, **507**, 131.
7. Kondo, M., Fukawa, M., Guo, L., and Matsuda, A. (2000) *J. Non-Cryst. Solids*, **266-269**, 84.
8. Hamers, E.A.G., Fontcuberta i Morral, A., Niikura, C., Brenot, R., and Roca i Cabarrocas, P. (2000) *J. Appl. Phys.*, **88**, 3674.
9. Hamers, E.A.G., Bezemer, J., and van der Weg, W.F. (1999) *Appl. Phys. Lett.*, **75**, 609.
10. Roca i Cabarrocas, P., Morin, P., Chu, V., Conde, J.P., Liu, J.Z., Park, H.R., and Wagner, S. (1991) *J. Appl. Phys.*, **65**, 2942.
11. Jun, S.-I., Rack, P.D., McKnight, T.E., Melechko, A.V., and Simpson, M.L. (2006) *Appl. Phys. Lett.*, **89**, 022104.
12. Kosku, N., Murakami, H., Higashi, S., and Miyazaki, S. (2005) *Appl. Surf. Sci.*, **244**, 39.
13. Jia, H.J., Saha, J.K., Ohse, N., and Shirai, H. (2006) *J. Phys. D: Appl. Phys.*, **39**, 3844.
14. Hajime, S., Nakamura, T., and Arai, T. (1998) *J. Non-Crystal. Solids*, **227-230**, 53.
15. Gordijn, A., Hodakova, L., Rath, J.K., and Schropp, R.E.I. (2006) *J. Non-Crystal. Solids*, **352**, 1868.
16. Lebib, S. and Roca i Cabarrocas, P. (2005) *J. Appl. Phys.*, **97**, 104334.
17. Kalache, B., Kosarev, A.I., Vanderhaghen, R., and Roca i Cabarrocas, P. (2002) *J. Non-Crystal. Solids*, **299-302**, 63.
18. Roca i Cabarrocas, P., Morin, P., Chu, V., Conde, J.P., Liu, J.Z., Fi Park, H., and Wagner, S. (1991) *J. Appl. Phys.*, **69**, 1.
19. De Vries, C.A.M. and Van Den Hoek, W.G.M. (1985) *J. Appl. Phys.*, **58**, 2074.
20. Zhang, X.D., Zhang, F.R., Amanatides, E., Mataras, D., and Zhao, Y. (2008) *Thin Solid Films*, **516**, 6829–6833.
21. Spiliopoulos, N., Mataras, D., and Rapakoulias, D. (1996) *J. Vac. Sci. Technol. A*, **14**, 2757.

22. Amanatides, E., Gkotsis, P., Syndrevelis, Ch., and Mataras, D. (2006) *Diamond Relat. Mater.*, **15**, 904.
23. Katsia, E., Amanatides, E., Mataras, D., Soto, A., and Voyiatzis, G.A. (2005) *Sol. Energy Mater. Sol. Cells*, **87**, 157.
24. Lyka, B., Amanatides, E., and Mataras, D. (2004) 19th European Photovoltaic Solar Energy Conference, June 7–11, 2004, Paris, p. 1395.
25. Amanatides, E., Hammad, A., Katsia, E., and Mataras, D. (2005) *J. Appl. Phys.*, **97**, 073303.
26. Johnson, E.V., Kherani, N.P., and Zukotynski, S. (2006) *J. Mater. Sci: Mater. Electron.*, **17**, 801–813.
27. Rath, J.K., Franken, R.H.J., Gordijn, A., Schropp, R.E.I., and Goedheer, W.J. (2004) *J. Non-Cryst. Solids*, **56**, 338–340.
28. van den Donker, M.N., Rech, B., Finger, F., Kessels, W.M.M., and van de Sanden, M.C.M. (2005) *Appl. Phys. Lett.*, **87**, 263503.
29. van den Donker, M.N., Rech, B., Finger, F., Kessels, W.M.M. and van de Sanden, M.C.M. (2007) *Prog. Photovoltaics*, **15**, 291–301.
30. van den Donker, M.N., Schmitz, R., Appenzeller, W., Rech, B., Kessels, W.M.M., and van de Sanden, M.C.M. (2006) *Thin Solid Films*, **511**, 562–566.
31. Nunomura, S., Kondo, M., and Akatsuka, H. (2006) *Plasma Sources Sci. Technol.*, **15**, 783–789.
32. Fukuda, Y., Sakuma, Y., Fukai, C., Fujimura, Y., Azuma, K., and Shirai, H. (2001) *Thin Solid Films*, **386**, 256–260.

22
Introduction to Diamond-Like Carbons

Masaru Hori

Diamond-like carbon (DLC) film including an sp^3 bonding structure has unique properties and has been applied to many industrial fields. The characteristics of DLC films are mainly classified into four types: tetrahedral amorphous carbon ta-C that is sp^3 rich and H free; ta-C including H (ta-C : H); amorphous carbon (a-C) that is less sp^3 and H free; and a-C with H (a-C : H). Generally, these typical DLCs have been synthesized by plasma technologies. In the industrial fields, the conventional DLC is an a-C : H, which has an sp^2 structure. To obtain ta-C and a-C that are H free, usually a vacuum arc evaporation or a sputtering technology is employed. However, it is very difficult to synthesize the a-C film by employing the sputtering method and recently the vacuum arc evaporation technology has been intensively focused on as a coating method. Thus, the ion-based processing with controlled energies will be promising for the synthesis of ta-C.

So far, there have been many reports on the synthesis of DLCs and their performances in industries. The main precursors and the related surface reaction mechanism to synthesize DLCs by the plasma processing have been discussed.

Generally, so far, the development of DLC films employing various kinds of plasmas has been performed by a trial-and-error method because the inside state of the plasma is unknown, similar to a black box. To establish the plasma process science for the synthesis of DLC films with a high quality, the precise design and control of species in the plasma basically become indispensable. In the last 20 years, various kinds of measurement methods for C-related radicals and ions and many models for the growth of DLCs have been proposed. However, there are a few reports on the synthesis of tailor-made DLC films for the industrial request on the basis of plasma science.

Now, industrial demands for the DLC films have been extraordinarily extended since the DLC films have been applied not only for coating on mechanical and optical parts but also in ultralarge integrated circuits (ULSIs) and biotechnology. According to the demand of each application, the various kinds of tailor-made DLC coating in one equipment are strongly desired. In these processings, it is vitally important to control the structure of DLCs at the atomic-size scale. In this equipment, therefore, the reaction of species not only in the gas phase but also on

Industrial Plasma Technology. Edited by Yoshinobu Kawai, Hideo Ikegami, Noriyoshi Sato, Akihisa Matsuda, Kiichiro Uchino, Masayuki Kuzuya, and Akira Mizuno

ISBN: 978-3-527-32544-3

the surface and/or subsurfaces is needed to be characterized by the densities and energies of species. In the following chapters, outstanding articles on science and technologies of DLC coating have been put together. The concepts to implement the plasma nanoprocessing for the synthesis of the tailor-made DLCs are expected to be found in these chapters. The chapter provides the important scientific principle as to how to control the plasma parameters for the synthesis of tailor-made DLC films, how to design and control properties of DLC precisely, and how to innovate the production equipment.

Consequently, it is considered that the establishment of useful plasma science, which enables to control specific species, will open new avenues for DLC films coating availability at the production level and thus the coating of hyperfunctional DLC films with only sp^3 bond structure on the complicated three-dimensional soft materials with an ultrahigh speed at a room temperature will be brought to realization.

23
Diamond-Like Carbon for Applications

John Robertson

23.1
Introduction

Diamond has a unique set of properties – it is a hard material with low thermal expansion coefficient, a wide bandgap, and high mobility semiconductor and is a biocompatible substrate. The cost of producing flat, large single crystals of diamond is still quite high. For many applications, a single-crystal diamond is not needed and a polycrystalline diamond is adequate. In many cases, a polycrystalline diamond surface is too rough, because of the grain size, and nanocrystalline diamond with its greater smoothness is preferable. The question then arises is, would amorphous diamond-like carbon (DLC) also be suitable for some of these applications. It is, therefore, of interest to compare the deposition and properties of DLC with that of the crystalline forms of diamond. The advantages of DLC are its high growth rate and reasonable mechanical properties; its disadvantages are its intrinsic stress and lack of complete semiconducting properties, as described below.

This chapter reviews applications of DLC; however, it does not provide a complete overview of diamond. Rather, it uses diamond as a benchmark to assess the growth, properties, and applications of DLC.

23.2
Growth Rates

Thin-film diamond is grown by chemical vapor deposition (CVD), typically by microwave-assisted CVD from a hydrogen-diluted methane plasma. The growth of diamond from hydrogen-diluted plasmas has been intensively studied. The plasma is a source of atomic hydrogen. The diamond surface is terminated by hydrogen. The basic growth mechanism involves three reactions: the abstraction by atomic hydrogen of the hydrogen from C–H bonds on the diamond surface to form carbon dangling bonds, the saturation of the dangling bonds by atomic hydrogen to create a quasi-equilibrium concentration of surface dangling bonds, and the addition of methyl radical (CH_3) growth species to the dangling bonds [1]. The growth rate,

Industrial Plasma Technology. Edited by Yoshinobu Kawai, Hideo Ikegami, Noriyoshi Sato, Akihisa Matsuda, Kiichiro Uchino, Masayuki Kuzuya, and Akira Mizuno

ISBN: 978-3-527-32544-3

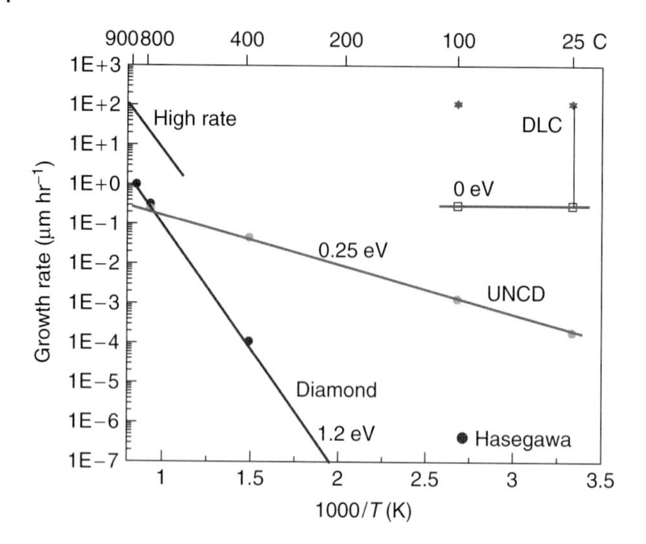

Figure 23.1 Typical growth rates of diamond, ultra-nanocrystalline diamond, and diamond-like carbon, showing the effect of their different temperature dependences.

therefore, depends on the surface concentration of dangling bonds, which in turn depends on the gas-phase atomic hydrogen concentration. The diamond phase forms, despite being metastable compared to graphite, because its surface is kept hydrogen-terminated, which stabilizes the sp^3 sites in the growing film [2].

The growth is thermally activated with an activation energy of 1.2 eV, resulting in a strong temperature dependence [3, 4], as seen in Figure 23.1. The rate-limiting step is the hydrogen abstraction [1]. The activation energy corresponds to that of the quasi-equilibrium concentration of dangling bonds. Figure 23.1 shows how this activation energy limits the growth rate of diamond, particularly at low temperatures. Early work searched for conditions to grow at lower temperatures using oxygen-containing plasmas [5, 6], but presently hydrocarbon plasmas still seem to dominate the field.

The most critical applications require single-crystal diamond. This requires homoepitaxial growth, as heteroepitaxial growth is not yet perfected. The growth rate of diamond depends on the face, with (110) being the fastest [4]. Generally growth on (100) is preferred as this leads to least defects such as stacking faults, which are prevalent in the (111) growth mode [7].

The material for many applications is a polycrystalline diamond, to avoid the need for homo- or heteroepitaxial growth. In this case, the grain size increases with the increase in thickness as in the van der Drift model, as faster growing faces overgrow slower growing faces [8]. This creates a rough surface. For some applications, there is a desire to decrease the grain size in order to decrease the surface roughness. *Nanodiamond* is defined as a material with a grain size below ~100 nm. This is just a small-grained polycrystalline diamond and is grown by the same mechanism. Its small grain size is controlled by starting with a very high nucleation density [9].

There is a related material grown from hydrogen sparse plasmas with a grain size of 4–10 nm, which is called *ultra-nanocrystalline diamond* (*UNCD*) to distinguish it from the larger grained material. This has been characterized by Birrell *et al.* [10, 11] and has recently been compared with nanocrystalline diamond developed by Williams *et al.* [12]. Its growth rate has a weaker temperature dependence than that of polycrystalline diamond, with the result that it can still be grown at 400 °C. The activation energy of the growth rate can be as low as 0.25 eV [13]. The growth mechanism is not fully settled, but the activation energy still limits the growth rate at low temperatures.

The early work on diamond CVD was constrained by the understanding that there is a trade-off between growth rate and crystalline quality, as in the model of Goodwin [14]. Higher methane contents in the plasma not only lead to higher growth rates but also lead to a higher incorporation of sp^2 sites, higher defect levels, and a loss of crystalline quality. Thus, many workers such as Okushi who desired the highest quality flat surfaces used very low methane fractions and very low growth rates [15], and this approach was also used when trying to incorporate a substitutional dopants such as phosphorus [16, 17].

It was then realized that high growth rates were possible at high methane fractions (6%), provided the plasma power and pressure were correspondingly increased [18, 19]. There was no loss of quality. This area is of great interest. Nevertheless, such growth still falls within the same thermally activated growth mechanism, so low temperature growth rates are still low.

Another factor in older work was that accidental nitrogen impurity could lead to higher growth rates, but this could be at the expense of crystal quality or changes in crystal faceting [20]. More recently, controlled nitrogen addition was used to obtain very high growth rates [21].

Hasegawa (private communication) has recently grown diamond films at very low temperatures on glass substrates using a high-power-distributed microwave plasma. The films are continuous. The substrate temperature has been reduced to an order 100 of °C, which is a remarkable achievement. Estimating its growth rate from the growth time and the need to make a continuous film show that this data point lies well above the trend for nanodiamond growth, but below that of UNCD (Figure 23.1).

DLC has a fundamentally different growth mechanism to that of diamond. DLC is an amorphous network solid, containing not only a high fraction of carbon sp^3 sites but also sp^2 sites and hydrogen. Its composition is represented by the ternary phase diagram [22, 23]. The presence of sp^2 sites means that it is never truly an "amorphous diamond". The maximum sp^3 content obtained so far in tetrahedral amorphous carbon (ta-C) is about 88% [24, 25] (Figure 23.2).

DLC is grown by plasma-enhanced chemical vapor deposition (PECVD), in which a large fraction of the film-forming flux must be an energetic species (ions) that have been accelerated by a positive DC bias of typically 10–300 V. This energy causes the ions to be "subplanted" (subimplanted) below the film surface in a

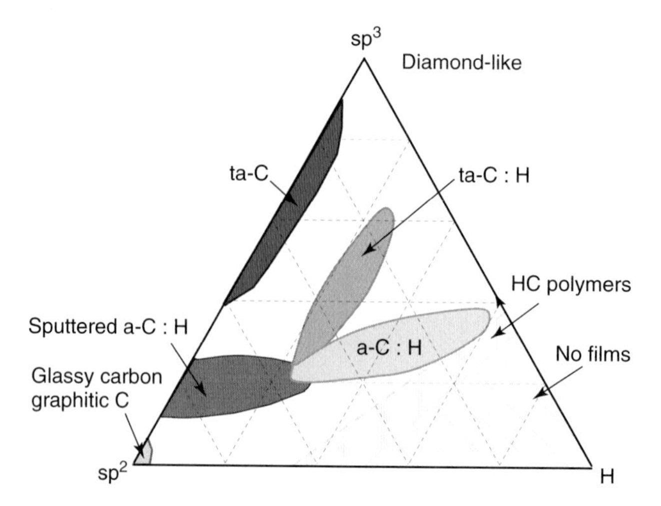

Figure 23.2 Ternary phase diagram of C, H system showing a-C : H and ta-C.

densified layer [26], where they tend to adopt the metastable sp^3 configuration instead of the more stable sp^2 configuration [27, 28].

Typically, the growth rates of DLC are 0.1–1 nm s^{-1} or 0.3–3 $\mu m\ h^{-1}$ (Figure 23.1) [23]. Being energetic, the subplantation process is independent of temperature, so the DLC growth rate is not thermally activated. The lack of thermal activation means that fundamentally DLC can be grown at high rates at room temperature. This is a tremendous practical advantage.

One should distinguish hydrogen-containing a-C : H from amorphous carbons with no hydrogen such as ta-C. Ta-C grows from C^+ ions beams, which are fully ionized. On the other hand, a-C : H is grown from hydrocarbon plasmas in which the film-forming flux is often only 10%. In that case, the growth rate strongly depends on the hydrocarbon source gas; higher growth rates will occur for gases with lower ionization potentials, with acetylene giving the highest rate (Figure 23.3) [23].

a-C : H can be grown by a range of PECVD processes, typically using 13.56 MHz RF plasmas rather than microwave at pressures of typically 10 Pa. The ion flux is then typically 10% of the total deposition flux. This leads to films of moderate sp^3 content and hydrogen contents of 25–55% (Figure 23.7).

The ion content of the film-forming flux can be increased by using low-pressure, high-density plasmas [30–33], such as a plasma beam source [30], electron cyclotron resonance systems [31], or electron cyclotron wave resonance (ECWR) systems [32].

High-density plasmas give growth rates up to 2–4 nm s^{-1}. Even higher growth rates are possible with a custom-designed microwave deposition system used to coat PET bottles with a-C : H, which reaches 20 nm s^{-1}, as seen in Figure 23.1, and is discussed later.

The optimum ion energy for deposition of the highest sp^3 content is about 100 eV. In ta-C, reasonable sp^3 contents can be obtained at ion energies as low as 30 eV, which is the self-bias voltage in some systems. In PECVD, again the

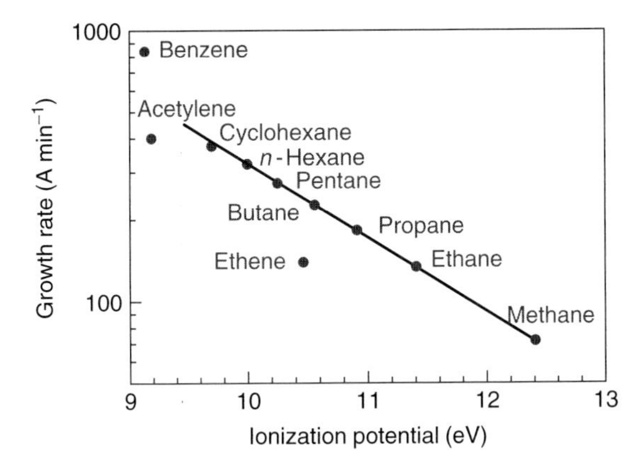

Figure 23.3 Typical growth rates of hydrogenated amorphous carbon by PECVD as a function of source gas.

optimum ion energy per C atom is about 100 eV. Note that the equivalent self-bias voltage can be higher due to ion scattering in the plasma.

23.3 Basic Properties

Crystalline diamond consists of only sp^3 sites, and so an ideal crystalline diamond has a clean bandgap with no gap states. Grain boundaries in polycrystalline diamond will introduce some sp^2 sites and these will create gap states. The sp^3 bonds of crystalline diamond are very strong and are the source of diamond's large Young's modulus, high-mechanical hardness, high Debye temperature, high carrier mobility, and small thermal expansion coefficient.

DLC always contains both sp^3 and sp^2 sites, and this makes its electronic structure fundamentally different from that of an ideal diamond [34]. The sp^2 sites will tend to form clusters to lower the free energy, and their π states form the band edge

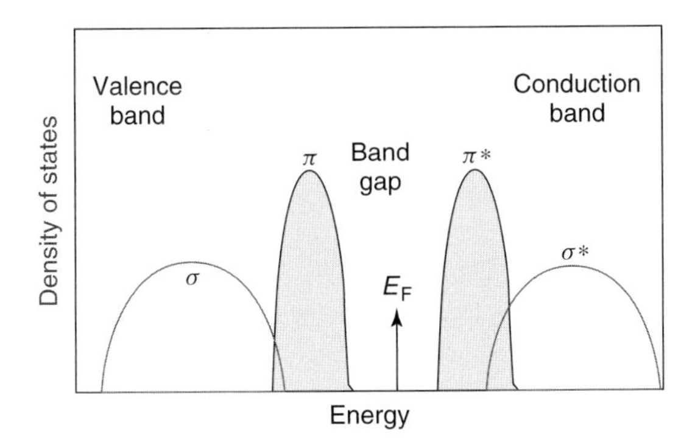

Figure 23.4 Schematic density of states of amorphous carbon.

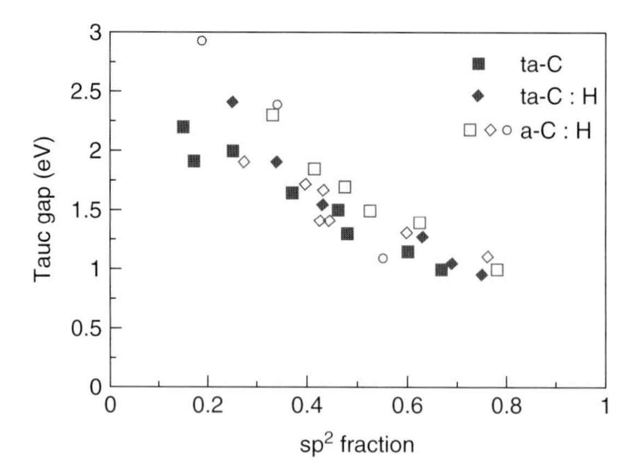

Figure 23.5 Variation of optical gap of amorphous carbon films deposited at room temperature as a function of their sp^2 content.

states [35, 36], as shown in Figure 23.4. Thus, the electronic structure depends on the configuration of the sp^2 sites. In as-deposited films, a strong correlation is found between the average configuration of sp^2 sites and the sp^2 concentration, so that the bandgap of ta-C and a-C : H depends on the average sp^2 fraction [35], as shown in Figure 23.5.

The gap states in nanocrystalline diamond are due to the π states of the sp^2 sites at the grain boundaries and, therefore, they follow the same band model [36]. Thus, the π states form midgap states and cause a subgap optical absorption.

The properties of ta-C depend, in a relatively simple way, on the sp^3 content and density, which are found to be linearly dependent [37], as shown in Figure 23.6.

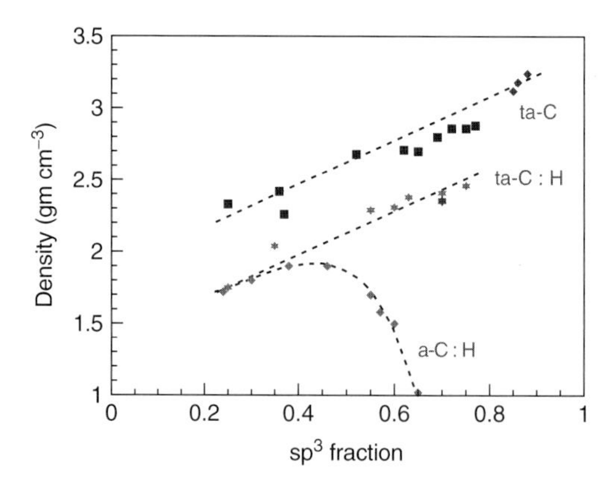

Figure 23.6 Density versus sp^3 fraction for the various types of amorphous carbon, showing the linear correlation for ta-C and the peak in density for a-C : H, with lower density for most sp^2-rich films and for polymeric highly sp^3-rich a-C : H.

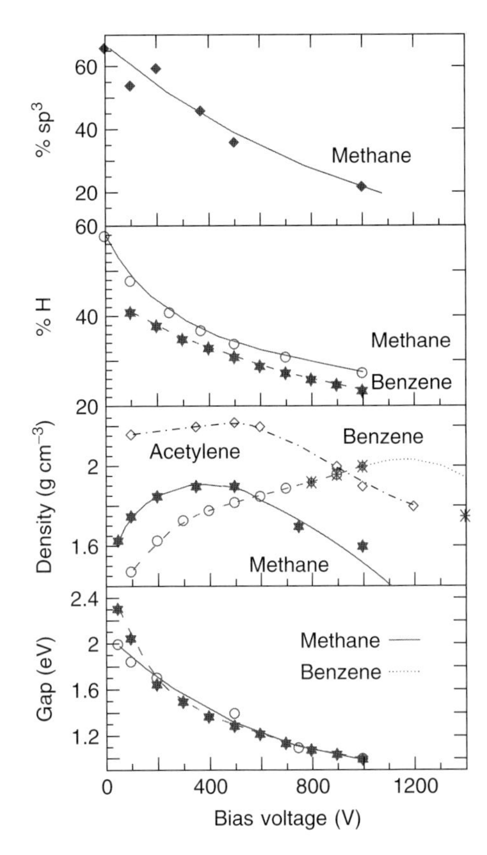

Figure 23.7 Properties of plasma-deposited a-C : H as a function of bias voltage (ion energy is equal to about 0.4 of bias voltage).

The properties of a-C : H depend, in a more complex manner, on their sp^3 fraction and hydrogen content, as summarized in Figure 23.7 for one type of a-C : H [29].

In contrast to the electronic properties, the mechanical properties of DLC depend on the mean C–C coordination and thereby on the fraction of C–C sp^3 sites. The mean coordination must exceed 2.4 for rigidity [23]. The C–H bonds add no rigidity to the network, so that Young's modulus depends only on the C–C coordination and thus on the sp^3 fraction, as seen experimentally [38] in Figure 23.8. This can also be related to the density. Thus, when the C–C sp^3 fraction is large, as in ta-C, Young's modulus and hardness approach that of crystalline diamond itself. The hardness is roughly 10% of Young's modulus. Similarly, the thermal expansion coefficient of ta-C will be similar to that of a diamond.

On the other hand, the thermal conductivity of amorphous materials is much lower than that of crystalline materials due to disorder-induced scattering, despite them having a similar Debye temperature. This is true for amorphous carbons, where the thermal conductivity is much lower than that of crystalline diamond. It is

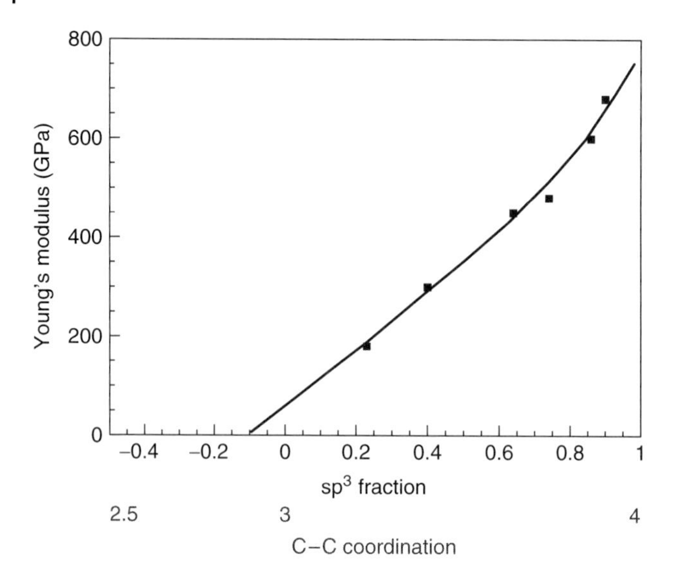

Figure 23.8 Variation of Young's modulus of ta-C films as a function of their sp^3 content and, thus, C–C coordination number [38].

found empirically that the thermal conductivity varies in proportion to the density of a-C [39].

23.4 Stress

In diamond, despite the generally low growth rates, diamond films can be grown to any thickness, with care and attention to stress generated by thermal mismatch to the substrate. On the other hand, DLC films have, to date, fundamentally limited thickness due to their compressive stress [23]. The compressive stress is created by the ion deposition process, the same process that creates the sp^3 bonding [40, 41]. It is, therefore, *intrinsic* to the subplantation growth mechanism. While there may be an absolute one-to-one correspondence between sp^3 fraction and stress [42], there is a general trend that high modulus films have a higher stresses at deposition [42–44].

DLC is mainly used for mechanical protective coating purpose; thus, it is coated on the substrate that is to be protected. The compressive stress limits the maximum thickness, h, of adherent films. If the stress is σ, the surface fracture energy is γ, and Young's modulus is E, then the maximum thickness of an adherent film is roughly given by [23]

$$h = 2\gamma E/\sigma^2 \tag{23.1}$$

both σ and E tend to be linearly correlated with the sp^3 fraction, $\sigma = k_1[sp^3]$ and $E = k_2[sp^3]$. While it is possible to reduce stress by lowering the sp^3 fraction, this is at the expense of the modulus (and thus hardness, as $H = 0.1E$). As γ is roughly constant, from Equation 23.1 the harder films have a lower maximum adherent thickness, $h = k_3/H$. Paradoxically, in this situation, making a harder film is of no use.

There are a number of strategies to reduce the compressive stress and it effects and to increase the adhesion. First, for ta-C, to lower stress, the film can be annealed to 500–600 °C [44–46]. It is found that this removes stress in ta-C, while only slightly lowering the sp^3 content. However, this method is not useful for a-C : H, because the hydrogen is lost first, so the film is no longer diamond-like. Second, alloying with Si or various metals will reduce stress [47–49]. Si is recommended if an optical gap has to be retained. Metals relieve stress by introducing a ductile component and generally by creating a nanocomposite [48, 49]. In some cases, carbide-forming metals such as Cr or Ti are used, and in others softer metals such as Al have been used. Other methods to reduce stress are by forming multilayers and graded layers, particularly of a-C : H and carbides in the latter case. Graded layers are particularly used in commercial films [50, 51].

So far, the compressive stress is intrinsic, and thus it was considered to be unavoidable. Recently, Bilek and McKenzie [52] have noted that this form of stress could be reduced by using high-energy-pulsed ion implantation, that is, for ion energies of order 1000 eV, well over the 100 eV optimum for maximum sp^3 content. Interestingly, pulsed implantation does not cause a complete loss of sp^3 bonding. This group only implemented it for C^+ ions, which is of limited relevance. Recently, Lusk *et al.* [53] have implemented a high-voltage-pulsed PECVD source, which can produce adherent layers of at least 50 μm thickness while retaining hardness of 15–20 GPa. The measured stress in these films is indeed 10 times lower than that for conventionally deposited films of the same hardness (Figure 23.9). In general, this development will have considerable importance for wear resistance coatings.

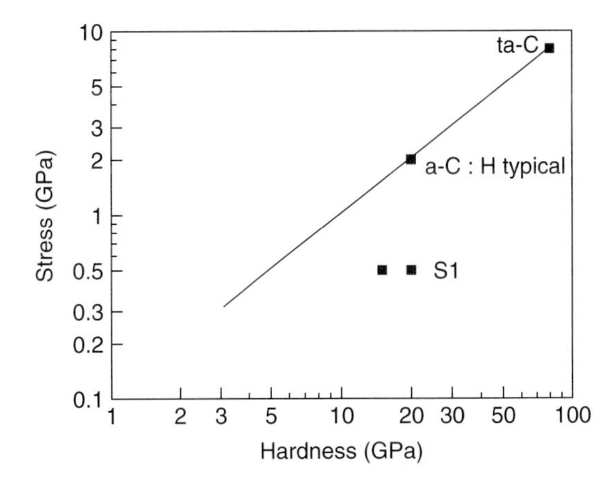

Figure 23.9 Compressive stress of a-C : H films deposited by Kusk *et al.*'s process.

The coatings often need to be wear resistant. Wear rates vary inversely as hardness. DLC wears by the adhesive mechanism. Wear rates for a-C : H are of order 10^{-7} mm^3 N^{-1} m^{-1}. Rates for ta-C can be as low as 10^{-9} mm^3 N^{-1}m^{-1} [23].

23.5 Applications of DLC

The main applications of DLC are as a wear-resistant coating material. This is used on bar-code scanners and recently in car components. There are numerous cases in the automotive sector where low-friction wear-resistant surfaces are needed (C Donnet, private communication). DLC provides a hard, smooth, low-friction surface. The coating will maintain its properties provided that its temperature does not exceed about 400 °C. Thus, DLC is useful for coating components such as cam-shaft bearings, plungers, and diesel injectors, which is now being implemented, as shown in Figure 23.10. It is not suitable for components such as piston rings, which are exposed to high piston temperatures. A key criterion for its usage in the automotive sector is its very low cost.

DLC is also used as a coating on the edges of razor blades. Major manufacturers have used this coating. Figure 23.11a shows a transmission electron microscopy image through a cross-sectional slice of a razor blade edge. The DLC lies on the steel blade. There is an outer layer of Teflon or a similar coating and an adhesive underlayer. The objective of using DLC is to keep the blade sharp. There is a minimum radius of curvature for ductile materials such as metals. Usage will blunt the blade edge. Carbon is less ductile and has a smaller comminution radius

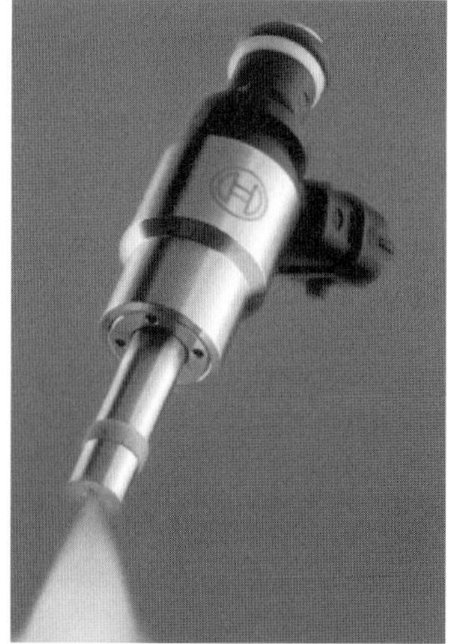

Figure 23.10 Images of DLC-coated motor components, after Donnet (private communication).

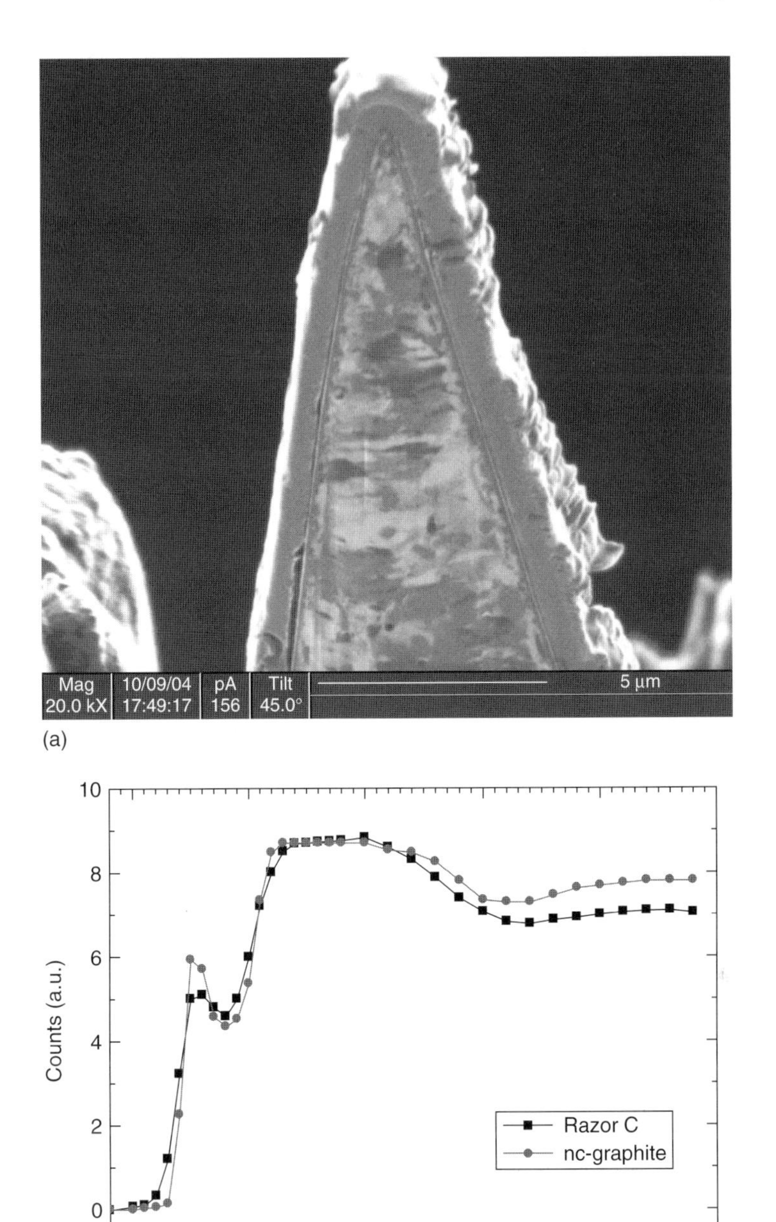

Figure 23.11 (a) Transmission electron microscopy image of a diamond-like-carbon-coated razor blade. (b) Electron energy loss spectrum of its carbon coating, compared to that of a microcrystalline graphite standard.

than that of steel due its large modulus [54]. Thus, it will help to retain its sharpness better.

The type of carbon used was analyzed. It is found to be ~80% sp^2 and hydrogen free, from the electron energy loss spectroscopy (size of the 285 eV peak and absence of a 287 eV peak from C–H bonding), see Figure 23.11b. This is consistent with a sputtered coating. Thus, it is not strictly "diamond" as shown in some advertisements.

The most important use of DLC economically is its use as a protective coating material on magnetic storage disks and their read-heads [55]. Magnetic disks store

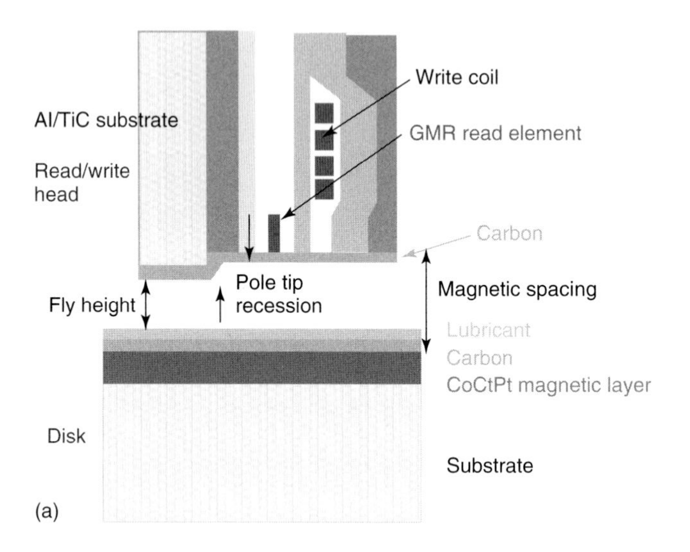

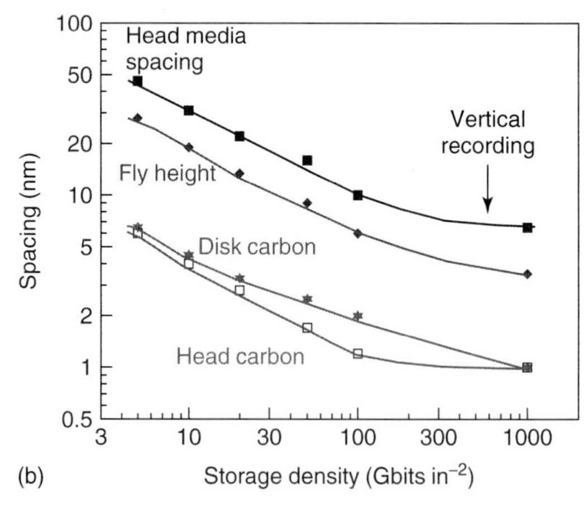

Figure 23.12 (a) Schematic of a head-disk interface of a hard-disk drive. (b) Scaling of vertical fly height, magnetic separation, and diamond-like carbon thickness versus storage density for hard-disk drives.

the bits of information as horizontally magnetized domains in a Co-based layer on a disk (Figure 23.12a). The information is read by a read-head using giant magnetoresistive or spin-valve sensors. Both disk and head are coated by a DLC layer. On top of this, there is a layer of perfluoro-polyether, which acts as a lubricant.

The storage densities have increased rapidly with year, by scaling down the size of the domains allowed for storage. This requires a parallel scaling of all dimensions of the disk and head device, such as the head–disk separation (Figure 23.12b) to keep the read sensitivity. Presently, this magnetic spacing has reduced to below 10 nm and has come down to close to its natural limit of 6.5 nm [56, 57]. For this spacing, the DLC layer can be at most 1.5 nm thick. The natural limit can be 1 nm, if it can be manufactured. Continued scaling is possible, because the technology is switching over to vertical bits [56], which allow a larger area storage density for a given magnetic spacing (Figure 23.12b).

For many years, the films were a-C : H produced by reactive sputtering [58]. More recently, this was replaced by sputtered a-CN_x, as thinner continuous films could be made. There is now a transition in the industry toward using ion beam deposition (ta-C) for deposition on heads [59] and PECVD for deposition on disks.

It is interesting to consider why DLC is so good for this usage [58]. Originally, the DLC layer acted as a wear-resistant coating. For today's disks, wear protection is no longer an issue for such thin films. The films must be pinhole free, to act as an efficient corrosion barrier, against aqueous oxidation of the read-head and disk metals (Fe and Co). For this, the films must be amorphous – no grain boundaries – and be as smooth as possible. This is equivalent to having the maximum nucleation density. DLC is able to achieve this because it nucleates wherever C ions impact the surface.

The film smoothness has been measured by atomic force microscopy (AFM), as a function of DLC type, for a range of deposition ion energies [60] and also versus film thickness [63–65] (Figures 23.13 and 23.14). It is found that most DLCs, both a-C : H and ta-C, are highly smooth, with RMS roughness of order 0.1 nm. They

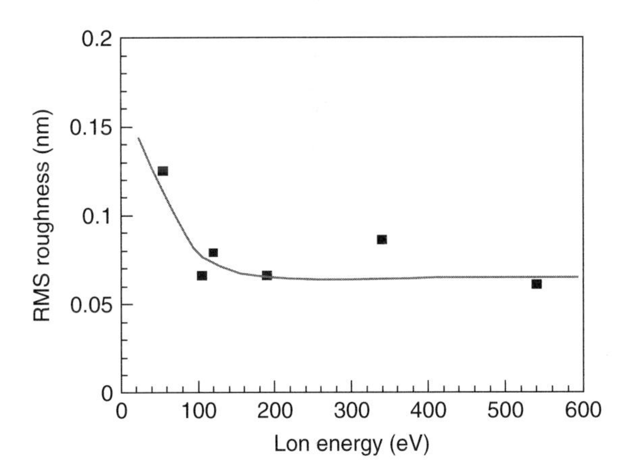

Figure 23.13 Variation of RMS roughness versus deposition ion energy for a-C : H films.

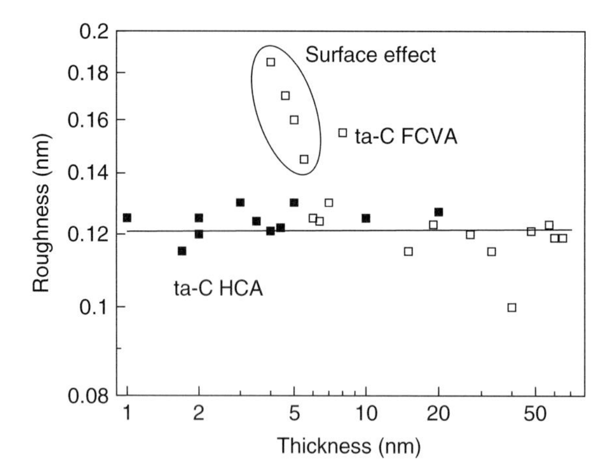

Figure 23.14 Variation of RMS roughness versus film thickness for ta-C films [61].

can achieve this, provided the deposition ion energy lies in the range 20–200 eV (Figure 23.13).

A very interesting fact is that the evolution of roughness with film thickness shows no increase in roughness with increasing thickness (Figure 23.14). Thus, expressing RMS roughness, r, as a function of average film thickness, h, we find

$$r = r_0 h^{\beta} \tag{23.2}$$

with $\beta \sim 0$. For most material systems, first, there is a nucleation barrier, giving an initial hump in roughness [58], and then for larger h we usually get $\beta \sim 0.5$, due to random growth. $\beta \sim 0$ has been identified as due to downhill deposition (due to energetic ions) [61].

Another useful coating application is as a gas membrane barrier on the surface of PET bottles used for drink and foodstuffs [64]. PET is a useful transparent plastic used in the beverage industry. However, it has a slight permeability to gases; oxygen will oxidize the contents (tea, cola, wine, etc.) while CO_2 can escape, leading to a loss of fizz. Previously, PECVD-deposited SiO_2 had been used as a transparent gas barrier, as it was a relatively easy process. Diamond is known to have the highest atomic density among solids [65]. DLC has a reasonable density, giving it gas diffusion barrier properties [66] and it can make a cost-effective sealant. The interesting thing is to make it transparent and economically viable.

The optical bandgap of an average hard DLC is only 1.5 eV, and it also has a broad optical absorption edge [23]. This would be dark brown and an unsuitable coating. It is necessary to increase the bandgap to over 2.5 eV to make a transparent coating. This means using a more polymeric a-C : H film with a higher content of CH_x groups to open up the bandgap. However, the C atom density decreases as the CH_x content increases. Thus, its gas impermeability character varies inversely with its transparency. We must optimize the bonding of the a-C : H film, in terms of precursor gas and growth conditions, to maximize the C atom density and also the bandgap. Many factors will narrow the bandgap or widen the Urbach tail and, thus,

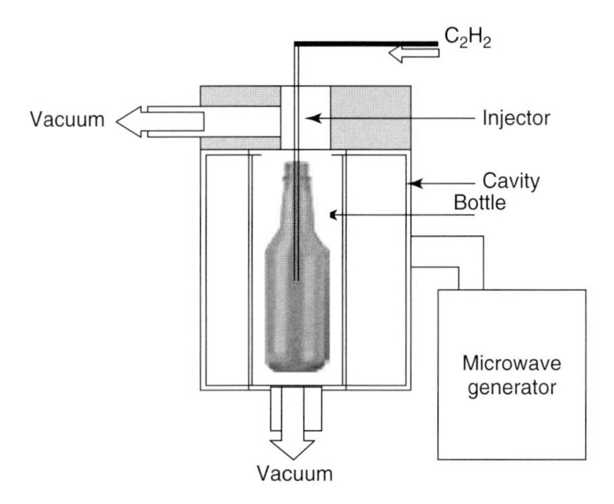

Figure 23.15 Schematic of the plasma deposition system to grow a-C : H gas barrier films inside PET bottles [65].

lose transparency. An optimization has been carried out for the related application of DLC for coating DVD disks [67]. The optimization for the case of PET bottles was carried out commercially, and the resulting films have been characterized [64].

The other key requirement is deposition rate. In a low-cost industry such as packaging, a DLC coating is useful only if it can be applied economically. This requires a coating of about 50 nm thickness to be grown in a cycle of ~4 seconds. Allowing 2 secs for pumping, this allows a total cycle time of 6 secs. The industrial process of Sidel [64] uses a microwave plasma and C_2H_2 as the source gas, as shown schematically in Figure 23.15.

A recent advance in this field is the development of atmospheric plasmas for a-C : H deposition [68]. This would remove one of the cost constrains.

23.6 MEMs

Microelectromechanical systems (MEMSs) are micromechanical systems fabricated using the methods of microelectronics. The standard MEMS materials are silicon and polysilicon, as they are adequately stiff and refractory, and their manufacturing routes are known. However, Si surface has a few disadvantages: it could be stiffer and its surfaces (usually SiO_2 terminated) are hydrophilic. At these small length scales, stiction or surface-to-surface adhesion becomes a problem, in that the working surfaces of a new device cannot be easily separated after manufacturing. DLC is hydrophobic. This problem can be overcome by coating the Si device with DLC with its hydrophobic surface.

Carbon-only MEMS has already been developed. Many of these use nanodiamond or UNCD for their smooth surfaces [69–72]. Nanodiamond has Young's modulus

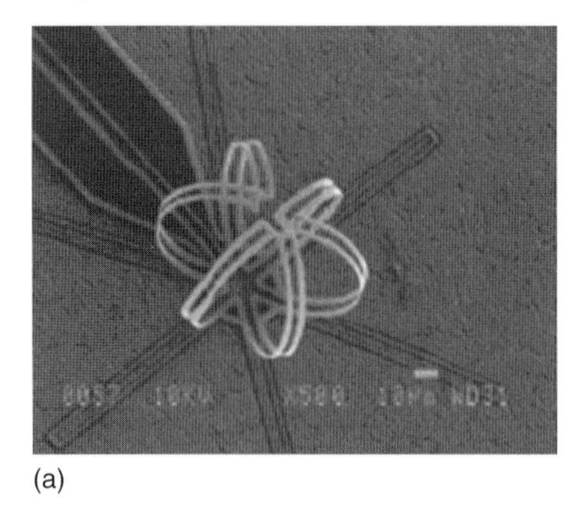
(a)

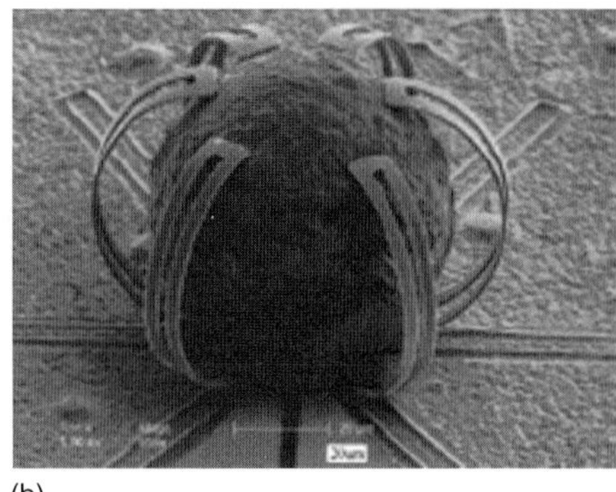
(b)

Figure 23.16 Micrograph of the bimorph gripper, (a) closed and (b) closed around a ball [75].

Table 23.1 Mechanical properties of ta-C, after [39]

Youngs modulus	750 GPa
Bulk modulus	330 GPa
Shear modulus	337 GPa
Poisson's ratio	0.12
Mass density	3.26 gm cm^{-3}

(stiffness) similar to that of diamond, if the grain size is optimized [71]. UNCD has a lower stiffness and may also have a mechanical loss [11].

Young's modulus of DLC has been measured [38, 45]. The mechanical parameters of a ta-C are listed in Table 23.1. Sullivan and Friedmann [74] were the first to develop ta-C for MEMS such as resonators. The main drawback of DLCs for MEMS is their intrinsic stress, which prevents the making of a flat device. The removal of stress by, for example, the annealing method [44] is critical for this application. MEMS devices using DLC have recently been reviewed by Luo *et al.* [75]. A novel device that uses the stress is a ta-C/metal bimorph gripper, as shown in Figure 23.16. The tensile stress makes it closed at rest. Passing a current through it heats it and opens up the gripper, using the differential expansion of Ni and C.

23.7 Electronic Applications

As noted, diamond is a wide bandgap semiconductor with a slightly indirect gap of 5.5 eV, high carrier mobilities [76], and the highest breakdown field of any semiconductor. It can easily be doped p-type by boron and less easily n-type by Phosphorus. Homopolar devices are possible; however, heteropolar devices

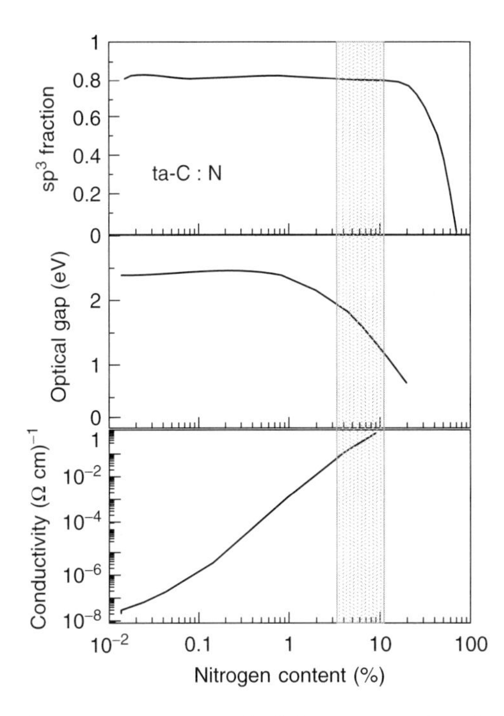

Figure 23.17 Variation of sp^3 fraction, optical bandgap, and conductivity for N-doped ta-C versus N content in the solid [79].

still need to be developed [77]. Micro- and nanocrystalline diamonds are also a heteropolar dopable semiconductor.

In UNCD, there can be a change in conductivity over many orders of magnitude by N, but mainly the N increases the width of grain boundaries and it is not true doping [11].

The electronic applications of DLC have been intensively probed, but ultimately this is not a successful electronic material. The band edges are formed from the π states from sp^2, which lie within the bandgap of the σ states (Figure 23.4). As in amorphous Si, any doping is resisted by autocompensation. The most-studied dopant is N. In a-C : H, as in UNCD, the effect of N is to narrow the bandgap and, thus, the conductivity does increase, but not by a true doping effect [23, 78].

In ta-C, N does provide partial doping. The conductivity does increase before the bandgap begins to decrease, or the sp^3 character is lost, as found by Kleinsorge *et al.* [79, 80]. This is shown in Figure 23.14. Nevertheless, there is only a narrow range of compositions at 3–10% N (shaded in Figure 23.17), where a useful doping effect occurs. The lack of doping was confirmed by the measurement of the Fermi level energy by the Kelvin probe. The work function is found to remain close to 4.7 eV irrespective of N concentration [81] (Figure 23.18). True substitutional n-type doping would cause the Fermi level to move up to the donor level, near the conduction band edge.

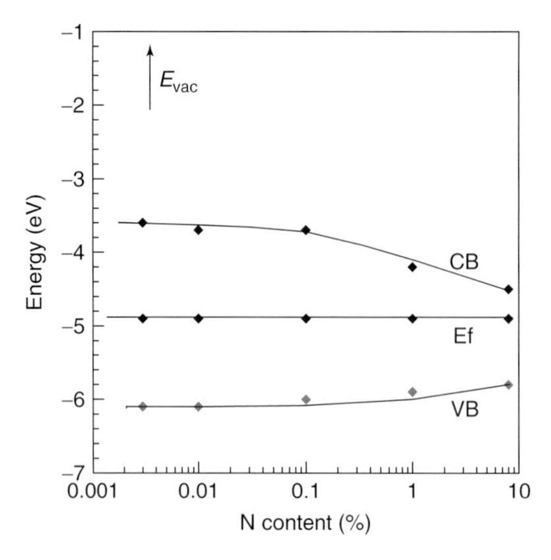

Figure 23.18 Variation of the Fermi energy and band edge energies of N-doped ta-C as a function of N content [81].

The third problem in DLC is that the band edge states are of π character. Because the effect of disorder in π states is strong [82], the carrier mobility is always very low, of order 10^{-6} cm^2 V^{-1} s^{-1}, as seen by photocurrent measurements [83]. The result of low mobility and poor doping is that ta-C and a-C : H do not make useful electronic materials.

A final hope for useful electronic applications of DLC was field emission. The hydrogen-terminated diamond surface can have a negative electron affinity, so that electrons are easily emitted into the vacuum, if somehow they can be injected into the conduction band. It was considered that ta-C might have a similar property, which would be useful for low-cost field emission devices. However, the lower bandgap of ta-C actually means that it has a sizeable positive electron affinity, as measured or estimated. Thus, this is not true [23, 84]. Furthermore, the field emission mechanism involves emission from graphitic tracks through the film, due to localized electrical breakdown. Fundamentally, a good field emitter should be a good conductor; DLC is not, so breakdown occurs to make it one. In practice, sharp objects make better field emitters than problematic semiconductors – carbon nanotubes are preferred.

23.8 Bioactive Surfaces

There has been intensive interest in trying to use diamond and DLC as biocompatible surfaces. The use of diamond depends on three factors. First, doped diamond is an electrode with an extremely wide potential window, due to its wide bandgap, which allows it to be used to oxidize or reduce species in solution without itself

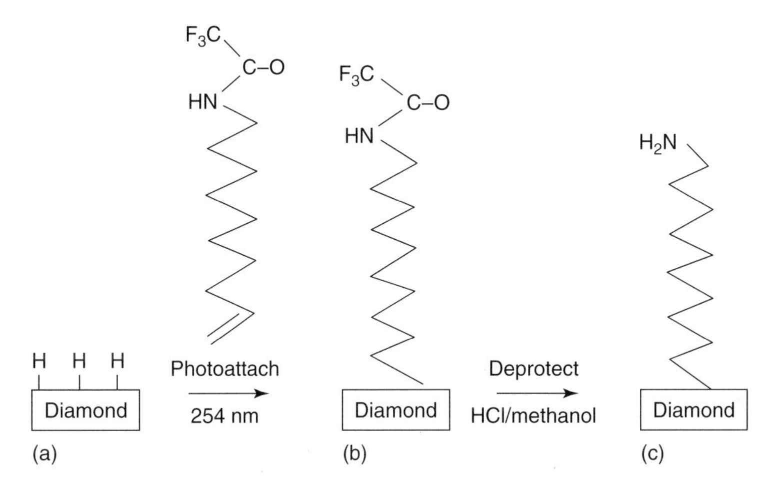

Figure 23.19 Process of attachment of a linker molecule to the diamond surface, after [85].

being oxidized or reduced. As long as the diamond is conductive, the exchange current is sufficiently high. Second, it is biocompatibe and chemically inert, which is an advantage over Si. Third, its band edges lie in a useful energy range.

Diamond surfaces are of great interest for possible biosensors [87–93]. This application requires that the diamond be conductive, for sensing, and also that it is possible to attach particular biomolecules for sensing activity. The target molecule is an amide, tri-fluoro-acetamide terminated 10-aminodec-1-ene (TFAAD). It can be attached to a diamond surface by 254-nm UV illumination [85]. The amide group can then be substituted by other groups such as amino acids (Figure 23.19). The photoattachment mechanism involves creating a hole in the diamond valence band. This is favored for hydrogen-terminated diamond surface, with its negative electron affinity. It is also favored for p-type.

The first experiments were carried out on UNCD. Interestingly, Yang *et al.* [85] observed that the molecules could also be attached to glassy carbon surfaces using the same process. Glassy carbon is not only conductive, but also a bulk material. It suggests that a form of DLC could be used, provided that it is conductive. Ta-C doped with either N or B would be a useful substrate. Ta-C has already been studied in terms of its use as an electrode, where it has a moderate potential window [91] (Figure 23.20). DLC has been widely used in terms of its biocompatibility [92].

Stutzmann *et al.* [90] have compared diamond and other wide gap semiconductors for their usefulness as biomaterials in terms of their band edge energies. They noted that if the band offset between the semiconductor substrate and the LUMO (lowest unoccupied molecular orbital) or HOMO (highest occupied molecular orbital) of the linker molecule allows charge to transfer onto the substrate, then this sets up a potential barrier and also a screening charge, which will inhibit the action of sensing. Thus, it is advantageous to have a relatively wide gap semiconductor with band edges outside the LUMO–HOMO gap of target molecules. Diamond is useful, but the best substrate is AlN.

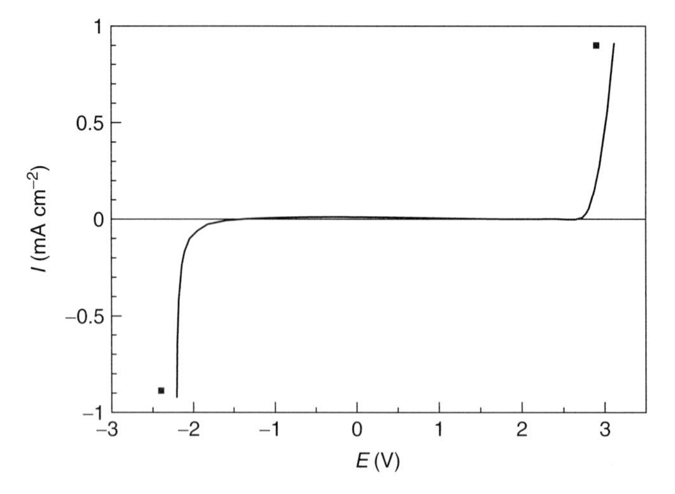

Figure 23.20 Potentiostatic diagram for N-doped ta-C, (after [91].)

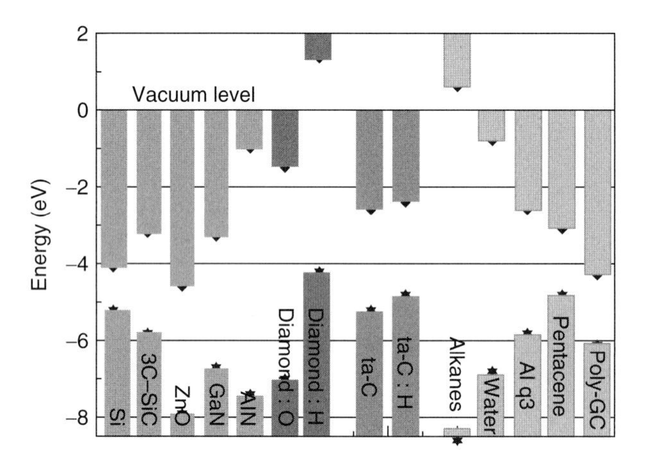

Figure 23.21 Band line ups for diamond, DLC, various other semiconductors, and some biomolecules. Poly-GC (polyguanine-cytosine) is synthetic DNA.

Figure 23.21 shows the respective alignment of various semiconductors, diamond terminated by H or by O, some target molecules, taken from Stuzmann *et al.* [90], plus various DLCs. It is seen that ta-C is a suitable substrate.

23.9 Conclusions

DLC has the advantage of high deposition rate at low temperatures. Its applications have been discussed. It has a number of applications such as a protective coating material, in magnetic disk technology, car components, food containers, and razor blades. It has future applications in MEMS. It might be of use in biosensors, on the basis of its surface properties and band structure.

Acknowledgments

The author would like to acknowledge discussion with C. Nebel, M. Stutzmann, O. Williams, and also the work of C. Casirashi and A.C. Ferrari.

References

1. Butler, J.E. and Woodin, R.L. (1994) *Philos. Trans. R. Soc. London*, **342**, 209.
2. Anthony, T.R. (1990) *Vacuum*, **41**, 1356.
3. (a) Kondoh, E., Ohta, T., Mitomo, T., and Ohtsuka, K. (1993) *J. Appl. Phys.*, **73**, 3042; (b) Kondoh, E., Ohta, T., Mitomo, T., and Ohtsuka, K. (1991) *Appl. Phys. Lett.*, **59**, 488.
4. Hu, C.J., Hauge, R.H., Margrave, J.L., and D'Evelyn, M.P. (1992) *Appl. Phys. Lett.*, **61**, 1393.
5. (a) Muranaka, Y., Yamashita, H., and Miyadera, H. (1994) *Diamond Relat. Mater.*, **3**, 313; (b) Muranaka, Y., Yamashita, H., and Miyadera, H. (1990) *J. Appl. Phys.*, **67**, 6247.
6. Stiegler, J. *et al.* (1996) *Diamond Relat. Mater.*, **5**, 226.
7. (a) Angus, J.C., Sunkara, M., Sahaida, S.R., and Glass, J.T. (1992) *J. Mater. Sci*, **7**, 3001; (b) Butler, J.E. and Oleynik, I. (2007) *Philos. Trans. R. Soc. Ser. A*, **366**, 295.
8. Wild, C., Kohl, R., Herres, N., Muller-Sebert, W., and Koidl, P. (1994) *Diamond Relat. Mater.*, **3**, 373.
9. Sumant, A.V., Gilbert, P.U.P.A., Grierson, D.S., Konicek, A.R., Abrecht, M., Butler, J.E., Feygelson, T., Rotter, S.S., and Carpick, R.W. (2007) *Diamond Relat. Mater.*, **16**, 718.
10. Birrell, J., Gerbi, J.E., Auciello, O., Gibson, J.M., Johnson, J., and Carlisle, J.A. (2005) *Diamond Relat. Mater.*, **14**, 86.
11. Xiao, X., Birrell, J., Gerbi, J.E., Auciello, O., and Carlisle, J.A. (2004) *J. Appl. Phys.*, **96**, 2232.
12. Williams, O., Nesladek, M., Daenen, M., Michaelson, S., Hoffman, A., Osawa, E., Haenen, K. and Jackman, R.B. (2008) *Diamond Relat. Mater.*, **17**, 1080.
13. McCauley, T.G., Gruen, D.M., and Krauss, A.R. (1998) *Appl. Phys. Lett.*, **73**, 1646.
14. Goodwin, D.G. (1993) *J. Appl. Phys.*, **74**, 6888.
15. Okushi, H. (2001) *Diamond Relat. Mater.*, **10**, 281.
16. Koizumi, S. (2000) *Diamond Relat. Mater.*, **9**, 935.
17. (a) Kato, H., Yamasaki, S., and Okushi, H. (2005) *Appl. Phys. Lett.*, **86**, 222111; (b) Kato, H., Yamasaki, S., and Okushi, H. (2007) *J. Phys. D*, **40**, 6189.
18. Achard, J., Silva, F., Bonnin, X., Lombardi, G., Haasouni, K., and Gicquel, A. (2007) *J. Phys. D*, **40**, 6175.
19. Silva, F. (2009) *Diamond Relat. Mater.*, **18**, 683.
20. Muller-Sebert, W., Worner, E., Fuchs, F., Wild, C., and Koidl, P. (1996) *Appl. Phys. Lett.*, **68**, 759.
21. Yan, C.S., Vohra, Y.K., Mao, H.K., and Hemley, R.J. (2002) *Proc. Natl. Acad. Sci. U.S.A.*, **99**, 12523.
22. Robertson, J. (1994) *Pure Appl. Chem.*, **66**, 1789.
23. Robertson, J. (2002) *Mater. Sci. Eng. R*, **37**, 129.
24. Polo, M.C., Andujar, J.L., Hart, A., Robertson, J., and Milne, W.I. (2000) *Diamond Relat. Mater.*, **9**, 663.
25. Xu, S., Tay, B.K., Tan, H.S., Zhong, L., Tu, Y.Q., Silva, S.R.P., and Milne, W.I. (1996) *J. Appl. Phys.*, **79**, 7234.

26. Lifshitz, Y., Kasi, S.R., and Rabalais, J.W. (1989) *Phys. Rev. Lett.*, **68**, 620.
27. McKenzie, D.R., Muller, D., and Pailthorpe, B.A. (1991) *Phys. Rev. Lett.*, **67**, 773.
28. (a) Robertson, J. (1993) *Philos. Trans. R. Soc.*, **342**, 277; (b) Robertson, J. (1994) *Diamond Relat. Mater.*, **3**, 361.
29. Koidl, P., Wagner, C., Dischler, B., Wagner, J., and Ramsteiner, M. (1990) *Mater. Sci. Forum*, **52**, 41.
30. Lieberman, M.A. and Lichtenberg, A.J. (1994) *Principles of Plasma Discharges and Materials Processing*, John Wiley & Sons, Inc., New York.
31. Weiler, M., Sattel, S., Giessen, T., Jung, K., Ehrhardt, H., Veerasamy, V.S., and Robertson, J. (1996) *Phys. Rev. B*, **53**, 1594.
32. Morrison, N.A., Rodil, S., and Milne, W.I. (1999) *Phys. Status Solidi A*, **172**,79.
33. Yoon, S.F., Yang, H., Rusli (1998) *Diamond Relat. Mater.*, **7**, 70.
34. Robertson, J. (1986) *Adv. Phys.*, **35**, 317.
35. Robertson, J. (1997) *Diamond Relat. Mater.*, **6**, 212.
36. Nesladek, M., Meykens, K., Stals, L.M., Vanacek, M., and Rosa, J. (1996) *Phys. Rev. B*, **54**, 5552.
37. Fallon, P.J., Veerasamy, V.S., Davis, C.A., Robertson, J., Amaratunga, G., Milne, W.I., and Koskinen, J. (1993) *Phys. Rev. B*, **48**, 4777.
38. Ferrari, A.C., Robertson, J., Beghi, M.G., Bottani, C.E., Ferulano, R., and Pastorelli, R. (1999) *Appl. Phys. Lett.*, **75**, 1893.
39. Shamsa, M., Liu, W.L., Balandin, A.A., Casiraghi, C., Milne, W.I., and Ferrari, A.C. (2006) *Appl. Phys. Lett.*, **89**, 161921.
40. Davis, C.A. (1993) *Thin Solid Films*, **226**, 30.
41. Tamor, M.A. and Vassel, W.C. (1994) *J. Appl. Phys.*, **76**, 3823.
42. Ferrari, A.C., Rodil, S.E., Robertson, J., and Milne, W.I. (2002) *Diamond Relat. Mater.*, **11**, 994.
43. Lau, D.W.M., McCulloch, D.G., Taylor, M.B., Partridge, J.G., McKenzie, D.R., Marks, N.A., Teo, E.H.T., and Tay, B.K. (2008) *Phys. Rev. Lett.*, **100**, 176101.
44. Sullivan, J.P., Friedmann, T.A., and Baca, A.G. (1997) *J. Electron. Mater.*, **26**, 1021.
45. Friedmann, T.A., Sullivan, J.P., and Knapp, J.A. (1997) *A ppl. Phys. Lett.*, **71**, 3820.
46. Ferrari, A.C., Kleinsorge, B., Morrison, N.A., Hart, A., Stolojan, V., and Robertson, J. (1999) *J. Appl. Phys.*, **85**, 7191.
47. Oguri, K. and Arai, T. (1991) *Surf. Coat. Technol.*, **47**, 710.
48. Zehnder, T. and Patscheider, J. (2000) *Surf. Coat. Technol.*, **133**, 138.
49. Voevodin, A.A., Prasad, S.V., and Zabinski, J.S. (1997) *J. Appl. Phys.*, **82**, 855.
50. Antilla, A., Lappalainen, R., Tiainen, V.M., and Hakovirta, M. (1997) *Adv. Mater.*, **9**, 1161.
51. Voevodin, A.A., Capano, M.A., and Laube, S.J.P. (1997) *Thin Solid Films*, **298**, 107.
52. Bilek, M.M.M. and McKenzie, D.R. (2006) *Surf. Coat. Technol.*, **200**, 4345.
53. Lusk, D., Gore, M., Boardman, W., Casserly, Y., Boinapally, K., Oppus, M., Upadhyaya, D., Tudhope, A., Gupta, M., and Cao, Y. (2008) *Diamond Relat. Mater.*, **17**, 1613.
54. Robertson, J. and Manning, M.I. (1990) *Mater. Sci. Technol.*, **6**, 81.
55. (a) Goglia, P., Berkowitz, J., Hoehn, J., Xidis, A., and Stover, L. (2001) *Diamond Relat. Mater.*, **10**, 271; (b) Bhushan, B. (1999) *Diamond Relat. Mater.*, **8**, 1985; (c) Robertson, J. (2001) *Thin solid Films*, **383**, 81.
56. Gui, J. (2003) *IEEE Trans. Mag.*, **39**, 716.
57. Mate, C.M., Toney, M.F., and Leach, K.A. (2001) *IEEE Trans. Mag.*, **37**, 1821.
58. Robertson, J. (2003) *Tribiol. Int.*, **36**, 405.
59. Druz, B., Yevtukhov, Y., and Zaritskiy, I. (2005) *Diamond Relat. Mater.*, **14**, 1508.
60. Peng, L., Barber, Z.H., and Clyne, T.W. (2001) *Surf. Coat. Technol.*, **138**, 23.
61. Casiraghi, C., Ferrari, A.C., Ohr, R., Flewitt, A.J., Chu, D.P., and Robertson, J. (2003) *Phys. Rev. Lett.*, **91**, 226104.
62. Patsalas, P., Logothetidis, S., and Kelires, P.C. (2005) *Diamond Relat. Mater.*, **14**, 1241.

63. Pisana, S., Casiraghi, C., Ferrari, A.C., and Robertson, J. (2005) *Diamond Relat. Mater.*, **15**, 898.
64. Moseler, M., Gumbsch, P., Casiraghi, C., Ferrari, A.C., and Robertson, J. (2005) *Science*, **309**, 1545.
65. Boutroy, N., Pernel, Y., Rius, J.M., Auger, F., von Bardeleben, H.J., Cantin, J.L., Abel, F., Zeinert, A., Casiraghi, C., Ferrari, A.C., and Robertson, J. (2006) *Diamond Relat. Mater.*, **15**, 921.
66. Jansen, F. and Angus, J.C. (1988) *J. Vac. Sci. Technol. A*, **6**, 1778.
67. Wild, C. and Koidl, P. (1987) *Appl. Phys. Lett.*, **51**, 1506.
68. Piazza, F., Grambole, D., Schneider, D., Casiraghi, S., Ferrari, A.C., and Robertson, J. (2005) *Diamond Relat. Mater.*, **14**, 994.
69. Suzuki, T., Kodama, H., Takano, K., Suemitsu, S., and Hotta, A. (2009) *Diamond Relat. Mater.*, **18**, 990.
70. Cho, S., Chasiotis, I., Friedmann, T.A., and Sullivan, J.P. (2005) *J. Micromech. Microeng.*, **15**, 728.
71. (a) Sekaric, L., Parpia, J.M., and Butler, J.E. (2002) *Appl. Phys. Lett.*, **81**, 4455; (b) Philips, J., Hess, P., Feygelson, T., and Butler, J.E. (2003) *J. Appl. Phys.*, **93**, 2164.
72. Hutchinson, A.B. *et al.* (2004) *Appl. Phys. Lett.*, **84**, 972.
73. (a) Auciello, O., Birrell, J., Carlisle, J.A., Gerbi, J.E., Peng, B., and Espinosa, H.D. (2004) *J. Phys. Conden. Mat.*, **16**, R552; (b) Krauss, A.R. (2001) *Diamond Relat. Mater.*, **10**, 722.
74. Sullivan, J.P. and Friedmann, T.A. (2001) *MRS Bull.*, **26**, 309. April
75. Luo, J.K. *et al.* (2007) *J. Micromech. Microeng.*, **17**, S147.
76. Nesladek, M. (2008) *Diamond Relat. Mater.*, in press, **17**, 1235.
77. Makino, T., Tokuda, N., Kato, H., Ogura, M., Watanabe, H., Ri, S.G., Yamasaki, S., and Okushi, H. (2006) *Jpn J. Appl. Phys.*, **45**, L1042.
78. Robertson, J. and Davis, C.A. (1995) *Diamond Relat. Mater.*, **4**, 441.
79. Kleinsorge, B., Ferrari, A.C., Robertson, J., and Milne, W.I. (2000) *J. Appl. Phys.*, **8**, 14905.
80. Waidmann, S., Knupfer, M., Fink, J., Kleinsorge, B., and Robertson, J. (2001) *J. Appl. Phys.*, **89**, 3783.
81. Ilie, A., Hart, A., Flewitt, A.J., Robertson, J., and Milne, W.I. (2000) *J. Appl. Phys.*, **88**, 6002.
82. Chen, C.W. and Robertson, J. (1998) *J. Non-Cryst. Solids*, **230**, 602.
83. Ilie, A., Conway, N.M.J., Kleinsorge, B., Robertson, J., and Milne, W.I. (1998) *J. Appl. Phys.*, **84**, 5575.
84. Robertson, J. (1999) *J. Vac. Sci. Technol. B*, **17**, 659.
85. Yang, W., Auciello, O., Butler, J.E., Cai, W., Carlisle, J.A., Gerbi, J., Gruen, D.M., Knickerbocker, T., Lassiter, T.L., Russell, J.N., Smith, L.M., and Hamers, R.J. (2002) *Nat. Mater.*, **1**, 253.
86. Hamers, R.J., Butler, J.E., Lasseter, T., Nichols, B.M., Russell, J.N., Tse, K.Y., and Yang, W. (2005) *Diamond Relat. Mater.*, **14**, 661.
87. Hartl, A., Schmich, E., Garrido, J.A., Hernando, J., Catharino, S., Walter, S., Feulner, P., Kromka, A., Steinmuller, D., and Stutzmann, M. (2004) *Nat. Mater.*, **3**, 730.
88. Rubio-Retama, J. *et al.* (2006) *Langmuir*, **22**, 5837.
89. Grierson, D.S. and Carpick, R.W. (2007) *Nano Today*, **2**, 12.
90. Stutzmann, M., Garrido, J.A., Eickhoff, M., and Brandt, M.S. (2006) *Phys. Status Solidi A*, **203**, 3424.
91. Yoo, K.S., Miller, B., Kalish, R., and Shi, X. (1999) *Electrochem. Solid State Lett.*, **2**, 233.
92. Hauert, R. (2003) *Diamond Relat. Mater.*, **12**, 583.

24
Applications of DLCs to Bioprocessing

Tatsuyuki Nakatani and Yuki Nitta

24.1
Introduction

Diamond-like carbon (DLC) films are amorphous carbonaceous films with both graphite and diamond bonds. It has interesting characteristics, such as a low-friction coefficient, high wear resistance, and a wide bandgap. It is already widely used for industrial purposes, such as in dies of various types and in automobile parts [1]. In addition, since the DLC films deposited by plasma-enhanced chemical vapor deposition (PECVD) methods have characteristics such as biocompatibility and antithrombogenicity, it is expected to be one of the biocompatible coatings on medical devices that prevent a reaction of biomaterial with the surface [2, 3], and studies aimed at the clarification of the deposition mechanisms [4]. Surface conditions of such films have been extensively pursued in recent years [5, 6].

Myocardial infarction, a disorder arising from occlusion or spasm of the coronary arteries, is one of the three major lifestyle-related diseases in addition to cancer and stroke. A technique for its treatment that has flourished in recent years is stent placement by means of a catheter, which is classified as a specially controlled medical device. Figure 24.1 shows a photograph of a coronary stent. A coronary stent mounted on a balloon catheter is delivered to coronary arteries to open blood vessels that are occluded or afflicted with stenosis. As for the medical devices, the character of its component was realized as a combination of various materials such as metals and polymers. However, there is the problem of biocompatibility for most medical materials when a device is inserted into the human body as a foreign substance. Concerning the stents available on the commercial market, a number of reports are currently being produced by academic societies on the adhesion of drug coatings and the biocompatibility of the stents after releasing the drug. Thus, the development of a new medical material that has superior adhesion to the base material and high biocompatibility is viewed as an urgent requirement.

Therefore, we aimed at biocompatible DLC films that have both the biocompatibility and durability required for a coronary stent. From the viewpoint of plasma surface treatment technology, the deposition process of DLC films provides high

Industrial Plasma Technology. Edited by Yoshinobu Kawai, Hideo Ikegami, Noriyoshi Sato, Akihisa Matsuda, Kiichiro Uchino, Masayuki Kuzuya, and Akira Mizuno

ISBN: 978-3-527-32544-3

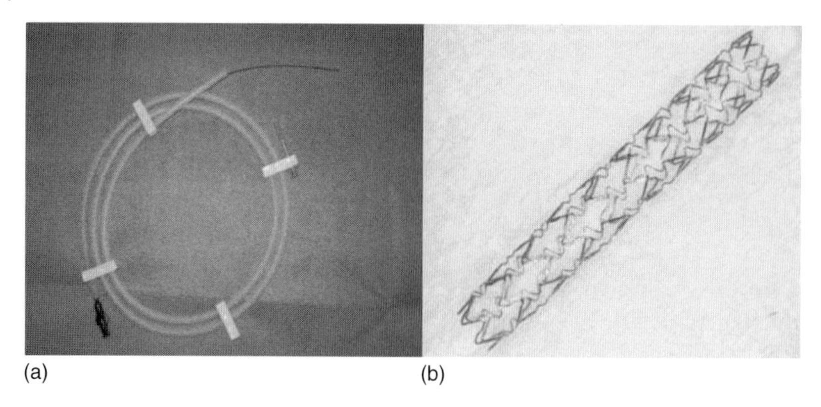

(a) (b)

Figure 24.1 Photograph of the coronary stent: (a) stent delivery system and (b) expanded stent.

adhesion and high biocompatibility that will resolve the problems of medical devices.

We have prepared new DLC films with increasing concentrations of Si doped during the DLC deposition process. These new DLC films have the superior adhesion that prevents cracking on the surface of the coronary stent during expansion. These new DLC films have been deployed on new domestic coronary stents, which approved the European CE marking in 2008 and it is now under clinical evaluation in Europe. Using plasma surface treatment techniques on these new DLC film, we can improve the adhesive property of the drug-containing biodegradable polymer layer and the antithrombotic drug layer and inhibit any allergic action after the complete release of the drug into the blood vessel.

Additionally, we have been developing next-generation multifunctional stents concurrently. We are focusing particularly on the interaction between the DLC surface and the biomaterial surface. We have investigated new biomimetic DLC films with the zeta potential of their surfaces controlled by plasma modification techniques for the purpose of improving cytocompatibility with the tissue cells, which are generally negatively charged. As a result, blood and tissue cell compatibility of the new biomimetic DLC was confirmed, and it was suggested that the optimization of zeta potential on the zwitterionic structural surface could be realized with the new DLC films. The functionality of biomimetic DLC films on medical devices have been improved by plasma modification techniques, and these DLC films will provide the opportunity to develop innovative and new medical devices in the near future.

In this chapter, we present the application and experience of DLC films on a coronary stent to introduce the DLC films specifically demanded for biocompatible medical devices and the plasma processing required for biocompatible DLC films. Furthermore, we propose the future prospects concerning the possibility of the DLC films used for medical devices, which will be realized through precise control of plasma processing.

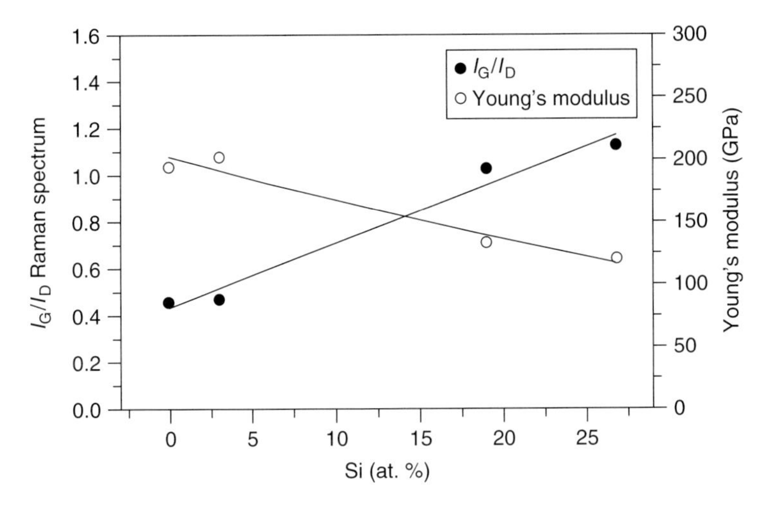

Figure 24.2 Relationship between the integrated intensity ratios of the Raman spectra (I_G/I_D) ratio and Si content in DLC films, and the relationship between Si content and Young's modulus obtained by a nanoindentation measurement.

24.2
High-Tenacity DLC Tin Films for Stents

A coronary stent can open an occluded coronary artery, being expanded by a balloon catheter to the plastic deformation region; thus, the biocompatible coating on the stents must follow the changing surface of the stents without any cracking. Therefore, we have examined the formation of high-tenacity concentration-gradient-type DLC films that were made to imitate the plastic deformation of the stent base material by adjusting the amount and concentration gradient of the Si doped into the DLC [7].

The DLC films used for the stents were prepared by ionization-assisted deposition. The operational gases used were benzene (C_6H_6) including tetramethylsilane ($Si(CH_3)_4$) gases. The film thickness was about 50 nm. Considering the Si content, the peeling resistance of the DLC films on the stents surface was evaluated in terms of hardness, crystalline structure, and adhesion. Figure 24.2 shows the relationship between the integrated intensity ratios of the Raman spectra (I_G/I_D) ratio and Si content in films, and the relationship between Si content and Young's modulus obtained by a nanoindentation measurement. When I_G is larger than I_D, the DLC films are partially graphitic and, consequently, are elastic. Therefore, when the Si content is changed depending on the depth in DLC films, the chemical structure can be controlled freely.

On the basis of the above results, to investigate the peeling resistance of the DLC films, two different 50-nm-thick coatings were applied to the surfaces of stents made of Co–Cr alloy. Specifically, one of the stents was coated with

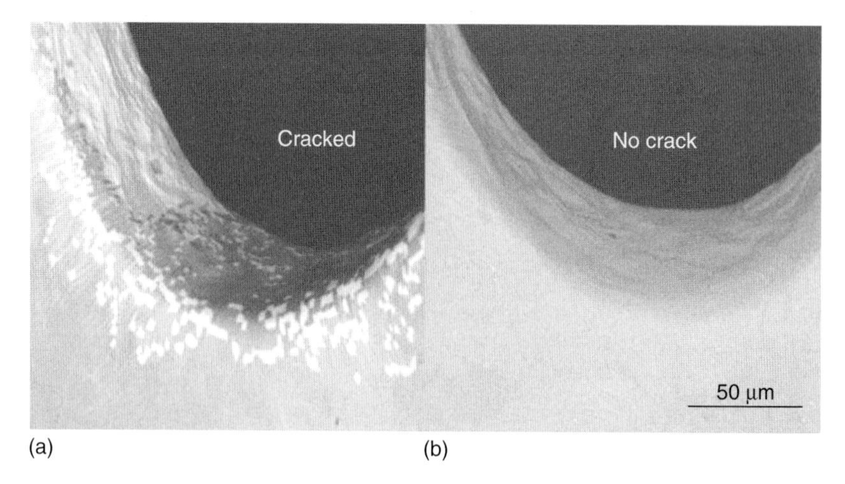

Figure 24.3 SEM image of the DLC-coated stent surface after expansion: (a) conventional DLC (0% Si) and (b) newly developed Si concentration-gradient-type DLC (26.8% Si).

DLC films (0%-Si) having high Young's modulus and the other was coated with DLC films (26.8%-Si) formed with a Si concentration gradient such that Young's modulus was lower for the base material. These stents were expanded from 1.55 to 3.0 mm in diameter with a balloon catheter, and the locations of the greatest strain were observed in an electron image. A Hitachi High-Tech Science Systems Corporation (TM-1000) desktop microscope was used for the reflected electron imaging. Figure 24.3 shows reflected electron images of the stent surfaces. The conventional DLC-coated stent shows a crack on the surface after expansion, as shown in Figure 24.3a, but the Si concentration-gradient-type DLC-coated stent has no crack on the surface, as shown in Figure 24.3b, because of its superior adhesion. Furthermore, newly developed DLC coating technologies enable the desired surface modification of three-dimensional pieces with complex shapes to be more completely accomplished compared with conventional methods.

Additionally, medical device testing services confirmed that 10-years postimplantation pulsatile fatigue and durability meet the requirements of the US Food and Drug Administration (FDA) without any peeling or cracking. In this test, the gradient-type DLC-coated stent satisfied the durability criterion. Thus, in addition to its adhesion performance, the newly developed Si concentration-gradient-type DLC films were proved, by a third-party organization, to possess superior endurance.

The above experimental results demonstrated the utility of high-tenacity DLC films and suggested that the practical application of stents with Si concentration-gradient-type DLC films formed by the ionized deposition method would be possible.

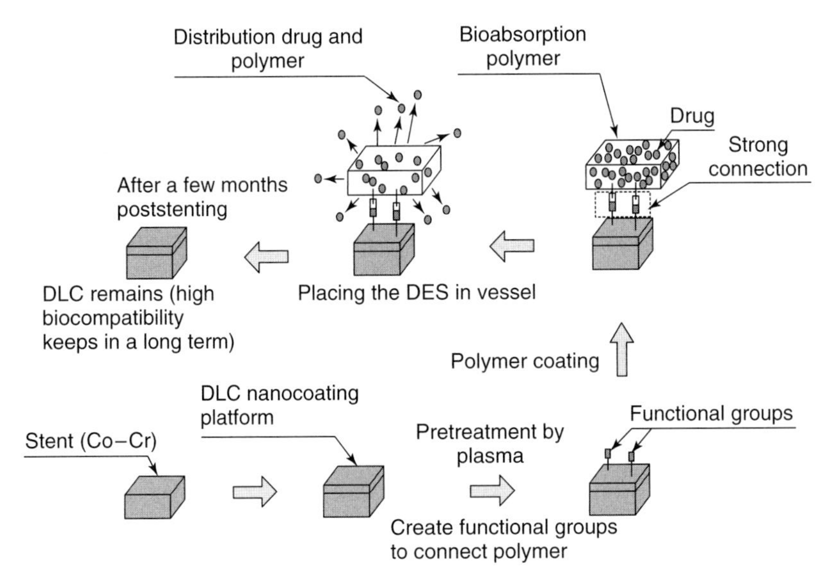

Figure 24.4 Schematic of the plasma surface engineering for coating of drug-eluting stent.

24.3 Applications of DLC Films to Coronary Drug-Eluting Stent

A drug-eluting stent slowly and continuously releases the drug inside the coronary vessel. After the drug release, the DLC-coated bare metal stent remains. To avoid injury or inflammation, the DLC-coated surface should not be cracked. To provide adhesion and biocompatibility, we attempted the fusion of (i) a biocompatible-polymer grafting technique employing DLC films – plasma irradiation–postgrafting polymerization techniques [8, 9, 10]; (ii) techniques for depositing Si-gradient-type DLC films [11]; and (iii) drug elution rate control employing biodegradable polymers.

Figure 24.4 shows the schematic for coating drug-eluting stents, which prevents restenosis over a long period of time. First, the Co–Cr alloy stent is coated with an Si concentration-gradient-type DLC film. Then its surfaces are irradiated with a radio-frequency plasma reactor, so as to bond functional groups to the surfaces. The process chamber is connected to an RF power supply with an excitation frequency of 13.56 MHz. The RF power of 30 W was injected to generate plasmas. Capacitively coupled plasma (CCP) was generated by means of two parallel-plate electrodes. Next, a biodegradable polymer containing a drug is fixed onto the surface. To coat the polymer, a solution containing the polymer and the drug is sprayed onto the DLC surface. This method is able to fix the polymer and functional groups firmly onto the DLC surface. Therefore, the polymer layer will not crack or peel away from the DLC surface when the stent is expanded with a balloon catheter.

Figure 24.5a shows a reflected electron image of the surfaces of a stent coated by the process as shown in Figure 24.4; it has been expanded with a balloon

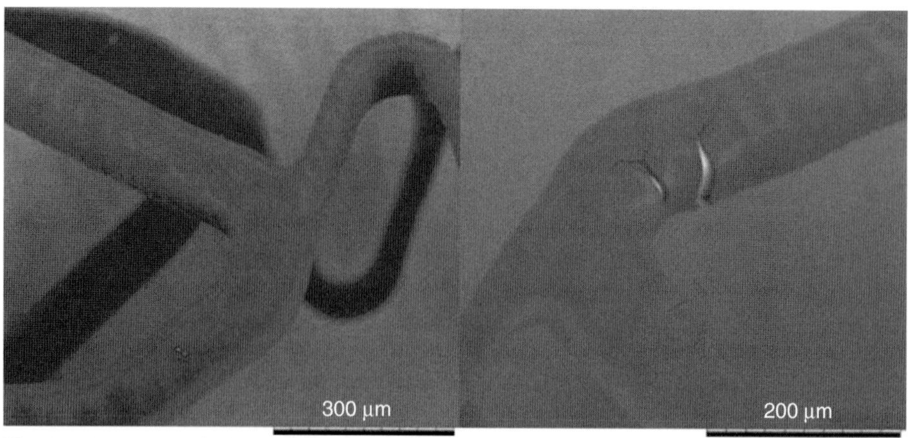

Figure 24.5 SEM images of the polymer coatings on stent surface: (a) DLC films + plasma irradiation and (b) non-DLC.

catheter. For comparison, Figure 24.5b shows a reflected electron image, taken after expansion, of a stent that was coated with the polymer directly, without a DLC film being applied to the stent base material. In this image, the white portions that are visible through the splits in the polymer layer are the Co–Cr alloy. It can be seen from this figure that the DLC films in addition to plasma modification improve the adhesion between the polymers and the base material, so that even under the deformation produced by stent expansion, no cracking or peeling of the polymer layer will occur [7].

In addition, animal studies and hemopathological analyses have been performed on more than 60 mini-pigs to confirm the safety and efficacy of DLC-coated coronary stents in the United States. As a result, we found that strut-associated inflammation was minimal with occasional macrophages. Endothelialization was nearly complete with mild subendothelial inflammation seen in each segment. Fibrin deposition was minimal.

The biocompatible-drug-eluting stent with the new DLC films is undergoing animal study and is expected to be realized in the near future.

24.4
DLC Films with Controlled Zeta Potential of Biomaterials

As the drug-containing biodegradable polymer is designed to elute the drug into the blood for several weeks after being implanted inside the artery, the DLC surface will ultimately come into contact with blood and endothelial cells (ECs). Therefore, imparting biocompatibility to the DLC surface ranks as an important task in functionalizing the surface of the stent.

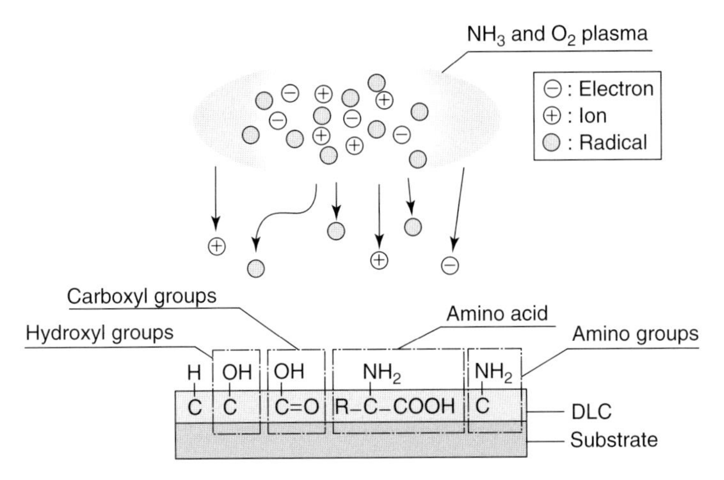

Figure 24.6 Schematic of the DLC films surface with introduced functional groups.

We have studied the possibility of imparting biocompatibility by controlling the functional groups that are generated on the surface of the DLC films by plasma surface treatment techniques. We believe that the creation of a new DLC film that is both durable and functional can be expected. We are focusing particularly on the interaction between the DLC surface and the biomaterial. Cells are generally negatively charged, and their surface potential varies depending on the individual cell. Hence, it is surmised that it will be possible to control the interaction with the DLC surface and cells such that they will either adhere the DLC surface as a result of having the same zeta potential or not be adsorbed. The zeta potential control method that we focused on involves the introduction of anionic and cationic groups by plasma surface treatment.

Figure 24.6 is a schematic of a DLC surface with functional groups introduced. Anions are negatively charged ions and cations are positively charged ions. If the composition ratio of these functional groups can be controlled at the DLC surface, it should be possible to control the zeta potential and thus to induce particular cells to selectively adhere or not be adsorbed by the DLC surface.

In this chapter, we report the relationship between the amounts of functional groups generated and also the zeta potential measurement results of the samples. The operational gases of plasma modifications used were O_2, Ar, NH_3, and C_2H_2. The zeta potential of the samples was measured with a zeta potentiometer (ELS-Z, Otsuka Electronics). The N/C ratio and O=C–O/C ratio were measured with an XPS (X-ray photoelectron spectroscope, JPS-9010MC).

Figure 24.7 shows the zeta potential measurement result for the O=C–O/C ratio. The N/C ratios for various cases are also shown in the figure. It was observed that the zeta potential decreased approximately twice as much as that of the untreated sample with increasing O=C–O/C ratio. It is known that when the carboxyl group –COOH is present, it will generally dissociate as $-COOH \rightarrow -COO^- + H^+$, and samples to which carboxyl groups have attached will be negatively charged owing

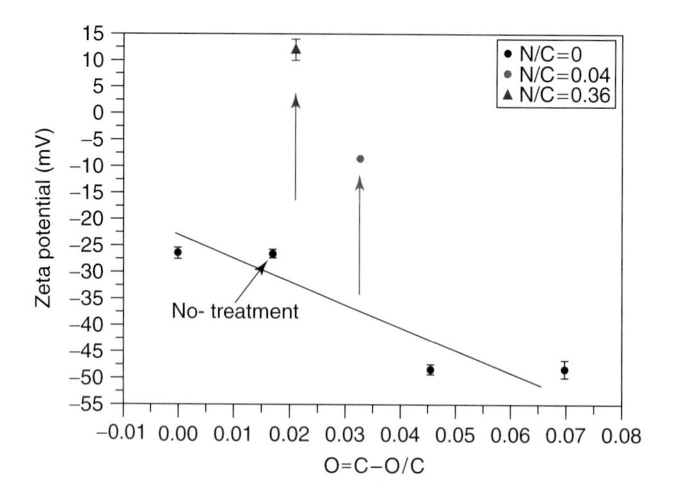

Figure 24.7 Dependence of the zeta potential on the O=C−O/C.

to the $-COO^-$. Therefore, it is considered that the zeta potential decreased with an increasing amount of carboxyl groups. It was also observed that when N/C ratio = 0.04 and 0.36, the zeta potential increases significantly compared with than that of untreated DLC films, particularly at the N/C ratio = 0.36, where the zeta potential turns positive. It is known that when the amino group $-NH_3^+$ with bonded protons is present on the surface of a sample, the sample will generally be positively charged. This is considered to account for the marked increase in the zeta potential of the DLC films when amino groups are introduced. This means that the zeta potential of the DLC films can also be controlled by controlling the amounts of these two functional groups [12].

The foregoing results indicate that we have succeeded, by using plasma surface treatment techniques, in developing a new multifunctional DLC film that does not use polymers and is suitable as a biocompatible material for stents.

24.5
Characterization of Biomimetic DLC and *In vitro* Biocompatibility Evaluation

We performed plasma surface treatment on the DLC films so as to examine the introduction of functional groups to its surface. We also measured the zeta potential of the DLC film with anionic and cationic groups introduced. Furthermore, by introducing anionic and cationic groups simultaneously onto a DLC surface, we created a biomimetic DLC film with the zwitterionic structure. Then, we investigated the relationship between the zeta potential and biocompatibility by varying the conditions on the DLC films. As a result, we confirmed the efficacy of the DLC film coating *in vitro* and concluded that it may be applied to medical devices, such as coronary stents, used in clinical treatment in the future.

24.5.1
Biomimetic DLC and Blood Compatibility

To assess *in vitro* blood compatibility, we assessed two properties of heparinized whole blood collected from a healthy person: platelet adhesion and blood coagulation. The amount of platelets adhering to the samples was measured by electron microscopy. The content of thrombin-antithrombin (TAT) III complexes in the plasma was evaluated using an ELISA kit (TAT-EIA, Enzyme Research Laboratories).

Figure 24.8 shows the electron microscope images of the representative platelet adhesion. The confirmed white dots are platelets adhering to the DLC surface. For the untreated DLC films and SUS 316L stainless steel sample, a large amount of activated platelet adhesion was observed. In the DLC film surface treated with $O_2 + NH_3$ plasma processing, a marked decrease in the amount of platelet adhesion was observed. It was found that the blood compatibility rates at the plasma-treated DLC surface and the untreated DLC surface were significantly different.

Figure 24.9 shows the amounts of TAT produced and platelets adhesion when the zeta potential of the DLC films was changed. As this figure shows, the greater the zeta potential, the higher the compatibility with regard to the blood coagulating factors. The correlative relationship between the N/O=C–O ratio and the zeta potential verified that the larger the N/O=C–O ratio, the greater is the zeta potential. Thus, it was found that an optimal compositional ratio also exists in the zwitterionic structure, resulting in the film showing superior blood compatibility if the zeta potential is high. However, it is not clear why the film exhibits superior compatibility with regard to both platelet adhesion and blood coagulation factors when the zeta potential is high if the O=C–O/C ratio is low and N/C is high. The O=C–O/C ratio is a functional group possessing a negative charge, while the N/C ratio is a functional group possessing a positive charge. It is possible that an effect on blood compatibility is exerted by the amounts of positive and negative charges and by the distances between the functional groups, on the surface of the carbonaceous films [13].

The foregoing results indicate that by optimizing the amounts of carboxyl and amino groups introduced to the DLC surface, we have succeeded in developing a zwitterionic structure. This new biomimetic DLC film may be promising as a new biocompatible material with high blood compatibility compared with conventional DLC films.

24.5.2
Biomimetic DLC and Cytocompatibility

Histocompatibility with human coronary artery ECs or smooth muscle cells (SMCs) requires the implanted stent surface with properties other than blood compatibility to come into contact with vascular tissue. After applying plasma surface treatment to DLC films for a stent, the DLC films become compatible with ECs. ECs in blood vessels are affected by external stimuli. The adjusting function was not

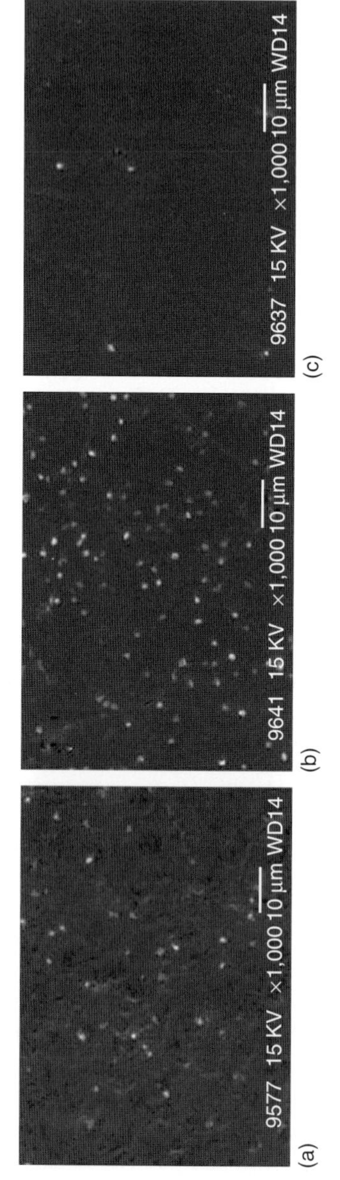

Figure 24.8 SEM images of the platelets adhesion on the various samples: (a) SUS 316L stainless steel, (b) DLC films, and (c) DLC films + plasma irradiation.

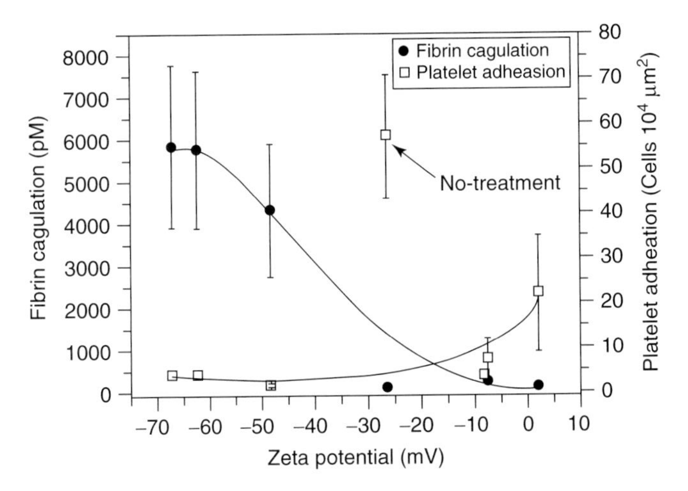

Figure 24.9 Blood coagulation and platelets adhesion characteristics as function of the zeta potential on the DLC films.

performed when an abnormality of ECs occurred. Then, thromboses with fibrin occur. Therefore, medical devices in contact with blood should be applied quickly to ECs and should not be recognized as foreign materials.

To investigate the *in vitro* cytocompatibility of functionalized DLCs, we measured the growth ratio of ECs and SMCs when the zeta potentials of the DLC-coated surface were changed under various plasma surface treatment conditions. ECs (HCAEC, Cell Applications) and SMCs (HCASMC, Cell Applications) from the coronary artery were seeded on cover glass at a density of 10^4 cells per well (500 μl). The cell growth ratio was measured with a cell counter (Cell counting kit8, Dojin Chemical) after four days of cultivation. The concentration of WST-8formazan (expressed in terms of absorbance) followed by a reduction in mitochondrial activity was measured with a microplate reader to compare cell growth ratios under various conditions of plasma surface treatment. The absorbance in the case of untreated DLC thin film was 100%, and cell growth ratios under various conditions were compared.

Figure 24.10 shows the measurement of EC and SMC growth ratios when the zeta potential of the DLC films was changed. Plasma surface treatment was performed under all conditions for all DLC films, except the untreated DLC film. The cell growth ratio in the case of the untreated DLC film was 100%, and the cell growth ratios under other conditions were measured. The results showed that the ECs growth ratio with the DLC films having surfaces treated with plasma was higher than that with untreated DLC. In contrast, there is little change in the SMC growth ratio with zeta potential. Platelet and TAT III complex compatibilities improved when the zeta potential reached 0 mV. At this value, the ECs' growth ratio was 140%. Therefore, our DLC films have high compatibility with ECs as well as with platelets and TAT III complexes. The ECs' growth ratio was lowest when the zeta potential was −26 mV (untreated DLC). Anionic functional groups,

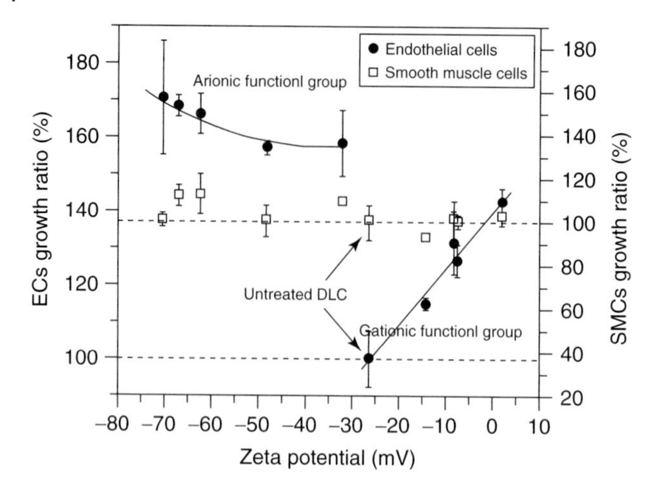

Figure 24.10 Growth ratios of ECs and SMCs as a function of the zeta potential on the DLC films.

such as carboxyl groups, were introduced into the DLC film surface at less than −26 mV using O_2 plasma. Cationic functional groups, such as amino groups, were introduced to the DLC film surface at more than −26 mV using NH_3 plasma. From the results of this study, we found that both anionic and cationic functional groups have high compatibilities with ECs [14].

The mechanisms behind the high compatibilities have not yet been determined. However, we considered that the amount and type of functional groups and the distance between anionic and cationic functional groups affect ECs compatibility. Details are expected to be presented in the near future.

24.6 Conclusion

We proposed future prospects concerning the possibility of DLC films used for medical devices. We aimed at biocompatible DLC films that have both the biocompatibility and durability required for a coronary stent. As a result, we developed a new Si concentration-gradient-type DLC-coated stent. This new DLC film has a superior adhesion feature that prevents cracking on the surface of the coronary stent during expansion. Applying plasma surface treatment techniques on this DLC film, we improved the adhesive property of the drug-containing biodegradable polymer layer and the antithrombotic drug layer such that no allergic action occurred after releasing the drug completely into the blood vessel. We also succeeded in obtaining zeta potentials spread over a wide range by varying the amounts of functional groups introduced to the DLC surface. As a result, we found that by optimizing the amounts of carboxyl and amino groups introduced to the DLC surface, its blood compatibility is enhanced compared with that of untreated DLC films. Additionally, we developed new DLC films having high compatibility with ECs as well as with platelets and TAT III complexes.

This new biomimetic DLC film on medical devices has improved functionality as a result of plasma modification techniques and will provide the opportunity to develop innovative and new medical devices in the near future.

Acknowledgments

The authors wish to thank Dr Shuzou Yamashita of Japan Stent Technology Co. Ltd., the promoter of the joint project on stents, for his multifaceted assistance in carrying out the research. The authors also gratefully acknowledge the kind advice they received on plasma processing from Emeritus Prof. Noriyoshi Sato of Tohoku University. In addition, the evaluation of the biocompatibility compilation is a result of collaboration with Prof. Akira Mochizuki of Tokai University.

References

1. Nakatani, T., Okamoto, K., Araki, A., and Washimi, T. (2006) *New Diamond Front. Carbon Technol.*, **16**, 187.
2. Maguire, P.D., McLaughlin, J.A., Okpalugo, T.I.T., Lemoine, P., Papakonstantinou, P., McAdams, E.T., Needham, M., Ogwu, A.A., Ball, M., and Abbas, G.A. (2005) *Diamond Relat. Mater.*, **14**, 1277.
3. Shirakura, A., Nakaya, M., Koga, Y., Kodama, H., Hasebe, T., and Suzuki, T. (2006) *Thin Solid Films*, **494**, 84.
4. Shinohara, M., Shibata, H., Cho, K., Nakatani, T., Okamoto, K., Matsuda, Y., and Fujiyama, H. (2007) *Appl. Surf. Sci.*, **253**, 6242.
5. Takabayashi, S., Motomitsu, K., Takahagi, T., Terayama, A., Okamoto, K., and Nakatani, T. (2007) *J. Appl. Phys.*, **101**, 103542.
6. Takabayashi, S., Okamoto, K., Shimada, K., Motomitsu, K., Motoyama, H., Nakatani, T., Sakaue, H., Suzuki, H., and Takahagi, T. (2008) *Jpn. J. Appl. Phys.*, **47**, 3376.
7. Nakatani, T., Okamoto, K., Omura, I., and Yamashita, S. (2007) *J. Photopolym. Sci. Technol.*, **20**, 221.
8. Nakatani, T., Abe, Y., Okamoto, K., and Shiraishi, K. (2004) Extended Abstracts of the 2nd International School of Advanced Plasma Technology, Varenna, 2004, p. 91.
9. Sheu, M.S., Hoffman, A.S., Terlingen, J.G.A., and Feijen, J. (1993) *Clin. Mater.*, **13**, 41.
10. Shiraishi, K., Ohnishi, T., and Sugiyama, K. (1998) *Macromol. Chem. Phys.*, **199**, 2023.
11. Okamoto, K., Nakatani, T., Yamashita, S., Takabayashi, S., and Takahagi, T. (2008) *Surf. Coat. Technol.*, **202**, 5750.
12. Nitta, Y., Okamoto, K., Nakatani, T., Hoshi, H., Homma, A., Tatsumi, E., and Taenaka, Y. (2008) *Diamond Relat. Mater.*, **17**, 1972.
13. Nakatani, T., Okamoto, K., Nitta, Y., Mochizuki, A., Hoshi, H., and Homma, A. (2008) *J. Photopolym. Sci. Technol.*, **21**, 225.
14. Nitta, Y., Okamoto, K., Nakatani, T., and Mochizuki, A. (2009) Proceedings of Plasma Science Symposium 2009 and the 26th Symposium on Plasma Processing, Nagoya, 2009, p. 50.

25
Plasma Processing of Nanocrystalline Semiconductive Cubic Boron Nitride Thin Films

Kenji Nose and Toyonobu Yoshida

25.1
Introduction

This chapter provides a brief review of developments in the syntheses of semiconductive cubic boron nitride (cBN) thin films in two decades. Many researchers in this field have been focusing on the mechanical properties of cBN for its use as a wear-resistant coating. On the basis of these studies, films with high cBN contents are fabricated in laboratories. On the other hand, it became evident through studies on bulk single crystals that electric properties of cBN are suitable for new electronic devices operating at higher power and temperature ranges. That is, cBN is one of the promising materials for these cutting-edge devices, which cannot be realized with existing semiconductors. From a purely scientific perspective, the III-group nitride composed of a pair of smallest atoms is a challenging subject in theoretical studies, and interesting reports about the electric states of cBN, hexagonal boron nitride (hBN), and BN nanotubes have recently been published. We consider that the time is now ripe to develop the processing of cBN thin films that can be applied to semiconductor devices. Coincidently, some research groups have begun to publish novel properties of prototype devices and doping experiments.

In this chapter, the state of the art in processing of cBN thin films with an aim to control their electronic properties is presented as follows. First, the ideal properties of cBN are briefly discussed in comparison with those of diamond and hBN in order to show why we choose cBN. Second, growth techniques for high-quality thin films are explained with models of nucleation and developing techniques of heteroepitaxial growth. *In situ* doping processes and modifications of electronic conductivities are explained at the end of this chapter. Finally, future trends of cBN processing are discussed on the basis of current achievements in semiconductive cBN thin films.

Industrial Plasma Technology. Edited by Yoshinobu Kawai, Hideo Ikegami, Noriyoshi Sato, Akihisa Matsuda, Kiichiro Uchino, Masayuki Kuzuya, and Akira Mizuno

ISBN: 978-3-527-32544-3

25.2 Fundamental Properties of cBN

25.2.1 Physical Properties

Basic physical properties of cBN have been clarified in studies of single crystals synthesized through a high-temperature and high-pressure (HTHP) method [1–5]. Basic physical properties of cBN are listed in Table 25.1. Mechanical properties of cBN and hBN, such as hardness and Young's modulus, are often discussed, with diamond and graphite as references, respectively. This is because the sp^3- and sp^2-bonded structures of BN and carbon allotropes renders the characteristics of these compounds similar. Cubic BN has the highest hardness, second to diamond [6]. Accordingly, sintered cBN, which is a composite of cBN and metal binders, is widely applied as a cutting tool for machining. The main advantages of a cBN cutting tool [7, 8] are its stability at high temperatures [9] and its high wear resistance during the machining of ferrous metals.

A hexagonal phase has a smaller hardness than a cubic phase because of the weak sp^2 bond in the hexagonal plane and the van der Waals force between adjacent planes [10, 11]. The macroscopic softness and chemical stability of hBN are utilized in an application of a solid-state lubricant [12]. Although hBN is sometimes referred to as a *white graphite*, its properties, except for mechanical ones – for example, optical reflectivity [13] and electrical conductivity – are far from those of graphite. This is explained by the ionic characteristic of the chemical bond of hBN [14], which accounts for the high chemical stability and electric insulation performance in the intrinsic state [15]. These properties indicate that hBN is a suitable crucible for chemical and metallurgical reactions and insulator of resistive heaters. It is widely known that hBN or turbostratic BN (tBN), which has disturbed alignments of basal planes of hBN, is inevitably grown before the nucleation of the cubic phase during thin-film deposition. This self-organized

Table 25.1 Lattice constants and basic physical properties of cBN, diamond, hBN, and graphite [97–100].

	cBN	Diamond	hBN (c/a)	Graphite (c/a)
Lattice constant (pm)	361.6	356.7	666.1/250.4	670.8/246.1
Atom density ($\times 10^{23}$ cm^{-3})	1.69	1.76	1.11	1.14
T_m(K)	3246	4073	–	>3773
E (GPa)	900	1220	22	10–20
Refractive index: n	2.17	2.4	2.13/1.65	–
ρ(g cm^{-3})	3.45	3.515	2.28	2.25
λ(W m^{-1}K^{-1})	1300	2000	63	50–130
a($\times 10^{-6}$K^{-1})	1.9	1.18	40.5/−2.9	27/−1.0

Table 25.2 Basic electronic properties of cBN, diamond, and hBN [3, 4, 97–99, 101].

	cBN	Diamond	hBN (c/a)
ΔE(eV)	6.4	5.45	5.2
μ_e(cm^{2}V^{-1}s^{-1})	1	2200	–
μ_h(cm^{2}V^{-1}s^{-1})	3, [4, 107]	1600	–
ε (static)	7.1	5.7	5.06/6.85
ε (dynamic)	4.5	5.6 (1 MHz)	4.1/4.95

initial layer often hinders the adhesion of cBN on substrates because sp^3- and sp^2-bonded BNs have far different mechanical properties, such as linear expansion coefficients.

25.2.2 Electronic Properties

The basic electronic properties of cBN are shown in Table 25.2 along with those of diamond and hBN. cBN has unique properties, such as wide bandgap (6.4 eV), high thermal conductivity (1300 W mK^{-1}), and high breakdown field. These fundamental properties are promising for high-power and high-temperature electronics. Another advantage of cBN is found in its dopability. That is, it is possible to dope cBN as both p- and n-types. It is often observed that carrier type is difficult to be controlled in wide-bandgap semiconductors. For example, n-type diamond and p-type zinc oxide had not been achieved for a long time, as seen in the literature. Unlike these wide-bandgap semiconductors, bulk single crystals of cBN exhibit both p- and n-type conductions with the appropriate choice of dopants. Electronic devices such as p–n junction diodes and ultraviolet light emitters fabricated with bulk single crystals [1, 3, 4] are good devices exhibiting the high potential of cBN. Electronic properties of intrinsic cBN and dopants are described below.

25.2.3 Intrinsic Electronic State

Theoretical calculations of energy bands of cBN were carried out by Gubanov *et al.* [16], MacNaughton *et al.* [17], and Xu and Ching [18]. These results showed an identical electronic structure, that is, the indirect energy bandgap of cBN. On the other hand, in hBN, there is almost no difference between the direct and indirect gap energies, suggesting that optical properties such as an efficient light emission could be observed in experimental measurements [19, 20]. The orbital-resolved and total densities of states (DOSs) simulated by these different studies show good agreement with those obtained by X-ray absorption spectroscopy (XAS) and soft

X-ray emission spectroscopy (XES) [17]. They indicate that N2p mainly contributes to the valence band (VB), and the conduction band (CB) originates from B2p. The sp^3 hybridization of boron atoms was also observed at the conduction band. On the other hand, the upper VB exclusively consists of N2p. Replacing boron by IV-group elements is expected to cause n-type conduction, on the basis of this result.

25.2.3.1 **Self-Defects**

The ratio of III/V elements is essential in controlling electric properties of III- and V-group semiconductors, because crystal defects such as interstitials, vacancies, and antisites, often affect carrier generation and doping efficiency. Orellana and Chacham reported the energy levels of nitrogen and boron vacancies (V_N and V_B) [14, 21]. They showed that V_N, the most stable defect in boron-rich crystals, can exist only in the donor state as a cation in the crystal. However, the formation energy of V_N under n-type condition is higher than that of a boron antisite, and it was concluded that V_N cannot serve as an electron source. That is, a theoretical study indicates that V_N is ineffective for carrier generation, although some researchers attributed p-type conduction in nondoped cBN to V_N in their discussions. According to the theory, actually, the V_N stably exists in p-type condition and compensates other acceptors. On the other hand, V_B behaves as a triple acceptor and exists with a low formation energy when the Fermi level is located near the conduction band. This suggests that V_B is a donor compensator in cBN. These formation energies and electric states of self-defects showed good agreements with those obtained by Gubanov *et al.* [16].

N_B and B_N have higher formation energies than other defects except for a case of B_N in an intrinsic or n-type conduction. Orellana and Chacham suggested that potential defect that contributes to carrier generation is a B_N acceptor under B-rich condition [14]. The ionicity of BN is lower than those of other III-group nitride semiconductors because of the small differences in electron orbital characteristics between boron and nitrogen. This fact explains low formation energies of antisites and predicts the electronic conductivity by nonstoichiometric compositions. However, how the electric properties are affected by defects in cBN is an open question at this stage.

25.2.3.2 **Dopants**

Dopants for III–V-group semiconductors are not as simple as those for IV-group semiconductors, in which the valence number exclusively determines the conduction type [22, 23]. To examine potential dopants for cBN, an *ab initio* simulation was performed by Gubanov *et al.* [24]. Energy levels of impurities and directions of lattice relaxations were calculated for Be, Mg, and Si. It was concluded in their study that beryllium and magnesium substituting for a B site contribute to the p-type conduction, and Si in the B site results in the n-type conduction.

Only a limited number of experimental results are published in the field of cBN thin-film doping, as described in Table 25.3. It is often difficult to identify the carrier type, mobility, and concentration directly by Hall measurement because of

Table 25.3 Experimental results of doping of BN thin films.

Dopant	Phase	Carrier type	Electrode	Method	Reference
Si	hBN	N	–	Field emission	Ronning *et al.* [89]
Si	cBN	N	Au/Cr	Hall effect	Yin *et al.* [92]
C	cBN	N	Ag	Seebeck effect	Phani *et al.* [102]
C	cBN	N	–	Field emission	Pyror *et al.* [103]
X[a]	cBN	N	Al	HJ[b] with Si	Zhang *et al.* [104]
S	hBN, cBN	N	Au, Al	HJ with Si	Szmidt *et al.* [105]
S	hBN	N	Au, Al	HJ with Si	Sugino *et al.* [106]
Zn	cBN	P	Ag	Seebeck effect	Nose *et al.* [88]
Be	cBN	P	Al, W	Hall effect	Liao *et al.* [93]
Be	cBN	P	Ag	Hall effect	He *et al.* [94]
Mg	hBN, cBN	P	In	Hall effect	Lu *et al.* [95]
Mg	cBN	P	Ag	Hall effect	Kojima *et al.* [96]
Intrinsic	cBN	P	In	Hall effect	Litvinov *et al.* [107]
X	hBN	P	Ni	HJ with Si	Kimura *et al.* [108]

[a] Dopant X denotes unidentified impurity.
[b] HJ indicates that carrier type was estimated through a device property of a heterojunction diode.

the limited thicknesses of the films and low Hall voltages. Accordingly, electrical properties of doped films were analyzed indirectly, for example, through the device performance of heterojunction diodes [25, 26] and the Seebeck effect. Low mobilities of cBN thin films are often explained by lattice imperfections, such as grain boundaries. Low doping efficiency and trap formation can also account for the small conductivity. However, the nature of nanocrystalline films indicates that the effect of grain boundaries on the electric conduction is different from those in microcrystalline films because of its macroscopic homogeneity. It should be noted that the crystal size of cBN thin films is sometimes less than 10 nm. For this reason, much attention should be paid to crystalline sizes when we compare the electric conduction of cBN thin films with that of micro- or single-crystalline semiconductors.

25.3 Growth of cBN Thin Films

25.3.1 Process and Structure

25.3.1.1 Plasma Processing of cBN Thin Film

It has been revealed that a strong nonequilibrium condition is essential for the nucleation and growth of cBN thin films [27]. That is, the deposition of cBN is characterized by a requirement of ion irradiation on the growing surface [28]. Accordingly, plasma-enhanced chemical and physical vapor depositions (PE-PVD

[29–31] and -CVD [32, 33]) and ion-beam-assisted deposition (IBAD) [28, 34–37] are applied to cBN syntheses. Ion energy is typically chosen from 30 to 1000 eV, and the threshold of momentum transfer per film atom is often described in the equation of $P/a = 200\ (\mathrm{eV\,amu})^{1/2}$ [38]. This equation is consistent in different deposition techniques and ion species. On the other hand, the threshold of substrate temperature exists at around 400 K [39, 40], which is much lower than those of other nitride semiconductors. Because of the limitations of flux and area of ion irradiation from ion beam sources, it is necessary to utilize plasma-enhanced processes when we use cBN thin films in industrial applications.

25.3.1.2 **Initial Layer Growth**

Figure 25.1 shows a typical structure of cBN thin films composed of amorphous/hexagonal/cubic layers observed by a cross-sectional transparent electron microscope (TEM). As widely accepted, this layered structure is formed independently of deposition processes and substrate materials [41, 42]. The thickness of the sp^2-bonded initial layer, that is, the sum of amorphous and hexagonal layer thickness ranges from 20 to 30 nm, and only a few reports about the direct growth of cBN on substrates have been published [34, 43]. In other words, the physics of the formation of the initial layer is still unclarified, and some researchers

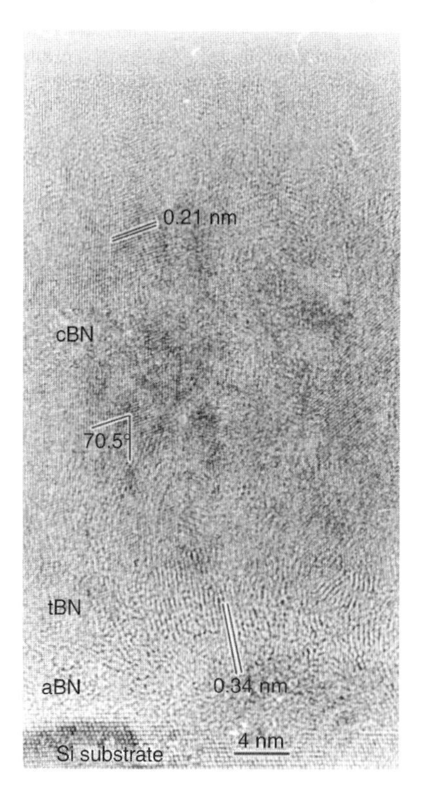

Figure 25.1 TEM image of a typical layered structure of cBN thin film grown on Si [57].

are trying to build a model for cBN nucleation and growth based on widely known experimental findings, that is, the requirement of ion irradiation and the self-organized sp^2-bonded BN. The epitaxial growth of cBN is discussed in detail in Section 25.3.2.

25.3.1.3 Nucleation and Growth of Cubic Phase

The proposed models of cBN nucleation were categorized into four groups by Yoshida [44] and Mirkarimi *et al.* [38]. These four models are typically referred to as *selective sputter*, *subplantation*, *compressive stress*, and *momentum transfer* or *thermal spike models*. In this decade, some new findings related to chemical etching, phases of the top surface, and stress evolution during film growth were reported in this field. They were closely related to model constructions and briefly introduced in this subsection.

The selective sputter model was proposed by Reinke *et al.* [45] and based on the selective etching [46] of the sp^2 phase by ion irradiation. This model does not explain how sp^3 bonds are generated but accounts for the elimination of the sp^2 phase during film growth. It is assumed in this model that the etching rate of sp^3 is lower than that of sp^2 [47, 48], and noncubic phases are removed consistently. In addition to this physical etching model, the chemical etching of the sp^2 bond by fluorine atoms is proposed in a CVD using BF_3 [28, 32, 49]. Stable and thick cBN films with low stresses were formed in this method, indicating a potential way to realize a process without highly energetic species.

It is considered in the subplantation model [50, 51] that high-energy ions penetrate the top surface of BN, resulting in the formation of local dislocations. These dislocations are assumed to function as the seeds of the sp^3-bonded domain in the sp^2 phase. This model is also used to explain the synthesis of diamond-like carbon (DLC) [52] and diamond nucleation on Si [53, 54]. If a local density is the driving force of sp^3 transition in this model, the top surface of cBN is supposed to be capped by an sp^2-bonded layer; however, conflicting results about the surface structures of cBN thin films were reported. Park *et al.* reported that the top surface of cBN was covered by an sp^2-bonded phase [55] by X-ray photoelectron spectroscopy (XPS) [56]. In contrast, TEM observation by Yang *et al.* [57] and the XPS analyses by Nose *et al.* [58] showed no sp^2 bond on the surface. These findings indicate that the density theory in the subplantation model cannot completely explain nucleation.

In the compressive stress model [59], ion irradiation causes inward stress, which results in the transformation of a turbostratic phase to a cubic phase. This theory is often associated with the high residual compressive stress (a few giga pascals) [60] of cBN thin films. *In situ* stress measurements [61, 62] are performed to clarify stress evolution during deposition [63–67]. It was observed in these experiments that the compressive stress increases during nucleation and then decreases with the progress of film growth [62, 68]. Pressure-temperature phase diagrams of BN in HTHP experiments support that the stress threshold of an equilibrium transformation from cBN to hBN is around a few giga pascals [69–72] at around 770 K. However, it is possible to argue that the evolution of film stress is only a result of cubic-phase nucleation and is not the driving force of nucleation.

High-purity cBN thin films are now deposited by multiple deposition methods on the basis of fundamental studies mentioned above. However, a clear description of cBN formation is still difficult to achieve. In addition to these *in situ* measurements and atomic scale analyzes, simulational approaches, such as a molecular dynamics method [73], are essential for describing each stage in nucleation.

25.3.2 Approaches to Epitaxial Growth

It has been an objective of this field to achieve epitaxial cBN by eliminating the initial layer for a long time. This is because direct growth on large crystalline substrates such as silicon, sapphire, and SiC is important for device fabrication. However, there had been no evidence about the direct or epitaxial growth of cBN thin films until recently. In this subsection, two approaches to cBN epitaxy are discussed in detail.

25.3.2.1 For Direct Growth on Silicon

An approach to direct growth on silicon has been developed by Yang *et al.* [74, 75]. They successfully eliminated the initial amorphous layer by a method named the *time-dependent bias technique*. In this method, the substrate bias is tuned continuously from sputtering conditions to deposition conditions. When the thickness of the initial layer is reduced to less than a few nanometers, it is almost impossible to identify tBN by Fourier-transform infrared (FT-IR) absorbance spectroscopy even though this method is exclusive for quantifying sp^2/sp^3 ratio nondestructively. They precisely examined interface structures between cBN and Si [76] by TEM, and proved that tBN can grow directly on silicon substrates without an amorphous layer. Additionally, they showed the direct growth of cBN films on Si at an atomic step composed of (111)–(100) planes later [77]. At this stage, there is no clear report of heteroepitaxy in a large area, but nucleation at the atomic step may lead to the heteroepitaxial growth of cBN on silicon.

25.3.2.2 Epitaxy on Diamond

In 2003, Zhang *et al.* published a novel result of cBN epitaxy on oriented diamond thin films [34]. As shown in Table 25.1, the lattice mismatch between cBN and diamond is approximately 1.4%, which is much smaller than that between cBN and Si boundary, and the deposition on diamond is a promising path for achieving heteroepitaxial growth. Boron, carbon, and nitrogen can form sp^3 bonds in a ternary compound, as expected in the BCN material [12, 78–81], indicating the formation of a stable boundary between BN and diamond. Zhang *et al.*, prepared 10-μm-thick oriented diamond thin films [82] on silicon by microwave-plasma CVD and deposited cBN on them by IBAD. Direct growth is confirmed by FT-IR, Rutherford backscattering (RBS), and electron energy loss spectroscopy (EELS [59, 83]) using an ultrahigh-voltage (1.25 MeV) TEM. The orientation of cBN was examined from TEM images and X-ray diffraction (XRD) spectra; these images and spectra showed the orientation relationship of cBN (001)//diamond (001). IBAD is often performed for cBN growth, and deposition parameters, such as substrate temperature (1173 K) and ion energy (280 eV), are also typically reported. Therefore, it is expected that the ideal surface of diamond is important for realizing epitaxial growth. A growth model of cBN on diamond was established on the basis of the experimental requirements of Zhang *et al.* for epitaxy [84]. They concluded that epitaxy occurs only at a temperature of around 1173 K and showed the necessity for removing the amorphous structure of diamond surfaces and maintaining the crystal orientation before the beginning of deposition. No other research group has reproduced the epitaxy on diamond yet, presumably because large and oriented diamond is not widely available at this stage. The growth of the epitaxial cBN with high orientation and large crystal grain on diamond is an essential method in purely scientific research studies as well as in various applications, such as heterojunction rectification diodes and hard coatings with high adhesion, in the near future.

25.4 Doping Processes and Electrical Characterization

25.4.1 In situ Doping Process

In previous studies on doping, impurity elements were introduced by the thermal evaporation of metals or as gases, such as SiH_4 and H_2S. The method of doping third elements to cBN thin films strongly affects the doping efficiency because substitutional sites, defects, and deposited phases depend on the doping method. This fact has been an open issue and no clear increase in electrical conductivity had been achieved until 2000 even though there are some reports of electric conduction in doped cBN, as listed in Table 25.3. In this section, a novel *in situ* doping method based on sputtering and controlled electronic properties is discussed.

25.4.1.1 Doping by Sputtering of Bulk Sources

Figure 25.2 shows a schematic diagram of an apparatus for sputter doping. In this method, a bulk dopant in the shape of a thin rod is inserted into the plasma during deposition. The typical dimensions of the dopant rod are 3 mm diameter and 120 mm length. Negative DC bias voltages were applied to the rod to enhance sputtering. The dopant concentration was controllable by changing the position of the rod and the applied bias voltages. The distance from the center of the substrate to the tip of the rod was denoted by L (mm) in this section, as shown in Figure 25.2.

The ion current of the rod was measured at various bias voltages at two different positions, as shown in Figure 25.3. That is, a type of single-probe analysis during the deposition of an insulating material was performed. It was shown in (a) that the dopant rod was sputtered continuously in the case of −200 V bias even when

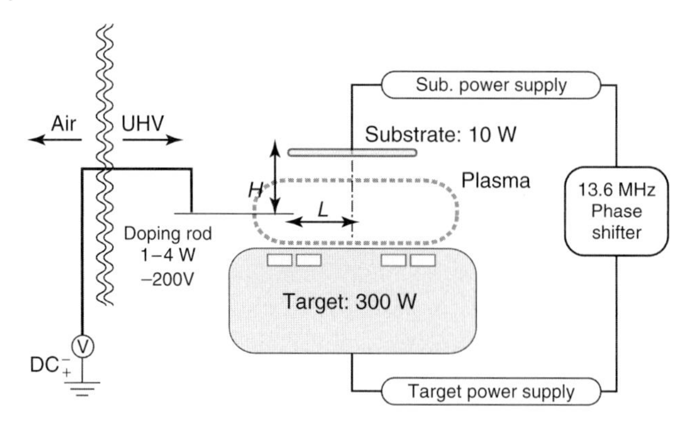

Figure 25.2 Schematic diagram of the sputter doping method. Areas of the target, substrate, and dopant source in this figure are approximately proportional to their applied electronic powers.

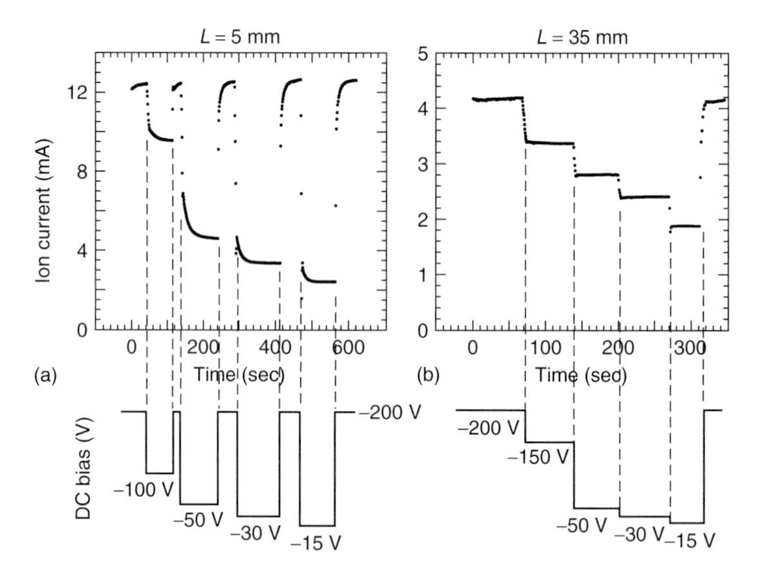

Figure 25.3 Dynamic current-voltage analyses using a Si doping rod as an electric probe. Probe currents are plotted against time at bias voltages from −15 to −200 V. (a) At L = 5 mm, the BN film deposited on the rod surface shields ion acceleration when the negative bias is lower than 100 V. (b) At L = 35 mm, almost no shielding effect is observed at any bias voltages.

the rod was inserted into the area of the maximum deposition rate (L = 5 mm). The surface of the rod remained to be metallic silicon under this condition. The lower bias voltage at L = 5 mm caused a gradual reduction in ion current because of the shielding effect of the deposited BN against ion acceleration. On the other hand, (b) the condition of L = 35 mm, resulted in constant currents independent of time, indicating that no BN film is deposited on the surface even at the bias voltage of −15 V. In conclusion, the surface of the dopant remained to be bare metal during sputter doping by increasing the bias voltage of the rod.

When the dopant source is located near the substrate, inhomogeneous dopant distributions are formed in the substrate plane. The change in dopant concentration with varying *L* shows us how dopant atoms travel in the plasma. Figure 25.4 shows the silicon concentration distributions measured by XPS and theoretical simulation using a Monte Carlo code to describe ion motions in the plasma [85, 86]. It is shown that the random scattering of dopant atoms on the basis of collisions on Ar in the plasma clearly explains the transport of dopant atoms from the source to the film surface. It is also demonstrated that the distribution of a metal with high vapor pressure at the substrate temperature, such as zinc, is predicted by reducing sticking probability from unity in this simulation. The disadvantage of the inhomogeneous distribution of dopant atoms can be overcome by changing the configuration of three electrodes, particularly by increasing the distance between the rod and the substrate.

25.4.1.2 **Effects of Doping on Nucleation and Growth**

Effects of Si [87] and Zn [88] doping on cBN nucleation in sputter doping were examined by controlling dopant concentration and the start time of doping. IR absorbances of sp^2 and sp^3 bonds in Si-doped films were plotted against the rod position in Figure 25.5. The silicon rod was inserted only after the beginning of the deposition ((a) 0 seconds) and after the nucleation of the cubic phase ((b) 120 seconds) in this experiment. That is, the layered structures dependent on whether

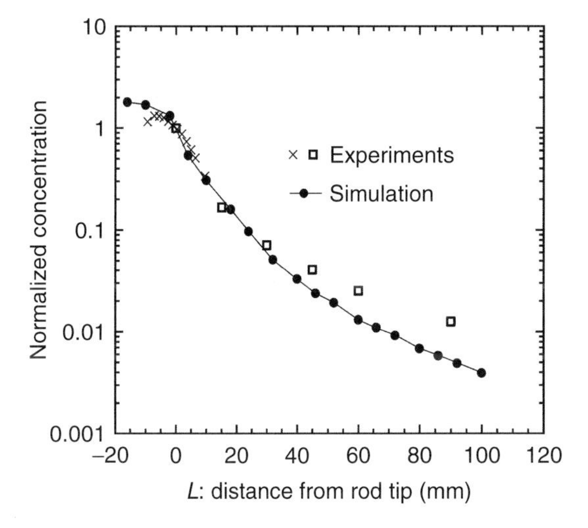

Figure 25.4 Silicon concentrations plotted as a function of the distance from the tip of the dopant rod. Crosses and squares indicate distributions measured on the surface of one film and centers of multiple samples, respectively. The simulational distribution denoted by dots is based on a Monte Carlo code.

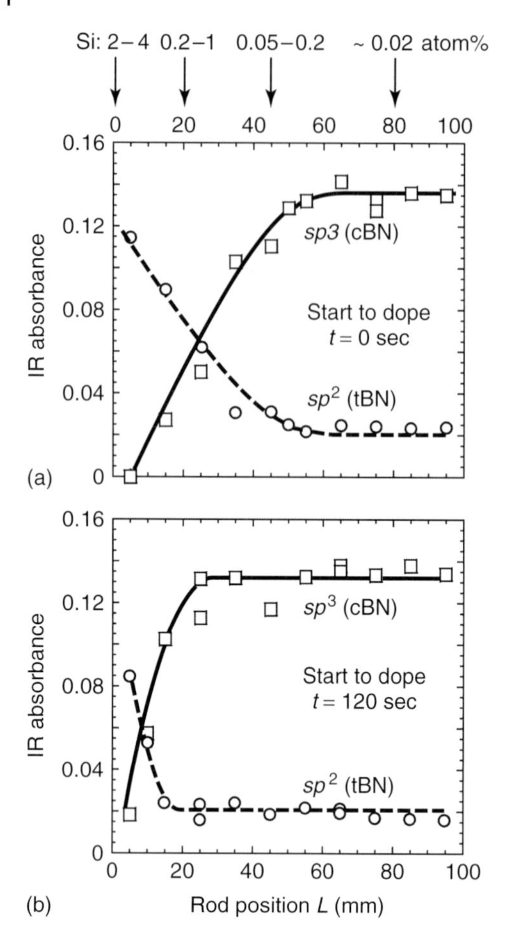

Figure 25.5 IR absorbances of sp^3 and sp^2 plotted against the dopant rod position; L, which is roughly proportional to the dopant concentration shown in Figure 25.4. Doping was started at (a) 0 seconds and (b) 120 seconds after the beginning of the deposition. The conditions of (a) and (b) correspond to the doping before and after the nucleation of cBN, respectively.

doping is started before or after the nucleation of the cubic phase were analyzed. In the case of (a), the silicon rod was inserted up to $L = 50$ mm without causing sp^2 transition on the top surface. The concentration of Si under this condition was about 0.05–0.2 atom %. On the other hand, once the cubic phase was nucleated, phase-pure cubic layers were continuously grown even when the dopant rod was inserted into the position $L = 25$ mm (b), corresponding to Si = 0.2–1 atom %. This fact indicates that the solubility of Si in the cubic phase is higher in the continuous growth stage than in the nucleation stage. In other words, cBN is difficult to nucleate from the Si-doped tBN layer. The transition of sp^3 to sp^2 phase by Si impurity was also reported by Ronning *et al.* [89]. They showed this phenomenon by silane gas or the evaporation of metal silicon in IBAD.

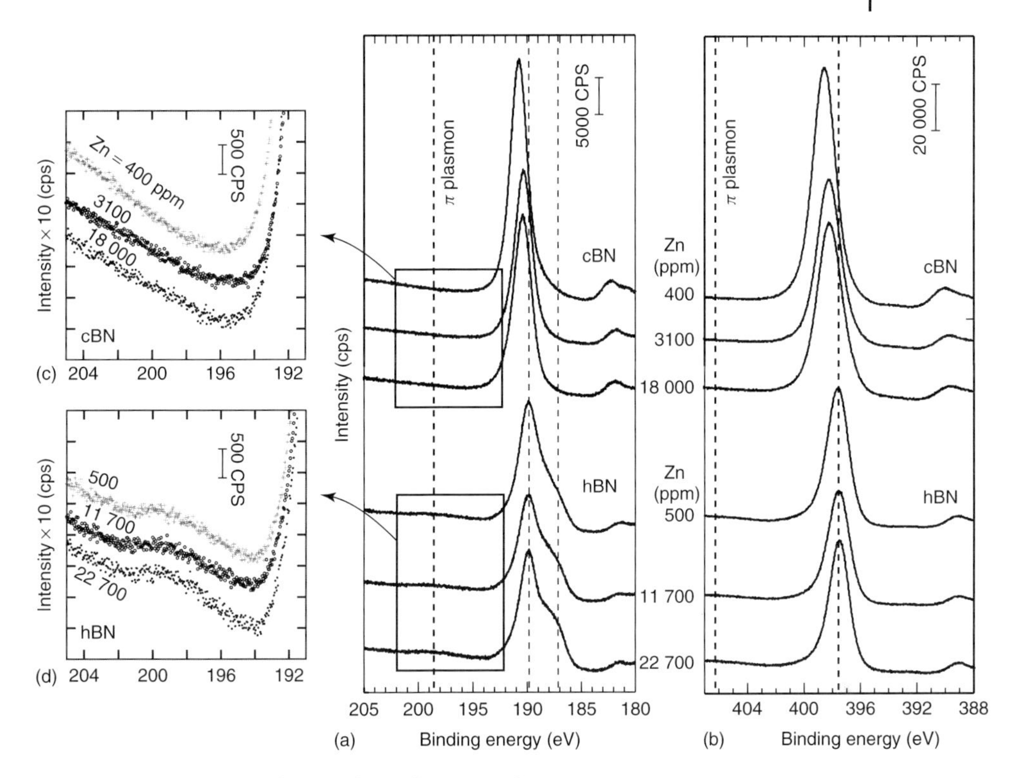

Figure 25.6 XPS spectra of B1s and N1s for cBN and hBN thin films with various zinc concentrations deposited on quartz substrates. Deposited phases were controlled by changing the substrate bias. Binding energies were calibrated by the position of O1s = 532.0 eV. (c) and (d) show the enlarged views of the π-plasmon peaks [88].

Zinc is doped to cBN by the same method. Figure 25.6 shows XPS spectra of B1s and N1s of cubic and hBN thin films [58] with different zinc concentrations grown on quartz substrates. The deposited phases were chosen by controlling substrate biasing. When we started doping after the nucleation of cBN, no transformation of the cubic phase to the turbostratic phase was observed. In this method, Zn-doped phase-pure cBN was deposited on insulating substrates, which is required for evaluating the electric conductivity depending on the dopant concentration.

25.4.2 Electrical Properties of Doped cBN Films

25.4.2.1 **Silicon-Doped Films**

It has been reported that Si-doped single-crystalline cBN exhibits n-type conduction [1, 90, 91]. The specific resistance and electron activation energy were evaluated to be 0.1 to 10^3 Ω cm and 1.1 eV, respectively. That is, Si impurity has a deep energy level, although it can act as an electric donor. Recently, Yin *et al.* have shown the n-type conduction of cBN thin films grown heteroepitaxially on (100) diamond [92]. In their experiments, Si^+ ions were implanted at acceleration energies from 0.5 to 180 keV after the film growth, and the electric conduction by the silicon donor was clearly distinguished from that by crystalline damage in comparison with Ar^+ implantation. They also showed that heavily doped films exhibit metallic conduction as in the case of boron-doped diamond. This is the first report of a controlled n-type conduction of cBN thin films by intentional doping, and this success is attributed to the growth methods for high-crystallinity thin films on diamond substrates. On the other hand, Ronning *et al.*, concluded that silicon cannot function as an electric donor in cBN [89] on the basis of their *in situ* doping experiments. Si-doped films deposited by sputter doping as mentioned above also showed an insulating property. This discrepancy indicates that grain sizes and unintended impurities strongly affect the doping effect in cBN thin films.

25.4.2.2 **Beryllium-Doped Films**

In studies of bulk single crystals, it was shown that beryllium functions as an acceptor in cBN [1, 4]. Liao *et al.* performed beryllium doping to cBN-containing films by thermal evaporation of beryllium powder [93]. Hole concentrations and mobilities as a function of the filament temperature were evaluated in this study, although silicon substrate without an insulating layer could affect the electrical analyses. Beryllium ion implantation was performed by He *et al.*, and a clear increase of electric conductivity of cBN thin film was achieved [94]. It was shown

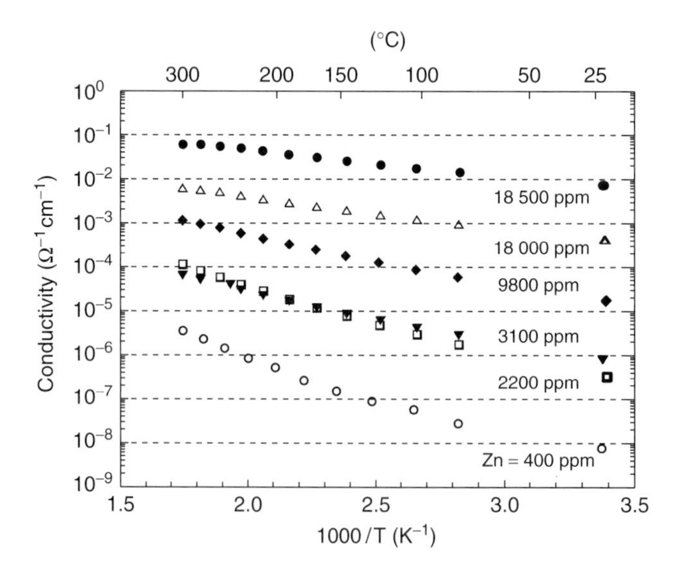

Figure 25.7 Electric conductivities of cBN thin films plotted against the reciprocal temperature as a function of zinc concentration. (From [88].)

that the electric conductivity is enhanced by a factor of 10^5 by applying a rapid thermal annealing after the implantation. The carrier mobility was relatively small ($3\ cm^2V^{-1}s^{-1}$), but a unique activation energy of 0.2 eV, which is close to that evaluated in bulk single crystals, was achieved in the sheet resistance. Beryllium is difficult to be used in industrial applications, but it is a candidate acceptor to achieve p-type as theoretically predicted by Gubanov [24].

25.4.2.3 **Zinc-Doped Films**

It was reported that the electric conductivity of cBN thin films was greatly increased by the sputter doping of zinc. Figure 25.7 shows the conductivity plotted against the reciprocal temperature as a function of zinc concentration [88]. Undoped films showed a conductivity less than $1 \times 10^{-8}\ \Omega^{-1}cm^{-1}$ at room temperature, and the conductivity increased to approximately $1 \times 10^{-2}\ \Omega^{-1}cm^{-1}$ in the film containing 2 atom % zinc. The temperature dependence of the conductivity was positive, indicating a semiconductive property. It is found that $\log(\sigma)$ increases linearly against $1/T$ for each zinc concentration. The conductivity at $1/T = 0$ was estimated from the least-squares lines shown in Figure 25.8, suggesting a constant doping efficiency independent of zinc concentration. Hole conduction was confirmed by evaluating the Seebeck effect in high conductive films, although the carrier type was not reproducibly identified by evaluating the Hall effect. On the other hand, in lightly doped or undoped films, it was impossible to detect thermoelectric current mainly because of their high interelectrode resistance. The

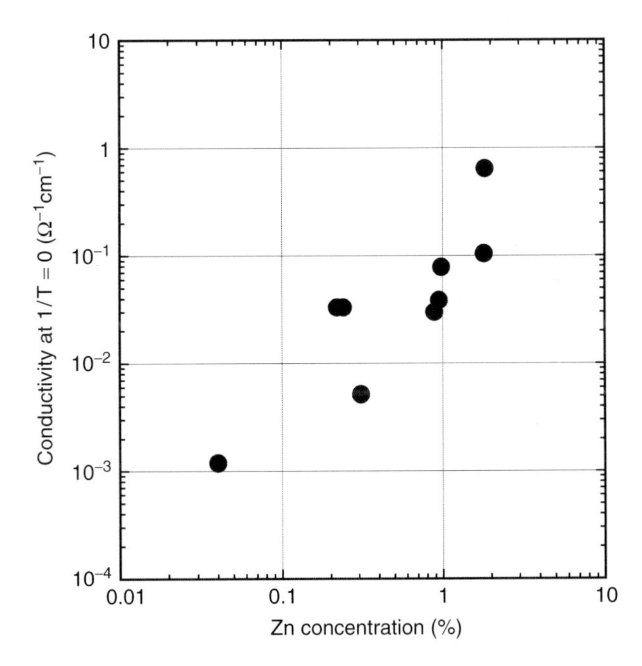

Figure 25.8 Conductivities at $1/T = 0$ estimated from least-squares lines of σ-$1/T$-relations in Figure 25.7.

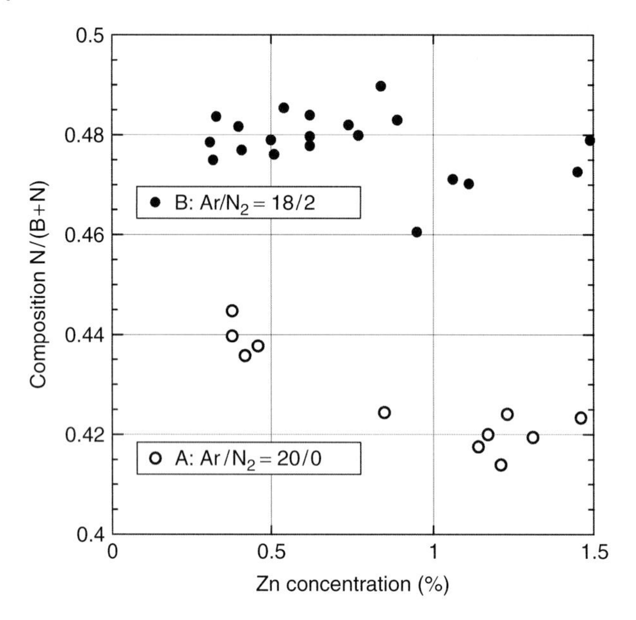

Figure 25.9 N/(B + N) ratios of films deposited with (A) pure Ar and (B) Ar/N_2 = 18 : 2 mixture plotted against zinc concentration. The composition was measured by XPS and calibrated using a commercial pyrolytic BN (Shinetsu). The data are referred from the B/(B + N) ratio in [88] and the converted N/(B + N) ratio to discuss the nitrogen vacancy in the films.

specific resistance of a film with 0.65 atom % zinc was approximately 10^5 Ωcm at room temperature. This is equivalent to the resistance of 2×10^{10} Ω between two electrodes in these films, making it difficult to detect currents generated by thermoelectric voltages less than a few hundred millivolts in these lightly doped films.

It is interesting to note that the effect of zinc doping on the electric conductivity was not observed when the gas composition was changed from (A) pure Ar to (B) a mixture gas of Ar/N_2=18/2 in the sputtering. The compositional change of the deposited films is shown in Figure 25.9. The N/(B + N) ratio decreased as the zinc concentration increased in the case of (A), indicating that doped zinc replaces the nitrogen site. The change in electric conductivity was observed only in these nonstoichiometric films. On the other hand, the composition of films (B) remained constant for zinc concentrations from 0.3 to 1.5%. In these stoichiometric films, the electric conductivity was unchanged from that of undoped films even if the zinc concentration was increased up to 2 atom %. It is unclear whether the zinc at the nitrogen site in cBN causes p-type conduction at this stage, but we expect that the control of the B/N ratio will be essential for yielding the electric conductivity in cBN films. Further studies involving theoretical band calculations should be conducted to clarify the relationship between the substitutional site and carrier generation in cBN.

25.4.2.4 **Magnesium-Doped Films**

Magnesium was doped to boron nitride thin films by Lu *et al.* [95], but no effect of the magnesium dopant on the electric conduction was observed presumably because of high concentrations of contaminations and the low concentration of the cubic phase, which was less than 33% in IR absorption. Kojima *et al.* applied sputter doping to magnesium and reached a high electric conductivity, which was dependent on magnesium concentration [96]. It is worth noting that they confirmed phase-pure cubic growth by in-plane XRD, indicating that magnesium was successfully doped to cBN by sputter doping when its concentration is less than 1 atom %. In a Hall-effect measurement of a heavily doped film, hole concentration and mobility were 4×10^{14} cm^{-3} and $6\ cm^2V^{-1}s^{-1}$, respectively, even though the Hall voltage was not reproduced at different temperatures and magnesium concentrations because of the low S/N ratio caused by the high resistance between two electrodes. Activation energies of electric conductivity depended on the magnesium concentration and were relatively lower than those of zinc. On the other hand, the doping effect was observed only in the films with boron-rich composition as was the case of zinc doping. It was mentioned in this report that magnesium replacement in the nitrogen site is a possible explanation for hole doping on the basis of the change of $N/(B + N)$ ratio of electric conductive films. Calculations of impurity states by Gubanov *et al.* clearly showed the p-type conduction of magnesium in the boron site [24]. Thus, it is interesting to calculate electronic structure of magnesium and zinc in the nitrogen site in a future study.

25.5 Conclusion

Deposition and doping, and electronic properties of doped cBN thin films were reviewed in this chapter. cBN has promising physical properties for use as electronic devices in the near future, and considerable attention is paid to controlling conductivity, carrier type, and mobility, which is based on the deposition of high-quality thin films. It is almost evident that the high electric conductivity of thin films is achieved by various doping methods. Silicon ion implantation on heteroepitaxial films on diamond has been one of the remarkable methods developed in recent years. These films can be applied to fabricate heterojunction pn-diodes between Si-doped cBN and boron-doped diamond, which can cover the low conductivity of n-type diamond.

It was shown that the *in situ* sputter doping of zinc and magnesium successfully results in high electric conductivities. This method is applicable to all plasma-based deposition processes, and it is challenging to apply this method to the deposition processes except for sputtering. In addition, it is an interesting topic to examine the effect of B/N composition on the doping efficiency, which should be approached through the evaluations of chemical bonds and electric conductivities of high-quality cBN bulks and thin films in the near future.

References

1. Mishima, O., Tanaka, J., Yamaoka, S., and Fukunaga, O. (1987) *Science*, **238**, 181.
2. Mishima, O., Yamaoka, S., and Fukunaga, O. (1987) *J. Appl. Phys.*, **61**, 2822.
3. Taniguchi, T., Teraji, T., Koizumi, S., Watanabe, K., and Yamaoka, S. (2002) *Jpn. J. Appl. Phys.*, **41**, L109.
4. Taniguchi, T., Koizumi, S., Watanabe, K., Sakaguchi, I., Sekiguchi, T., and Yamaoka, S. (2003) *Diamond Relat. Mater.*, **12**, 10983.
5. Nistor, L.C., Nistor, S.V., Dinca, G., Georgeoni, P., Landuyt, J.V., Manfredotti, C., and Vittone, E. (2002) *J. Phys.: Condens. Matter*, **14**, 10983.
6. Yang, H.S. and Yoshida, T. (2005) *Surf. Coat. Technol.*, **200**, 984.
7. Keunecke, M., Wiemann, E., Weigel, K., Park, S.T., and Bewilogua, K. (2006) *Thin Solid Films*, **515**, 967.
8. Remadna, M. and Rigal, J.F. (2006) *J. Mater. Process. Technol.*, **178**, 67.
9. Tokoro, H., Fujii, S., and Oku, T. (2004) *Diamond Relat. Mater.*, **13**, 1139.
10. Bosak, A., Serrano, J., Krisch, M., Watanabe, K., Taniguchi, T., and Kanda, H. (2006) *Phys. Rev. B*, **73**, 041402.
11. Iwamoto, C., Yang, H.S., Watanabe, S., and Yoshida, T. (2003) *Appl. Phys. Lett.*, **83**, 4402.
12. Hasegawa, T., Yamamoto, K., Kakudate, Y., and Ban, M. (2003) *Surf. Coat. Technol.*, **169**, 270.
13. Cappellini, G., Satta, G., Palummo, M., and Onida, G. (2001) *Phys. Rev. B*, **64**, 035104.
14. Orellana, W. and Chacham, H. (2001) *Phs. Rev. B*, **63**, 125205.
15. Nose, K., Yang, H.S., Oba, H., and Yoshida, T. (2005) *Diamond Relat. Mater.*, **14**, 1960.
16. Gubanov, V.A., Lu, Z.W., Klein, B.M., and Fong, C.Y. (1996) *Phys. Rev. B*, **53**, 4377.
17. MacNaughton, J.B., Moewes, A., Wilks, R.G., Zhou, X.T., Sham, T.K., Taniguchi, T., Watanabe, K., Chan, C.Y., Zhang, W.J., Bello, I., Lee, S.T., and Hofsäss, H. (2005) *Phys. Rev. B*, **72**, 195113.
18. Xu, Y.N. and Ching, W.Y. (1991) *Phys. Rev. B*, **44**, 7787.
19. Watanabe, K., Taniguchi, T., and Kanda, H. (2004) *Nat. Mater.*, **3**, 404.
20. Watanabe, K., Taniguchi, T., Kuroda, T., and Kanda, H. (2006) *Appl. Phys. Lett.*, **89**, 141902.
21. Orellana, W. and Chacham, H. (1999) *Appl. Phys. Lett.*, **74**, 2984.
22. Sze, S.M. (1981) *Physics of Semiconductor Devices*, 2nd edn, John Wiley & Sons, Inc., ISBN: 0-471-05661-8.
23. Morkoç, H. (1999) *Nitride Semiconductors and Devices*, Springer, ISBN: 4-621-04481-8.
24. Gubanov, V.A., Pentaleri, E.A., Fong, C.Y., and Klein, B.M. (1997) *Phys. Rev. B*, **56**, 13077.
25. Nose, K., Tachibana, K., and Yoshida, T. (2003) *Appl. Phys. Lett.*, **83**, 943.
26. Nose, K., Yang, H.S., and Yoshida, T. (2005) *Diamond Relat. Mater.*, **14**, 1297.
27. Weissmantel, C., Bewilogua, K., Dietrich, D., Erler, H.J., Hinneberg, H.J., Klose, S., Nowick, W., and Reisser, G. (1980) *Thin Solid Films*, **72**, 19.
28. Kester, D.J. and Messier, R. (1992) *J. Appl. Phys.*, **72**, 504.
29. Abendroth, B., Gago, R., Kolitsch, A., and Möller, W. (2004) *Thin Solid Films*, **447**, 131.
30. Mieno, M. and Yoshida, T. (1990) *Jpn. J. Appl. Phys. Part 2*, **29**, L1175.
31. Ulrich, S., Scherer, J., Schwan, J., Barzen, I., Jung, K., Scheib, M., and Ehrhardt, H. (1996) *Appl. Phys. Lett.*, **68**, 909.
32. Zhang, W.J. and Matsumoto, S. (2000) *Chem. Phys. Lett.*, **330**, 243.
33. Ichiki, T. and Yoshida, T. (1994) *Appl. Phys. Lett.*, **64**, 851.
34. Zhang, X.W., Boyen, H.G., Deyneka, N., Ziemann, P., Banhart, F., and Schreck, M. (2003) *Nat. Mater.*, **2**, 312.
35. Fitz, C., Kolitsch, A., Möller, W., and Fukarek, W. (2002) *Appl. Phys. Lett.*, **80**, 55.

36. Litvinov, D. and Clarke, R. (1997) *Appl. Phys. Lett.*, **71**, 1969.
37. Hofsäss, H., Ronning, C., Griesmeier, U., Gross, M., Reinke, S., and Kuhr, M. (1995) *Appl. Phys. Lett.*, **67**, 46.
38. Mirkarimi, P.B., McCarty, K.F., and Medlin, D.L. (1997) *Mater. Sci. Eng. R*, **21**, 47.
39. Hofsäss, H., Feldermann, H., Sebastian, M., and Ronning, C. (1997) *Phys. Rev. B*, **55**, 13230.
40. Hofsäss, H., Eyhusen, S., and Ronning, C. (2004) *Diamond Relat. Mater.*, **13**, 1103.
41. Inagawa, K., Watanabe, K., Ohsone, H., Saitoh, K., and Itoh, A. (1989) *Surf. Coat. Technol.*, **39**, 253.
42. Jimenez, I., Jankowski, A.F., Terminello, L.J., Sutherl, D.G.J., Carlisle, J.A., Doll, G.L., Tong, W.M., Shuh, D.K., and Himpsel, F.J. (1997) *Phys. Rev. B*, **55**, 12025.
43. Feldermann, H., Ronning, C., Hofsäss, H., Huang, Y.L., and Seibt, M. (2001) *J. Appl. Phys.*, **90**, 3248.
44. Yoshida, T. (1997) *Diamond Films Technol.*, **7**, 87.
45. Reinke, S., Kuhr, M. and Kulisch, W. (1994) *Diamond Relat. Mater.*, **3**, 341.
46. Chapman, B. (1980) *Glow Discharge Processes*, John Wiley & Sons, Inc., ISBN: 0-471-07828-X.
47. Harris, S.J., Doll, G.L., Chance, D.C., and Weiner, A.M. (1995) *Appl. Phys. Lett.*, **67**, 2314.
48. Harris, S.J., Weiner, A.M., Doll, G.L., and Meng, W.J. (1997) *J. Mater. Res.*, **12**, 412.
49. Zhang, W.J., Chan, C.Y., Chan, K.M., Bello, I., Lifshitz, Y., and Lee, S.T. (2003) *Appl. Phys. A*, **76**, 953.
50. Dworschak, W., Jung, K., and Ehrhardt, H. (1995) *Thin Solid Films*, **254**, 65.
51. Lifshitz, Y., Kasi, S.R., and Rabalais, J.W. (1990) *Phys. Rev. B*, **41**, 10468.
52. Robertson, J. (1993) *Diamond Relat. Mater.*, **2**, 984.
53. Lifshitz, Y., Kasi, S.R., and Rabalais, J.W. (1989) *Phys. Rev. Lett.*, **62**, 1290.
54. Lifshitz, Y., Meng, X.M., Lee, S.T., Akhveldiany, R., and Hoffman, A. (2004) *Phys. Rev. Lett.*, **93**, 056101.
55. Park, K.S., Lee, D.Y., Kim, K.J., and Moon, D.W. (1997) *Appl. Phys. Lett.*, **70**, 315.
56. Trehan, R., Lifshitz, Y., and Rabalais, J.W. (1990) *J. Vac. Sci. Technol. A*, **8**, 4026.
57. Yang, H.S., Iwamoto, C., and Yoshida, T. (2002) *Thin Solid Films*, **407**, 67.
58. Nose, K., Oba, H., and Yoshida, T. (2006) *Appl. Phys. Lett.*, **89**, 112124.
59. McKenzie, D.R., Cockayne, D.J.H., Möller, D.A., Murakawa, M., Miyake, S., Watanabe, S., and Fallon, P. (1991) *J. Appl. Phys.*, **70**, 3007.
60. Cardinale, G.F., Howitt, D.G., McCarty, K.F., Medlin, D.L., Mirkarimi, P.B., and Moody, N.R. (1996) *Diamond Relat. Mater.*, **5**, 1295.
61. Friesen, C. and Thompson, C.V. (2002) *Phys. Rev. Lett.*, **89**, 126103.
62. Fitz, C., Fukarek, W., Kolitsch, A., and Möller, W. (2000) *Surf. Coat. Technol.*, **128**, 474.
63. Klett, A., Malave, A., Freudenstein, R., Plass, M.F., and Kulisch, W. (1999) *Appl. Phys. A*, **69**, 653.
64. Klett, A., Freudenstein, R., Plass, M.F., and Kulisch, W. (2001) *Diamond Relat. Mater.*, **10**, 1875.
65. Kim, H.S. and Baik, Y.J. (2004) *J. Appl. Phys.*, **95**, 3473.
66. Klett, A., Freudenstein, R., Plass, M.F., and Kulisch, W. (1999) *Surf. Coat. Technol.*, **116**, 86.
67. Djouadi, M.A., Mortet, V., Khandozhko, S., Jouan, P.Y., and Nouet, G. (2001) *Surf. Coat. Technol.*, **142**, 899.
68. McKenzie, D.R., McFall, W.D., Sainty, W.G., Davis, C.A., and Collins, R.E. (1993) *Diamond Relat. Mater.*, **2**, 970.
69. Fitz, C., Fukarek, W., Kolitsch, A., and Möller, W. (2000) *Surf. Coat. Technol.*, **128**, 292.
70. Solozhenko, V.L. (1994) *Diamond Relat. Mater.*, **4**, 1.
71. Corrigan, F.R. and Bundy, F.P. (1975) *J. Chem. Phys.*, **63**, 3812.
72. Fukunaga, O. (2000) *Diamond Relat. Mater.*, **9**, 7.
73. Koga, K., Nakamura, Y., Watanabe, S., and Yoshida, T. (2001) *Sci. Technol. Adv. Mater.*, **2**, 349.

74. Yang, H.S., Iwamoto, C., and Yoshida, T. (2003) *J. Appl. Phys.*, **94**, 1248.
75. Iwamoto, C., Yang, H.S., and Yoshida, T. (2002) *Diamond Relat. Mater.*, **11**, 854.
76. Zhang, X.W., Boyen, H.-G., Yin, H., and Banhart, F. (2005) *Diamond Relat. Mater.*, **14**, 1474.
77. Yang, H.S., Iwamoto, C., and Yoshida, T. (2007) *Diamond Relat. Mater.*, **16**, 642.
78. Okada, K., Kimura, C., and Sugino, T. (2006) *Diamond. Relat. Mater.*, **15**, 1000.
79. Kimura, C., Shima, H., Okada, K., Funakawa, S., and Sugino, T. (2005) *J. Vac. Sci. Technol. B*, **23**, 1948.
80. Gao, J. and Hou, Q.R. (1997) *Mod. Phys. Lett. B*, **11**, 749.
81. Kurapov, D., Neuschütz, D., Cremer, R., Pedersen, T., Wutting, M., Dietrich, D., Marx, G., and Schneider, J.M. (2003) *Vacuum*, **68**, 335.
82. Shiomi, H., Tanabe, K., Nishibayashi, Y., and Fujimori, N. (1990) *Jpn. J. Appl. Phys.*, **29**, 34.
83. Widmayer, P., Ziemann, P., and Boyen, H.-G. (1998) *Diamond Relat. Mater.*, **7**, 385.
84. Zhang, X.W., Boyen, H.-G., Ziemann, P., Ozawa, M., Banhart, F., and Schreck, M. (2004) *Diamond Relat. Mater.*, **13**, 1144.
85. Nakano, T. and Baba, S. (2006) *Vacuum*, **80**, 646.
86. Ito, T. and Cappeli, M.A. (2007) *Appl. Phys. Lett.*, **90**, 101503.
87. Oba, H., Nose, K., and Yoshida, T. (2007) *Surf. Coat. Technol.*, **201**, 5502.
88. Nose, K. and Yoshida, T. (2007) *J. Appl. Phys.*, **102**, 063711.
89. Ronning, C., Banks, A.D., McCarson, B.L., Schlesser, R., Sitar, Z., Ward, B.L., and Nemanich, R.J. (1998) *J. Appl. Phys.*, **84**, 5046.
90. Wang, C.X., Yang, G.W., Zhang, T.C., Liu, H.W., Han, Y.H., Luo, J.F., Gao, C.X., and Zou, G.T. (2003) *Appl. Phys. Lett.*, **83**, 4854.
91. Wang, C.X., Zhang, T.C., Liu, H.G., Gao, C.X., and Zou, G.T. (2002) *J. Phys. Condens. Matter*, **14**, 10989.
92. Yin, H., Pongrac, I., and Ziemann, P. (2008) *J. Appl. Phys.*, **104**, 023703.
93. Liao, K.J., Wang, W.L., and Kong, C.Y. (2001) *Surf. Coat. Technol.*, **141**, 216.
94. He, B., Zhang, W.J., Zou, Y.S., Chong, Y.M., Ye, Q., Ji, A.L., Yang, Y., Bello, I., Lee, S.T., and Chen, G.H. (2008) *Appl. Phys. Lett.*, **92**, 102108.
95. Lu, M., Bousetta, A., Bensaoula, A., Waters, K., and Schultz, J.A. (1996) *Appl. Phys. Lett.*, **68**, 622.
96. Kojima, K., Nose, K., Kambara, M., and Yoshida, T. (2009) *J. Phys. D*, **42**, 055304.
97. Karim, M.Z., Cameron, D.C., and Hashmi, M.S.J. (1993) *Surf. Coat. Technol.*, **60**, 502.
98. Siklitsky, Vadim (2002) New Semiconductor Materials. Characteristics and Properties. Electronic archive. Ioffe Physico-Technical Institute. 3 Sept. 2002. Web. *http://www.ioffe.ru/SVA/NSM/Semicond/BN/index.html*. 18 Dec. 2009
99. National Astronomical Observatory (1999) *Rika Nenpyo, Chronological Scientific Tables*, Maruzen, ISBN: 4-621-04519-9.
100. Lide, D.R. (2003) *CRC Handbook of Chemistry and Physics*, 83rd edn, CRC Press LLC, ISBN: 0-8493-0483-0.
101. Mohammad, S.N. (2002) *Solid-State Electron.*, **46**, 203.
102. Phani, A.R., Manorama, S., and Rao, V.J. (1995) *Semicond. Sci. Technol.*, **10**, 1520.
103. Pryor, R.W. (1996) *Appl. Phys. Lett.*, **68**, 1802.
104. Zhang, X.W., Zou, Y.J., Yan, H., Wang, B., Chen, G.H., and Wong, S.P. (2000) *Mater. Lett.*, **45**, 111.
105. Szmidt, J., Werbowy, A., Jarzebowski, L., Gebicki, T., Petrakova, I., Sokolowska, A., and Olszyna, A. (1996) *J. Mater. Sci.*, **31**, 2609.
106. Sugino, T., Tanioka, K., Kawasaki, S., and Shirafuji, J. (1998) *Diamond Relat. Mater.*, **7**, 632.
107. Litvinov, D., Taylor, C.A.II, and Clarke, R. (1998) *Diamond Relat. Mater.*, **7**, 360.
108. Kimura, C., Yamamoto, T., and Sugino, T. (2001) *Diamond Relat. Mater.*, **10**, 1404.

26
Fundamentals on Tribology of Plasma-Deposited Diamond-Like Carbon films

Julien Fontaine and Christophe Donnet

26.1
Introduction

Increasing demands for improved performance, longer durability, and greater efficiency in mechanical systems are pushing current materials to their limits. To achieve such goals, protective and/or solid lubricant coatings are becoming prevalent. Such coatings can be divided into two families (Figure 26.1): "soft coatings," with low friction but some wear (or solid lubricant coatings); and "hard coatings," with low wear but high friction (or protective coatings) [1].

Solid lubricant coatings include polymers, soft metals, or lamellar solids (graphite, MoS_2, etc.), which are easily sheared, while antiwear coatings are made of hard ceramics. The threshold between these two families is about 10 GPa in hardness – close to the hardness of ball-bearing or tool steels, usual structural materials for mechanical systems – and about 0.3 in friction – close to lower friction values for uncoated and unlubricated metals.

Combining low friction and high wear resistance in "dry" tribological contacts is indeed a great challenge. However, some carbon-based materials combine to some extent not only in low friction but also high wear resistance. These materials are widely known as *diamond* and *diamond-like carbon* (*DLC*) coatings. They are usually harder than most metals and/or alloys, thus, affording very high wear resistance and, at the same time, impressive friction coefficients, generally between 0.05 and 0.22. Friction values lower than 0.01 have even been reported [2, 3], a sliding regime often referred to as "*superlubricity*" or "*superlow friction.*"

26.2
Special Case of DLC Coatings

To account for such exceptional tribological behavior, some fundamental processes leading to energy dissipation in dry sliding must be considered (Figure 26.2):

Industrial Plasma Technology. Edited by Yoshinobu Kawai, Hideo Ikegami, Noriyoshi Sato, Akihisa Matsuda, Kiichiro Uchino, Masayuki Kuzuya, and Akira Mizuno

ISBN: 978-3-527-32544-3

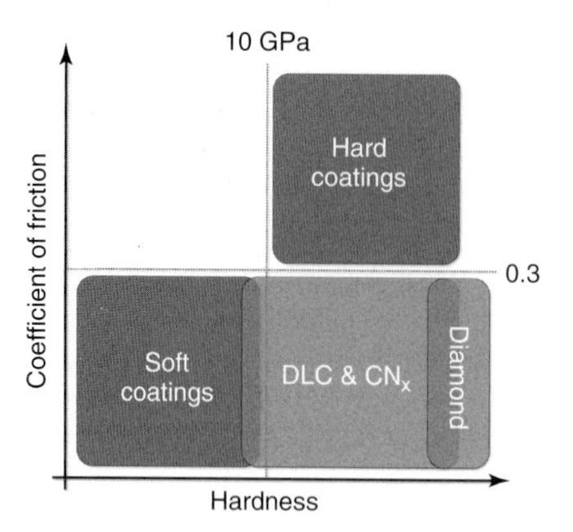

Figure 26.1 Classification of coatings with respect to hardness and coefficient of friction, highlighting the special case of carbon-based coatings.

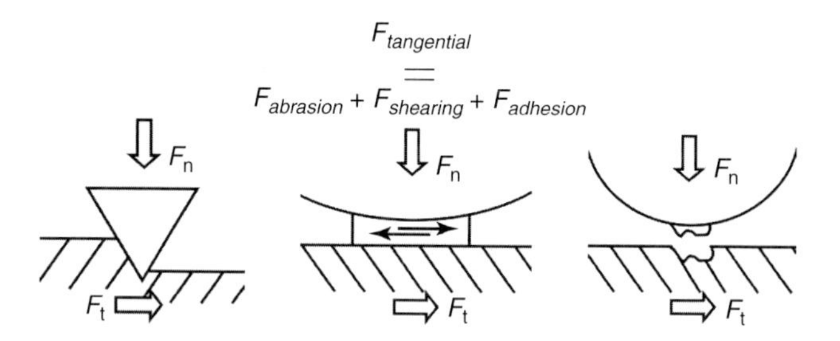

Figure 26.2 Fundamental contributions to tangential force F_t under a normal force F_n.

- **Abrasion:** plowing or scratching of surfaces by asperities or debris is not likely to occur, thanks to smoothness and hardness of DLC films;
- **Shearing:** plastic flow of asperities or interfacial material is limited, due to high hardness of DLC and relatively small amounts of transfer material (compared to other solid lubricants);
- **Adhesion:** release of adhesive junction between sliding counterfaces could be the major contribution to friction of DLC coatings. Surface nature is thus prevalent, and the weak van der Waals interactions provided by –H or –OH passivated surfaces account for the low friction of these materials in most environments.

26.3
Superlubricity of DLC Coatings

Hydrogenated amorphous carbon films, deposited by plasma-enhanced CVD processes, may indeed provide extremely low friction coefficients (<0.01) under ultrahigh vacuum, thanks to hydrogen passivation of surfaces (Figure 26.3) [4, 5].

However, adhesive phenomena are more complex and seem to control the friction evolution of DLC coatings under vacuum. Indeed, thanks to adhesion at

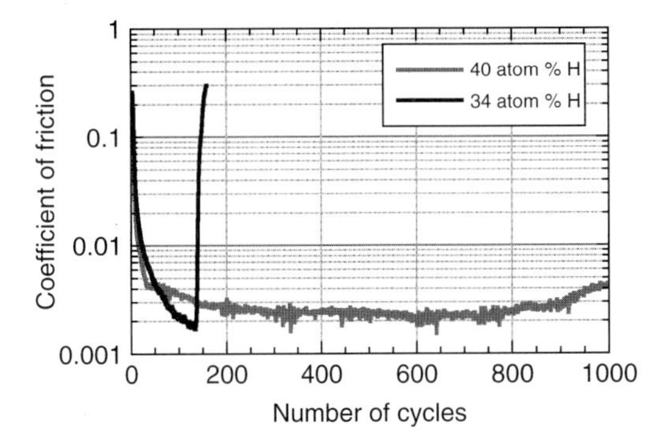

Figure 26.3 Friction coefficient in ultrahigh vacuum for DLC coatings with different hydrogen content.

the early stages of sliding, a very thin tribofilm is formed on metallic counterface, allowing weak interactions and thus friction decrease.

The loss of superlow friction for the less hydrogenated films is also related to adhesion. However, the progressive wear of nanoscale surface roughness – controlling the size of adhesive junctions – appears to be critical. When the surface becomes smoother, large adhesive junctions appear and have to be released. Depending on the mechanical properties of both counterfaces, these junctions might be released in softer DLC (sample with 40 atom % H) – increasing wear but allowing to maintain low friction, thanks to weak interactions – or in steel pin in case of harder DLC (sample with 34 atom % H), damaging both counterfaces and leading to the drastic friction increase observed in Figure 26.3.

Nevertheless, in the presence of hydrogen gas during friction, it is possible to restore or even avoid such dramatic adhesive phenomena, thanks to friction-induced reaction with DLC surface, which lead to passivation of sliding surfaces.

26.4 Conclusion

Combination of high mechanical properties with weak surface interactions accounts for the low friction and for the high wear resistance of DLC coatings. Such combination would indeed minimize the three fundamental contributions to friction force: abrasion, shearing, and adhesion. Adhesion appears to control friction, both through the nature of surface interactions, and also through surface roughness and size of adhesive junctions.

References

1. Holmberg, K. and Matthews, A. (1994) *Coatings Tribology – Properties, Techniques and Applications in Surface Engineering*, Elsevier, Amsterdam.
2. Donnet, C. and Erdemir, A. (eds) (2008) *Tribology of Diamond-Like Carbon Films – Fundamentals and Applications*, Springer, New York.

3. Erdemir, A. and Martin, J.M. (eds) (2007) *Superlubricity*, Elsevier, Amsterdam.
4. Fontaine, J., Belin, M., Le Mogne, T., and Grill, A. (2004) *Tribol. Int.*, **37**, 869.
5. Fontaine, J., Le Mogne, T., Loubet, J.L., and Belin, M. (2005) *Thin Solid Films*, **482**, 99.

27
Diamond-Like Carbon Thin Films Grown in Pulsed-DC Plasmas

Enric Bertran, Miguel Rubio-Roy, Carles Corbella, and José-Luis Andújar

27.1
Introduction

RF discharges have been used for years as synthesis technologies to produce amorphous carbon and hydrogenated amorphous carbon coatings for diverse applications. This technology is based on capacitive plasma-enhanced chemical vapor deposition (PECVD), where carbon films are usually grown on the cathode. The deposition conditions can be chosen according to several useful criteria. For example, negative self-bias induces changes on the a-C microstructure and properties. Also, a low self-bias provides polymeric and transparent films, whereas a high-impact energy of ions reaching the substrate induces graphitization of the films (optically opaque). Only intermediate values of impact ion energy can provide extreme amorphous carbon mechanical properties (hardness and elasticity as main). Hence, self-bias becomes an important deposition parameter which can be controlled through electrode asymmetry and rf power. However, in spite that the technology based on RF discharges has been studied in the past and was used to deposit films as protective coatings in a huge amount of industrial applications, requiring low processing temperatures, it shows significant limitations such as low deposition rate, high mechanical stress, and low adhesion to the substrate, especially on polymers.

The industrial interest for pulsed-DC technology arises from anytime major availability of high-power pulsed-DC supplies generating pulsed-DC plasmas at higher-power levels.

From the industrial point of view, the advantages of this technology are related to the possibility of operation without impedance matching networks and the easy implementation to plasma processes, both for chemical vapor deposition (CVD) as for physical vapor deposition (PVD) processes.

Focusing on the possibilities, pulsed-DC technology adds new extra control parameters such as frequency, peak width, and peak voltage, which can increase the plasma density, the electronic temperature, and the energy of ion bombardment.

Among the pulsed-power supplies, asymmetric bipolar pulsed-DC power is an expanding technology to increase the ion density and bombardment energy in the

Industrial Plasma Technology. Edited by Yoshinobu Kawai, Hideo Ikegami, Noriyoshi Sato, Akihisa Matsuda, Kiichiro Uchino, Masayuki Kuzuya, and Akira Mizuno

ISBN: 978-3-527-32544-3

processes of PECVD and reactive magnetron sputtering. The aim of the present study is to evidence the possibilities of this technology when applied to grow diamond-like carbon (DLC) thin films. For that, we present the effects of varying the pulsed-DC parameters on the film growth and on the structural properties of DLC coatings. These effects have been discussed and compared to those of films produced by conventional PECVD in the same deposition system.

Asymmetric bipolar pulsed-DC power increases the ion density and the bombardment energy of deposition processes using plasma. Reactive magnetron sputtering and PECVD expand their possibilities because asymmetric bipolar pulsed-DC technology provides extra control parameters and growing conditions with a higher power density than conventional RF plasma processes. Particularly, in RF plasma processes, the reactor geometry influences the self-bias voltage. Thus, a greater asymmetry between the powered and the grounded electrodes leads to a higher developed self-bias voltage. As a consequence, the coating of large area objects would require processing chambers of huge dimensions, which complicates the industrial implementation of RF–PECVD processes [1]. On the contrary, asymmetric bipolar pulsed-DC plasmas allow achieving higher bias values without the need of geometrical changes. When applied to grow DLC thin films, new possibilities for modulating their characteristics appear. In particular, the performances of DLC thin films and their multifunctional properties can be extended for ultrathin antifriction and antiwear protective coatings. The adaptation of asymmetric bipolar pulsed plasma technology to magnetron sputtering and PECVD processes becomes a real alternative to conventional methods based on RF discharges to obtain films of different microstructures. The aim of the present study is to evidence the possibilities of the asymmetric bipolar pulsed-DC plasma technology when applied to grow DLC thin films showing low friction and reduced wear.

27.2
Experimental Details

DLC films, over 600 nm thick, were deposited on c-Si substrates by pulsed DC-reactive magnetron sputtering in a spherical stainless steel vessel. Prior to each procedure, the chamber was evacuated with a turbomolecular pump until a base pressure of 2×10^{-5} Pa. The absolute pressure in the chamber was monitored by a capacitive sensor during the deposition process and kept constant by an automatic throttle valve of variable conductance. The chamber of plasma used in this work (Figure 27.1) has two faced ports in which diverse and exchangeable anodes or cathodes can be assembled.

Two experiments were designed; the first one consists of a magnetron sputtering head (cathode) with a grounded substrate holder (anode), and the second one of a capacitively coupled cathode. Both experiments can be powered by the pulsed-DC power supply, and can operate alternately in the same chamber using the same vacuum system. The magnetron sputtering cathode supports a graphite target of electronic grade in front of a water cooled substrate holder. The samples were

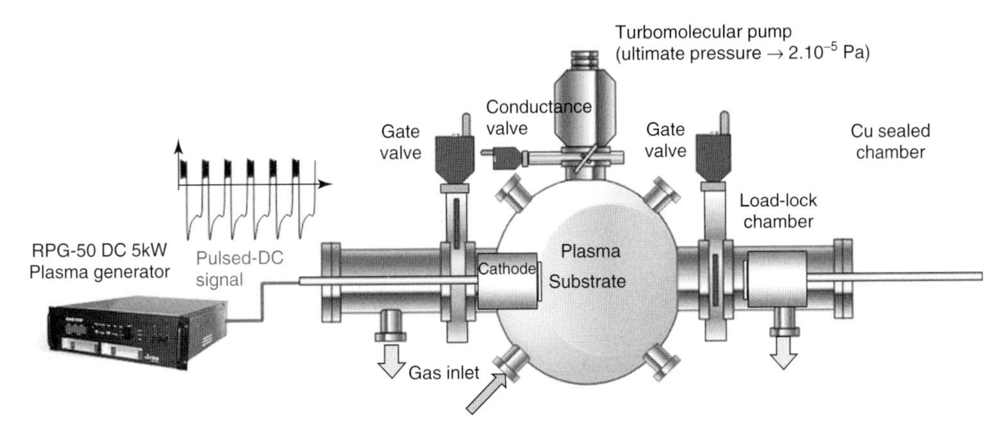

Figure 27.1 Asymmetric pulsed-DC plasma deposition system.

placed on the holder at 1.3 Pa of Ar with 7.5% of CH_4 during the deposition processes.

The magnetron sputtering cathode was operated at a fixed power density of 4.4 W/cm^2 using an asymmetric bipolar pulsed-DC power supply (ENI RPG-50). The characteristic trend of the asymmetric bipolar pulsed-DC power excitation is a deep negative peak voltage up to −5 kV, supplied at a frequency in the range of hundreds of kilohertz. The asymmetric pulsed-DC power works at a fixed positive pulse of +37.5 V, followed by a longer and variable negative pulse. The peak amplitude of the sputtering experiment was varied from −600 to −700 V and the pulse frequency was kept at 150 kHz at a constant positive pulse of 2016 ns, resulting in a duty cycle of 70%. On the other hand, substrate bias voltage was varied between 0 and −300 V.

Methane-glow discharges were produced using a cooled cathode without magnetron and powered by the asymmetric bipolar pulsed-DC power supply described above. The DLC films were grown on c-Si substrates placed on the cathode, as usual in DLC grown by PECVD. After achieving the lowest residual pressure, the chamber was filled with pure methane at a constant flow rate of 30 sccm. During the deposition process, the total pressure was kept constant at 10 Pa by the automatic throttle valve.

In the PECVD experiment, the pulsed-DC source was operated in the range 100–200 kHz of pulse frequency. The negative voltage peak of the pulsed-DC source ranged from −600 to −1400 V, whereas the positive voltage was kept at +37.5 V. As above, the positive pulse duration was set at 2016 ns in all the experiments.

To compare the characteristics of the DLC films produced by pulsed-DC sputtering and PECVD, the stress and thickness of the DLC samples were determined by profilometry (Dektak 3030 profilometer) working with a high vertical resolution in the order of 0.1 nm. In addition, some of the DLC films were deposited by RF–PECVD and used as comparative samples. For that, DLC films were deposited on the cathode powered at 12 W of RF (13.56 MHz) and −200 V of bias voltage. The experimental set-up is described in detail elsewhere [1].

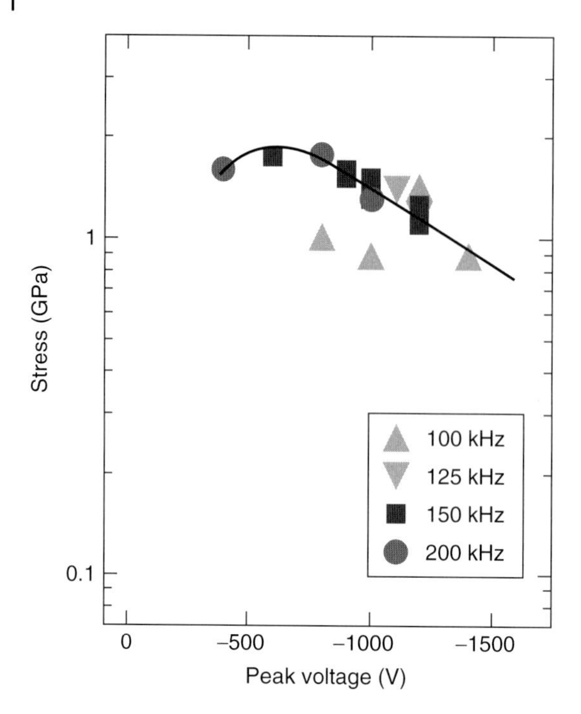

Figure 27.2 Variation of DLC film compressive stress as a function of negative pulse voltage amplitude in pulsed-DC PECVD.

27.3 Results and Discussion

The negative peak voltage affects the compressive stress of the DLC films deposited by pulsed-DC PECVD. Figure 27.2 shows the dependence of the compressive stress versus the negative pulse voltage amplitude for pulse frequencies in the range 100 –200 kHz [1, 2].

The main trend is a reduction in compressive stress with increasing negative peak voltage amplitude. Pulsed-DC PECVD films showed lower compressive stress values (between 1 and 1.5 GPa) than those of RF–PECVD films (up to 2.5 GPa in Figure 27.3 [3]), thus enabling the growth of well-adhered 1000 nm thick DLC films without using any adhesion interlayer deposited on silicon substrates. It should be noted in Figure 27.2 that the DLC films grown at the lowest pulse frequency (100 kHz) showed low and nearly constant compressive stress values below –1 GPa.

Figure 27.3 shows the dependence of stress from profilometry measurements with the substrate DC bias of carbon films deposited by pulsed-DC magnetron sputtering beside those grown by RF–PECVD at 100 W. The lower ion bombardment of pulsed-DC samples and the relaxation of carbon atoms and ions adsorbed onto the film surface during the low-voltage positive pulses might account for this

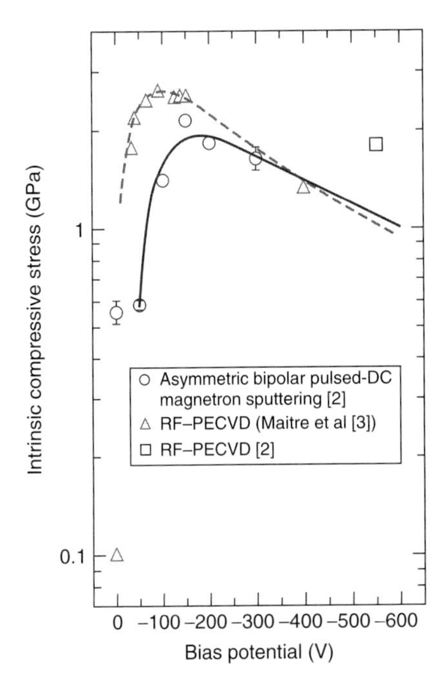

Figure 27.3 Intrinsic compressive stress dependence with substrate DC bias of DLC films grown at 100 W of pulsed-DC power by magnetron sputtering and RF–PECVD processes. (○) Experimental stress measurements of DLC films grown by asymmetric bipolar pulsed-DC reactive sputtering have been fitted with subplantation model (solid line). (Δ) Stress measurements of DLC films grown by RF–PECVD. Dashed line corresponds to the fitting using the Davis subplantation model [4].

considerable lowering in the stress. Using the Davis subplantation model [4]:

$$\sigma = \frac{c\sqrt{E}}{R/j + 0.016p(E - E_0)^{5/3}}$$

where c is a constant, E is the ion energy, E_0 is the activation energy of the relaxation process, j is the ion flux, R is the depositing flux, and p is a material-dependent parameter which is of the order of 1. Compared to typical DLC stress values [3], the obtained films show a low intrinsic stress with a maximum for substrate bias voltage of −180 V. The adjustment of the data to the model provides parameters of the same order as those obtained by Maître *et al.* [3]: $c = 0.33, p = 0.78,\ E_0 = 3.3eV,\ V_p = -44.2V$ (being $E \approx V_p - V_{\text{bias}}$) except for the R/j ratio, which is significantly higher for our experiment: 1.4 versus 0.86 from Maître *et al.* [3], but similar to 1.59 according to Lacerda and Marques [5], who proposed that the rigidity of the films was basically provided by a matrix of dispersed cross-linked sp^2 sites, in addition to the contribution of the sp^3 sites.

To proceed with the adjustment, the criterion $E \approx V_p - V_{\text{bias}}$ has been considered, on the basis that in low-pressure discharges the mean free path is of the order or larger than the plasma sheath (collisionless regime). V_p is the mean plasma potential for each particular type of discharge. The fact that R/j is high implies an important part of the deposition process is carried out primarily by neutrals, which are not susceptible to being controlled by substrate bias voltage. These results evidence a possible mechanism of subplantation used to explain the formation of ta-C : H films, acting also during the formation of a-C : H films deposited by methane decomposition in pulsed-DC reactive sputtering. This gives account of the high energetic species generated in pulsed processes.

Other results sensitive to pulsed-DC technology are related to tribologic behavior. These results have shown a higher wear rate of the sputtered films, compared to those deposited by the other techniques. The particular waveform of the pulsed-DC signal, for the used parameters, induces a strong ion bombardment during negative pulse and an almost total trapping of electrons during the positive pulse promoting the inclusion of no energetic species to the substrate.

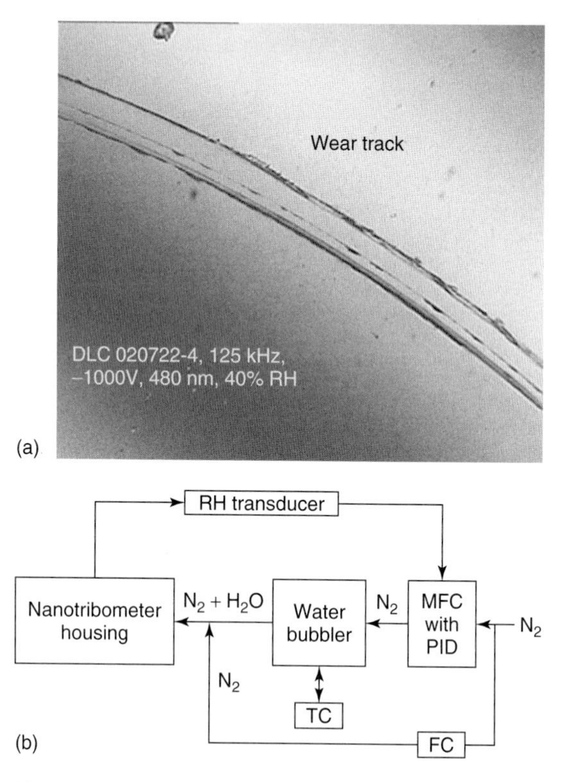

Figure 27.4 (a) Wear track of DLC films grown on c-Si after the friction test and using an optical microscope at a 100× magnification. The nanotribometer was set to a normal load up to 100 mN and the measurements were carried out under controlled relative humidity (40% RH). Few debris formation appears at normal forces of 50 mN. (b) Schematic representation of relative humidity controller using a water bubbler and dry nitrogen as dragging gas controlled by a PID mass flow controller and a RH sensor.

The friction coefficient was evaluated with a nanotribometer. The main advantages of this instrument are the low-load operation and the relatively reduced track diameter. The measurements were performed in ball-on-disk configuration using a 3 mm-diameter tungsten carbide ball, which sliced on the DLC surface at 1 mm/s of linear speed and applying 50 mN of normal load. This resulted in 6250 cycles. Each experiment was conducted in a $N_2 + H_2O$ atmosphere, with precisely controlled relative humidity (RH). Figure 27.4a shows the surface image of the wear track, provided by an optical microscope at a 100× magnification, corresponding to a DLC film grown on c-Si after the friction test. In our experiments, the nanotribometer was set to a normal load up to 100 mN and the measurements were carried out under a controlled relative humidity of 40%. Few debris formations appear at normal forces of 50 mN. The self-made RH controller is schematized in Figure 27.4b.

Initial point and stable friction coefficient as a function of pulse frequency and peak voltage have been studied (Figure 27.5). The continuous friction force recording not only provides numerical values of friction coefficient (μ), but also

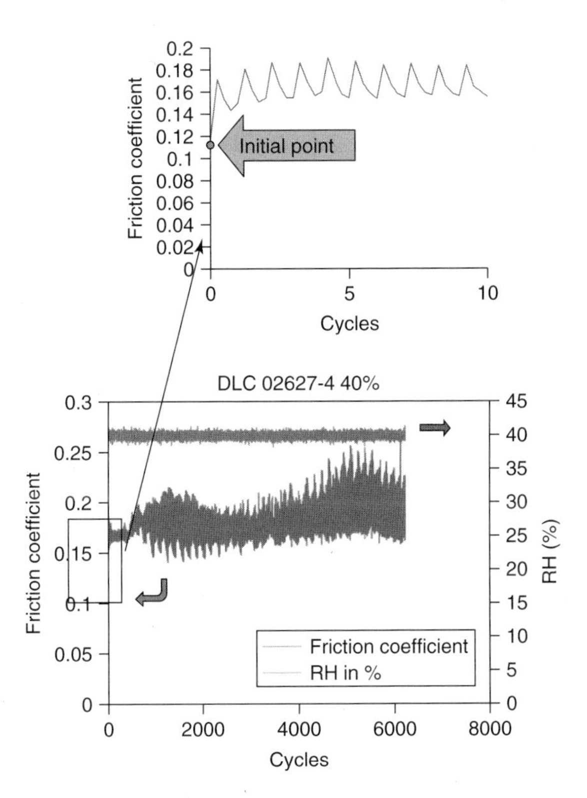

Figure 27.5 Evolution of friction coefficient of DLC films grown on c–Si by pulsed-DC PECVD. The experimental conditions of friction were kept at (40 ± 0.5)% of RH and (25.0 ± 0.1)°C.

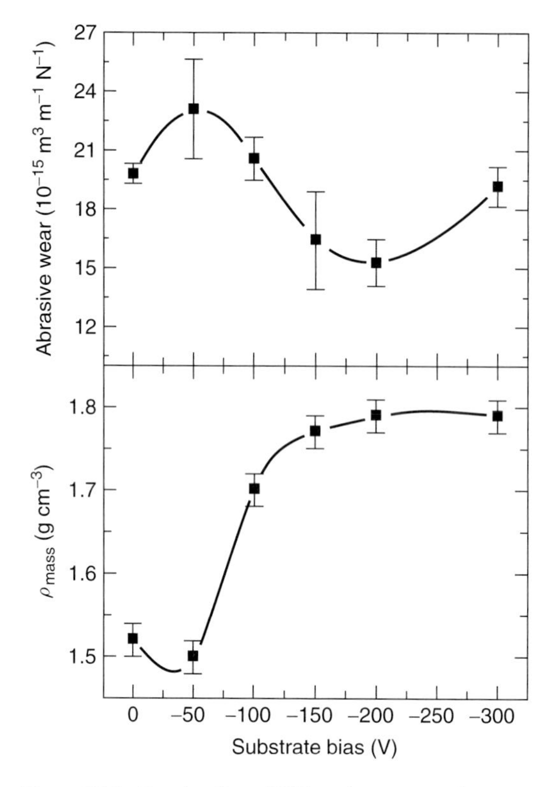

Figure 27.6 Density (from XRR) and wear rate (from [6]) dependence with substrate bias. These magnitudes show opposite behaviors.

monitors the changes in sliding behavior. The oscillations in μ give partial account of the holder's balancing during the measurements.

After the wear test, the examination of the wear tracks was carried out by optical microscopy at a magnification of 100×. From these observations, we detected little debris formation at normal forces of 50 mN. Wear is not appreciable after 6000 cycles, since the remaining tracks could be wiped out with ethanol. Indeed, such traces consisted of an accumulation of debris produced during the ball-on-disk operation.

The abrasive wear resistance of the pulsed-DC PECVD was characterized with a Calotest equipment. The corresponding values of the wear rate were calculated from the crater depth after abrasion. Then, the obtained rates ranged between 15 and $23.10^{-15}\frac{m^3}{m.N}$, indicating that the deposited films have diamond-like properties. Figure 27.6 shows these results, where a dependence of the wear rate on substrate bias was adduced. The film mass density was studied by the means of X-ray reflectivity (XRR). The results (Figure 27.6) show a correlation with the substrate bias and the wear rate. In particular, higher wear rates appear associated to reduced mass densities, which points to an improved mechanical stability for samples exhibiting low porosity.

27.4 Conclusions

Asymmetric bipolar pulsed DC technique appears as a promising alternative for the production of DLC films by PECVD and sputtering. This technology is being extensively implemented in industrial systems and it offers a new window of additional process parameters (pulse frequency, time intervals, and amplitudes of positive and negative voltages).

Pulsed plasmas show a greater complexity than those of DC or RF discharges. The special conditions exhibited by pulsed plasmas at low pressure of the sputtering processes seem to indicate a possible mechanism of subplantation, explaining the growing mechanism of this low-stressed DLC. Weak but remarkable effects have been found on stress, friction, and wear, in the short studied interval of parameters (pulse frequency and peak voltage). These effects are useful for modulating DLC properties basically for surface applications, needing low friction and wear rate and specific surface free energies.

Acknowledgments

This study was partially supported by the Generalitat de Catalunya (Project No. 2005SGR00666) and the MICINN of Spain (Project No. DPI2007-61349). The authors thank Serveis Cientifico-Tècnics of the Universitat de Barcelona (SCT-UB) for measurement facilities. M.R.-R. acknowledges financial support from CSIC of Spain (Grant No: UAC-2005-0021). C.C. acknowledges financial support from MICINN of Spain for a Juan de la Cierva contract.

References

1. (a) Andújar, J.L., Vives, M., Corbella, C., and Bertran, E. (2003) *Diamond Relat. Mater.*, **12**, 98–104; also in: (b) Corbella, C., Vives, M., Oncins, G., Canal, C., Andújar, J.L., and Bertran, E. (2004) *Diamond Relat. Mater.*, **13**, 1494–1499.
2. Rubio-Roy, M., Corbella, C., Garcia-Céspedes, J., Polo, M.C., Pascual, E., Andújar, J.L., and Bertran, E. (2007) *Diamond Relat. Mater.*, **16**, 1286–1290.
3. Maître, N., Girardeau, Th., Camelio, S., Barranco, A., Vouagner, D., and Breelle, E. (2003) *Diamond Relat. Mater.*, **12**, 988–992.
4. Davis, C.A. (1993) *Thin Solid Films*, **226**, 30–34.
5. Lacerda, R.G. and Marques, F.C. (1998) *Appl. Phys. Lett.*, **73**, 617–619.
6. Rubio -Roy, M., Pascual, E., Polo, M.C., Andújar, J.L., and Bertran, E. (2008) *Surf. Coat. Technol.*, **202**, 2354–2357.

28
Plasma Deposition of N-TiO_2 Thin Films

Pablo Romero-Gómez, Angel Barranco, José Cotrino, Juan P. Espinós, Francisco Yubero, and Agustín R. González-Elipe

28.1
Introduction

Plasmas containing N_2 and H_2 are extensively studied because of their widespread applications in the processing of materials [1]. Despite their extensive use, only limited information is available on the role of the different nitrogen species of the plasma in the nitridation processes [2]. Thus, for example, although it is known that NH_x radicals play an important role as intermediate precursors of N and H atoms in plasma-driven reactions [3], little is known about the plasma–surface interactions involving this kind of species.

This situation is clearly exemplified by the studies aiming at preparing N-doped TiO_2 thin films. The quest for N-doped TiO_2 thin films and powders has been fostered by the discovery that this doped titanium dioxide may present photoactivity in the visible region [4–6]. At present there is a vivid controversy about the reasons for this photoactivity [7, 8] and how it is mediated by the amount of nitrogen incorporated into the films, its chemical state, and other characteristics of the samples such as their structure, microstructure, optical properties, and so on. Several methods, including magnetron sputtering using nitrogen gas [9] or plasma-enhanced chemical vapor deposition (PECVD) [10], have been reported for the preparation of N-doped TiO_2 films. However, in most cases, the plasma conditions used by these preparations are not well described and the reproducibility of the results is not always straightforward.

The present chapter presents characterization of data of a series of plasmas used for the synthesis of N-doped TiO_2 thin films by PECVD. For this purpose, the composition (i.e., nitrogen or hydrogen content) and other characteristics of the plasma used for decomposition of the titanium precursor have been systematically changed. The plasma characteristics have been analyzed by optical emission spectroscopy (OES) and Langmuir probe measurements. These plasma characteristics have been correlated with the amount of nitrogen incorporated within the structure of the film, a key issue for the control of the visible photoactivity of this oxide.

Industrial Plasma Technology. Edited by Yoshinobu Kawai, Hideo Ikegami, Noriyoshi Sato, Akihisa Matsuda, Kiichiro Uchino, Masayuki Kuzuya, and Akira Mizuno

ISBN: 978-3-527-32544-3

28.2
Experimental Setup and Diagnostic Techniques

N-doped TiO_2 thin films were prepared by PECVD in a plasma reactor with a remote configuration. The system, supplied with a microwave plasma source (SLAN, from Plasma Consult, GMBh, Germany) has been described in previous works [11, 12]. The plasma source is powered by a 2.45-GHz microwave generator working at 400 W and was separated from the reaction chamber by a grid to avoid the microwave heating of the substrates.

Titanium tetra isopropoxide (TTIP) was used as titanium precursor. For dosing the TTIP in a controlled way, the precursor was placed in a stainless steel recipient through which oxygen was bubbled while heating at 305 K. Both the bubbling line and the shower-type dispenser used to dose the precursor into the chamber were heated at 373 K to prevent any condensation on the tube walls. The plasma source was operated with a mixture of gases containing nitrogen, basically N_2, $N_2 + H_2$, $N_2 + O_2$, and $N_2 + H_2 + O_2$. During deposition, oxygen was always present as a carrier gas of the precursor. In other plasma characterization experiments (e.g., Langmuir probe measurements), neither oxygen nor TTIP was fed into the discharge. During operation, N_2, H_2, and O_2 gas flows and total pressure in the chamber were controlled by using mass flow controllers and a throttle valve, respectively. The total pressure during deposition and the other plasma characterization experiments was 4×10^{-3} torr. A detailed description of this experimental setup can be found in a previous publication [13].

The emission spectrum was analyzed using a 0.5 m CVI/Digikrom DK480 monochromator (CVI Laser Corporation, Albuquerque, NM) with a 1200 grooves/mm grating, spectra resolution of 0.2 nm, and spectral sensitivity ranging from 200 to 900 nm (Hamamatsu R928). The light was collected by an optical fiber that was fixed to the monochromator and positioned 4 cm downstream from the source center. The light was collected through a hole in the air-cooling ring. It should be noted that the spectroscopic intensities are integrated over the line of sight, thus preventing any spatial resolution of the measurements.

The plasma parameters were assessed by a single Langmuir probe movable along the vacuum chamber diameter at an axial distance of about 8 cm from the grid that avoids microwave heating of the substrates. The probe tip was made of tungsten, had a radius of 0.125 mm, and a length of 7 mm. The nonactive part of the probe wire was protected from the plasma by a coaxial ceramic tube of length 10 cm. The stainless steel walls of the vacuum chamber were used as a reference electrode. The measurements were performed by a computer-controlled probe system which has been described in detail in [14]. Briefly, the system offers the possibility of single- and floating double-probe measurements. The algorithm to derive the ion density from the single-probe characteristics is described in detail in [14]. Langmuir's orbital-motion limited (OML) theory is used to analyze the results [15, 16]. We found a total standard uncertainty of the ion density of 25%, which results from the uncertainty of the probe surface area (systematic error $\pm 8\%$) and of the statistical uncertainty of 17% when running the discharge at different times under identical conditions of the process parameters.

TiO_2 thin films were prepared at two substrate temperatures, 298 and 523 K. Nitrogen incorporation into the TiO_2 thin films was assessed by X-ray photoemission spectroscopy (XPS). XP spectra were recorded on an ESCALAB 210 spectrometer working under energy transmission constant conditions. The Mg Kα line was used for excitation of the spectra. They were calibrated in binding energy (BE) by referencing to the C 1s peak taken at 284.6 eV owing to the surface contamination. Quantification was done by calculating the area of the peaks and by correcting then with the sensitivity factor of each element/electronic level.

28.3
Results and Discussion

Table 28.1 summarizes the values of the ion density, electron temperature, and plasma potential determined with the Langmuir probe for plasmas of pure N_2, H_2, and a mixture (50%) of hydrogen plus nitrogen. The analysis of the radial dependence of these magnitudes showed that their values decrease for measurements made far from the central position (i.e., about 15 cm) of the chamber where the deposition experiments were carried out. The values reported in Table 28.1 correspond to this central position where the plasma characteristics were homogeneous. A significant difference is that, for all radial positions, the ion density follows the order $N_2 + H_2 > H_2 > N_2$, while the plasma potential has an opposite trend and the electron temperature is minimum for the $N_2 + H_2$ plasma (cf. Table 28.1). The enhancement of the ion density for the $N_2 + H_2$ mixture can be attributed to an increase in the intensity in the tail of the electron energy distribution function [17] and the lower ionization energy of H_2 with respect to N_2. Besides this observation, it is also worthy of note the high ion density values characteristic of these plasmas. These conditions ensure a high reactivity and the total decomposition of the TTIP precursor used for the deposition of the films.

Figure 28.1 shows, as examples, OES spectra corresponding to $O_2/N_2/H_2$ (i.e., 65%/12%/23% in volume) and O_2/N_2 (88%, 12% in volume) plasmas. In this

Table 28.1 Ion density, electron temperature, and plasma potential of different plasmas used in the present work.

Plasma composition	Ion density (m^{-3})	Electron temperature (eV)	Plasma potential (V)
N_2	6×10^{16}	5.6	28
H_2	9.5×10^{16}	5.6	27
$N_2 + H_2$	10.5×10^{16}	5.0	25

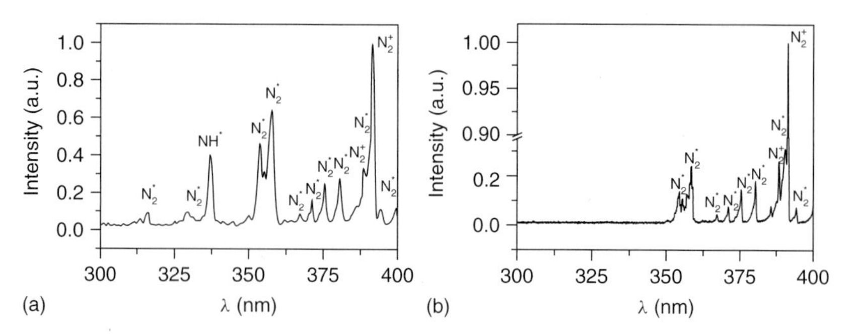

Figure 28.1 OES spectrum recorded in the region from 300 to 400 nm for a $N_2/O_2/H_2$ (65, 12, 23%) and N_2/O_2 plasmas.

figure, it is possible to detect the presence of N_2^*, NH^*, and $N_2{}^+$ species. It is important to remark that no NH^* species are detected in the absence of hydrogen in the plasma. Similar spectra were also recorded for other spectral regions under these and other plasma compositions, with and without the TTIP precursor. A full account of the species detected under these different working conditions is reported in Table 28.2, where we also include the wavelengths of these peaks/bands used for assignation according to previous works from literature and data basis [18–20]. It is important to note that the type of species detected did not change when the precursor was added to the plasma (i.e., under deposition conditions), although the intensity of the bands attributed to O^* in the plasmas of mixtures O_2/N_2 decreased or disappeared and that of the bands/peaks of H^* significantly increased. It is also worthy of note that although no O^* species were detected in the plasma of mixtures $O_2/N_2/H_2$, this does not mean that no oxygen-reactive species are formed. The formation of (N-doped, see below) titanium oxide under these conditions proves the reactivity of the precursor with active species of oxygen in the plasma.

An important evidence from the data in Table 28.2 for the purpose of incorporating nitrogen within the TiO_2 lattice is the detection of $N_2{}^*$ and $N_2{}^+$ species for mixtures of $N_2 + O_2$ as plasma gas. The observed bands correspond to the bands of the second positive (i.e., $C^3\Pi_u - B^3\Pi_g$) and the first positive (i.e., $B^3\Pi_g - A^3\Sigma_u^+$) systems of molecular nitrogen and to the first positive systems of $N_2{}^+$ molecular ion (i.e., $B^2\Sigma_u^+ - X^2\Sigma_g^+$). It is also worth noting the development of a series of peaks attributed to NH^* species and the detection of the hydrogen Balmer α (656 nm) and Balmer β (486 nm) lines in the $N_2 + H_2 + O_2$ mixtures. From the work of Dood *et al.* [21], the formation of NH^* species can be accounted for by the reaction $N(^2D) + H_2 \rightarrow NH + H$, which implies an important $N(^2D)$ state population.

To examine the influence of hydrogen in the characteristics of the plasma, different mixtures of nitrogen and hydrogen with increasing amounts of this latter gas were also analyzed. We found that the intensities of the lines of atomic hydrogen, H_α and H_β located at 656.3 and 486.1 nm, increased and that of the head bands $N_2(0 \rightarrow 2)$ and $N_2^+(0 \rightarrow 0)$ decreased with the amount of hydrogen in the

Table 28.2 Plasma species detected by OES.

Plasma composition (%N_2, %O_2, %H_2)	Species	Wavelength (s) (nm)	Transition(s)
(88,12,0)	N_2*	250–500 600–900	Molecular bands SPS[a] ($C^3\Pi, v \rightarrow B^3\Pi, v'$) FPS[b] ($B^3\Pi, v \rightarrow A^3\Sigma, v'$)
	N_2^+	391.4[d]	Molecular bands FNS[c] ($B^2\Sigma, v \rightarrow X^2\Sigma, v'$)
(17,83,0)	N_2*	250–500 600–900	Molecular bands SPS[a] ($C^3\Pi, v \rightarrow B^3\Pi, v'$) FPS[b] ($B^3\Pi, v \rightarrow A^3\Sigma, v'$)
	N_2^+	391[d]	Molecular bands FNS[c] ($B^2\Sigma, v \rightarrow X^2\Sigma, v'$)
	O*	436.8, 532.9, 615.6, 645.6, 777.4, 844.6	Atomic lines involving the 3s and 3p states
(65,12,23)	H*	656, 486	H_α 2p-3d; H_β 2p-4d
	NH*	337.0	Molecular band (1,1) ($^3\Pi, v \rightarrow ^3\Sigma, v'$)
	N_2*	250–500 600–900	Molecular bands SPS[a] ($C^3\Pi, v \rightarrow B^3\Pi, v'$) FPS[b] ($B^3\Pi, v \rightarrow A^3\Sigma, v'$)
	N_2^+	391[d]	Molecular bands FNS[c] ($B^2\Sigma, v \rightarrow X^2\Sigma, v'$)
	H*[e]	H_α (656 nm) H_β (486 nm)	$2p-6d$; $2p-5d$; $2p-4d$; $2p-3d$

[a] SPS, second positive system.
[b] FPS, first positive system.
[c] FNS, first negative system.
[d] Only the band at 391 nm is clearly visible in the spectra, the other bands overlaps with those due to sthe N_2 SPS.
[e] Visible when the precursor is added to the plasma even for gas mixtures without hydrogen.

plasmas. It was also found that the increase in the H_2 flow rate had a strong impact on the intensity of the emission lines of the radical NH that relatively increased with the proportion of this gas. The fact that the maximum incorporation of nitrogen into the TiO_2 is reached for plasmas containing hydrogen (see Table 28.3) proves that the incorporation of nitrogen into the titanium oxide lattice is favored under chemical reduction conditions and, very likely, the presence of NH* species in the plasma.

The TiO_2 thin films prepared under different conditions of work were characterized by XPS, with the purpose of estimating both the amount of nitrogen incorporated into the films and its chemical state. Table 28.3 summarizes the value of the N/Ti ratios determined by XPS for a series of thin films prepared by changing the plasma conditions and substrate temperature. It is apparent from this table

Table 28.3 Ratio N/Ti and percentage of the two types of nitrogen species detected in N-doped TiO_2 thin films prepared with different plasmas and substrate temperatures.

Preparation conditions (N_2,O_2,H_2 in %) and (T substrate in K)	N/Ti	$N_{400\ eV}$ (%)	$N_{396\ eV}$ (%)
(12,88,0)(523)	0.01	100	0
(65,12,23)(523)	0.11	28	72
(88,12,0)(523)	0.04	67	33
(88,12,0)(298)	0.11	100	0

that the amount of nitrogen incorporated into the films increases for the plasmas containing hydrogen or when the deposition was carried out at room temperature.

To get a deeper insight into the characteristics of the incorporated nitrogen, we have looked in more detail at the shape of the N1s spectra. Figure 28.2 shows two examples corresponding to the films obtained with the N_2/O_2 and $N_2/O_2/H_2$ plasmas. Besides the quite different intensity of the two spectra (note the multiplication factor in the spectrum of the sample prepared with the N_2/O_2 plasmas), the most noticeable difference is that the peaks in the two cases appear at quite different BEs, that is, at 400.0 and 396.1 eV. According to the literature the second peak can be attributed to nitrogen species bonded to titanium and forming a kind of nitride species [22]. The attribution of the first peak is less straightforward, although it has been tentatively assigned to nitrogen species interacting simultaneously with titanium and oxygen atoms in interstitial positions [23].

An overall assessment of the data in Figure 28.2 and Table 28.3 in relation with the plasma characteristics shows that the formation of the nitrogen species at 400.0 eV is favored when no hydrogen has been added to the plasma or when

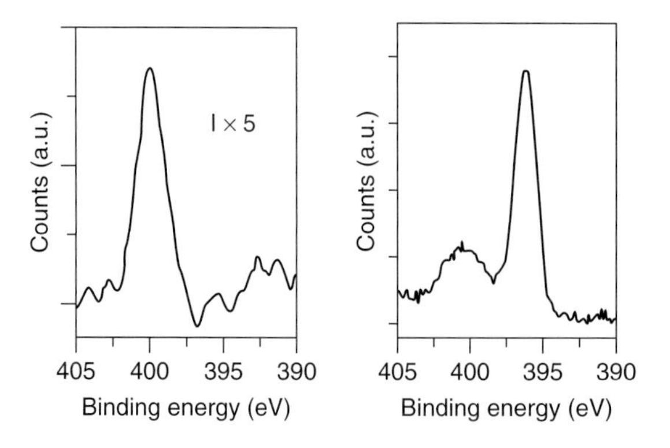

Figure 28.2 N 1s photoemission spectra recorded for two N-doped TiO_2 thin films prepared with (a) a N_2/O_2 (12, 88%) and (b) a $N_2/O_2/H_2$ (65, 12, 23%) plasma at 523 K.

working at 523 K. By contrast, the maximum peak of 396.1 eV is formed with plasmas containing hydrogen and/or when performing the deposition at 523 K. It is likely that the presence of NH* species detected in the plasma under these conditions plays a favorable role in the doping processes of the films. Under more oxidant conditions as those existing in the N_2/O_2 plasmas or by decreasing the temperature of deposition, less nitrogen is incorporated within the films. In this case, the nitrogen is characterized by a BE of 400.0 eV. Although the attribution of these species is still controversial in the literature, recent papers attribute the photo-activity in the visible region of the N-doped TiO_2 to its presence as a doping species [23]. Additional work is being carried out in our laboratory to clearly establish a correlation between the detection of these species and the visible photoactivity of N-doped TiO_2.

In conclusion, our work has shown that it is possible to incorporate nitrogen into the structure of anatase prepared by PECVD and that the characteristics of this nitrogen incorporation depend on the plasma conditions for mixtures $N_2 + O_2 + H_2$. The analysis of the plasma by OES has provided important clues to understand this phenomenology.

Acknowledgments

We thank the Ministry of Science and Education of Spain (projects MAT 2007 65764/NAN2004-09317, the CONSOLIDER INGENIO 2010-CSD2008-00023) and the Junta de Andalucia (projectsTEP2275 and P07-FQM-03298) for financial support. Part of this work has been carried out within the EU project NATAMA (contract n° 032583).

References

1. Hueso, J.L., Espinós, J.P., Caballero, A., Cotrino, J., and González-Elipe, A.R. (2007) *Carbon*, **45**, 89.
2. Suraj, K.S., Bharathi, P., Prahlad, V., and Mukherjee, S. (2007) *Surf. Coat. Technol.*, **202**, 301.
3. Gordiets, B., Ferreira, C.M., Pinheiro, M.J., and Ricard, A. (1998) *Plasma Sources Sci. Technol.*, **7**, 363.
4. Asahi, R., Morikawa, T., Ohwaki, T., Aoki, K., and Taga, Y. (2001) *Science*, **293**, 269.
5. Sakthivel, S., Janczarek, M., and Kirsch, H. (2004) *J. Phys. Chem. B*, **108**, 19384.
6. Yates, H.M., Nolan, M.G., Sheel, D.W., and Pemble, M.E. (2006) *J. Photochem. Photobiol. A: Chem.*, **179**, 223.
7. Livraghi, S., Paganini, M.C., Giamello, E., Selloni, A., Valentin, C.D., and Pacchioni, G. (2006) *J. Am. Chem. Soc.*, **128**, 15666.
8. Lee, J.-Y., Park, J., and Cho, J.-H. (2005) *Appl. Phys. Lett.*, **87**, 011904.
9. Wong, M.-S., Chou, H.P., and Yang, T.-S. (2005) *Thin Solid Film*, **494**, 244.
10. Maeda, M. and Watanabe, T. (2006) *J. Electrochem. Soc.*, **153**, C186.
11. Borrás, A., Cotrino, J., and González-Elipe, A.R. (2007) *J. Electrochem. Soc.*, **154**, 152.
12. Gracia, F., Holgado, J.P., and González-Elipe, A.R. (2004) *Langmuir*, **20**, 1688.
13. Cotrino, J., Palmero, A., and Rico, V. (2001) *J. Vac. Sci. Technol. B*, **19**, 1071.

14. Brockhaus, A., Borchardt, A., and Engeman, J. (1994) *Plasma Sources Sci. Technol.*, **3**, 539.
15. van Helden, J.H., van den Oever, P.J., Kessels, W.M.M., van de Sanden, M.C.M., Schram, D.C., and Engeln, R. (2007) *J. Phys. Chem. A*, **111**, 11460–11472.
16. Jauberteau, J.L. and Jauberteau, I. (2008) *Plasma Sources Sci. Technol.*, **17**, 015019.
17. Loureiro, J. and Ricard, A. (1993) *J. Phys. D: Appl. Phys.*, **26**, 163.
18. Pearse, R.W.B. and Gaydon, A.G. (1963) *The Identification of Molecular Spectra*, John Wiley & Sons, Inc., New York .
19. Wiese, W.L., Smith, M.W., and Glennon, B.M. (1966) *Atomic Transition Probabilities, National Bureau of Standards*, Gaithersburg.
20. Boudam, M.K., Saoudi, B., Moisan, M., and Ricard, A. (2007) *J. Phys. D: Appl. Phys.*, **40**, 1694.
21. Dood, J.A., Lipson, S.J., Flanagan, D.J., Blumberg, W.A.M., Person, J.C., and Green, B.O. (1991) *J. Chem. Phys.*, **94**, 4301.
22. Qiu, X., Zhao, Y., and Burda, C. (2007) *Adv. Mater.*, **19**, 3995.
23. Asahi, R., Morikawa, T., Hazama, H., and Matsubara, M. (2008) *J. Phys.: Condens. Matter*, **20**, 064227.
24. Cong, Y., Zhang, J., Chen, F., and Anpo, M. (2007) *J. Phys. Chem. C*, **111**, 6976.

29
Investigation of DLC and Multilayer Coatings Hydrophobic Character for Biomedical Applications

Rodica Vladoiu, Aurelia Mandes, Virginia Dinca, Mirela Contulov, Victor Ciupina, Cristian Petrica Lungu, and Geavit Musa

29.1
Introduction

Diamond-like carbon (DLC) has drawn been of considerable interest to scientists as a potential surface-coating material, owing to its remarkable properties such as wear resistance, electrical properties, hardness, roughness, smoothness, low friction coefficients, chemical inertness, and high refractive index [1].

Besides these, biocompatibility is one of the main characteristics of DLC coatings, which can be used inside the human body for medical purposes without generating an allergic reaction. All these properties match well with the criteria of a good biomaterial for applications in orthopedics, cardiovascular devices, contact lenses, or dentistry [2–4].

Knowledge about the surface free energy of solid materials, in particular, information on hydrophobicity or hydrophilicity of the surface is crucial in biomedical applications.

For example, in the case of the DLC coating, a *hydrophilic* character (meaning a low value of the water contact angle) could be oriented on the simulation of cell growth or on the reduction of the inflammatory reactions, while a *hydrophobic* character – given by higher water contact angle values – could be focused on prevention of unwanted cell growth.

At the same time, multilayers based on carbon (i.e., carbon–metal) are known to have excellent tribological properties against various materials. Many researches have been focused on the microstructure effects on the tribological behavior of multilayers films but few have been reported on the surface properties. However, because of its low adhesion strength to a substrate and/or the difficulty of preparing the required film, thin films have not been as extensively applied to tools as expected and different outstanding tribological properties can be obtained [5]. Also, the free surface energy of the films is an important factor because of the fact that it can give some clues to the surface properties of the film.

DLC as well as multilayers based on carbon films have been prepared by several methods which provide a wide range of mechanical, physical, and tribological

Industrial Plasma Technology. Edited by Yoshinobu Kawai, Hideo Ikegami, Noriyoshi Sato, Akihisa Matsuda, Kiichiro Uchino, Masayuki Kuzuya, and Akira Mizuno

ISBN: 978-3-527-32544-3

properties [6, 7]. In this study, the method used for DLC and multilayers coatings synthesis was thermionic vacuum arc (TVA), which might become one of the most suitable technologies to significantly improve the quality of the surfaces.

29.2 Experimental Setup

The TVA deposition method consists of an externally heated cathode surrounded by a Wehnelt cylinder that concentrates the high-voltage accelerated electrons on the anode used (a graphite rod) (Figure 29.1).

The symmetry of the cathode–anode arrangement allows a perpendicular bombardment of the electron beam on the anode surface. Owing to the high applied voltage, the continuously evaporated material from the anode ensures the formation within the interelectrodes space of steady-state carbon vapors with enough density in order to ignite and to maintain a bright discharge [8].

The main experimental parameters involved in this study were d – the distance between samples and the point of the ignition of the discharge; I_f – the intensity of the heating filament; U_{arc} – the applied voltage of the arc [9].

It is to be emphasized that, in the case of TVA technology, the deposition takes place in high vacuum conditions, so plasma does not fill the whole chamber. In this way, the substrate can be protected against the thermal heat of the plasma

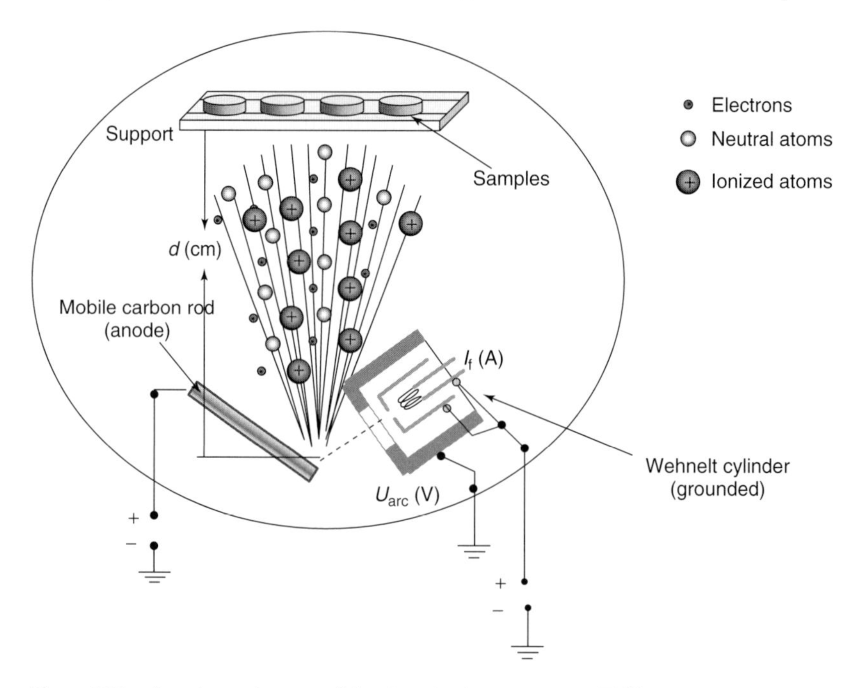

Figure 29.1 Experimental setup of the thermionic vacuum arc (TVA).

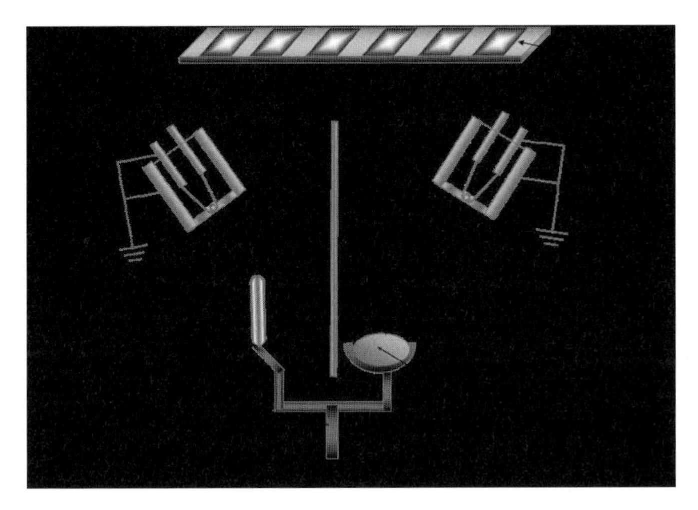

Figure 29.2 Experimental setup for simultaneous deposition of two different materials in the two-gun configuration.

by placing it away from the core of the localized plasma. This is the reason that the distance d plays a very important role for determining the properties of the condensing film.

To grow unstressed carbon–metal films (C—Ag, C—Al, and C—Cu), we used a two electron gun TVA plasma arrangement.

The experimental setup is shown in Figure 29.2. The electron guns are symmetrically arranged with respect to the substrate mounted at a distance of 400 mm on the central line. The synthesis of multilayer films is not a trivial task. Both metal (Ag, Al, or Cu) and carbon were evaporated one after the other.

The cathode of each of the guns consisted of a heated tungsten filament surrounded by molybdenum Wehnelt cylinder, which focuses the electrons on the anode surface. A hydrogen-free carbon rod was used as anode in the case of carbon discharge, and a carbon crucible filled with metal grains in the case of metal discharge. The role of the panel is to prevent the interaction between the two plasmas.

One of the most significant characteristics of the TVA method is the presence of energetic ions in the pure carbon and metal vapor plasmas. Moreover, the energy of the ions can be fully controlled from outside the discharge vessel via the control of the arc voltage drop value [10]. The angle ϕ between the Wehnelt cylinder axis and a virtual perpendicular line on the crucible surface is an important geometrical parameter for the control of the ion energy. Owing to the high energy dissipated in unit volume of the plasma, the material is totally dispersed and is completely free of droplets and the bombarding ions are just the ions of the depositing material (carbon and metal).

The contact angles and free surface energy of the DLC films by TVA method were determined by means of surface energy evaluation (SEE) system [11]. It is based on the usage of a charge coupled device (CCD) camera, which captures a photo of the liquid drop spread on the studied surface. The contact angle of a sessile drop on a solid surface, defined by mechanical equilibrium of the drop under the action of three interfacial tensions – solid/vapor, solid/liquid, and liquid/vapor – could be fitted with more than three basic points: two on the liquid–solid interface – the base line – and one on the drop contour.

There are several models for evaluation of the surface free energy based on contact angle measurement, depending on the testing liquid. In this chapter, the surface free energy of the deposited films has been calculated by mean of Owens and Wendt (Lifshitz–van der Waals) method, which enables to determine the electron-acceptor and electron-donor parameters of the surface energy [12]. Owens and Wendt proposed the division of the total surface energy into two components: the dispersive component γ^d and polar component γ^p.

The surface energy is a combination of these two separate components and is an extension of the equation given by Fowkes [13]:

$$\gamma_{SL} = \gamma_S + \gamma_L - 2\left(\sqrt{\gamma_S^d \gamma_L^d} + \sqrt{\gamma_S^p \gamma_L^p}\right) \tag{29.1}$$

From the Young–Dupre equation, it is possible to obtain the work of adhesion W_{SL} between the liquid and the thin film at atmospheric pressure:

$$W_{SL} = \gamma_L(1 + \cos\Theta) \tag{29.2}$$

Combining the expression of the surface energy given by Equation 29.1 with Dupre's formula:

$$\gamma_{SL} = \gamma_S + \gamma_L - W_{SL} \tag{29.3}$$

we obtain Equation 29.4:

$$(1 + \cos\Theta)\gamma_L = 2\left(\sqrt{\gamma_S^d \gamma_L^d} + \sqrt{\gamma_S^p \gamma_L^p}\right) \tag{29.4}$$

The data obtained with this method depends on testing parameters such as temperature, drop volume, quality of the surface, drop size, drop environment, liquid evaporation, and so on.

29.3
Results and Discussions

The results on contact angle Θ and the free surface energy γ_{LS} measurements of the deposited DLC films calculated by mean of Owens–Wendt (Lifshitz–Van der Waals) method are summarized in Table 29.1, next to the deposition parameters visualized in Figure 29.1, where d is the distance between the arc ignition point and the sample and I_f is the intensity of the filament heating.

As can be seen that the contact angle has values ranging between 51 and 62° for ethylene glycol as testing liquid, while for water the values were between 77 and 91°. The highest value of the contact angle, which is corresponding to the lowest

Table 29.1 Deposition conditions, measured contact angles Θ_i, and the surface free energy for DLC samples.

Sample	d (cm)	I_f (A)	Θ (°) water	Θ (°) ethylene glycol	γ_{total} (mJ m^{-2})
B3	20	50	80.67	53.93	31.15
B5	29	50	77.68	55.28	29.56
B7	28	51	79.46	51.53	32.55
B8	30	50	77.44	51.27	32.01
B10	31	45	83.25	54.62	31.98
B11	35	53	85.37	60.39	27.86
B12	35	45	90.34	61.42	30.52

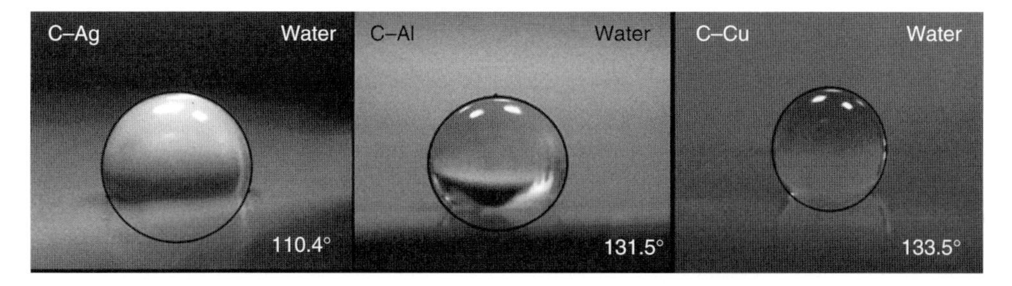

Figure 29.3 Pictures of the droplets on different samples (C–Ag, C–Al, C–Cu).

free total surface energy (30 mJ m^{-2}) has been obtained in the case of the greatest distance from the ignition of the carbon plasma (sample B12).

Figure 29.3 presents three pictures taken during the measurement of the contact angle on different multilayers based on carbon samples.

The recorded pictures, which were quite extensive, were earlier analyzed manually by the user for the full control of the contact angle determination. This evaluation method is now integrated in the software and is fully automatically carried out by the computer. Selection of several points of solid–liquid and liquid–vapor interface makes it possible to fit the drop profile and to calculate the tangent angle of the drop with the solid surface. Small droplets of 4-μl constant volume of liquids (water and ethylene glycol) were used for the measurements in order to minimize gravitational effects. From all the evaluated carbon-based multilayers, the C–Cu film presents the highest value for the contact angle (133.3°) obtained for analyses made with water, which means the lowest value of the free surface energy polar component.

The variations of the free surface energy, obtained by mean of Owens Wendt method are shown in Figure 29.4. The obtained data for free surface energy, showed the lowest value of $\gamma_{total} = 46.4$ mJ m^{-2} for C–Al, while the C–Cu revealed the highest value ($\gamma_{total} = 52.6$ mJ m^{-2}).

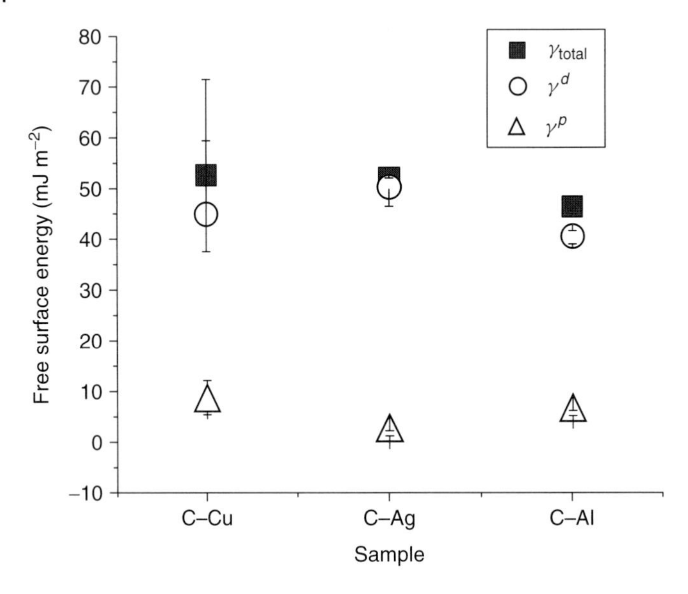

Figure 29.4 Diagram of the free surface energy values.

The total surface energy value of the C–Ag multilayer has been evaluated as intermediate between the other multilayers taken into account ($\gamma_{total} = 51.9$ mJ m^{-2}), but it could be also included in the nonwettable materials.

29.4 Conclusions

The surface free energies calculated with Owens–Wendt method reveal a good hydrophobic character of the DLC thin films especially in the higher values of the distance from the ignition point of the carbon plasma. The wettability of the C–Al, C–Cu, and C–Ag films obtained by means of the TVA method in the electron-gun configurations was investigated with drop-shape measurement. The high values of the contact angle prove the hydrophobic character of the multilayer film surface especially in the case of C–Cu and C–Al multilayers.

In this way, for multilayers composed of C coatings with very thin buffer layers, metals like Al or Cu have become a major area of interest not only for tribological but also for biomedical applications. Moreover, as silver is known to be a potent antibacterial agent, C–Ag, having both good blood compatibility and antimicrobial characteristics, could be useful biomaterials in biomedical engineering.

We conclude that TVA method can be used successfully for obtaining DLC and multilayers based on carbon-film deposition and might be an alternative to current material for biomedical applications.

Acknowledgments

This work was supported by Romanian Ministry of Education and Research, under project ID_230/2007 and project CEx 237/2006.

References

1. Zykova, A., Safonov, V., Virva, O., Luk'yanchenko, V., Walkowich, J., Rogowska, R., and Yakovin, S. (2008) *Journal of Physics: Conference Series*, **113**, 012029.
2. Mathew Mate, C. (2008) *Tribology on the Small Scale*, Oxford University Press.
3. Yang, W., Thordarson, P., Justin Gooding, J., Ringer, S.P., and Braet, F. (2007) *Nanotechnology*, **18**, 412001/12.
4. Dearnaley, G. and Arps James, H. (2005) *Surf. Coat. Technol.*, **200**, 2518–2524.
5. Hauert, R. (2004) *Tribol. Int.*, **37**, 991–1003.
6. Roy, R.K. and Lee, K.R. (2007) *J. Biomed. Mater. Res. B: Appl. Biomater.*, **83** (1), 72–84.
7. Lu, X., Li, M., Tang, X., and Lee, J. (2006) *Surf. Coat. Technol.*, **201** (3–4), 1679–1684.
8. Vladoiu, R., Ciupina, V., Surdu-Bob, C., Lungu, C.P., Janik, J., Skalny, J.D., Bursikova, V., Bursik, J., and Musa, G. (2007) *J. Optoelectron. Adv. Mater.*, **9** (4), 862–866.
9. Musa, G., Mustata, I., Ciupina, V., Vladoiu, R., Prodan, G., Vasile, E., and Ehrich, H. (2004) *Diamond Relat. Mater.*, **13**, 1398–1401.
10. Surdu-Bob, C., Vladoiu, R., Badulescu, M., and Musa, G. (2008) *Diamond Relat. Mater.*, **17** (7–10), 1625–1628.
11. Bursikova, V., Stàhel, P., Navratil, Z., Bursik, J., and Janca, J. (2004) *Surface Energy Evaluation of Plasma Treated Materials by Contact Angle Measurement* Brno Masaryk University Press
12. Navratil, Z., Bursikova, V., Stahel, P., Sira, M., and Zverina, P. (2004) *Czech. J. Phys.*, **54** (Suppl. C), 877.
13. Fowkes, F.M. (1983) in *Physicochemical Aspects of Polymer Surfaces*, vol. 2 (ed. K.L. Mittal), Plenum Press, New York.

30
Creation of Novel Electromagnetic and Reactive Media from Microplasmas

Kunihide Tachibana

30.1
Introduction

Recently, small-scale plasmas of sub-millimeter size called *microplasmas* have been attracting much attention because of their properties that are different from traditional plasmas and a variety of potential applications [1, 2]. As an example, Figure 30.1 shows the territory of microplasmas on the map of plasma (electron) density n_e and characteristic size d [1]. Since the operating pressure p of a microplasma is generally high in accordance with Paschen's law, n_e becomes higher than 10^{12} cm^{-3} even though its ionization degree stays low. Therefore, an unexplored territory is opening up for the study of microplasmas. As a representative of microplasmas, one can think of a unit discharge cell of a plasma display panel (PDP), whose characteristic dimensions are 700 μm long × 200 μm wide × 150 μm deep, with a typical reported value of n_e on the order of 10^{13} cm^{-3} [3]. The PDP discharge cell is located at the entrance of the area known as the *meso-exotic area*" The intention is to promote performance within the territory in the desired direction according to the requirement of individual application technology.

Interesting features of a microplasma can be seen in Figure 30.2, where electron plasma frequency $\nu_{pe}(= \omega_{pe}/2\pi)$, free space wavelength λ corresponding to ω_{pe}, collision frequency of electrons ν_m, and Debye length λ_D are taken along the vertical axes as functions of n_e [4]. Let us take an example of n_e at 3.2×10^{14} cm^{-3}. Then, ν_{pe} becomes 2×10^{11} Hz (200 GHz) and the corresponding λ is 1.5 mm. If we assume the ionization degree α is about 10^{-5}, the corresponding gas density N becomes 3.2×10^{19} cm^{-3}, which coincides with the atmospheric pressure gas density at room temperature. In He gas, for instance, ν_m is estimated to be 1.2×10^{12} Hz at a mean electron energy of 1 eV. The Debye length λ_D corresponding to the value of n_e becomes 4×10^{-4} mm, which is much smaller than the typical size of a microplasma. These data suggest, in the first place, that a microplasma of our concern is collision dominant. Secondly, the plasma frequency can easily be on the order of 100 GHz (or sub-terahertz) range, which is very attractive for the study of the interaction of electromagnetic waves with microplasmas of millimeter to sub-millimeter sizes with the above-mentioned properties.

Industrial Plasma Technology. Edited by Yoshinobu Kawai, Hideo Ikegami, Noriyoshi Sato, Akihisa Matsuda, Kiichiro Uchino, Masayuki Kuzuya, and Akira Mizuno

ISBN: 978-3-527-32544-3

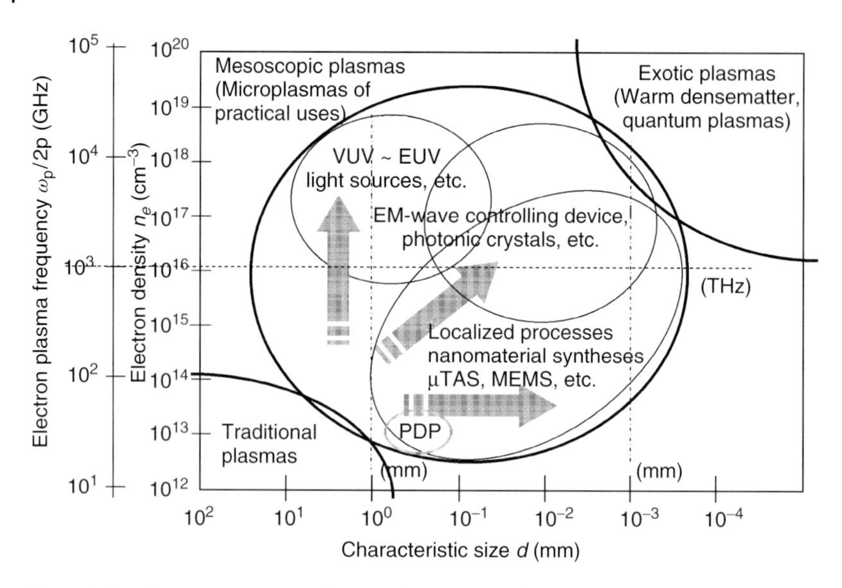

Figure 30.1 Characteristic area of microplasmas in a plane of spatial size d and electron density n_e.

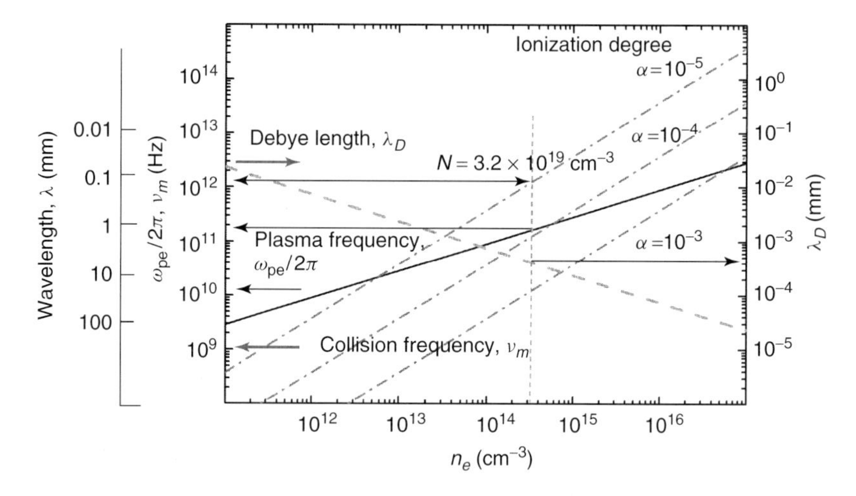

Figure 30.2 Relations of n_e with electron plasma frequency ω_{pe}, collision frequency (ν/m), and Debye length λ_D.

The electron temperature T_e, which is another important plasma parameter, is not in equilibrium with the gas temperature T_g even though the gas pressure is high as shown in Figure 30.3 [1]. This is due to the extremely short residence time within the small plasma volume even under a usual gas-flow condition and also due to the short duration of pulsed discharge mostly used for the generation of microplasmas. The degree of the nonequilibrium state can be designed by adjusting these conditions as illustrated in the figure.

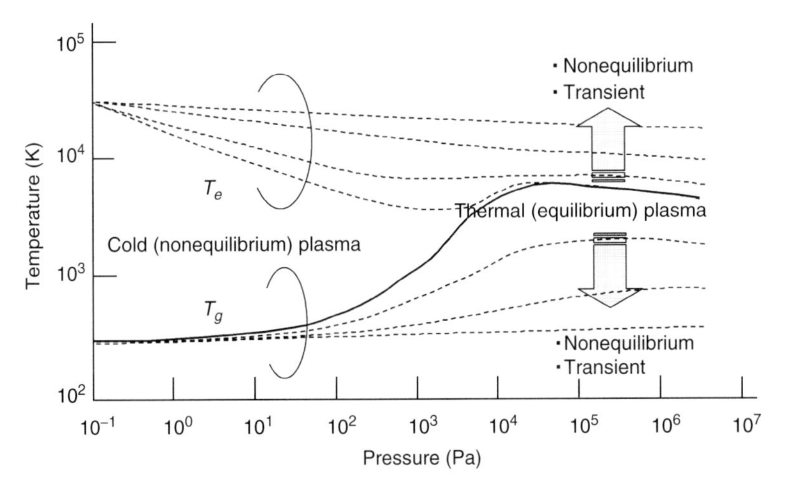

Figure 30.3 Pressure dependence of electron temperature T_e and gas temperature T_g.

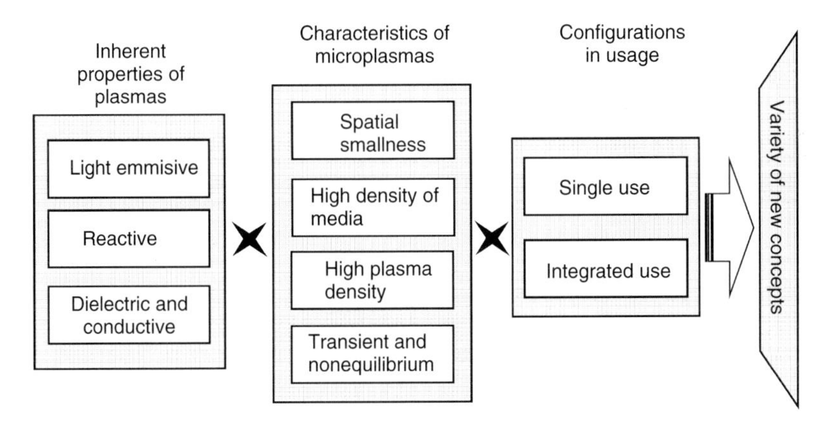

Figure 30.4 Scheme for the creation of new concepts by combinations.

The properties of microplasmas can be combined with the inherent properties of usual plasmas to create a variety of new concepts with the additional choice of the configuration in single or integrated use as indicated in Figure 30.4 [4]. In this article, some examples of new concepts created with microplasmas are described, such as on-demand reactive media for material-processing applications and variable electromagnetic media for applications to electromagnetic-wave-controlling devices.

30.2
Microplasma in Single Use as Plasma Jet

There are a variety of configurations that can be considered for the generation of microplasmas using power sources in the frequencies from DC to GHz range. In

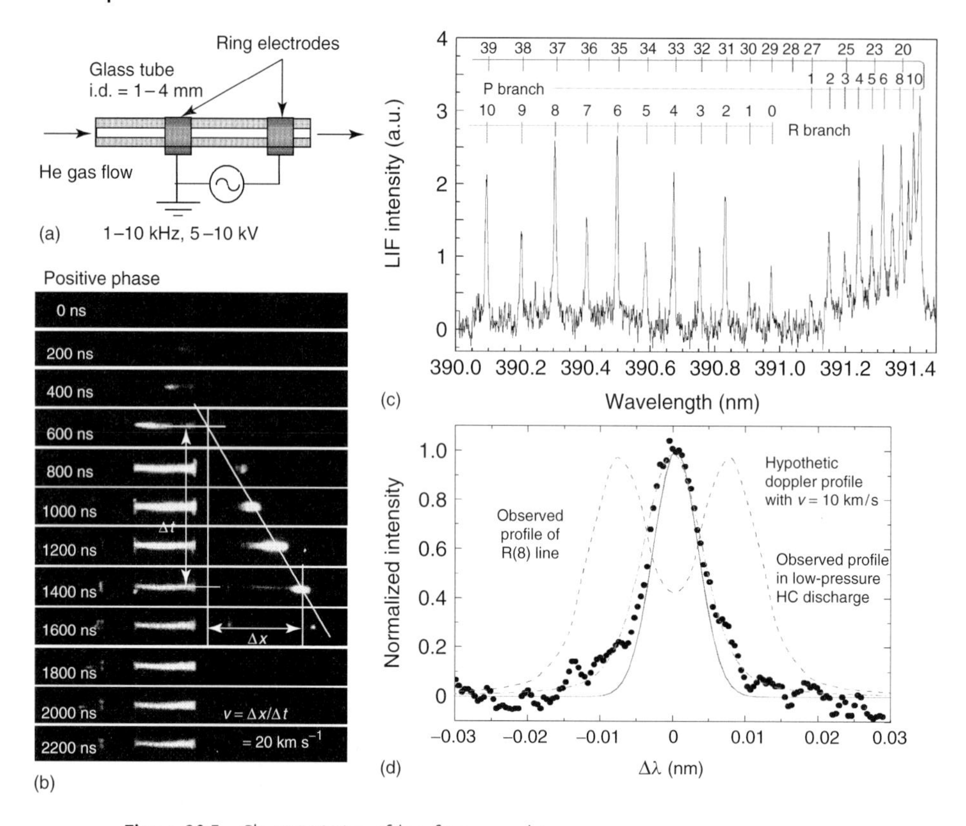

Figure 30.5 Characteristics of low-frequency-driven microplasma jet: (a) schematic structure, (b) time-resolved image of plasma jet, (c) rotational components of the first negative (0, 0) band of N_2^+, and(d) line profile of R(8) component.

most of these, microplasmas are confined in bounded spaces, but in some of them, plasma plumes are extended to free spaces. Here, as an example of the latter case, a typical microplasma jet with a simple design is introduced [5], since this has been attracting considerable interest in recent years because of its high potential for applications in various fields including biomedicine.

The structure of the microplasma jet is shown in Figure 30.5a [4]. A glass capillary tube of one to a few millimeters bore diameter was equipped with a couple of ring electrodes wound on its outside, which were connected to a low-frequency power source provided with sinusoidal or square-wave voltage. Rare gas such as He or Ar was fed through the capillary. This is actually a kind of DBD (dielectric barrier discharge) in a tubular configuration. Once the discharge was ignited, a long plasma plume a few centimeters long was ejected into ambient air. When the jet was observed by a fast gated ICCD (intensified charge coupled device) camera, it was seen that a series of bunched emissions were effused synchronously with the driving frequency in the positive phase of the applied voltage to the front electrode as seen in Figure 30.5b [4]. The apparent traveling speed was on the order of $10\,km\,s^{-1}$, which is larger than the gas-flow speed by 3 orders of magnitudes. The gas temperature T_g monitored by an LIF (laser induced fluorescence) method using the rotational spectrum of the first negative (0–0) band of N_2^+ ions remained around room temperature as shown in Figure 30.5c [6]. There was no noticeable Doppler shift observed corresponding to the apparent velocity; if the bunch is traveling with the apparent speed, split peaks should be observed by a 45° tilted configuration of the laser beam as indicated in Figure 30.5d [6]. From these observations and also theoretical models, the driving mechanism of these bulletlike bunches is attributed to the ionization waves similar to the case of positive corona discharge along the rare gas channel, where the electron drift velocity and the ionization coefficient are larger than in ambient air. Since the counterelectrode is at an infinite distance, the electric field is induced near the positive ion core in the bunch, because electrons are attracted toward the front electrode with positive potential.

As an example of its application, the use of this type of a plasma jet was attempted in the deposition of SiO_2 films using vaporized TEOS (tetraethoxysilane) as the source material onto Si substrates [7]. We employed two configurations; in the first one, TEOS, with He carrier gas, was supplied coaxially to the plasma jet, and in the second one, the TEOS flow was crossed with the plasma jet near the substrate surface as shown in Figure 30.6a. As a result, the deposition rate in the second case was about twice as large as in the first case, so only the data in the latter case is shown in Figure 30.6b. The reason for this rate difference may be attributed to difference in the transport of deposition precursors toward the substrate; in the first case, the precursors are produced in the jet, and some part of them is lost through the formation of powders during the transport. There are two points to be

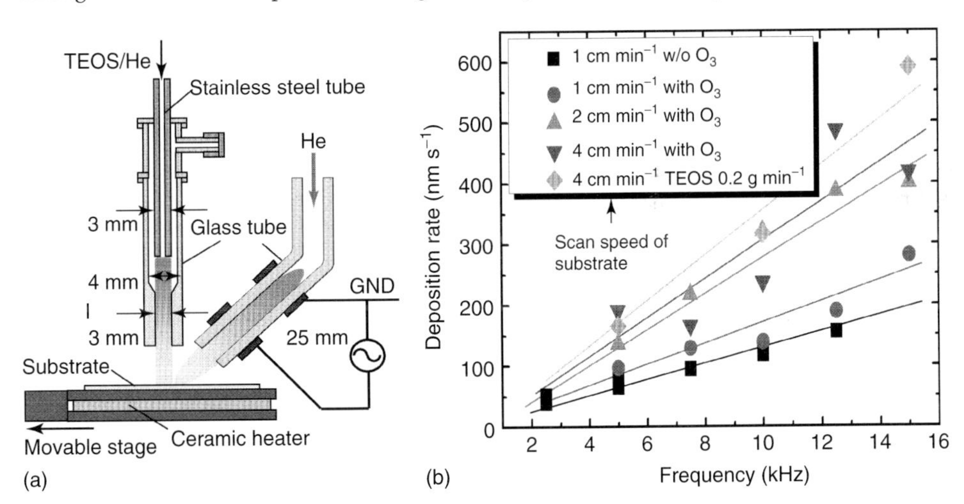

Figure 30.6 (a) Configuration for the deposition of SiO_2 film using microplasma jet with TEOS source. (b) Deposition rate at different scan speeds of the stage with or without O_3 supply condition.

noted here. First, when a small amount of ozone was admixed with the TEOS flow, a slight increase was observed in the deposition rate. Second, the deposition rate strongly depended on the translational speed of the substrate. This is probably due to the charging effect on the surface of deposited insulating film: the jet tends to spread wider avoiding the charged-up area on the film.

Other interesting applications are in environmental and biomedical fields, such as sterilization, dental, and dermatological treatments as well as in material-processing for biosensor devices [4]. High-performance microplasma jets for local treatments are beneficial for most of these applications. Improvements in instrumentation are taking place simultaneously with application technologies to lower the operating voltage by replacing one of the pair of electrodes by an internal electrode [8] or by using ultrahigh frequency (100-MHz range) power sources [9, 10]. Integrated assemblies of microplasma jets have also been developed for larger area processing [11, 12].

30.3
Microplasma Integration for Large-Area Material Processing

Microplasma-integrated devices will be useful for material processing over much larger areas, but simple and robust structures are desirable. For this purpose, we developed a mesh-type electrode device, where two metal meshes were covered with insulating material such as Al_2O_3 and stacked together [13]. By applying a sinusoidal or square-wave shaped bipolar pulse voltage between the metal electrodes, the surface DBD was ignited along the inner surface of each aligned opening of sub-millimeter size as shown in Figure 30.7a. Since each discharge is confined within the opening and there is no strong interaction between adjacent microdischarges in this scheme, the scale of the electrode assembly over several tens of square centimeters can be enlarged while retaining good uniformity.

As a modified structure of the mesh-type device, we developed also a fabric-type device, where insulated metal wires were weaved to form a flat electrode assembly as shown in Figure 30.7b [14]. We used a kind of plastic tubes as the insulating material. The DBD was generated between the warp and weft to cover uniformly the whole area of the fabric electrode. The merits of this structure are its easy fabrication and flexibility although there is a demerit of possible degradation of insulating material, causing contaminations depending on the kinds of applications.

As for the characterization of plasma parameters in these devices, we applied a millimeter-wave transmission technique at a frequency range up to 100 GHz [15]. For this purpose, the size of each opening was designed to be rectangular, with the longer side being larger than the wavelength of the transmitting millimeter-wave. When the electric vector of the electromagnetic wave is aligned vertical to the longer side, it can be transmitted through the opening even though the length of the shorter side is less than the wavelength just as in the case of a grid polarizer. An example of the attenuated transmission signal observed in the fabric-type device operated in He gas at atmospheric pressure is shown in Figure 30.7c after

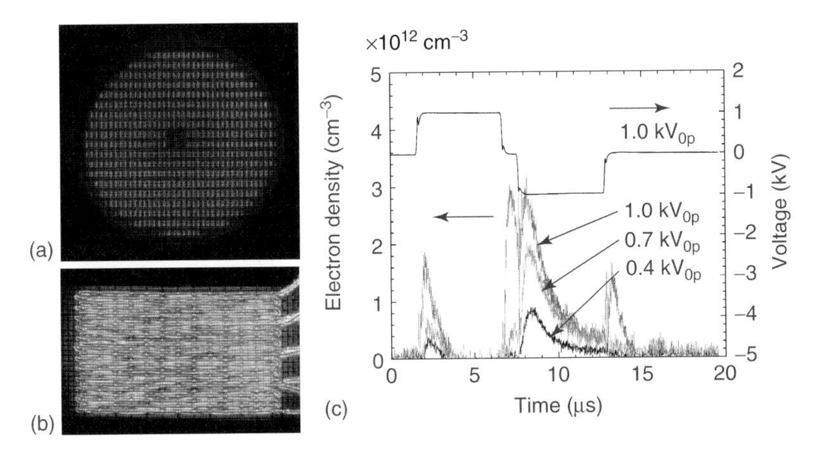

Figure 30.7 Examples of microplasma-integrated devices: (a) mesh-electrode type and (b) fabric-electrode type. (c) Temporal behavior of electron density by measured millimeter-wave transmittance in fabric-type device.

conversion into the electron density n_e together with the driving bipolar pulsed voltage waveform. In this example, the peak value varies according to the driving voltage on the order of 10^{12} cm^{-3} [14]. Similar results have also been obtained in the case of the mesh-type device [15].

Both of these electrode structures are fitted for the source gas flow through the plane vertically toward the treated substrate set parallel to the plane. As an example, we tried to treat a substrate with ITO (indium tin oxide) film in the down flow of Ar discharge with a small amount of water vapor transported through the fabric-type device. We noticed an increase in the work function of ITO film due to the additional oxidation probably by OH radicals produced in the discharge [4].

Another interesting feature of the fabric-type device is that it can be used in underwater conditions. As shown in Figure 30.8a, the device was set in an aqueous solution containing a small amount (3 wt%) of electrolyte (Na_2CO_3) together with a metal plate as the third electrode [16]. In this application, the warp was made of bare wire while the weft was insulated wire. By applying a DC voltage between the warp and the third electrode, hydrogen- or oxygen-containing microbubbles were successfully produced by electrolysis depending on the polarity of the applied voltage. Those bubbles were held on the surface of the fabric electrode. When a bipolar pulsed voltage of about 3-kV peak was applied between the warp and the weft at a certain growth stage of the bubble size, the occurrence of discharge at the inside of bubbles was noticed as seen in Figure 30.8b. From the Lissajous curve between the voltage and the accumulated charge, the net input power could be estimated. The observed optical emission spectrum in the H_2-containing bubbles were composed of atomic hydrogen H_α (656 nm) and H_β (486 nm) lines and a continuous bandlike structure due to H_2 transitions as seen in Figure 30.8c, while in O_2-containing bubbles, only a strong emission at 777 nm was observed. In the case of H_2 bubbles, the hydrogen activity was estimated from the change in pH

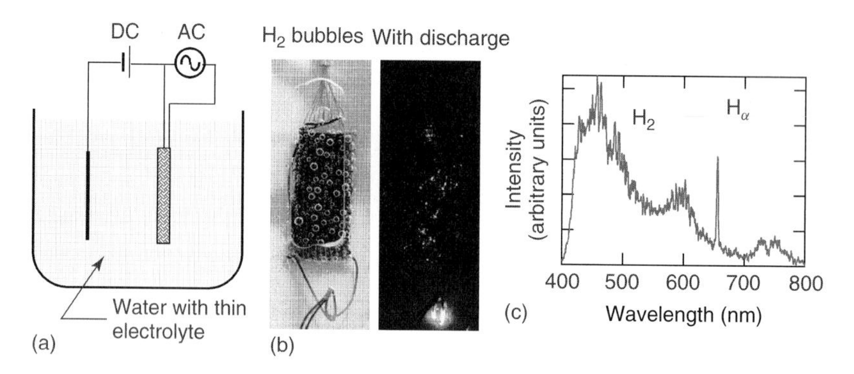

Figure 30.8 (a) Fabric-type electrode assembly used in aqua solution of electrolyte (Na_2CO_3, 3 wt.%). (b) H_2 bubbles generated by electrolysis and image of discharge in bubbles. (c) Observed spectra in the discharge.

number. It decreased according to the discharge time, and the net transport of H atoms and/or H^+ (probably in the form of H_3O^+) into the solution was confirmed [16]. In this situation, bubbles were not destroyed by the discharge, so that the transport was performed through the gas–liquid interface of bubbles.

This technique can be applied to the liquid treatment such as purification, sterilization, and disinfection by using reductants or oxidants produced in the bubbles. However, in order to make the process more realistic, the supplying speed of those agents should be enhanced to a practical level. Thus, the development of a cyclic process is essential for the production of bubbles, discharge in the bubbles, and destruction of the bubbles by discharge or forced flow, and efficient release of the produced reagents into the liquid.

30.4 Microplasma Integration for Electromagnetic Media

It is well known that a plasma exhibits dielectric property for high-frequency electromagnetic waves. According to the Drude model, its relative dielectric constant ε_p is given as a function of the frequency of electromagnetic waves by the following equation:

$$\varepsilon_p = 1 - \left(\frac{\omega_{pe}}{\omega}\right)^2 \frac{1}{1 - i(\nu_m/\omega)} \tag{30.1}$$

where $\omega_{pe} = (n_e e^2/\varepsilon_0 m)^{1/2}$ is the electron plasma (angular) frequency given by the electron charge e, mass m, and the dielectric constant in vacuum ε_0. Let us consider a situation where the medium is composed of a two-dimensional periodic array of columnar microplasmas. Then, the dispersion relation between the frequency ω and the wave number k depends on the propagation direction of electromagnetic waves. In this case, we can represent ε_p at an arbitrary position $r_{//}$ on the plane perpendicular to the columnar array by using Fourier expansion with reciprocal

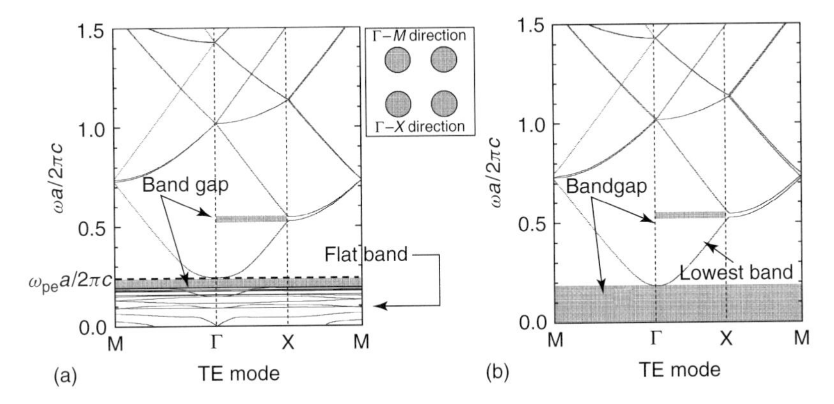

Figure 30.9 Dispersion relation (photonic band structure) for electromagnetic waves propagating through square array of columnar microplasmas with lattice constant a in (a) TA mode and (b) TE mode.

lattice vectors $\mathbf{G}_{//}$ as follows [17]:

$$\varepsilon(\mathbf{r}_{//}|\omega) = \sum \hat{\varepsilon}(\mathbf{G}_{//})\exp\left(i\mathbf{G}_{//}\cdot\mathbf{r}_{//}\right) \tag{30.2}$$

where $\hat{\varepsilon}(\mathbf{G}_{//})$ is given by

$$\hat{\varepsilon}\left(\mathbf{G}_{//}\right) = \varepsilon_d - f_v\left\{1 - \varepsilon_d - \frac{\omega_{pe}^2}{\omega(\omega - i\nu_m)}\right\} \quad \text{for } \mathbf{G}_{//} = 0$$

$$= f_v\left\{1 - \varepsilon_d - \frac{\omega_{pe}^2}{\omega(\omega - i\nu_m)}\right\}\frac{2J_1\left(\mathbf{G}_{//}\mathbf{R}\right)}{\mathbf{G}_{//}\mathbf{R}} \quad \text{for } \mathbf{G}_{//} \neq 0$$

Here J_1 is the first-order Bessel function, ε_d is the dielectric constant of the medium around the plasma columns of radius R, and f_v is the volumetric ratio of the plasma columns. It is noted here that this expression is based on the approximation of superposed plane-waves for the electromagnetic waves propagating through the periodic structure.

This situation is quite similar to the photonic crystals in the optical frequency range. However, the corresponding frequency range shifts toward sub-terrahertz range according to the lattice constant of our periodic structure of millimeter to sub-millimeter scales. This is also due to the corresponding frequency range of ω_{pe} for our microplasmas. As a typical example, we calculated the dispersion relation in the case of a two-dimensional square lattice with the lattice constant a under the following conditions: $\omega_{pe}a/2\pi c = 0.25$, $\nu_m = 0.5\omega_{pe}$, and $f_v = 0.3$. The results are shown in Figure 30.9 in two different situations where the electric vector of the electromagnetic waves is set perpendicular to the plasma columns in the TE mode (a) and parallel in the TM mode (b) [18]. It is noted that in both cases a bandgap appears in the propagation along $\Gamma - \mathrm{X}$ direction at around $\omega_a/2\pi c = 0.5$ (where c is the velocity of light) even above the cutoff frequency range. (The flattened band structures seen in the TE mode below the cutoff frequency are attributed to the propagation of surface waves around the plasma columns.)

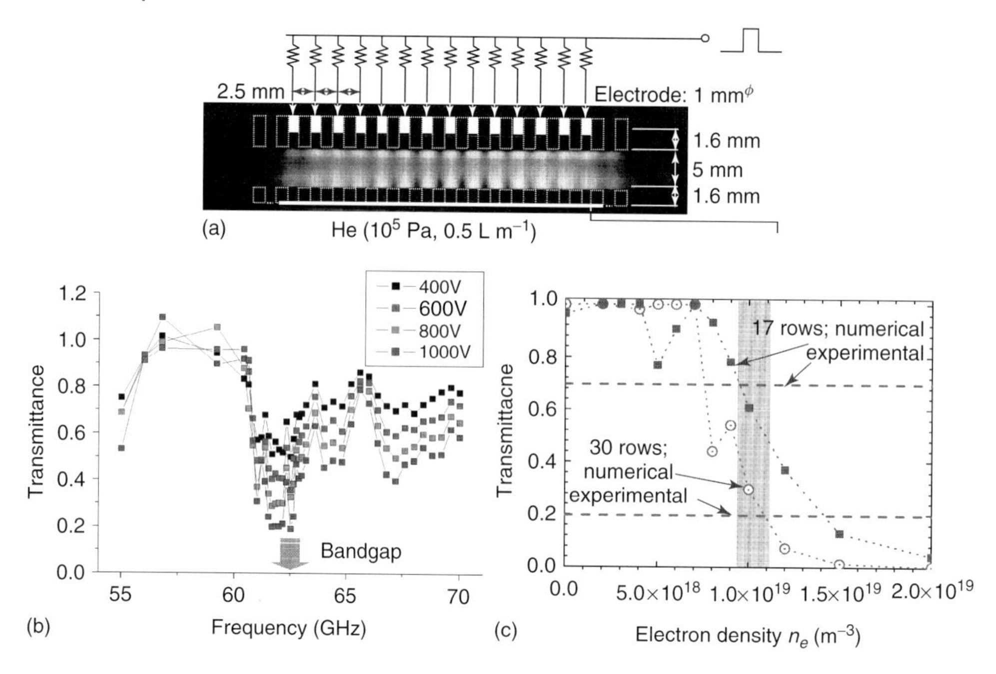

Figure 30.10 (a) Structure of microdischarge device for generation of periodic array of columnar microplasmas. (b) Measured frequency-dependent transmittance. (c) Determination of n_e by comparison of experimental values of transmittance with calculated results at two different numbers of rows of the microplasma array.

In order to verify experimentally the presence of the bandgap, we constructed a discharge device constituted of a square lattice array of columnar microplasmas as shown in Figure 30.10a with a lattice constant of 2.5 mm [19]. Each discharge unit had a pair of capillary electrodes of 1-mm bore and 1.6-mm depth separated by 5-mm distance. Pulsed bipolar voltage was applied between the pair through a series resistor. The relative transmittance T measured as a function of ω is plotted in Figure 30.10b. A sharp dip is seen at the frequency range around 62 GHz, which corresponds to the bandgap mentioned above. The location is mostly determined by the lattice constant and independent of the change in the applied voltage, that is, the electron density n_e, while the depth and the width depend on n_e. The depth depends also on the number of rows in the array along the propagation direction. We calculated T numerically by changing n_e at two values for the number of rows (see [18] for the calculation method and the assumptions). The results are plotted in Figure 30.10c. In comparison with the experimental results of T measured in those two cases, it is seen that coincidence between the experimental results and the calculated results is obtained at the same value of electron density, that is, $n_e = 1.0 \times 10^{13}\text{cm}^{-3}$. From this result, we can say that n_e is successfully determined by this method.

30.5 Concluding Remarks and Perspectives

We have demonstrated that a microscale plasma operated in small space at higher pressure can have different properties than those of a traditional large-scale plasma usually operated at lower pressure. Taking the advantage of these properties, we can create new reactive media for a variety of material-processing technologies under nonequilibrium and transient conditions. In recent years, microplasmas have been successfully applied to the syntheses of various nanoparticles [20] and nonequilibrium materials [21]. Other interesting directions are the applications to environmental and biomedical issues.

An integrated scheme of microplasmas enables material processing over larger areas. We have shown some examples of the integrated structures suitable for the purpose. In a more sophisticated application, a novel idea of "plasma stamps" has been developed recently for designed pattern processing [22]. Unfortunately, other interesting applications of microplasma-integrated structures cannot be discussed here due to the limited space, but their usefulness has been shown for light sources [23] and photosensors [24].

A new concept of electromagnetic medium has also been developed by using a periodic structure of microplasmas. It exhibits a property similar to that of a photonic crystal, and can be applied to the control of electromagnetic waves in the range from a few tens of gigahertz to 1 THz. The merit of using plasma as an electromagnetic medium is that the structure (lattice constant) and the dielectric constant can be varied by the operation conditions, leading to an idea of dynamic control devices. It has also been suggested that in a combination with inductance components such as helically shaped electrodes both the permittivity and permeability can be tuned for the realization of metamaterials which may have negative refractive indexes [4, 25].

Acknowledgments

The author acknowledges O. Sakai, T. Shirafuji, Y. Ito, and K. Urabe for their collaboration in the experimental research work presented in this article. The work has been partially supported by Grant-in-Aids for Scientific Research on Priority Areas from Ministry of Education, Culture, Sports, Science and Technology, Japan.

References

1. Tachibana, K. (2006) *IEEJ Trans.*, **1**, 145.
2. Becker, K.H., Schoenbach, K.H., and Eden, J.G. (2006) *J. Phys. D: Appl. Phys.*, **39**, R55.
3. Hassaballa, S., Sonoda, Y., Tomita, K., Kim, Y.-K., Uchino, K., Hatanaka, H., Kim, Y.-M., Park, C.-H., and Muraoka, K. (2005) *Jpn. J. Appl. Phys. Pt. 2*, **44**, L442.
4. Tachibana, K., Ishii, S., Terashima, K., and Shirafuji, T. (2009) *Microplasmas – Fundamentals and Applications*, Ohm-sha, Tokyo (in Japanese).

5. Teschke, M., Kedzierski, J., Finantu-Dinu, E.G., Korzec, D., and Engemann, J. (2005) *IEEE Trans. Plasma Sci.*, **33**, 310.
6. Urabe, K., Ito, Y., Tachibana, K., and Ganguly, B.N. (2008) *Appl. Phys. Exp.*, **1**, 066004.
7. Ito, Y., Urabe, K., Takano, N., and Tachibana, K. (2008) *Appl. Phys. Exp.*, **1**, 067009.
8. Lu, X.-P., Jiang, Z.-H., Xiong, Q., Tang, Z.-Y., and Pan, Y. (2008) *Appl. Phys. Lett.*, **92**, 151504.
9. Ito, T. and Terashima, K. (2002) *Appl. Phys. Lett.*, **80**, 2648.
10. Ichiki, T., Koidesawa, T., and Horiike, Y. (2003) *Plasma Sources Sci. Technol.*, **12**, s16.
11. Cao, Z., Walsh, J.L., and Kong, M.G. (2009) *Appl. Phys. Lett.*, **94**, 021501.
12. Tachibana, K. (2009) *J. Inst. Electrost. Jpn.*, **33**, 61 (in Japanese).
13. Sakai, O., Kishimoto, Y., and Tachibana, K. (2005) *J. Phys. D: Appl. Phys.*, **38**, 431.
14. Sakai, O. and Tachibana, K. (2007) *J. Phys. Conf. Ser.*, **86**, 012015.
15. Sakai, O., Sakaguchi, T., Ito, Y., and Tachibana, K. (2005) *Plasma Phys. Control. Fusion*, **47**, B617.
16. Sakai, O., Kimura, M., Shirafuji, T., and Tachibana, K. (2008) *Appl. Phys. Lett.*, **93**, 231501.
17. Kuzmiak, V. and Maradudin, A.A. (1997) *Phys. Rev. B*, **55**, 7427.
18. Sakai, O., Sakaguchi, T., and Tachibana, K. (2007) *J. Appl. Phys.*, **101**, 073304.
19. Sakaguchi, T., Sakai, O., and Tachibana, K. (2007) *J. Appl. Phys.*, **101**, 073305.
20. Sankaran, R.M., Holunga, D., Flagan, R.C., and Giapis, K.P. (2005) *Nano Lett.*, **5**, 537.
21. Shimizu, Y. (2008) *Oyobutsuri*, **77**, 404 (in Japanese).
22. Lucas, N., Ermel, V., Kurrat, M., and Buttgenbach, S. (2008) *J. Phys. D: Appl. Phys.*, **41**, 215202.
23. Park, S.-J., Kim, K.S., and Eden, J.G. (2005) *Appl. Phys. Lett.*, **86**, 221501.
24. Eden, J.G., Park, S.-J., Ostrom, N.P., McCain, S.T., Wagner, C.J., Vajak, B.A., Chen, J., Liu, C., von Allen, P., Zenhausern, F., Sadler, D.J., Jensen, C., Wilcox, D.L., and Ewing, J.J. (2003) *J. Phys. D: Appl. Phys.*, **36**, 2869.
25. Sakai, O., Shimomura, T., Lee, D.-S., and Tachibana, K. (2008) Proceedings of the 2nd International Congress on Advanced Electromagnetic Materials in Microwaves and Optics (Metamaterials 2008), Pamplona.

31
Nanoblock Assembly Using Pulse RF Discharges with Amplitude Modulation

Shinya Iwashita, Hiroshi Miyata, Kazunori Koga, and Masaharu Shiratani

31.1
Introduction

Nanomaterials have made themselves attractive for an increasing number of applications, including electronics, medical components, fillers, catalysts, and fuel cells [1–6]. There are two approaches to fabricate nanomaterials – one is the top-down approach such as etching and deposition processes used in semiconductor manufacturing [7–11] and the other is the bottom-up one such as fabricating DNA-coated nanowires [12–15]. Recent progress in nanomaterials having assembled-structures and -components which exhibit novel and significantly improved physical, chemical, and biological properties, because of their nanoscale size, has required the development of the bottom-up approach. There have been many different bottom-up methods for assembly of nanobuilding blocks. Charged aerosol nanoblocks are deposited on a silicon-oxide surface patterned with lines of induced surface charges [16]. Metallic nanoblocks are manipulated on an atomically flat graphite surface by atomic force microscopy techniques [17]. Micro- and nanoblocks in liquids are manipulated and immobilized both by electroosmosis and dielectrophoresis for biological applications [18]. A wide variety of bottom-up methods for assembly of nanoblocks, however, are required to be developed more precisely to realize "molecular manufacturing," which is claimed by K. Eric Drexler [19].

We have proposed a novel one-step method for assembly of nanoblocks using reactive plasmas [20–22]. Our method consists of production of nanoblocks and radicals (adhesives) in reactive plasmas, their transport toward a substrate, and their arrangement on the substrate using pulse radio frequency (RF) discharges with the amplitude modulation (AM) of discharge voltage. For the success of this method, control of the size of nanoblocks as well as their accurate assembly in gas phase by controlling their agglomeration is important.

In this chapter, we discuss formation mechanisms of nanoblocks and their agglomeration by pulse RF discharges in Section 31.2, and describe their assembly in gas phase by controlling their agglomeration by pulse RF discharges with AM in Section 31.3.

Industrial Plasma Technology. Edited by Yoshinobu Kawai, Hideo Ikegami, Noriyoshi Sato, Akihisa Matsuda, Kiichiro Uchino, Masayuki Kuzuya, and Akira Mizuno

ISBN: 978-3-527-32544-3

31.2 Formation of Nanoblocks and Their Agglomeration

Nanoblocks are formed in 13.56 MHz RF discharges of $Si(CH_3)_2(OCH_3)_2$ diluted with Ar as shown in Figure 31.1 [20–26]. During the discharging period, nanoblocks are nucleated in the plasma-sheath boundary region around the powered electrode, and they tend to grow there resulting from the balance between ion drag force which pushes nanoblocks toward the powered electrode and electrostatic force which repels them toward plasma bulk [21, 22]. After nucleation of nanoblocks, there are two possible growth mechanisms: agglomeration and CVD process. The size (diameter) of nanoblocks is determined by the discharging period T_{on}. The pulse RF discharge, therefore, is a promising method for controlling their size [20–22].

Figure 31.2 shows time evolution of the size and density of nanoblocks after turning off a pulse RF discharge of $T_{on} = 0.3$ s [27]. After turning off the discharge, the size of nanoblocks increases, whereas their density decreases owing to their agglomeration. The decrease in number density is given by

$$\frac{\partial n_p}{\partial t} = D\frac{\partial^2 n_p}{\partial x^2} - Kn_p^2 - \frac{n_p}{\tau_p} \tag{31.1}$$

where D, K, and τ_p are the diffusion coefficient of nanoblocks, their agglomeration coefficient, and the characteristic time of their transport to a substrate, respectively. Assembly of nanoblocks in gas phase using their agglomeration needs to control their transport time from their generation region toward a substrate. Such control is described in the next section.

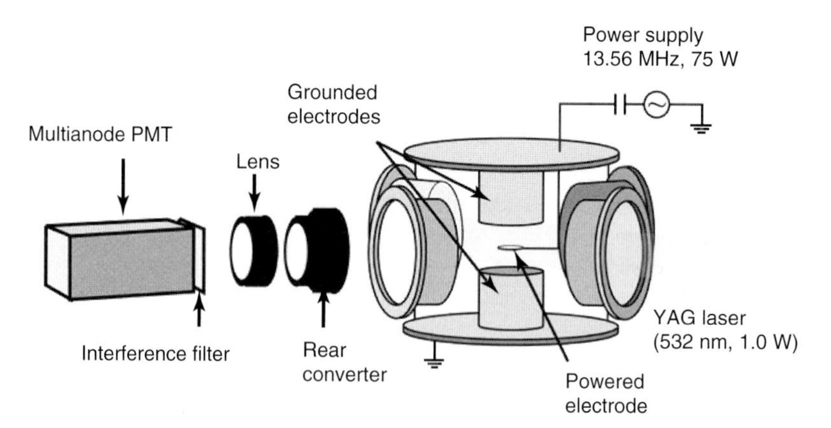

Figure 31.1 Experimental setup of two-dimensional photon-counting laser-light-scattering (2DPCLLS) system together with the reactor. We have proposed the high sensitivity 2DPCLLS method to detect nanoblocks down to 1 nm in size formed in CVD plasmas $\frac{\partial n_p}{\partial t} \cong -Kn_p^2$.

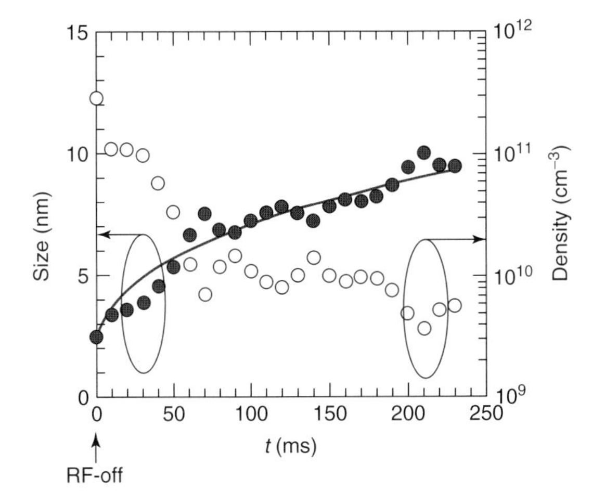

Figure 31.2 Time evolution of size and density of nanoblocks after turning off pulse RF discharge having $T_{on} = 0.3$ seconds without AM. Ar 40 sccm, $Si(OCH_3)_2(CH_3)_2$ 0.2 sccm, 1.0 torr, $V_{pp} = 828$ V, and 75 W. Nanoblocks grow due to agglomeration after turning off the discharge. When the nanoblock dynamics is dominated by agglomeration, Eq. (31.1) can be simplified as $\frac{\partial n_p}{\partial t} \cong -Kn_p^2$. To determine the agglomeration coefficient K, the data are fitted using equation in [27] as shown by the solid line. The coefficient K is $7.1 \times 10^{-10} cm^3 s^{-1}$.

31.3
Assembly of Nanoblocks in Gas Phase Using Agglomeration

In this section, we describe transport control of nanoblocks by pulse RF discharges without and with AM as shown in Figure 31.3 [21, 22]. Figure 31.4 shows the cross-sectional view of the reactor and the observation area using the 2DLLS method. After turning off unmodulated discharges, nanoblocks are transported at a velocity of 6–9 cm s^{-1} from their generation region around the powered electrode toward the upper grounded electrode due to temperature gradient, and they are agglomerated with each other during the transport as shown in Figure 31.5a. The transport is determined by the balance between thermophoretic force and gas viscous force [28]. Their velocity v_p is given by

$$v_p = \frac{3p\lambda \nabla T}{\pi n_g m_g v_g T} \tag{31.2}$$

where n_g is the number density of gas molecules, m_g and v_g their mass and thermal velocity, p the gas pressure, λ the mean free path of gas molecules, and T and ∇T the gas temperature and its gradient. The experimental results described elsewhere [21, 22] show the velocity of nanoblocks can be well expressed by Eq. (31.2).

With AM, nanoblocks are transported at a velocity more than 60 cm s^{-1} from their generation region around the powered electrode toward the upper grounded electrode during the modulation period as shown in Figure 31.5b. Just after the initiation of the modulation, electrostatic force drives nanoblocks, and then

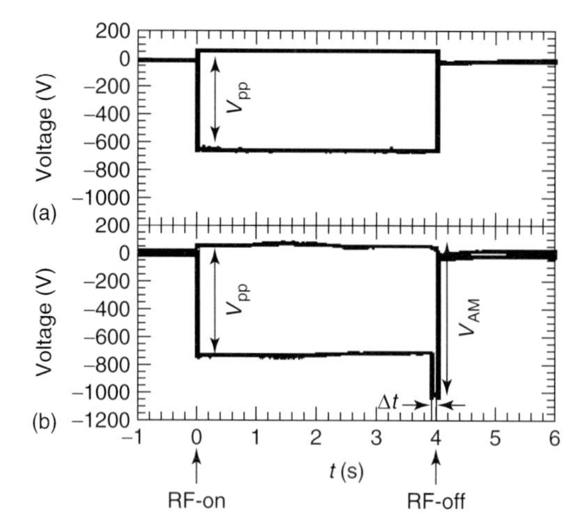

Figure 31.3 Envelope of discharge voltage of single pulse discharge without (a) and with (b) AM. $V_{pp} = 816\,\text{V}$, $V_{AM} = 1193\,\text{V}$, $T_{on} = 4.0\,\text{s}$, and $\Delta t = 100\,\text{ms}$.

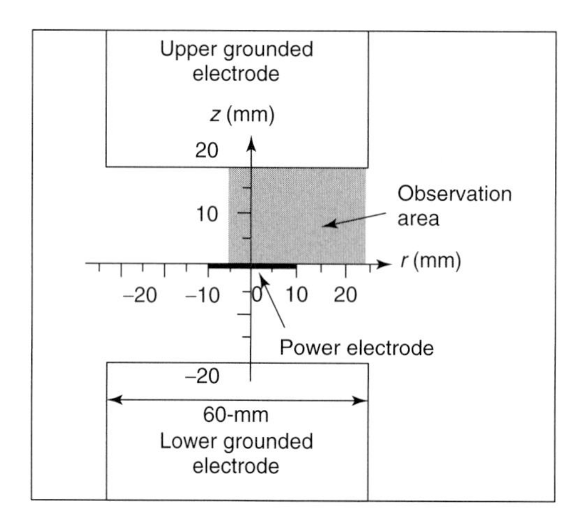

Figure 31.4 Cross-sectional view of the reactor and observation area of 2DLLS measurements with an ICCD camera. Shaded area is the observation area in this study. The scattered light from nano-blocks is detected at right angles with the ICCD camera with an interference filter of wavelength of 532 nm and FWHM of 1 nm.

ion drag force drives them toward the upper grounded electrode as shown in Figure 31.6. The larger nanoblocks (longer T_{on}) need the longer Δt and higher V_{AM} for their rapid transport because of their large inertia. The key parameters for the rapid transport of nanoblocks are the period and voltage of the modulation and asymmetry of the discharges, which is characterized by the DC self-bias voltage [21, 22]. It should be noted that although most nanoblocks are neutral, some

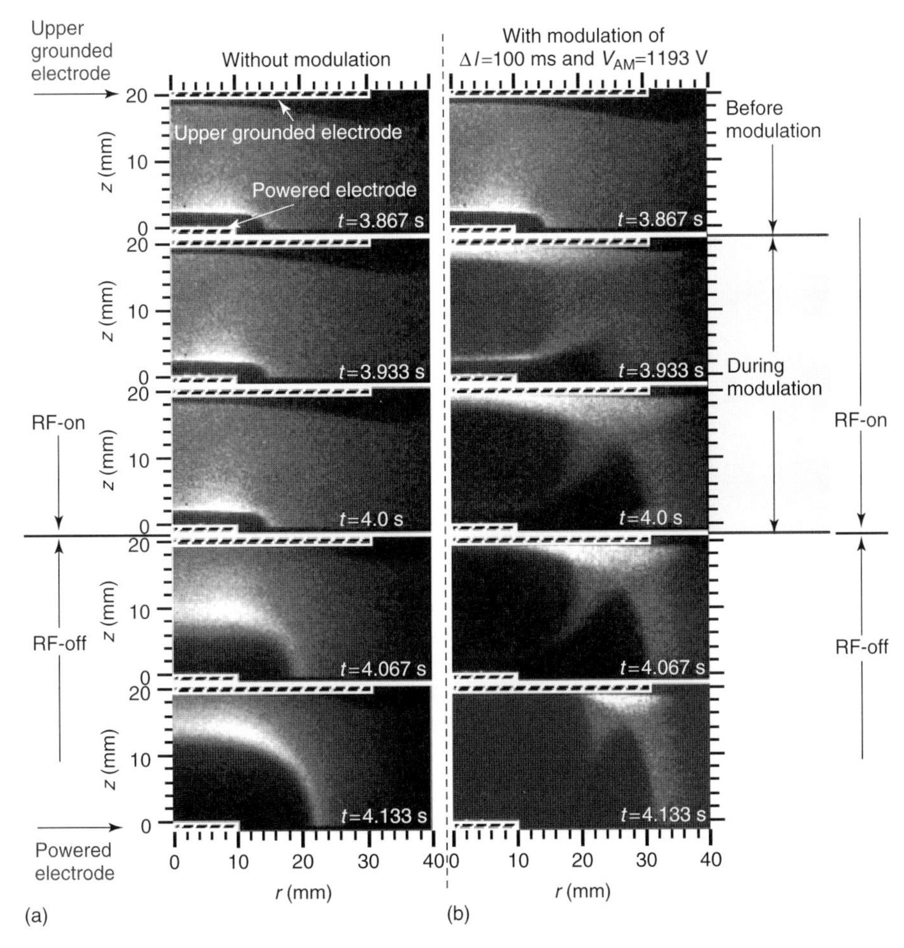

Figure 31.5 Two-dimensional spatial images of LLS intensity without (a) and with (b) AM as a parameter of time *t*. Ar 40 sccm, $Si(CH_3)_2(CH_3)_2$ 0.2 sccm, 1.0 torr, 75 W, $V_{pp} = 816\,V$, $V_{AM} = 1193\,V$, $V_{dc} = -350\,V$, $T_{on} = 4.0\,s$, and $\Delta t = 100\,ms$. Without AM, nanoblocks of 26 nm in size are transported at a velocity of $6-9\,cm\,s^{-1}$ due to temperature gradient after turning off discharges, while with AM they are transported rapidly at a velocity more than $60\,cm\,s^{-1}$ during the modulation period.

of them turn negatively charged due to charge fluctuation and such negatively charged nanoblocks are driven by electrostatic and ion drag forces [21–22, 29, 30]. The agglomeration frequency with AM is probably much lower than that without AM because the velocity of nanoblocks during the modulation period is at least seven times as high as that after turning off unmodulated discharges, and some of them are charged negatively and such partial charging suppresses significantly their agglomeration [21–22, 31]. We have also realized three dimensional transport of nanoblocks by AM discharges [32]. In short, the method utilizing AM discharges is a promising one for assembly of nanoblocks in gas phase using agglomeration.

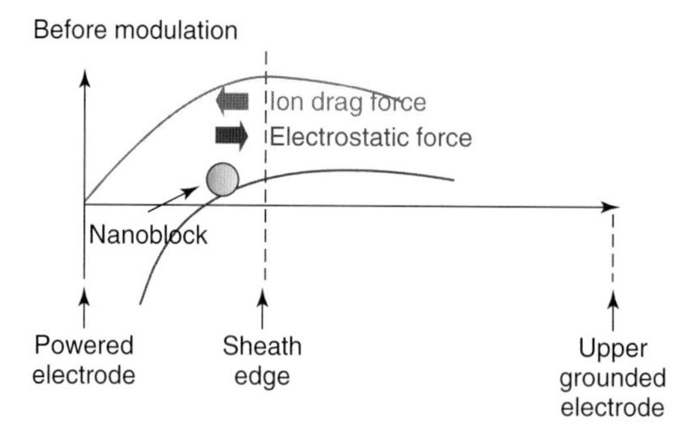

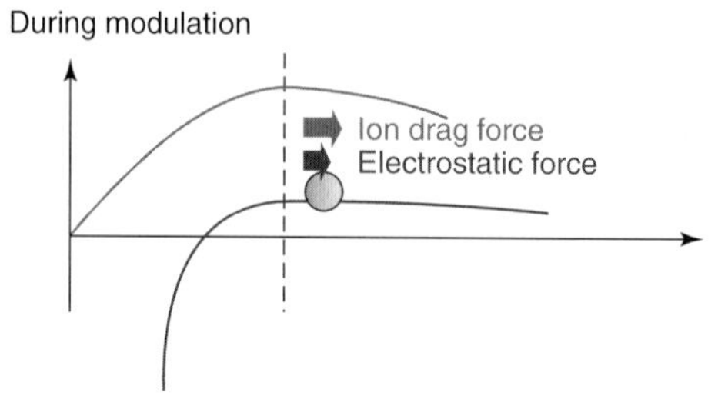

Figure 31.6 Mechanisms of rapid transport of nanoblocks during modulation. Immediately after the initiation of the modulation, electrostatic force drives nanoblocks, and then ion drag force drives them toward the upper grounded electrode.

31.4 Conclusions

In this chapter, we have proposed a novel one-step method for assembly of nanoblocks using reactive plasmas. The following conclusions are obtained.

1) Nanoblocks are generated and grow via agglomeration and CVD during the discharging period. The size of nanoblocks is controlled by the duration of pulse RF discharges.
2) Nanoblocks are agglomerated with each other during their transport. The method utilizing AM discharges is a promising one for assembly of nanoblocks in gas phase using agglomeration because transport velocity of nanoblocks is realized by the AM discharges.

Acknowledgments

We are grateful to Prof. Y. Watanabe and Dr S. Nunomura for valuable discussions. We also would like to acknowledge the assistance of Messers T. Kinoshita and H. Matsuzaki who contributed greatly to the preparation of the experimental setup. This work was partially supported by a grant from the Japan Society of the Promotion of Science and from New Energy and Industrial Technology Development Organization.

References

1. Joachim, C., Gimzewski, J.K., and Airam, A. (2000) *Nature*, **408**, 541.
2. Rutkevych, P.P., Ostrikov, K., and Xu, S. (2007) *Phys. Plasmas*, **14**, 043502.
3. Gill, V., Guduru, P.R., and Sheldon, B.W. (2008) *Int. J. Solids Struct.*, **45**, 943.
4. Ennen, I., Hoink, V., Weddemann, A., Hutten, A., Schmalhorst, J., Reiss, G., Waltenberg, C., Jutzi, P., Weis, T., Engel, D., and Ehresmann, A. (2007) *J. Appl. Phys.*, **102**, 013910.
5. Khanduja, N., Selvarasah, S., Chen, C.L., Dokmeci, M.R., Xiong, X., Makaram, P., and Busnaina, A. (2007) *Appl. Phys. Lett.*, **90**, 083105.
6. Nel, A., Xia, T., Madler, L., and Li, N. (2006) *Science*, **311**, 622.
7. Spears, K.G., Robinson, T.J., and Roth, R.M. (1986) *IEEE Trans. Plasma Sci.*, **PS14**, 179.
8. Watanabe, Y., Shiratani, M., Kubo, Y., Ogawa, I., and Ogi, S. (1988) *Appl. Phys. Lett.*, **53**, 1263.
9. Selwyn, G.S., Singh, J., and Benne, R.S.J. (1989) *J. Vac. Sci. Technol.*, **A7**, 2758.
10. Watanabe, Y. (2006) *J. Phys. D*, **39**, R329.
11. Lieberman, M.A. and Lichtenberg, A.L. (2005) *Principles of Plasma Discharges and Materials Processing*, 2nd edn, John Wiley & Sons, Inc., p. 571 .
12. Morrow, T.J., Li, M., Kim, J., Mayer, T.S., and Keating, C.D. (2009) *Science*, **323**, 352.
13. Ostrikov, K. and Murphy, A.B. (2007) *J. Phys. D*, **40**, 2223.
14. Zhang, S. (2003) *Nat. Biotechnol.*, **21**, 1171.
15. Nunomura, S., Kita, M., Koga, K., Shiratani, M., and Watanabe, Y. (2005) *Jpn. J. Appl. Phys.*, **44**, L1509.
16. Krinke, T.J., Fissan, H., Deppert, K., Magnusson, M.H., and Samuelson, L. (2001) *Appl. Phys. Lett.*, **78**, 3708.
17. Dietzel, D., Mönninghoff, T., Jansen, L., Fuchs, H., Ritter, C., Schwarz, U.D., and Schirmeisen, A. (2007) *J. Appl. Phys.*, **102**, 084306.
18. Heeren, A., Luo, C.P., Henschel, W., Fleischer, M., and Kern, D.P. (2007) *Microelectron. Eng.*, **84**, 1706.
19. Drexler, K.E. (1992) *Nanosystems*, John Wiley & Sons, Inc., p. 411.
20. Nunomura, S., Kita, M., Koga, K., Shiratani, M., and Watanabe, Y. (2006) *J. Appl. Phys.*, **99**, 083302.
21. Koga, K., Iwashita, S., and Shiratani, M. (2007) *J. Phys. D: Appl. Phys.*, **40**, 2267.
22. Shiratani, M., Koga, K., Iwashita, S., and Nunomura, S. (2008) *Faraday Discuss.*, **137**, 127.
23. Lieberman, M.A. and Lichtenberg, A.L. (2005) *Principles of Plasma Discharges and Materials Processing*, 2nd edn, John Wiley & Sons, Inc., p. 649.
24. Miyahara, H., Iwashita, S., Miyata, H., Matsuzaki, H., Koga, K., and Shiratani, M. (2009) *Plasma Fusion Res., SERIES*, **8**, 700.
25. Matsuoka, Y., Shiratani, M., Fukazawa, T., Watanabe, Y., and Kim, K.S. (1999) *Jpn. J. Appl. Phys.*, **38**, 4556.
26. Koga, K., Matsuoka, Y., Tanaka, K., Shiratani, M., and Watanabe, Y. (2000) *Appl. Phys. Lett.*, **77**, 196.
27. Iwashita, S., Morita, M., Matsuzaki, H., Koga, K., and Shiratani, M. (2008) *Jpn. J. Appl. Phys.*, **47**, 6875.
28. Hinds, W.C. (1982) *Aerosol Technology*, John Wiley & Sons, Inc., p. 153.
29. Lieberman, M.A. and Lichtenberg, A.L. (2005) *Principles of Plasma Discharges and Materials Processing*, 2nd edn, John Wiley & Sons, Inc., p. 165.
30. Song, Y.P., Field, D., and Klemperer, D.F. (1990) *J. Appl. Phys.*, **23**, 673.
31. Nunomura, S., Shiratani, M., Koga, K., Kondo, M., and Watanabe, Y. (2008) *Phys. Plasmas*, **15**, 080703.
32. Iwashita, S., Miyata, H., Matsuzaki, H., Koga, K., and Shiratani, M. (2009) *Plasma Fusion Res., SERIES*, **8**, 582.

32
Thomson Scattering Diagnostics of Discharge Plasmas

Kiichiro Uchino and Kentaro Tomita

32.1
Introduction

Thomson scattering is the light scattering by free electrons in plasmas. Because the scattering cross-section is determined by the size of the electron, the probability of the light scattering is low. However, the use of a high intensity Q-switched laser makes it possible to detect scattered lights. Quantitative measurements of the intensity and the spectrum of the scattered lights give the values of electron density and electron temperature. In fact, the laser Thomson scattering (LTS) method has been well established as the most reliable diagnostic method of electron density and electron temperature in plasmas for fusion research.

Thomson scattering is characterized by a parameter α ($= 1/k\lambda_d$, where k is a differential wave-number and λ_d is the Debye length) [1]. In the limit of $\alpha \ll 1$ (incoherent regime), scattering is induced by electrons which are independently moving. The scattered light intensity is proportional to the electron density n_e, and therefore, the absolute value of n_e can be obtained by performing absolute calibration of the detection optics. Such calibration can be easily achieved by detecting Rayleigh scattering lights from the chamber filled at a certain pressure with a gas having a known Rayleigh scattering cross-section. The scattered spectrum has a Gaussian shape which is caused by Doppler broadening due to thermal motions of electrons, and then, the electron temperature T_e can be obtained from the measurement of the spectral spread of the scattered lights. If the electron energy distribution function (EEDF) does not follow the Maxwellian distribution, the spectral shape gives EEDF itself. In addition, because the detected lights are from the volume determined by the intersection of the laser beam and the observation sight line, obtained n_e and T_e are local values. On the other hand, for the case of $\alpha > 1$ (collective regime), the collective behavior of electrons concerns the scattering. The scattered spectrum consists of the ion term which is due to electron motions following ions and the electron term which is due to electron collective behavior (electron plasma wave). Precise measurements of both terms give n_e, T_e, and the ion temperature T_i.

Industrial Plasma Technology. Edited by Yoshinobu Kawai, Hideo Ikegami, Noriyoshi Sato, Akihisa Matsuda, Kiichiro Uchino, Masayuki Kuzuya, and Akira Mizuno

ISBN: 978-3-527-32544-3

The authors have been developing LTS as a diagnostic method for industrially applied plasmas. First, we have applied LTS to low-pressure processing plasmas, and recently, are extending to microdischarge plasmas. In this article, we will describe about these efforts of ours. For the application of LTS to industrially applied plasmas, we are often faced with following problems. First, signals are difficult to be detected due to low electron densities. In order to solve this problem, we adopt the signal accumulation over many laser shots because plasmas are maintained continuously or produced repeatedly at a high frequency. Second, the strong stray lights which are the laser lights scattered by windows and chamber walls can easily swamp the scattered signal lights. Because the spectral spread of the stray lights is determined by the instrumental function of the spectrometer used for the detection optics system, the problem can be resolved by using a triple-grating spectrometer (TGS) which has high stray-light rejection characteristics. Third, the high laser energy may change values of electron density and temperature. For this problem, we must examine the laser conditions (energy, wavelength, pulse width, and so on) which do not induce laser perturbations.

The application of LTS to low-pressure plasmas is reviewed in [2, 3] and described briefly in Section 32.2 For microplasmas which are recently being developed, LTS is the unique diagnostic method for some cases. As an example in the incoherent regime, measurements of microplasmas produced for the plasma display panel (PDP) is introduced in Section 32.3. As an example in the collective regime, LTS applied to high density plasmas produced for the extreme ultraviolet (EUV) light source which will be used for next generation lithography is described in Section 32.4.

32.2
Application to Low-Pressure Plasmas

First, we think about the LTS signal intensity from low-pressure discharge plasmas. Following setup can be considered. We use a frequency doubled Q-switched YAG laser (wavelength 532 nm, photon energy $E_{ph} = 3.7 \times 10^{-19}$ J, pulse width 10 nanoseconds) as a light source. The laser beam is focused into the plasma with a spot diameter less than 1 mm. The scattered lights from the volume determined by the laser spot diameter and the length $\Delta L = 10$ mm along the laser beam are collected by a lens with a solid angle of $\Delta\Omega = 0.01$ *sr*. The detection direction is 90° from the laser beam and the laser light polarization. The scattered lights are spectroscopically analyzed by a spectrometer (transmission efficiency $\eta = 0.4$) and detected by a photomultiplier tube (quantum efficiency $\varepsilon = 0.1$). The photoelectron number N_{pe} obtained at the surface of the photocathode of the photomultiplier tube can be expressed as follows.

$$N_{pe} = \left(E_L/E_{ph}\right) \cdot n_e \cdot \Delta L \cdot \left(d\sigma_T/d\Omega\right) \cdot \Delta\Omega \cdot \eta \cdot \delta \cdot \varepsilon \qquad (32.1)$$

Here, $d\sigma_T/d\Omega (= 8 \times 10^{-30}\ m^2 sr^{-1})$ is the Thomson scattering differential cross-section and $\delta (= 0.1)$ is the detected portion of the Thomson scattered spectrum.

When the electron density $n_e = 10^{17}\ \mathrm{m}^{-3}$, N_{pe} is estimated to be 0.1. If $n_e < 10^{17}\ \mathrm{m}^{-3}$, the photon-counting method should be adopted for the signal detection.

We first applied LTS to an ECR processing plasma produced in an argon gas [4]. Because the electron density was of the order of $10^{18}\ \mathrm{m}^{-3}$, the analog detection was adopted and the signals were observed using an oscilloscope. Since the plasma was continuously maintained, signal accumulations over many laser shots were demonstrated to be effective to obtain an enough signal-to-noise ratio. The photon-counting method was used to measure the detailed structure of the scattered spectrum. When EEDFs of inductively coupled plasmas were examined, it was found that argon plasmas have the Maxwellian EEDF and, on the other hand, EEDFs of plasmas produced in molecular gases such as methane and oxygen deviated remarkably from the Maxwellian [5]. For capacitively coupled plasmas, the lowest detection limit was pursued and the electron density as low as $5 \times 10^{15}\ \mathrm{m}^{-3}$ was proved to be detectable by LTS after a longtime signal accumulation [6].

In the reactive plasmas, the interaction between laser lights and plasma particles sometimes induces problems for LTS measurements. We must examine the particle composition of the plasma and evaluate the possible laser perturbations to the plasma. The strategy for such situation is reviewed in [3]. When the Raman scattering, whose spectrum overlaps with the LTS spectrum, was the problem, the LTS spectrum was separated from the Raman spectrum by observing both spectra from two different scattering angles [7]. When the negative ions exist in the plasma, electrons can be easily detached from the negative ions by the laser beam injection. Then, the photo-detached electrons can be observed as a hump on the LTS spectrum. Thus, LTS can be used to measure the negative ion density [8].

32.3 Diagnostics of PDP Plasmas

32.3.1 LTS Using a Visible System

For our first LTS system, we used the second harmonics beam of the Nd : YAG (wavelength 532 nm) laser as a light source. When we applied the LTS method to PDP microdischarge plasmas, there were two main difficulties, namely the small scattering volume and the high stray-light level. The small size of microdischarge plasmas implies a small scattering volume, which results in a very small Thomson scattering signal. This difficulty could be overcome by using the photon counting accompanied by data accumulation method.

The second difficulty is that the stray laser light is very strong because the wall of the discharge cell is very close to the discharge volume. In order to overcome this problem, we designed a special TGS. The stray-light rejection factor of the TGS was measured to be 10^{-8} at $\Delta\lambda = 1$ nm away from the laser wavelength. This high rejection allowed us to apply the LTS method to microdischarge plasmas [9].

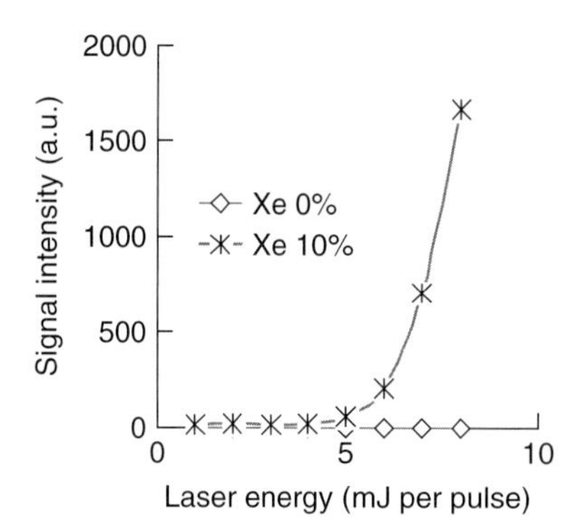

Figure 32.1 Signals observed for Ne/Xe gas.

By using this system the striation phenomenon which had been universally observed above the anode of ac-PDP [10] was investigated. The discharge was produced in the Ne/Ar (10%) gas mixture at a pressure of 200 torr. Results of the LTS measurements showed modulations in both n_e and T_e profiles above the anode, and these modulations had a similar trend to striations appeared in the optical emission images. Also, it was found that the modulations in n_e and T_e were out of phase [11].

When we injected the second harmonics beam of the Nd : YAG laser into discharges in gas mixtures containing the Xe gas, we observed abnormally large signals. The large signals were also observed even when there was no discharge plasma. Figure 32.1 shows signals observed when the second harmonic beam of the Nd : YAG laser at different energies were injected into a pure Ne gas at 200 torr and a gas mixture of Ne/Xe (10%) at 200 torr. For the pure Ne gas, no signal was observed. On the other hand, for the gas mixture of Ne/Xe, signals appeared and increased steeply above the laser energy E_L of 5 mJ. This implies that Xe atoms at the ground state (ionization potential 12.1 eV) were ionized by the laser with $E_L > 5$ mJ. In fact, the steep increase of the signal in the range 5 mJ $< E_L <$ 8 mJ is consistent with seven-photon absorption process (six photons for multiphoton ionization and one photon for Thomson scattering). The lower photon energy of the laser will be favorable to avoid the laser perturbation of the discharge.

32.3.2
Infrared LTS System

The configuration of the electrodes used in this study is shown in Figure 32.2. The structure was similar to the sustaining electrodes of a coplanar ac-PDP. The electrodes were built on a glass substrate which has a width of 2 mm and a thickness of 3 mm. The electrodes had a discharge gap of 0.1 mm. The electrodes were covered with a 15 μm glass layer, followed by a 0.5-μm MgO layer.

Figure 32.3 shows the schematic diagram of the experimental setup. The system was similar to the system using a visible light source. Differences were the laser

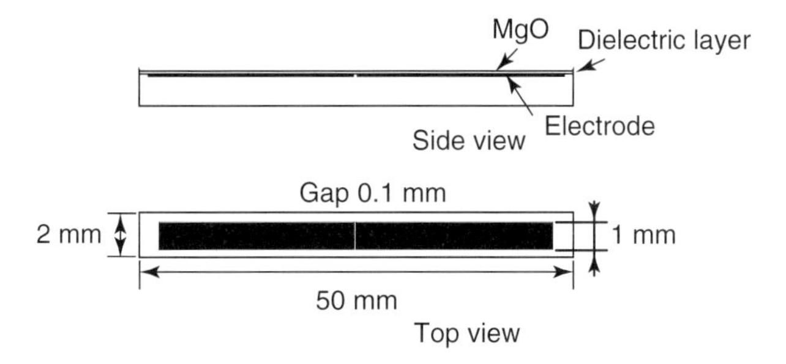

Figure 32.2 Schematic diagram of the geometrical structure of the electrodes.

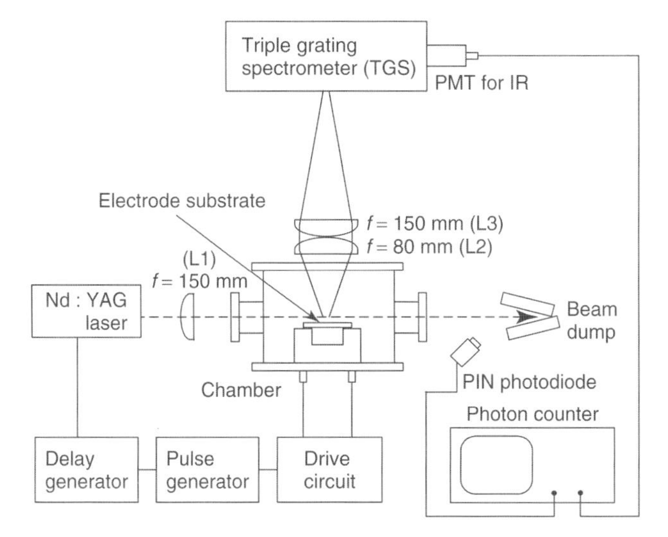

Figure 32.3 Experimental setup for the PDP plasma study.

light source having a wavelength of 1064 nm and the optical components, such as achromatic lenses and gratings used in the TGS, which were designed for 1064 nm. Also, the detector was for 1064 nm, namely an infrared photomultiplier tube (Hamamatsu, R5509) which had a quantum efficiency of ~5% at 1064 nm and could be used for the photon counting.

The electrode substrate was housed in the chamber. After evacuating the discharge chamber, the Ne/Xe gas mixture was introduced. Alternating voltage pulses with a square waveform were applied between the electrodes at a frequency of 20 kHz and duty ratio of 0.4. The peak value of the applied pulse voltage was 270 V.

Figure 32.4 shows the schematic diagram of the TGS for the infrared system. The observation wavelength of the system was tuned by rotating the three gratings of the homemade infrared TGS, and we confirmed that rotation angles of these gratings were just as calculated when optical components of TGS were properly arranged at designed positions. The most important design parameters were the

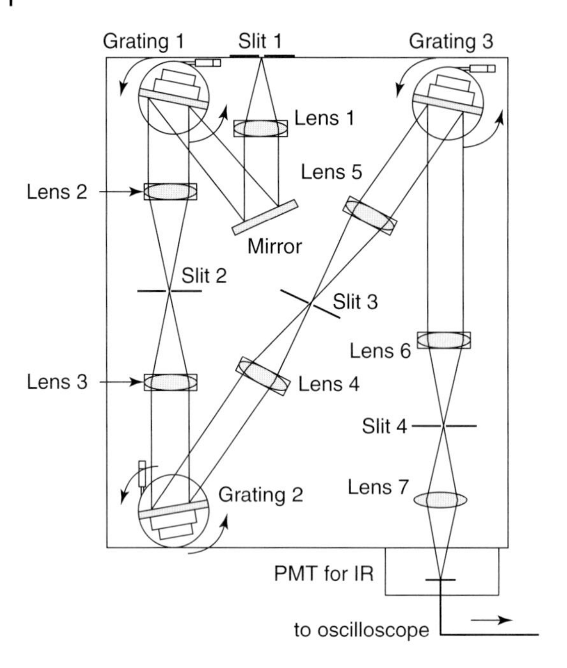

Figure 32.4 Schematic diagram of the triple-grating spectrometer for the infrared system.

grating constant 900 grooves/mm, focal lengths of lenses, and differential angles of incidence and reflection of gratings to be 30°. Focal lengths of lenses placed between the slit 1 and the slit 4 were all 150 mm. Three gratings were arranged so that dispersions of three gratings were additive. An inverse dispersion at a surface of the slit 4 was 5.2 nm mm^{-1}.

Measured spectral profiles of the stray light using the TGS for the infrared system are shown in Figure 32.5. To measure these profiles grating 1 and grating 2, in Figure 32.4, were adjusted at the central laser wavelength (1064 nm), and then at 2 nm away from the laser wavelength, respectively. After that, grating 3 was tuned to cover the spectral range of ±6 nm away from the laser wavelength. The rejection factor of the TGS is 10^{-8} at $\Delta\lambda = 2$ nm away from the laser wavelength. This means that the performance of the TGS for the infrared system is comparable to that of the TGS for the visible system.

32.3.3 Results and Discussion

32.3.3.1 Laser Perturbations

Detailed experimental measurements have been performed to investigate the influence of the fundamental output of the Nd : YAG laser on Ne/Xe gas mixtures and Ne/Xe plasmas. In case of Ne/Xe gas mixtures we could not observe any signals, indicating that there is no photo ionization from ground state Xe atoms. On the other hand, for Ne/Xe plasma the linearity of LTS signals against the laser

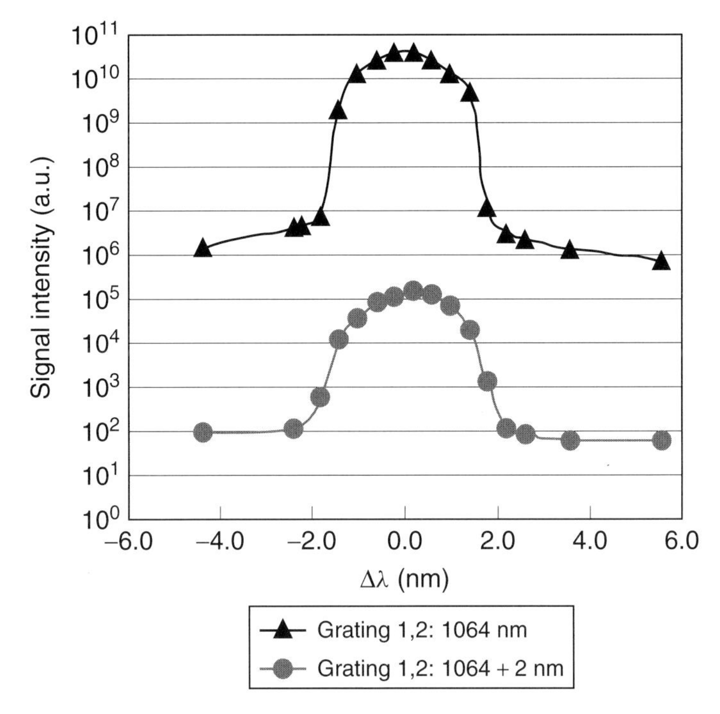

Figure 32.5 Measured spectral profiles of the stray light by the infrared system.

energy was investigated as shown in Figure 32.6 for (a) Ne/Xe (10%) and (b) Ne/Xe (20%) respectively. In Figure 32.6, the solid lines show that the signal is linearly proportional to the laser energy and dotted lines show that the signal is proportional to the fourth power of laser energy. Here, the laser energy of 4 mJ corresponds to the laser intensity of 10^{15} W m^{-2}. These measurements were carried out at the points where largest LTS signals were observed. Other conditions were fixed as follows. Gas pressure was 200 torr, observation time was 270 nanoseconds from the start of the discharge, and the observed wavelength was at $\Delta\lambda = 2.6$ nm from the laser wavelength.

It can be seen from Figure 32.6a that the LTS signal is linearly proportional to laser energy in the range <20 mJ. While for laser energy >20 mJ, the LTS signals deviate from the linearity and proportional to the fourth power of laser energy. This fourth power dependence can be due to the photoionization of metastable Xe atoms by three photon absorption. For higher Xe percentages, as shown in Figure 32.6b, LTS signals deviate from the linearity at lower laser energy. This can be due to the increase of the excited Xe density with the increase of Xe percentage.

Obviously, LTS measurements must be done with the laser energy in the linear ranges of Figure 32.6. Therefore, for new discharge conditions signal linearity against laser energy must be checked before the full measurements of scattered spectra.

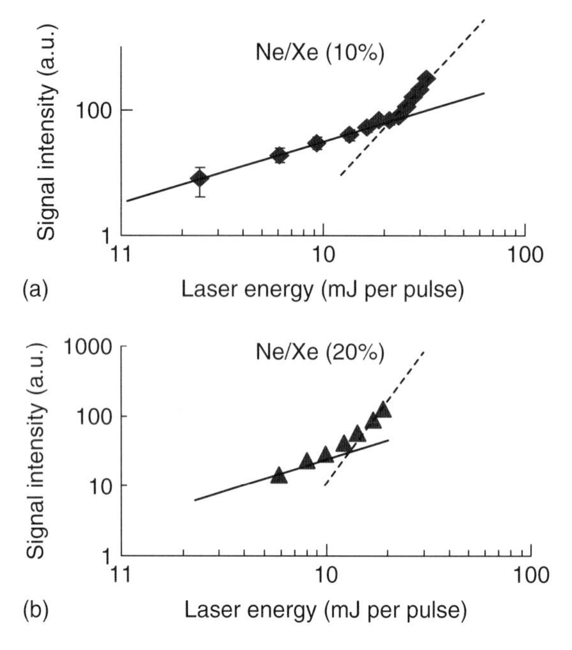

Figure 32.6 LTS signal intensity against laser energy for (a) Ne/Xe (10%) and (b) Ne/Xe (20%).

32.3.3.2 Results of LTS Measurements

Figure 32.7 shows distributions of the LTS signals along the electrode surface for different Xe percentages of 5, 10, and 30%. Measurements were performed at a fixed height of 100 μm above the electrode surface and at a fixed wavelength ($\Delta\lambda = 2.6$ nm). The signal intensity at this wavelength is relatively insensitive against T_e in the range $T_e = 1.5–3$ eV. Therefore, these distributions can be considered to indicate electron density distributions in the relative scale. It can be seen from Figure 32.7 that the LTS signals increase with the increase of the Xe percentage.

32.4 Diagnostics of EUV Plasmas

EUV lights are going to be used for semiconductor lithography after the 32 nm half-pitch technology node. Only a Mo/Si multilayer mirror has reasonable reflectance in the EUV region and it restricts the wavelength range to be at around 13.5 nm with a bandwidth of 2%. For the practical use, the power necessary in this bandwidth should be more than 115 W at the intermediate focal point [12, 13]. There are two candidates as the EUV light source. One is the discharge produced plasma (DPP) and the other is the laser produced plasma (LPP) [14, 15]. However, these sources have not yet succeeded to radiate the EUV lights in the required

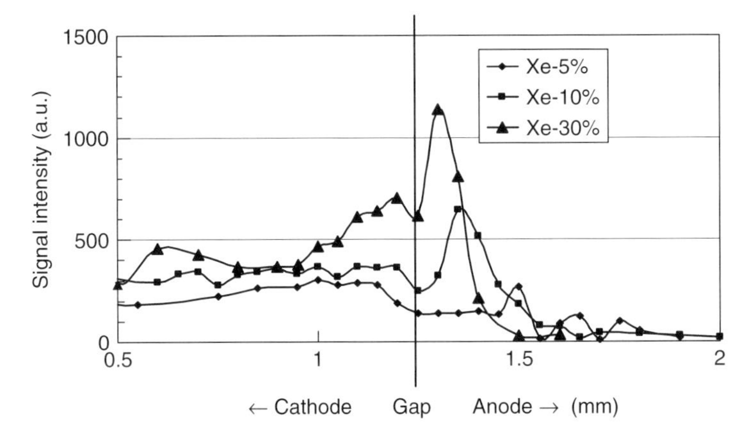

Figure 32.7 Distributions of LTS signals along the electrode surface.

band with the necessary power. For both methods, high density (electron density $n_e = 10^{24}$–10^{26} m^{-3}) and high temperature (electron temperature $T_e = 10$–30 eV) plasmas should be generated using Xe or Sn atoms [16, 17]. In order to produce required EUV lights efficiently, these plasma parameters should be optimized. A prerequisite for such optimization is the quantitative measurements of these parameters.

In the case of DPP, the high density and high temperature plasma is generated through the mechanism of magnetic implosion (pinch) of the preproduced plasma by applying a high current with a short rise time. The pinched plasma is alive during a short time (<50 nanoseconds) and the plasma size is small (radius <200 μm). Therefore, a high temporal resolution (<10 nanoseconds) and a high spatial resolution (<100 μm) are required for measurements of DPP sources. LTS can be expected to fulfill these requirements. The method can yield n_e and T_e values unambiguously.

When we use the second harmonics of Nd : YAG laser ($\lambda = 532$ nm) as the light source, the LTS spectra from the EUV plasmas having aforementioned plasma parameters are in the collective regime. The spectrum of the collective Thomson scattering consists of an ion term and an electron term. The ion term is present very close to the central wavelength of the probing laser, and its expected spectral spread for the EUV plasma is about 100 pm. On the other hand the spectral spread of the electron term is of the order of 10 nm. Taking account of the strong background radiation from the plasma, we determined to measure the ion term, for which we could expect enough SN ratios against the background radiation. One problem to measure the ion term is that the spectral resolution of 10 pm is needed, and the other problem is that the intense wall-scattered laser lights easily overwhelm the ion spectra. In order to overcome these problems, we constructed a newly designed LTS measurement system whose spectral resolution and stray-light rejection were enough to resolve fine feature of the ion term. Using the system, we succeeded to evaluate n_e, T_e, and effective ionic charge $\overline{Z}$ of a Z-pinch type DPP. Furthermore, temporal variations of n_e and T_e spatial profiles were measured and the results were discussed in connection with the evolution of the EUV emission.

32.4.1 Experiment

LTS measurements were performed for the EUV source plasma produced in a compact Z-pinch device. The device has a main capacitor bank of 42 nF. The capacitor bank is rapidly charged up to 24 kV by a magnetic pulse compressor and then a current pulse with an amplitude of 15 kA and duration of 120 nanoseconds is delivered to the Z-pinch discharge tube. The Z-pinch is generated by magnetic pressure due to the current flowing through the tube. More detailed explanations about the Z-pinch device used in this study are described in [14].

The experimental arrangement for LTS measurements is shown in Figure 32.8. The second harmonics of a Nd : YAG laser (Continuum Powerlite 9010 with an injection seeder; spectral spread <0.1 pm) was used as a light source of LTS measurements. The laser energy injected into the plasma was less than 10 mJ, and an achromatic lens (L1, $f = 170$ mm) was used to focus the laser beam. The laser spot size at the focusing point was measured by detecting Rayleigh scattering signals from a nitrogen gas at a pressure of 300 torr. For this experiment, the spatial resolution was set to be 20 μm. The evaluated laser spot diameter was 80 μm (FWHM). The timing between the laser pulse and the plasma generation was controlled by using a delay pulse generator (Stanford Research Systems Inc., DG535).

Scattered lights into the scattering angle of 150° from the plasma were focused onto the entrance slit of a TGS that is described below, and then dispersed by it. The signal lights passed through an exit slit of the TGS were guided by an optical fiber into an electromagnetically shielded room which was very effective to avoid electromagnetic noise produced by the discharge circuit for the plasma generation. Finally, the signal lights were detected by a photomultiplier tube (Hamamatsu, R943-02). Because the Z-pinch plasmas were generated in a ceramics tube having a diameter of 5 mm, wall-scattered laser lights were very strong and overwhelmed

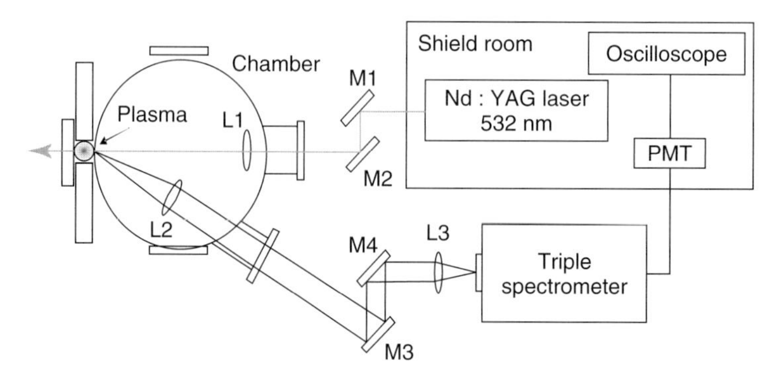

Figure 32.8 Experimental arrangement for the EUV plasma study.

the ion term if they were not eliminated carefully. To overcome this problem, we specially designed and fabricated the TGS to eliminate the wall-scattered laser lights. The structure of the TGS was almost the same as before [9], but its design was different from the previous one to achieve a high spectral resolution. The TGS has three gratings of identical specification (58 mm × 58 mm holographic gratings with 1800 Grooves/mm, blazed at 500 nm), four slits and six lenses (achromatic lenses of focal length $f = 250$ mm). Widths of four slits were around 10 μm. Three gratings were arranged to give additive dispersion at each stage. The overall inverse dispersion of the TGS was 0.57 nm mm^{-1}. In order to change the measurement wavelength, the two intermediate slits and the final lens were finely translated with a step of 1 μm by using stepping motors.

By this TGS, the spectral resolution of 16 pm (FWHM) and the stray-light rejection of 10^{-5} at $\Delta\lambda = 40$ pm were achieved. These performances of the TGS made us possible to measure the ion term spectra from EUV plasmas.

32.4.2 Results and Discussions

We used the new LTS system to observe the ion term spectra from the Z-pinch plasmas. Normally, Xe gas is used for the EUV source as a working gas. Therefore, at first we tried to observe ion term spectra from Xe plasmas. The detected spectra had two peaks at both shorter and longer wavelength sides apart from the laser wavelength. However, the widths of those peaks were much broader than those predicted by the Thomson scattering theory. We thought that this was possibly caused by the steep radial distributions of n_e and T_e of the pinched Xe plasma. Because the spatial resolution of the present LTS system would not be enough, the observed spectrum should be composed of contributions from different fractional scattering volumes with different n_e and T_e values. In order to observe clear ion term spectra using the present system, we changed the working gas of the discharge from Xe to Ar. Since the mass of Ar is three times smaller than that of Xe, we expected that the radius of the pinched plasma would become larger and the distributions of n_e and T_e should be less steep. At the same time, the measurement of the ion term spectrum of the Ar plasma would be much easier than that of the Xe plasma, because the width of the ion term spectrum is inversely proportional to the root of the atomic mass number [18]. For the production of the pinched Ar plasma, Ar was fed into the discharge tube at the flow rate of 300 cm^{3}min^{-1}. The discharges were generated at a frequency of 10 Hz.

Figure 32.9 shows an example of the LTS signal and the background radiation signal observed at $\Delta\lambda = 60$ pm collected from the Ar gas discharge. It is clear that the LTS signal is almost one order of magnitude larger than the background radiation signal. Figure 32.10 shows the ion term spectrum collected from the Ar gas discharge at 10 ns after the pinch. In order to measure this spectrum, the signals were averaged over a number of 50 laser shots, and these averaged measurements were repeated for 10 times at each wavelength. The error ranges shown in Figure 32.10 is the standard deviation of these data by 10 times measurements.

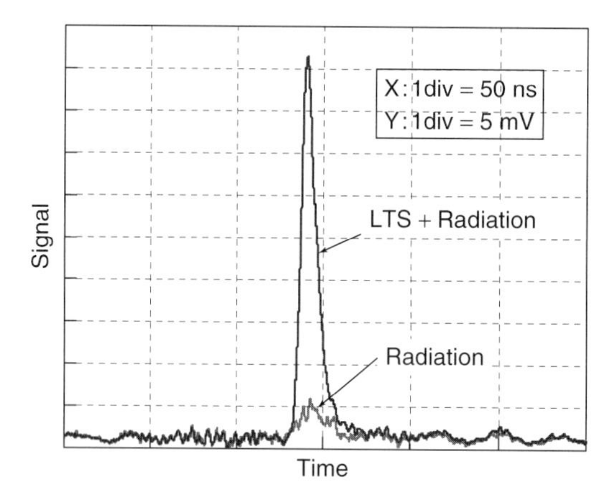

Figure 32.9 Examples of the LTS signal and the background radiation signal observed at $\Delta\lambda = 60$ pm for an Ar plasma.

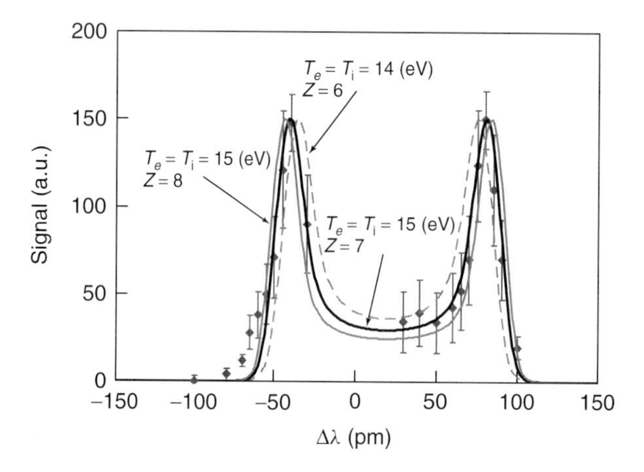

Figure 32.10 Observed ion term spectrum from an Ar gas discharge curve and a fitted theoretical curve.

In this case, the plasma was flowing to the chamber side along the axis of the discharge (plus side of the z-axis). Therefore, the ion term spectrum is Doppler shifted to the blue side by 20 pm. The measured position was $z = 500\ \mu$m. Here, we selected the origin of the z-axis ($Z = 0$) to be the center of the pinched plasma. From the width of the spectral shift, the plasma velocity can be estimated, and the velocity was $1.2 \times 10^4\ \text{ms}^{-1}$ to the z-direction. Because of the strong stray-light signals, the ion term spectrum couldn't be measured at $\Delta\lambda < 25$ pm. However, it is easy to reconstruct the spectral shape of $\Delta\lambda < 25$ pm from the fittings of theoretical curves with the measured part of the ion term spectrum.

The absolute value of LTS signal is also needed to evaluate plasma parameters. The calibration of the LTS signal was done by using the Rayleigh scattering signals from a nitrogen gas at a pressure of 300 torr. Since the shape, the spectral peak wavelength, and the absolute value of the ion term could be measured, n_e, T_e, and $\overline{Z}$ could be evaluated. It can be seen from Figure 32.10 that the solid line is most close to the measured spectrum. So the plasma parameters deduced from the spectrum were $n_e = 1.5 \times 10^{24}\ \mathrm{m}^{-3}$, $\overline{Z} = 7$, $T_e = T_i = 15\ \mathrm{eV}$. The error ranges for these values were less than $\pm 10\%$.

In conclusion, we have developed a collective LTS system for measurements of n_e, T_e, and $\overline{Z}$ of Z-pinch plasmas produced for EUV lithography. The system was successfully applied to the diagnostics of pinched plasmas produced in the argon gas. However, in order to apply the LTS system to the Xe plasmas, spatial and spectral resolutions of the system must be improved. Such improvements are now in progress.

References

1. Evans, D.E. and Katzenstein, J. (1969) *Rep. Prog. Phys.*, **32**, 207.
2. Muraoka, K., Uchino, K., and Bowden, M.D. (1998) *Plasma Phys. Control. Fusion*, **40**, 1221.
3. Muraoka, K., Uchino, K., Yamagata, Y., Noguchi, Y., Mansour, M., Suanpoot, P., Narishige, S., and Noguchi, M. (2002) *Plasma Sources Sci. Technol.*, **11**, A143.
4. Bowden, M.D., Okamoto, T., Kimura, F., Muta, H., Uchino, K., Muraoka, K., Sakoda, T., Maeda, M., Manabe, Y., Kitagawa, M., and Kimura, T. (1993) *J. Appl. Phys.*, **73**, 2732.
5. Bowden, M.D., Tabata, R., Suanpoot, P., Uchino, K., Muraoka, K., and Noguchi, M. (2001) *J. Appl. Phys.*, **90**, 2158.
6. Mansour, M., Bowden, M.D., Uchino, K., and Muraoka, K. (2001) *Appl. Phys. Lett.*, **78**, 3187.
7. Narishige, S., Suzuki, S., Bowden, M.D., Uchino, K., Muraoka, K., Sakoda, T., and Park, W.Z. (2000) *Jpn. J. Appl. Phys.*, **39**, 6732.
8. Noguchi, M., Ariga, K., Hirao, T., Suanpoot, P., Yamagata, Y., Uchino, K., and Muraoka, K. (2002) *Plasma Sources Sci. Technol.*, **11**, 57.
9. Hassaballa, S., Yakushiji, M., Kim, Y.K., Tomita, K., Uchino, K., and Muraoka, K. (2004) *IEEE Trans. Plasma Sci.*, **32**, 127.
10. Yoshioka, Y., Okigawa, A., Tessier, L., and Toki, K. (1999) IDW' 99, p. 603.
11. Hassaballa, S., Sonoda, Y., Tomita, K., Kim, Y.K., and Uchino, K. (2005) *J. Soc. Inf. Display*, **13**, 639.
12. Banine, V. and Moors, R. (2004) *J. Phys. D: Appl. Phys.*, **37**, 3207 p.
13. International Technology Roadmap for Semiconductor (2008) *http://www.itrs.net/Links/2007ITRS/2007_Chapters/2007_Lithography.pdf.* (accessed 22 December 2009]
14. Katsuki, S., Kimura, A., Kondo, Y., Hotta, H., Namihira, T., Sakukgawa, T., and Akiyama, H. (2006) *J. Appl. Phys.*, **99**, 013305.
15. Tao, Y., Nishimura, H., Fujioka, S., Sunahara, A., Nakai, M., Okuno, T., Nishihara, N., Miynaga, N., and Izawa, Y. (2005) *Appl. Phys. Lett.*, **86**, 201501.
16. Kieft, E.R., van der Mullen, J.J.A.M., and Banine, V. (2005) *Phys. Rev. E*, **72**, 026415.
17. Sasaki, A., Nishimura, K., Koike, F., Kagawa, T., Nishikawa, T., Fujima, K., Kawamura, T., and Furukawa, H. (2004) *IEEE J. Sel. Top. Quantum Electr.*, **10**, 1307.
18. Sheffield, J. (1975) *Plasma Scattering of Electro- magnetic Radiation*, Academic Press.

33
Crystallized Nanodust Particles Growth in Low-Pressure Cold Plasmas

Laïfa Boufendi, Marie Christine Jouanny, Marjory Cavarroc, Abdelaziz Mezeghrane, Maxime Mikikian, and Yves Tessier

33.1
Introduction

Observation of visible photoluminescence from silicon nanocrystals (nc-Si) of indirect gap at room temperature [1], nanostructured semiconductors such as porous Si, silicon nanostructured particles, Si nanocrystals embedded in thin layers such as silica (nc-Si/SiO_2) or amorphous silicon (nc-Si/a-Si : H), and so on has induced an intense research activity because of the potential optoelectronic applications of these materials [2–6] and also because of their unique mechanical properties making them useful as superhard nanocomposites [7]. Different systems have attracted the interests of the researcher. However, silicon nanocrystals embedded in a-Si : H or silicon oxide matrix constitute the most studied nanomaterial because of the fact that they are considered to be the most promising on a technological level. Different growth techniques are used to achieve good control of nanocrystal size, size distribution, and volume fraction.

Recently, however, silicon thin films grown with a significant contribution of nanoparticles coming from the plasma have been found to exhibit improved properties of transport and stability and high optical gap as compared to a-Si : H [8]. Knowledge of the powder formation pathway and the use of modulated radiofrequency (RF) plasma have permitted anew the selective incorporation of nanocrystallites of few nanometers into the growing thin film. These films have been described as a mixture of amorphous and ordered material and are called *polymorphous Si* (pm-Si : H) or *nanostructured Si* (ns-Si : H). It has been claimed that the unusual structure of these films, dominated by the ordered structure of the Si nanoparticles embedded therein, is responsible for the unusual properties of pm-Si : H. However, there is no clear picture on the atomic structure of these Si nanoparticles or clusters of about 2 nm. This question will be discussed in this chapter based on the results of high-resolution transmission electron microscopy (HRTEM) and selected area electron diffraction (SAED).

Among these techniques, one can cite high-temperature thermal annealing, usually above 1000 °C, to form nanocrystals by precipitation of excess of Si in Si

Industrial Plasma Technology. Edited by Yoshinobu Kawai, Hideo Ikegami, Noriyoshi Sato, Akihisa Matsuda, Kiichiro Uchino, Masayuki Kuzuya, and Akira Mizuno

ISBN: 978-3-527-32544-3

oxide matrix and the homogeneous nucleation and growth of single nanocrystals in the gas phase of a silane-based plasma.

In this contribution, we report on this last technique. The silicon nanocrystallites constitute the first phase of dust particle growth in low-pressure plasmas [9]. Many experimental and modeling studies have been devoted in the last two decades to the formation of particles in low-pressure silane-based plasmas. An exhaustive review of this research [10] shows that growth phenomena, from monomer precursors to large-sized particles, are rather well elucidated in the typical conditions of pressure, power density, and diffusion lengths relevant for PECVD technologies. This growth involves successive steps: nucleation leading to the formation and accumulation of few nanometer-sized (100–1000 Si atoms) dust particles; agglomeration of these "protoparticles" when critical densities are achieved (few 10^{11} cm^{-3}); further growth by a radical sticking process when charging effects prevent further aggregation. The first step and its limiting transition to the agglomeration phase are key points when nanostructured thin films are grown with inclusion of a significant fraction of nanocrystals in an amorphous matrix. It is a fast (10–100 ms) process leading to the formation of nanometer-sized (2–4 nm) particles, mainly neutral, with low hydrogen content and a monocrystal-like structure.

The first part will be devoted to the description of the experimental setup and the diagnostic techniques developed and used to follow the nucleation and the growth of the nanocrystallites. The second part will focus on the atomic structure of these nanoparticles. The gas temperature effect on growth mechanisms of dust particles will be addressed in the third part. The conclusion will give a summary of the most important aspects of this subject.

33.2 Description of the Experimental Setup

The experimental setup was described in detail in the previous articles [11]. The RF discharge is produced in a grounded cylindrical box (13-cm inner diameter) equipped with a shower-type RF powered electrode (Figure 33.1). The bottom of the box is closed with a 20% transparency grid. This allows a vertical laminar gas flow. The discharge structure is surrounded by a cylindrical oven. The gas temperature can be varied from room temperature up to 200 °C and is measured in the gas flow below the bottom grid by a thermocouple. Three vertical slits (2 mm wide, 4 cm high) allow optical access to the plasma at 0°, 90°, and 180° around the chamber. The whole system is enclosed in a vacuum vessel of 30 cm height and 30 cm inner diameter. Three optical viewports on the vacuum vessel (5 cm in diameter and 90° apart) are aligned with the slits.

The discharge and plasma are characterized by time evolution of the self-bias and electron density. These two parameters allow detection of the appearance of dust particles in the gas phase [12]. Self-bias is measured using a voltage probe. Electron density is measured using the discharge box as a microwave resonant cavity. Two antennae are placed inside the box on the grounded electrode. One

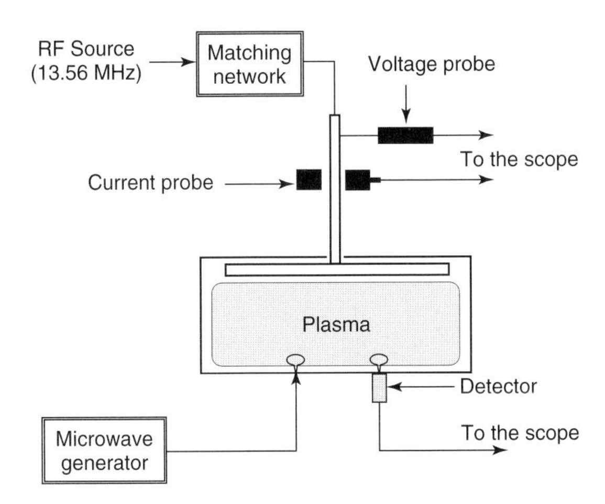

Figure 33.1 General schematics of the experimental set-up.

of them is connected to a microwave generator and the second one is connected to a digital scope through a detection diode. The resonance mode frequency shift, induced when the plasma is turned on, is used to determine the electron density from the expression given below.

$$n_e = \frac{2m_e\varepsilon_0(2\pi f)^2\Delta f}{e^2 f_0} \tag{33.1}$$

where f_0 and f are the resonance frequencies without and with a plasma respectively, Δf is the frequency shift, m_e is the electron mass, e is the elementary charge, and ε_0 is the permittivity of vacuum. This diagnostics allows us to follow the time evolution of the electron density during the formation of the negative ions which are the seeds for dust particle formation.

Ex situ measurements are performed in order to characterize the collected dust particles and the nanostructured thin films. Electron microscopy (TEM and MEB) is used for imaging and to study their morphology. Micro-Raman spectroscopy is used to determine the size and crystalline volume fraction of the thin layers.

33.3 Detection of Nanocrystallites Formation in the Plasma Gas Phase

To detect the formation of nanocrystallites in the plasma gas phase, we followed the time evolution of the electron density and self-bias occurring on the powered electron. These two parameters are very sensitive to the modification that the nanoparticle occurrence induces on the discharge and plasma properties [12]. With these two diagnostic tools, it is possible to detect the earlier dust particles in the nanometer size range. Figure 33.2 shows the time evolution of both the electron density and self-bias voltage. The first sharp drop of electron density

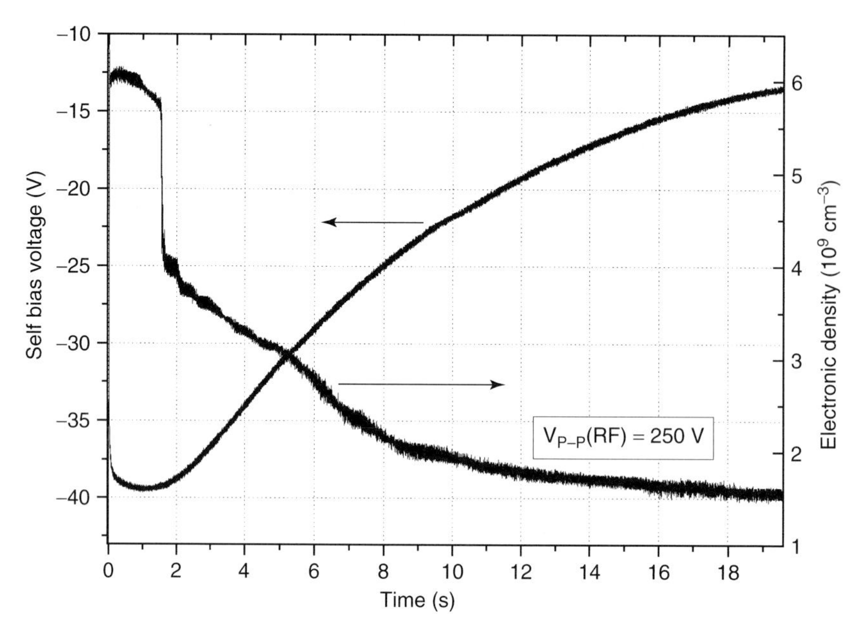

Figure 33.2 Time evolution of the electron density and the self-bias voltage during the dust particle formation.

corresponds to the formation of negation ions. We can emphasize here that self-bias is affected by the formation of the negative ions, which are the nuclei from which starts the formation of the dust particles. The slight slope that follows is related to the formation and accumulation of the nanocrystallites. Their charge cannot exceed one elementary charge. Most of these nanocrystallites are neutral. That is why the total charge of these nanoparticles is of the same order as that of the negative ions. In this phase, the drop in self-bias is due to the modification of the resistivity of the plasma. Indeed, the electron will have a collision with the nanoparticles.

The second electron density drop indicates the beginning of the agglomeration phase of the nanocrystallites to form bigger particles. As soon as they start growing up, they are able to attach more than one electron. The drop of the self-bias here is associated with the drop of the electron density. With this indication one can modulate the plasma in order to prevent the agglomeration and thus enable thin layers containing nanocrystallites to grow.

33.4
Atomic Structure of the Nanocrystallites

Figure 33.3 shows an example of a nanostructured Si thin film with a high concentration of ordered domains, which is evident from its dark-field and SAED images. This film corresponds to a ns-Si sample using a plasma on time of $T_{\text{on}} = 5$

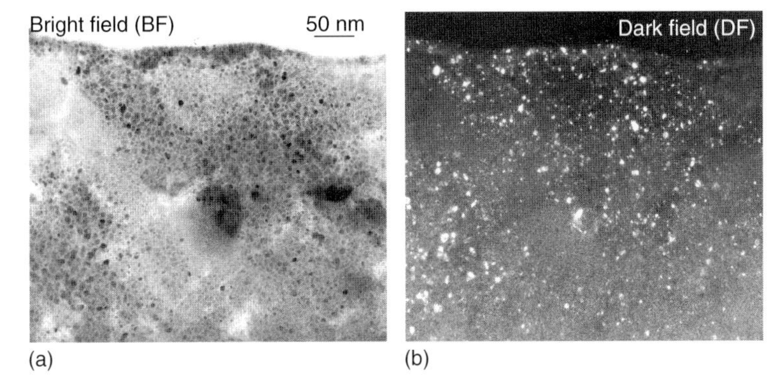

Figure 33.3 Bright field (a) and dark-field (b) TEM images of a nanostructured Si thin film deposited from modulated RF plasmas of Ar-diluted SiH_4 in dust-forming conditions (ns-Si with $T_{on} = 5$ seconds and $T_G = 100^\circ C$). The insert in the dark-field image shows the corresponding SAED pattern.

seconds and a gas temperature of $T_G = 100^\circ C$. The plasma conditions used here were chosen to attain powder formation. Selective incorporation of nanoparticles into the growing film was controlled through the square-wave modulation of the RF plasma. During the plasma on time (T_{on}) of the modulation cycle, an amorphous Si film is deposited onto the substrate and, at the same time, nanoparticles nucleate and grow in the plasma gas phase. During the afterglow periods (T_{off}), the particles leave the plasma and are deposited onto the amorphous Si film. Consequently, after a great number of cycles, the final structure will consist of Si nanoparticles embedded in an amorphous matrix.

The atomic structure of the films has been well resolved using the SAED technique. SAED images of polymorphous Si thin films and of isolated Si nanoparticles have revealed the existence of different cubic phases: fcc and bc8 (Table 33.1). These phases are commonly unstable under ambient conditions and can be obtained

Table 33.1 Summary of the structures observed in the electron diffraction patterns (see inset in Figure 33.3) during HRTEM analysis of the nanostructured thin films.

Sample	Ring	*dhkl*	*hkl*	*a* (Å)	Space group
ns-Si : H	1	2.11	111	3.659	fcc
	2	1.83	200	3.654	Fm3m (225)
	3	1.29	220	3.649	
	4	1.10	311	3.640	
	5	1.05	222	3625	
	6	0.84	331	3.647	
	7	0.81	420	3.616	

under pressure compression. The occurrence of such phases in low-temperature plasma processes has been explained in terms of the particular kinetics of development of the particles in the plasma and of the stability changes involving size effects. The fcc and bc8 phases are very similar to the hexagonal hcp one, which is currently reported in the literature for nanocrystalline Si.

33.5
Size and Crystalline Volume Fraction Measurements

Raman spectroscopy is recognized as a powerful technique for the characterization of Si structures. As is well known, the Raman spectrum for amorphous Si (a-Si : H) is characterized by four bands, the most intense peak of which is the TO mode located at 480 cm^{-1}. For crystalline Si (c-Si), the Raman peak is a sharp and nearly Lorentzian band centered at 520 cm^{-1} and with a full width at half maximum (FWHM) of about 3 cm^{-1}. For nanocrystalline or nanostructured Si (nc-Si or ns-Si, respectively, the crystalline Raman peak exhibits a frequency downshift and peak broadening caused by a phonon confinement effect. The Raman peak can also be modified by the presence of strain (or stress) in the films and/or by simple sample heating due to the laser irradiation during the Raman experiment, which causes disturbances in the Si lattice. As a matter of fact, Raman spectroscopy is an efficient probe that is very sensitive to the local atomic arrangements.

Figure 33.4 shows the Raman spectra of a standard a-Si : H and of three different ns-Si : H thin films in the range 200–800 cm^{-1}. The power density of the Raman laser was the same for the three samples and equal to 75 kW cm^{-2}. This laser power is normally considered sufficiently small for no thermal or crystallization effects to be induced on the samples. It is evident that the Raman spectrum for the a-Si : H film (Figure 33.4a) shows the features of amorphous silicon, that is, the TO mode around 480 cm^{-1} and a broadband due to the combination of the LO and LA modes around 340 cm^{-1}.

However, for the ns-Si : H films, the presence of very small crystals seems to arise by the appearance of a narrow Raman signal at around 494–496 cm^{-1} (Figure 33.4b), which is superimposed on the Raman signal of an amorphous Si structure. In addition, the presence of Si crystallites in the ns-Si : H films is found to be greater as the plasma-T_{on} increases, which could well be related to the fact that larger T_{on} leads to higher incorporation of nanoparticles during the film growth. The position of this crystalline-like peak is found to downshift by 25 cm^{-1} (as compared to the reference of c-Si). These Raman spectra are interpreted using Zi [13] and Balkansky [14] models in order to determine respectively the size, crystalline volume fraction, and the sample surface temperature during the Raman analysis. These data [15] give sizes around 2–3 nm, crystalline volume fractions of 20–30%, and surface temperatures around 700–800 °C.

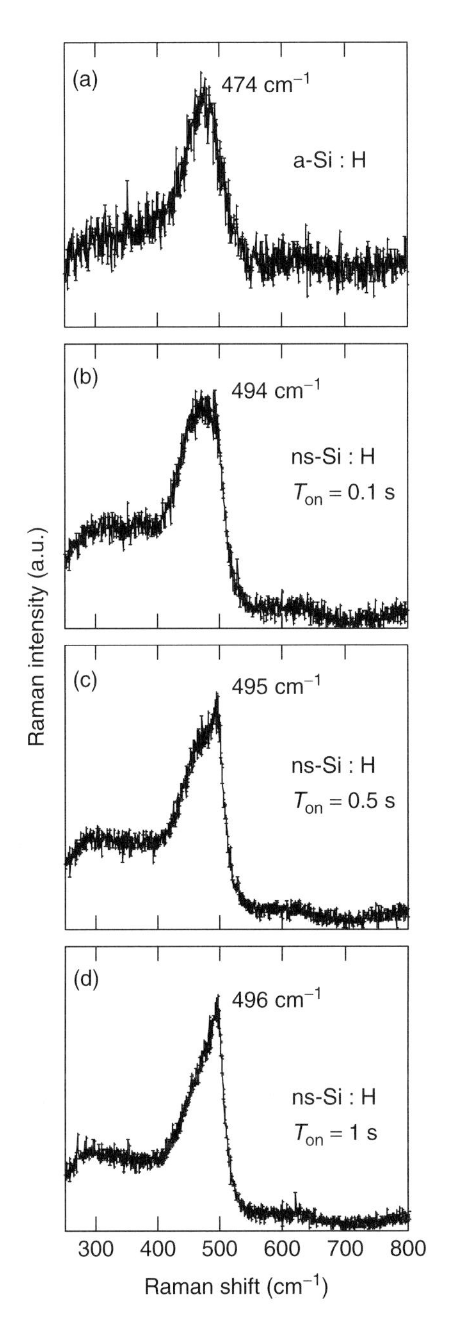

Figure 33.4 Raman spectra of an a-Si : H thin film and of ns-Si:H films deposited at 100 °C and at different T_{on}. The laser power density was set at 75 kW cm^{-2} for all the samples.

33.6
Gas Temperature Effects on Dust Nanoparticle Nucleation and Growth

One very interesting aspect is the sensitivity of the clustering process to the gas temperature. Evidence of this sensitivity was initially given using standard laser light scattering methods [16]. Using a more sophisticated technique, called *laser-induced particle explosive emission* (LIPEE), it was shown that this temperature sensitivity was related to the initial growth step of nanosized crystals. This thermal sensitivity has been attributed to two different processes implying vibrational excited states of SiH_4 [17] or gas temperature effects on the effective electron attachment on *n*-atom clusters [9] or radical loss by Brownian motion [18].

Figure 33.5 shows the effect of gas temperature on the nucleation and growth of dust particles through time evolution of the amplitude of the third harmonic of the discharge current for different gas temperatures. It is clear here that the first phase of the process becomes longer. However, as soon as the density of the nanocrystallites reaches its critical value (few 10^{11} cm^{-3}) the agglomeration phase starts. The decrease in amplitude during the first phase is nearly the same for all the temperatures investigated. To perform this experiment we kept the number density of the precursor molecules (SiH_4) constant. These curves also highlighted

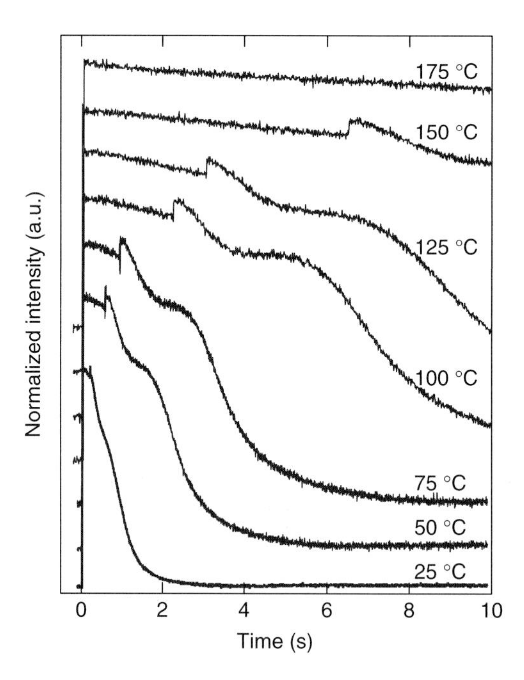

Figure 33.5 Time evolution of the amplitude of the third harmonic of the discharge current during the dust particle growth for different gas temperatures. It has been shown that the third harmonic and the self-bias exhibit the same trends.

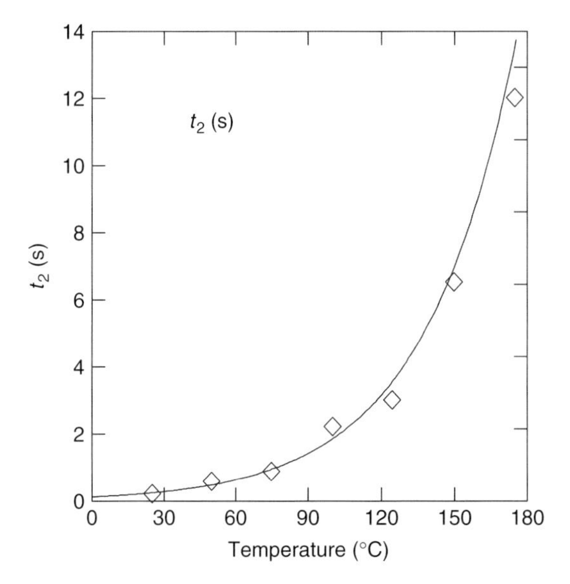

Figure 33.6 Evolution of the beginning time of the agglomeration phase versus the gas temperature.

a transient phase (a plateau), which becomes longer with the gas temperature. On the basis of our previous measurements it corresponds to a dust particle size of about 10 nm. When these particles are formed they accumulate in the plasma gas phase because of their charge and then restart agglomerating.

However, the study of the time evolution of electron density at these different gas temperatures shows that the formation of negative ions is not affected by this parameter. Only the phase corresponding to the formation of nanocrystallites is affected. It is drastically slowed down. This means that temperature has no effect on the electron dissociative attachment but has a big effect on the chemical reaction leading to the formation of the nanoparticles.

Figure 33.6 shows the evolution of time when the agglomeration phase begins versus the gas temperature. It has an exponential evolution comparable to the evolution of the nucleation rate in a supersaturated medium. This means that the dust particle growth in low-pressure plasma evolves in a thermodynamical manner, which has to be considered. Work is now underway to elucidate this question.

33.7 Conclusion

In this contribution, we focused our interest on the formation of silicon nanometer-sized crystallites in low-pressure cold silane-based plasma. Single nanocrystals constitute the first phase of the dust particle nucleation and growth mechanisms. They exhibit a metastable atomic structure, mainly FCC, which can be revealed through electron diffraction during TEM analysis of the collected dust. Their size is measured using micro-Raman scattering. The time evolution of the self-bias voltage, which is well correlated to the growth mechanisms, can be used to efficiently target the collection of the nanocrystallites formed in the plasma.

References

1. Takagi, H., Ogawa, H., Yamazaki, Y., Ishizaki, A., and Nakagiri, T. (1990) *Appl. Phys. Lett.*, **56**, 2379.
2. Canham, L.T. (1990) *Appl. Phys. Lett.*, **57**, 1046.
3. Roca i Cabarrocas, P., Gay, P., and Hadjadj, A. (1996) *J. Vac. Sci. Technol. A*, **14**, 655.
4. Baron, T. *et al.* (2000) *Appl. Surf. Sci.*, **164**, 29–34.
5. Khriachtchev, L. *et al.* (2001) *Appl. Phys. Lett.*, **79**, 1249.
6. Oda, S. (2003) *Mater. Sci. Eng. B*, **101**, 19.
7. Vepoek, S. (1996) *Pure Appl. Chem.*, **68** (5), 1023–1027.
8. Longeaud, C., Kleider, J.P., Roca i Cabarrocas, P., Hamma, S., Meaudre, R., and Meaudre, M. (1998) *J. Non-Cryst. Solids*, **227–230**, 96.
9. Boufendi, L. and Bouchoule, A. (1994) *Plasma Sources Sci. Technol.*, **3**, 262.
10. Bouchoule, A. (ed.) (1999) *Dusty Plasmas, Physics, Chemistry and Technological Impacts in Plasma Processing*, John Wiley & Sons, Inc., New York.
11. Boufendi, L., Hermann, J., Bouchoule, A., Dubreuil, B., Stoffels, E., Stoffels, W., and de Giorgi, M.L. (1994) *J. Appl. Phys.*, **76**, 148.
12. Boufendi, L., Gaudin, J., Huet, S., Viera, G., and Dudemaine, M. (2001) *Appl. Phys. Lett.*, **79**, 4301.
13. Zi, J., Buüscher, H., Falter, C., Ludwing, W., Zhang, K., and Xie, X. (1996) *Appl. Phys. Lett.*, **69**, 200.
14. Balkansky, M., Wallis, R.F., and Haro, E. (1983) *Phys. Rev. B*, **28**, 1928.
15. Viera, G., Huet, S., and Boufendi, L. (2001) *J. Appl. Phys.*, **90**, 4181.
16. Bouchoule, A., Plain, A., Boufendi, L., Blondeau, J.Ph., and Laure, C. (1991) *J. Appl. Phys.*, **70**, 1991.
17. Fridman, A., Boufendi, L., Hbid, T., Potapkin, B., and Bouchoule, A. (1996) *J. Appl. Phys.*, **79**, 1303.
18. Bhandarkar, U. *et al.* (2003) *J. Phys. D: Appl. Phys.*, **36**, 1399.

34
Collection and Removal of Fine Particles in Plasma Chambers

Noriyoshi Sato

34.1
Introduction

On the basis of experiments on fundamental behaviors of fine particles in plasmas [1], the NFP-collector (negatively charged fine-particle collector) was proposed and has proved to be useful for collection and removal of fine particles of size less than a few tens of micrometers, which are negatively charged, levitating in plasmas [2–4], although the collector is applicable to much larger particles levitating under the microgravity condition. The NFP-collector could be of crucial importance in various kinds of plasma application. The collector is just a simple electrode with hole(s), which is externally biased higher than the floating potential to collect fine particles levitating in plasmas. Fine particles pass through the hole(s) without impinging on the electrode surface into the electrode hole(s). When fine particles near the collector are collected, fine particles left away from the collector approach the collector because there is the Coulomb-force balance among fine particles confined by an external potential. They are pulled one after another into the collector, and finally, there is almost no fine particle left in plasmas.

When a metal plate is set in plasmas, fine particles levitate against the gravity in the horizontal plane above the plate in the presence of such a large sheath potential in front of the plate surface as to provide an upward force balancing the force due to the gravity. The horizontal spread of these particles depends on the electrostatic-potential structure in the horizontal plane. If the plate for particle levitation has the surface with ditches, negatively charged fine particles levitate above the ditches because the potential is higher in the region above the ditches than that in other regions on the same horizontal plane. Also found here is fine-particle collection and removal based on specially structured ditches formed on the plate for particle levitation [3–5]. They are shaped to provide such a potential profile that fine particles are guided to flow in the outward direction. If the ditch is in the direction toward the NFP-collector, the collector can be located at a position far away from the central plasma region, where the plasma disturbance caused by the collector is negligibly small. Even if the ditches are filled up with insulator or

Industrial Plasma Technology. Edited by Yoshinobu Kawai, Hideo Ikegami, Noriyoshi Sato, Akihisa Matsuda, Kiichiro Uchino, Masayuki Kuzuya, and Akira Mizuno

ISBN: 978-3-527-32544-3

the surface with ditches is covered with a plane insulator film to yield the plane surface, the fine-particle levitation and flow are quite similar to those in case of no insulator. Therefore we can have almost the same results for fine-particle collection and removal.

In the experiments mentioned above, fine particles are externally injected into plasmas in order to prove the principals of the NFP-collector and ditch guidance of fine particles. The NFP-collector and ditch guidance could also be applied to fine particles sputtered out of plasma chamber walls or dirty metal plates in the chambers [3–5]. In case of the sputtering method for cleaning plasma chambers, particles sputtered out of some parts of the chamber walls could be removed by the NFP-collector without accumulation on other parts of/in the chambers.

It is well known that dusty fine particles are often formed in reactive plasmas used for various kinds of material and device processing. We have applied the NFP-collector for removal of dusty particles formed in a silane plasma which has been employed in fabrication of amorphous and microcrystalline silicon films. Measurements have confirmed that the film quality is improved in the presence of the NFP-collectors surrounding a substrate on which amorphous silicon films are formed [6].

34.2 NFP-collector

In the presence of upward force provided, for example, by the electric potential slope in front of the metal plate, fine particles levitate against the gravity force on the horizontal plane in plasmas. When we have a small electrode in a thin layer of fine-particle clouds levitating in plasmas, there appear vortices of fine particles on the horizontal plane [1]. As shown (top view) in Figure 34.1, the directions of vortices depend on the polarity of electric potential applied. Here, we are interested in vortices produced by the electrode biased positively. Their size is found to be smaller than that in case of the negative potential. Fine particles are attracted

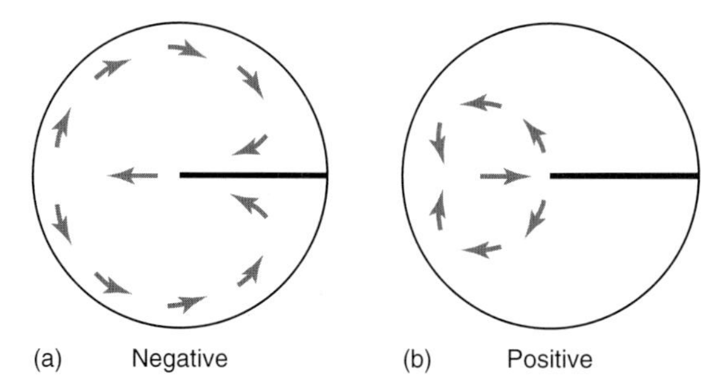

Figure 34.1 Schematic for vortices of fine particles, triggered by small electrodes biased negatively (a) and positively (b) (top view: view downward).

toward the electrode, but they do not touch on the electrode, flowing sideways in front of the electrode. This is because the electrode potential is smaller than the plasma potential even in this situation. As a result, there appear vortices in the fine-particle cloud.

Figure 34.2 shows vortices (top view) produced by one (left) and two (right) electrodes. In the latter case, the electrode separation is smaller than the size of vortices and then particles are guided to flow into a region between the two electrodes without hitting the electrodes. This behavior of vortices is typical of negatively charged particles. This is a basic principle of the NFP-collector. We cannot expect this feature of vortices for positively charged particles when, instead of the positive potential, a negative potential is applied to the two electrodes. Under this situation, fine particles are collected on the electrode wall, just as in case of usual dust collectors. These behaviors of particles are schematically illustrated for cylindrical geometry of the electrodes in Figure 34.3. An essential point of the NFP-collector is that fine particles are collected without impinging on the collector wall. This is possible in the presence of plasma surrounding the collector.

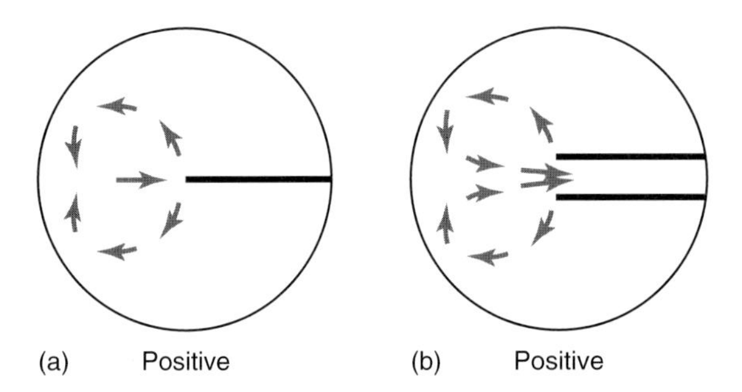

Figure 34.2 Schematic for vortices of fine particles, triggered by a positively biased small electrode (a) and two positively biased small electrodes separated closely (b) (top view).

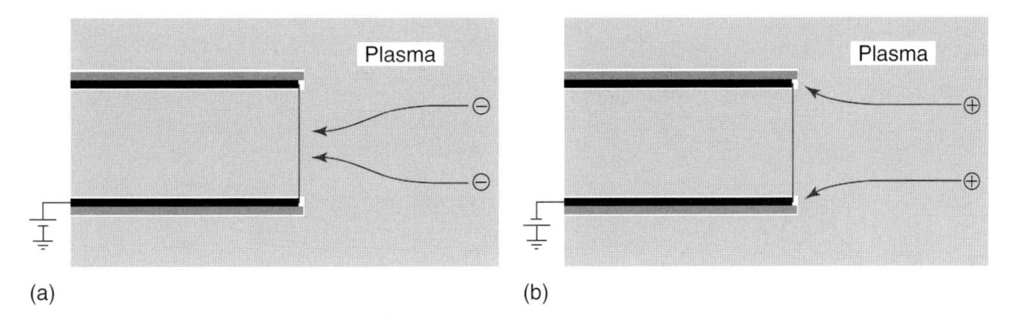

Figure 34.3 Collection of (a) negatively charged particles and (b) positively charged particles.

It has to be again emphasized here for a specific feature of the particle collection and removal by the NFP-collector. When fine particles near the collector are removed, fine particles left away from the collector approach the collector in the presence of the force balance among particles, being pulled one after another into the collector. This feature based on the force balance among fine particles enhances an efficiency of particle collection and removal. We can imagine a similarity between this collector action and a dynamics of water in a vessel inclined as illustrated in Figure 34.4. Therefore, in principle, it is not necessary to move the NFP-collector to collect and remove particles in plasmas. It is interesting to note that this point is in contrast to usual vacuum cleaners used at home, which has to be moved around to clean rooms.

A simple configuration of the NFP-collector is provided by a cylindrical metal tube. It is better to cover the outside wall with insulator (for example, ceramic tube). Figure 34.5 shows the collector used in the actual experiment. An action of the collector is demonstrated in Figure 34.6, where particles (methyl-methacrylate polymer) of 10 μm in diameter are employed in a dc Ar discharge plasma under the pressure of 100–300 mtorr. The laser irradiation is used to detect fine particles.

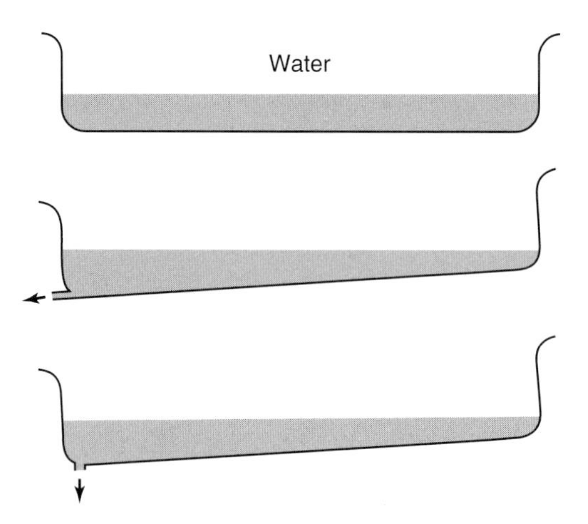

Figure 34.4 Water in inclined vessels (side view).

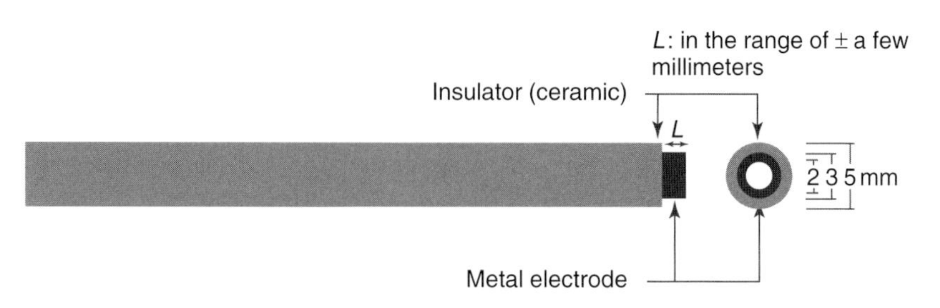

Figure 34.5 A tube-typed NFP-collector.

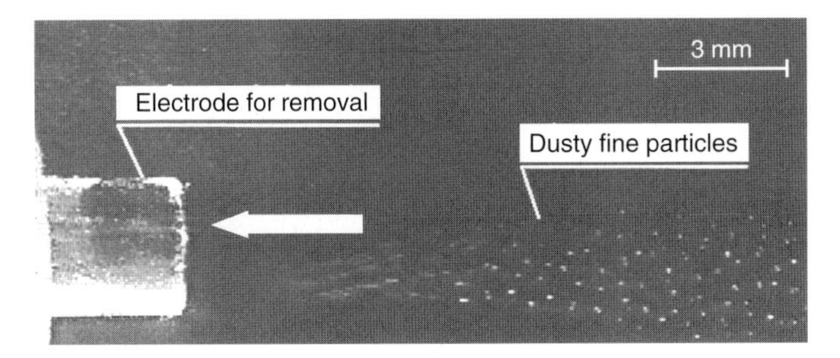

Figure 34.6 Particle removal (side view) by a tube-typed NFP-collector.

The collector collects fine particles, which are externally injected into the plasma and levitate above a metal plate for particle levitation. Fine particles are removed into the tube without impinging on the inside wall of the tube electrode. The result is consistent with the particle flow in Figure 34.3a. This mechanism could also be applied to the experiment on particle collection at the HEDRC in Russia [7], which followed our works on the NFP-collector.

A different configuration of the NFP-collector is schematically described in Figure 34.7. In this case, a duct surrounds fine-particle clouds levitating in the plasma. There are several holes at the inside wall of the duct, behind which electrodes are situated for fine-particle collection and removal. When the electrodes are biased, particles are pulled into the duct, as demonstrated in Figure 34.8. A pumping system could be connected with the duct to guide fine particles into a region outside the vacuum chamber.

The measurements mentioned above are on fine particles levitating before collection and removal by the NFP-collector. But, the collector works also on fine particles, which, for example, are falling down in plasmas.

Both of the collector configurations mentioned above are quite simple for manufacturing. Various ideas could be found for other configurations of the NFP-collector, depending on real situations for application. But, the collector itself is so simple that there might be no difficulty in finding a proper collector structure in any case. The electrode used for plasma production could be used also as the

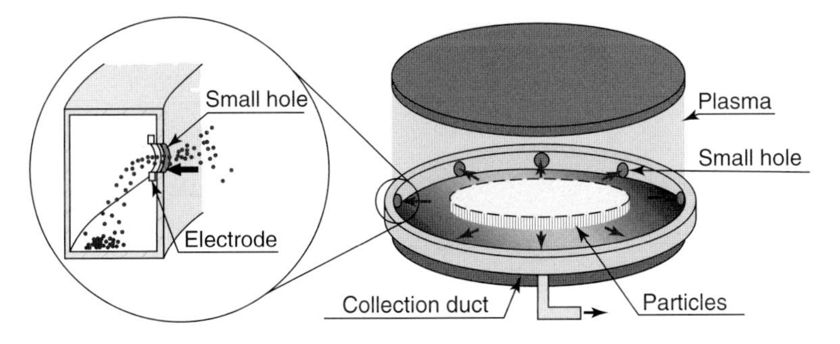

Figure 34.7 A duct-typed NFP-collector.

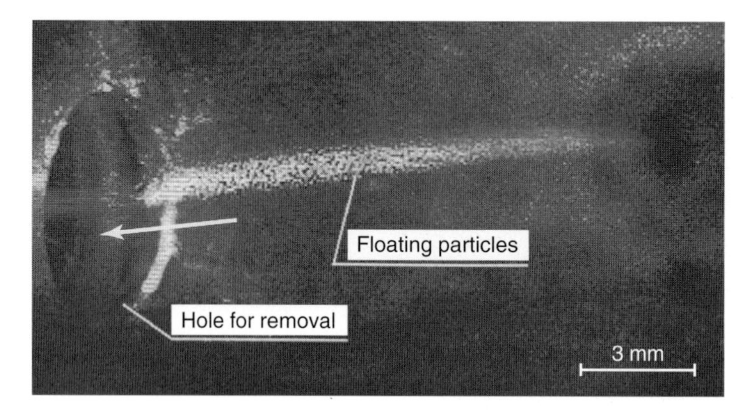

Figure 34.8 Particle removal by a duct-typed NFP-collector.

collector if the electrode has holes on its surface even though some ideas are necessary for its structure to be well matching with the discharge system.

34.3
Ditch Guidance

As introduced in Section 34.1, fine particles are horizontally confined in the region above a ditched region of the metal plate for particle levitation. The horizontal particle distribution is determined by the ditch figure. In Figure 34.9, where the plate diameter is 60 mm, the ditch figure is just the word "DUST." The laser irradiation for particle detection is found to figure out the "DUST" in a plasma.

There is a similarity between fine-particle flows and water flows described in Figure 34.4. This similarity also suggests us to form a ditch structure so as to induce a flow of fine particles along the ditch, for example, as shown in

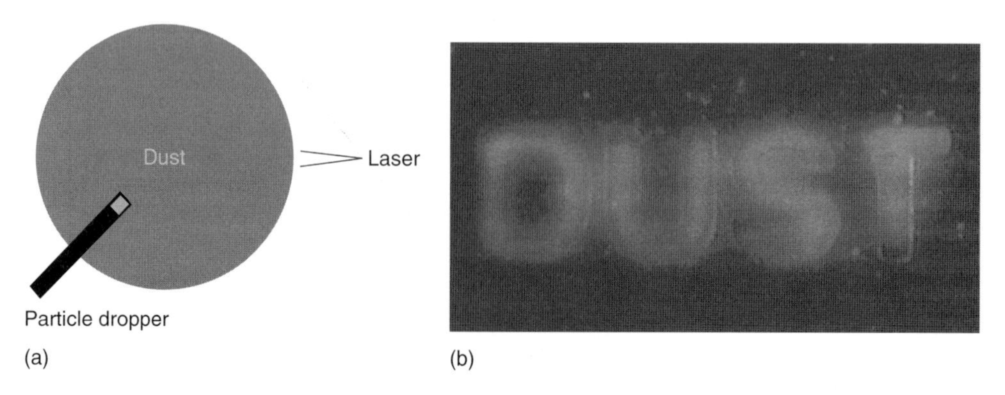

Figure 34.9 Metal plate with ditch "DUST" (a) and particle levitation above "DUST (b) (top view).

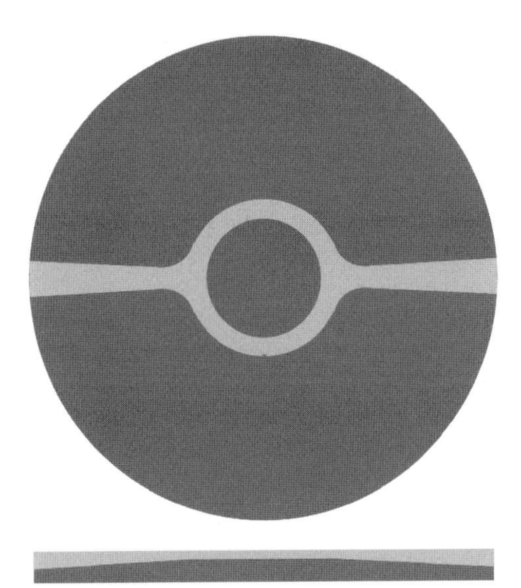

Figure 34.10 Example of ditch for particle guidance toward the radial edge (top view). The depth and width of radial parts of the ditch increase in the radial direction.

Figure 34.10. Here, the ditch has radial paths with depth and width increasing in the radial direction. Fine particles are trapped at first above the ditch and then flow along the ditch in the radial direction just as in case of water flow. When the NFP-collector is situated at the radial edge of the ditch, the collector action is enhanced even if the collector is located at the plasma edge far from the central region. Measurements show that some particles are left in the crossing regions of the circular and radial paths of the ditch when the size of cross regions are larger than the width of other parts of the ditch. This problem is solved by decreasing the cross area (for example, see Figure 34.11a). If we do not like a nonplane structure due to the ditch, we can put a thin plane insulator sheet above the plate, which guarantees a similar potential profile to that without the insulator on the horizontal plane, as already mentioned in Section 34.1.

Figure 34.11 demonstrates measurements on the ditch guidance of particle flow and the NFP-collector action at the ditch end. Here, the ditch is formed on the surface of metal plate of 6 cm in diameter, as shown in Figure 34.11 (upper, left). This is covered with a thin plane insulator sheet and the NFP-collector is situated at the ditch end (Figure 34.11 (upper, right)). When fine particles are externally injected into a plasma, they levitate and are guided along the ditch into its wide region in front of the collector. When the collector is not biased, the particle flow stops in this region (see Figure 34.11a). Immediately after the collector is biased, however, particles flow into the collector (see Figure 34.11b), and finally there is no particle left in the ditch (see Figure 34.11c) even if the bias potential is quite small.

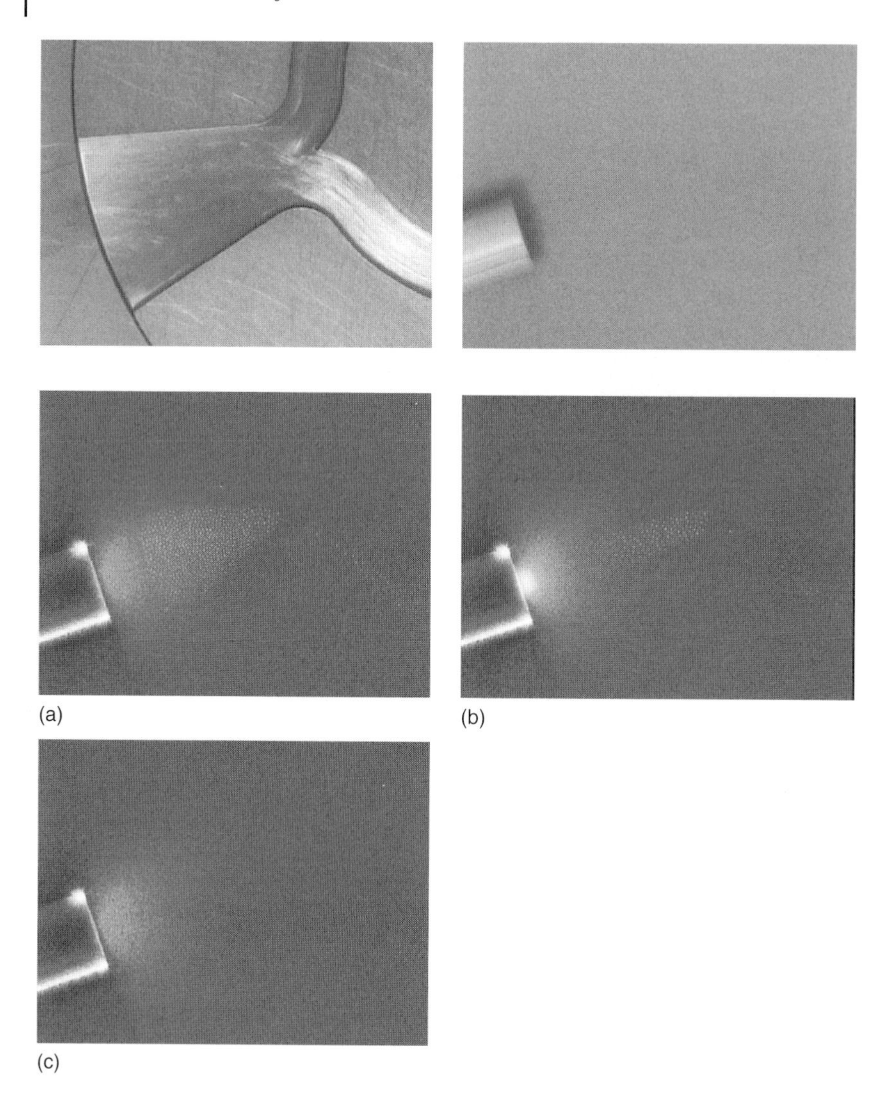

Figure 34.11 Ditch-guided particle removal by tube-typed NFP-collector (top view). A part of the ditch (upper, left) on the plate surface, which is covered with a thin insulator sheet (upper, right). Ditch-guided particles stopping in front of the collector, which is not yet switched on (a). Particles being pulled into the collector immediately after the collector is switched on (b). No particle left in the ditch after several seconds of switch-on (c).

Another shape of the ditch used is spiral as presented in Figure 34.12. The ditch depth and width increase along the spiral toward the plate edge. Figure 34.13a shows a particle flow in the absence of the collector. Particles flow along the spiral, as shown in Figure 34.13b. After some seconds, almost all particles flow out from the ditch, as shown in Figure 34.13c. This shape of the ditch is confirmed to

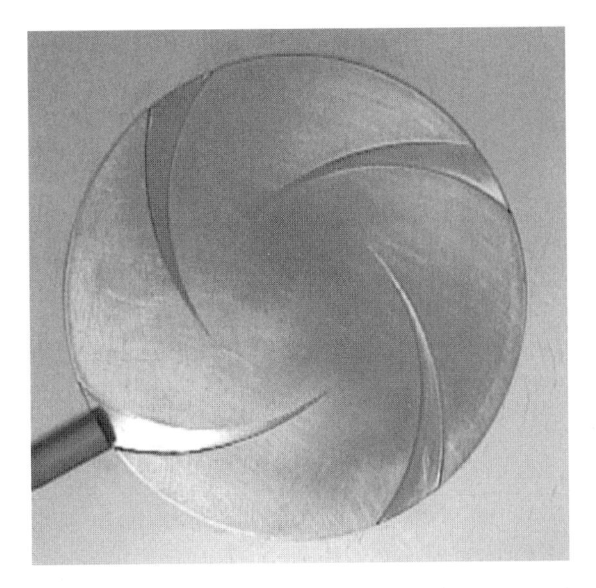

Figure 34.12 A spiral-typed ditch on a 6-cm-diameter metal plate for particle guidance (top view). Ditch depth and width increase in the outward direction.

be better than that in Figure 34.11 for fine-particle guidance and removal. Other shapes of the ditch could be found out, depending on real situations for application. A point of this particle guidance is that both of the ditch depth and width increase along the ditch path toward the outside edge of the ditch.

34.4
Plate Cleaning

Here is presented an important role of the NFP-collector for cleaning the vacuum chamber wall or dirty parts in the vacuum chamber. Low-pressure discharge or ion-bombardment cleaning method is often used for cleaning dirty surfaces. In this case, a lot of dusty particles are sputtered out into plasmas, as illustrated in Figure 34.14. These particles contaminate other parts of the surfaces. The NFP-collector could be used for collection and removal of dusty particles in plasmas in order to avoid this contamination.

In the experiments described above, spherical particles of 10 μm in diameter are externally injected into plasmas, where they are negatively charged up. In a region above the plate for particle levitation, there is an upward electric force yielding the particle levitation against the gravity force. If there is no plasma or if the particle size is quite large, however, fine particles fall down, contaminating the plate. Here, this dirty plate is employed to demonstrate how to use the NFP-collector for the low-pressure discharge or ion-bombardment cleaning.

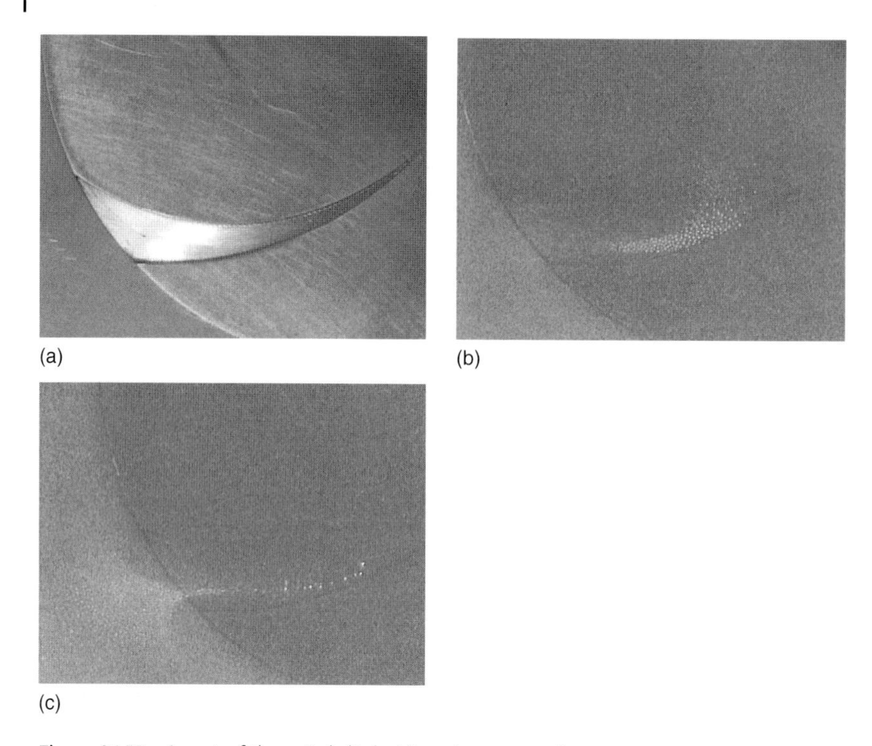

Figure 34.13 A part of the spiral ditch (a) and corresponding particle flow (top view): particles flowing along the ditch (b) and going finally out of the ditch (c).

In Figure 34.15a, there is a plate below a coarse grid (RF electrode for discharge). A region inside a circle is used for the particle levitation. Many experiments using this plate for particle levitation result in formation of dirty plate surface, as found in Figure 34.15b. A method of ion bombardment is employed to demonstrate a cleaning process using the NFP-collector. A dc negative potential in the range of 100–300 V is applied to the dusty plate with respect to the chamber wall in the presence of plasma. Then, many particles with different sizes (10 nm–10 μm) are sputtered out and levitate above the plate, as shown in Figure 34.16. The vertical position of particle levitation depends on the particle sizes. Now we use the NFP-collector to remove these particles. By moving the collector vertically, particles with different sizes are collected and removed, as already demonstrated in Figures 34.6 and 34.8. It is to be remarked that, in this method of cleaning, it is important to confine particles sputtered out of dirty plates in plasmas to avoid contamination of particles on other parts in the chambers.

This combination of the ion bombardment and the NFP-collector could be applicable, in principle, to cleaning of various kinds of plasma chambers including small-sized plasma processing devices and large-sized fusion devices.

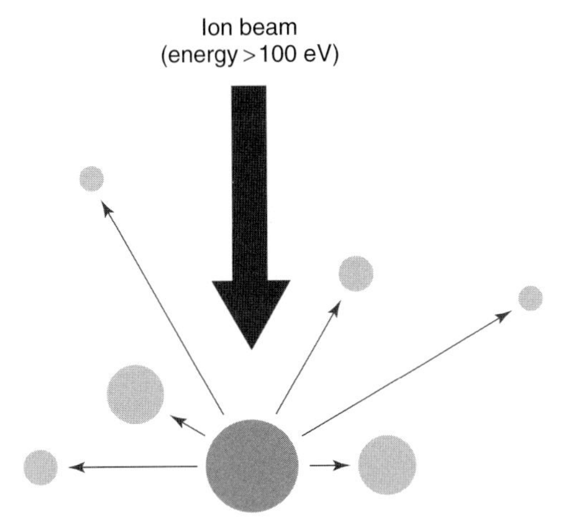

Figure 34.14 Schematic for ion bombardment.

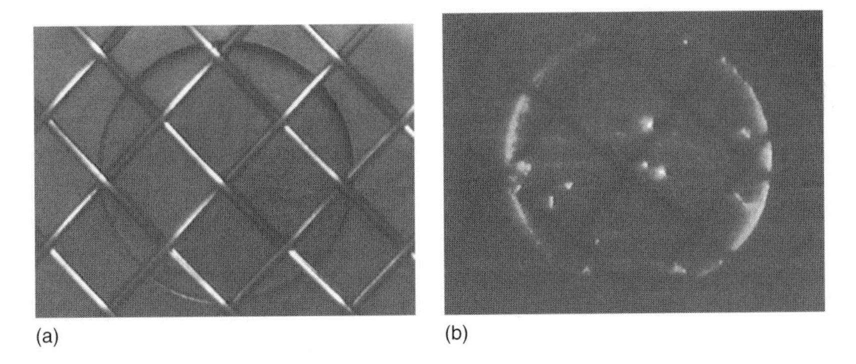

Figure 34.15 Surfaces of a metal plate (top view): clean (a) and dirty (b).

34.5 Effects on Particles Produced in Plasmas

Sections 34.2 and 34.3 are on fine particles injected externally into plasmas. Section 34.4 is on fine particles sputtered out of plates into plasmas. Fine particles are produced also in reactive plasmas [8, 9]. Now we are concerned with fine particles produced in plasmas.

It is well known that a lot of dusty particles are produced in silane plasmas. Here presented is an application of the NFP-collector to this plasma as an example of the NFP-collector efficiency [6]. Silane plasmas have been employed in fabrication of amorphous and microcrystalline silicon films [10]. But, it is of crucial importance now to solve problems caused by dusty fine particles produced in silane plasmas during the fabrication process.

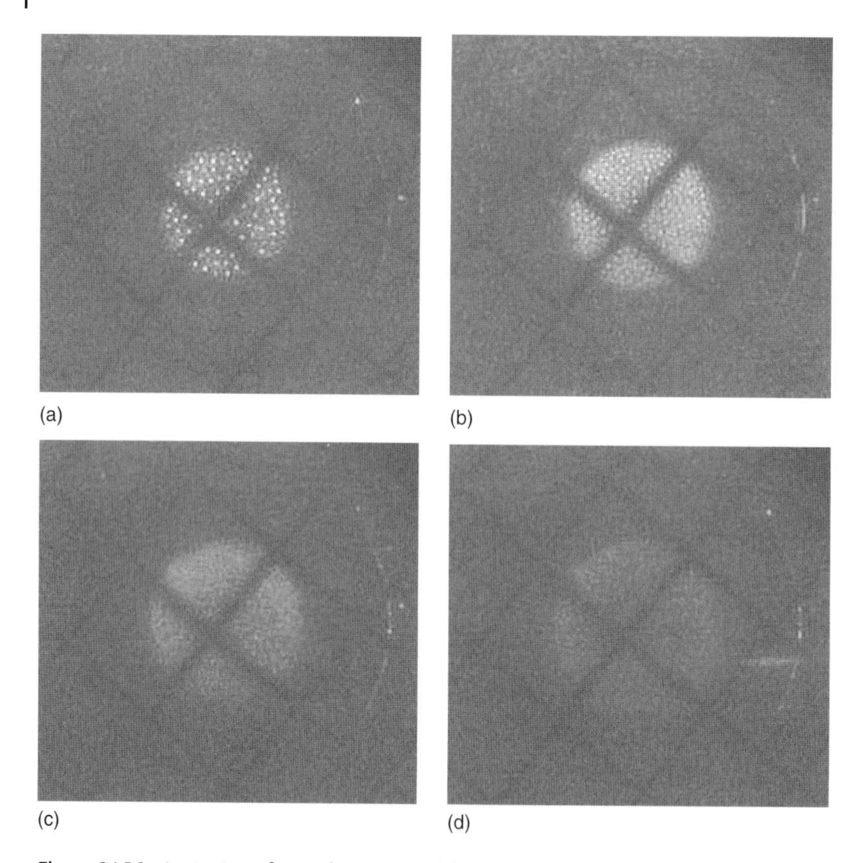

Figure 34.16 Levitation of particles sputtered by ion bombardment (top view): particle size depends on the vertical position (from 10 μm (left) to less than 1 μm (right)). The vertical position is changed upward by a step of 1 mm [(a) to (d)].

The plasma CVD device used is shown in Figure 34.17a, where a silane plasma is produced by modified magnetron-typed (MMT) RF discharge [4, 11] and amorphous silicon (a-Si : H) films are formed on a substrate (glass). As found in Figure 34.17b, the NFP-collectors are installed, surrounding the substrate, in order to find effects of particle removal on the films formed on the substrate.

Two different measurements performed are as follows. In one of them, the NFP-collector is switched on after particles grow up so large ($\geq$100 nm in diameter) in the RF silane plasma that they are clearly observed to levitate above the plate (substrate). The collector is found to collect and remove these particles as in the same way as explained till now. In the other measurement, the collector is switched on just after the silane plasma is triggered. Then, even for a long plasma operation, there appears no appreciable particle growth as far as the particle size is, roughly speaking, larger than 100 nm in the measurements. The results mean that the NFP-collector is effective not only in particle collection and removal but also in

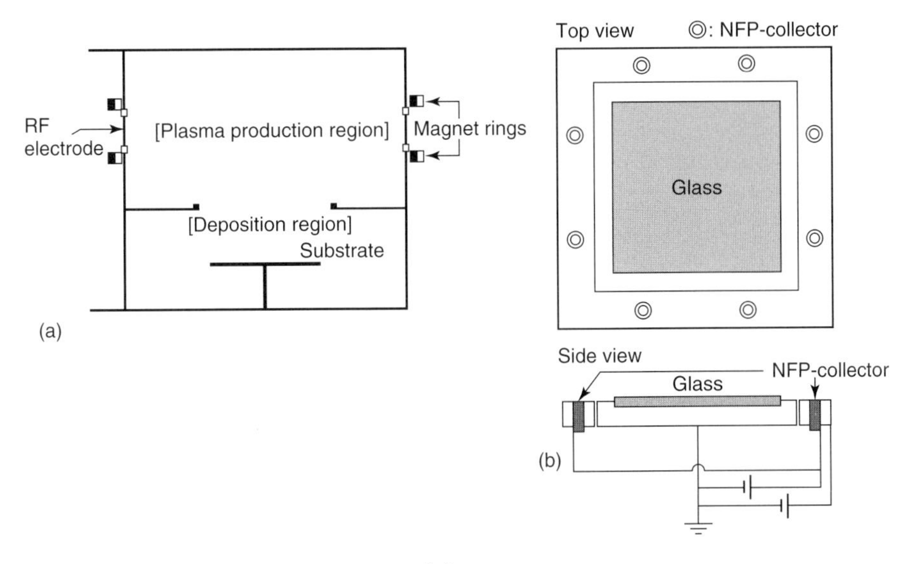

Figure 34.17 (a) Plasma CVD apparatus and (b) NFP-collectors arranged for a-Si : H deposition experiment.

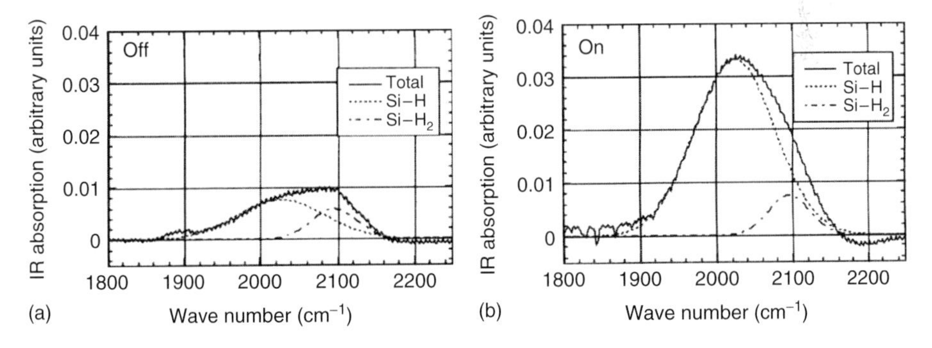

Figure 34.18 IR absorption spectra of a-Si : H film deposited (a) without and (b) with NFP-collectors.

suppressing particle growth in plasmas. The suppression is due to removal of fine particles much smaller than 100 nm.

Figure 34.18 demonstrates IR absorption spectra of the a-Si : H film deposited (a) without and (b) with NFP-collector. In the absence of dusty particles, the SiH component is found to be drastically increased. This component is known to reduce degradation of the films formed. This means that, under this condition, amorphous silicon films formed are improved for providing better solar-cell batteries. In fact, the measurements have proved a decrease of the photo-degradation of the films formed in the absence of dusty fine particles.

This suppression of particle growth would be also important in other reactive plasmas used for other material and device fabrication, where dusty fine particles degrade materials and devices fabricated.

34.6 Conclusions

The NFP-collector and ditch guidance are quite useful for collection and removal of fine -particles in plasmas. In addition to its simple principle, the collector is basically simple in its configuration. The ditch guidance of particles employed together with the NFP-collector decrease disturbances given by the collector, increasing the collection and removal efficiency. There is a proposal to employ the collector in the particle-transport method using low-frequency traveling plasma modulations in order to collect and remove particles [12]. Other possible methods of particle transport and guidance are also to be used with the NFP-collector for collection and removal of particles in plasmas.

The collector is expected to assist discharge (ion-bombardment) cleaning of vacuum chambers in its efficiency. In addition, the methods presented could be applied to remove dusty particles responsible for degrading products in various kinds of plasma application. The collector provides a new approach to removal of the air pollution due to various fine-particle dusts in the air.

Finally, the NFP-collector can be also used for collection and mass separation of useful and functional particles produced in plasmas. The mass separation is possible because of both the vertical particle levitation position and the horizontal distance flown by the particle from the collector entrance into the collector hole (when it is used as shown in Figure 34.6) depends on the particle mass.

Acknowledgments

The works presented here have been performed under the collaborations at Tohoku University, Hitachi Kokusai Electric Inc., and SHARP Corporation.

References

1. Sato, N., Uchida, G., Ozaki, R., and Iizuka, S. (1998) in *Physics of Dusty Plasmas* (eds M. Horanyi *et al.*), American Institute of Physics, pp. 239–246.
2. Uchida, G., Iizuka, S., and Sato, N. (2000) 17th Symposium on Plasma Processing, Japan, pp. 617–620.
3. Sato, N. (2005) *Ukr. J. Phys.*, **50** (2), 171.
4. Sato, N. (2008) in *Advanced Plasma Technology* (eds R. d'Agostino, *et al.*), Wiley-VCH Verlag GmbH, pp. 1–15.
5. Sato, N. and Koshimizu, T. (2003) 30th IEEE International Conference on Plasma Science, Korea, June 2–5, 2003, Abstracts p. 195.
6. Kurimoto, Y., Matsuda, N., Uchida, G., Iizuka, S., and Sato, N. (2004) *Thin Solid Films*, **457**, 285.
7. Winter, J., Fortov, V.E., and Nefedov, A.P. (2001) *J. Nucl. Mater.*, **90-293**, 509.
8. Watanabe, Y., Shiratani, M., and Koga, K. (2008) in *Advanced Plasma Technology* (eds R. d'Agostino *et al.*), Wiley-VCH Verlag GmbH, pp. 227–242.

9. Boufendi, L. (2010) Crystallized nanodust particles growth in low pressure cold plasmas, *Kawai, Industrial Plasma Technology*, Wiley-VCH Verlag GmbH, Weinheim.
10. Matsuda, A. (2008) in *Advanced Plasma Technology* (eds R. d'Agostino *et al.*), Wiley-VCH Verlag GmbH, pp. 197–210.
11. Li, Y., Iizuka, S., and Sato, N. (1994) *Appl. Phys. Lett.*, **65**, 28.
12. Li, Y., Konopka, U., Jiang, K., Shimizu, T., Hüfner, H., Thomas, H.M., and Morfill, G.E. (2009) *Appl. Phys. Lett.*, **94**, 081502.

Index

Industrial Plasma Technology. Edited by Yoshinobu Kawai, Hideo Ikegami, Noriyoshi Sato, Akihisa Matsuda, Kiichiro Uchino, Masayuki Kuzuya, and Akira Mizuno

ISBN: 978-3-527-32544-3

o

p

t

u

v

w